MARINE NATURAL PRODUCTS

Chemical and Biological Perspectives

Volume II

Contributors

J. C. BRAEKMAN

D. DALOZE

WILLIAM FENICAL

L. J. GOAD

RICHARD R. IZAC

M. KAISIN

SYNNØVE LIAAEN-JENSEN

ALLAN F. ROSE

JAMES J. SIMS

B. TURSCH

MARINE NATURAL PRODUCTS

Chemical and Biological Perspectives

Volume II

EDITED BY

PAUL J. SCHEUER

Department of Chemistry
University of Hawaii
Honolulu, Hawaii

ACADEMIC PRESS New York San Francisco London 1978

A Subsidiary of Harcourt Brace Jovanovich, Publishers

ACADEMIC PRESS, INC.
111 Fifth Avenue, New York, New York 10003

United Kingdom Edition published by
ACADEMIC PRESS, INC. (LONDON) LTD.
24/28 Oval Road, London NW1 7DX

Library of Congress Cataloging in Publication Data

Main entry under title:

Marine natural products.

Includes bibliographies and index.
1. Natural products–Addresses, essays, lectures.
2. Biological chemistry–Addresses, essays, lectures. 3. Marine pharmacology–Addresses, essays, lectures. I. Scheuer, Paul J.
QD415.M28 547'.7 77–10960
ISBN 0–12–624002–7

PRINTED IN THE UNITED STATES OF AMERICA

78 79 80 81 82 9 8 7 6 5 4 3 2 1

Dedication

Edgar Lederer, who celebrated his seventieth birthday June 5, 1978, is outstanding among the pioneers in marine natural product research. Many of his contributions over the past fifty years to our understanding of physiologically significant marine metabolites are only now experiencing full realization.

It is with great pleasure, therefore, that we dedicate this volume, which contains the first comprehensive treatment of marine carotenoids, to Professor Edgar Lederer.

Contents

List of Contributors

Numbers in parentheses indicate the pages on which the authors' contributions begin.

J. C. BRAEKMAN (247), Collectif de Bio-Ecologie, Unité de Chimie Bio-Organique, Faculté des Sciences, Université Libre de Bruxelles, 1050 Bruxelles, Belgium

D. DALOZE (247), Collectif de Bio-Ecologie, Unité de Chimie Bio-organique, Faculté des Sciences, Université Libre de Bruxelles, 1050 Bruxelles, Belgium

WILLIAM FENICAL (173), Institute of Marine Resources, Scripps Institution of Oceanography, University of California, San Diego, La Jolla, California 92093

L. J. GOAD (75), Department of Biochemistry, University of Liverpool, Liverpool L69 3BX, England

RICHARD R. IZAC (297), Department of Plant Pathology, University of California, Riverside, California 92521

M. KAISIN (247), Collectif de Bio-Ecologie, Unité de Chimie Bio-Organique, Faculté des Sciences, Université Libre de Bruxelles, 1050 Bruxelles, Belgium

SYNNØVE LIAAEN-JENSEN (1), Organic Chemistry Laboratories, Norwegian Institute of Technology, University of Trondheim, N-7034 Trondheim-NTH, Norway

ALLAN F. ROSE (297), Department of Plant Pathology, University of California, Riverside, California 92521

JAMES J. SIMS (297), Department of Plant Pathology, University of California, Riverside, California 92521

B. TURSCH (247), Collectif de Bio-Ecologie, Unité de Chimie Bio-Organique, Faculté des Sciences, Université Libre de Bruxelles, 1050 Bruxelles, Belgium

Preface

Research, publications, and conferences on marine natural products are continuing with undiminished vigor, and an ever increasing number of experienced scientists are being attracted to the field. I hope that the present volume remains in tune with our initial goal to provide "timely and critical reviews that are representative of major current researches and that, hopefully, will also foreshadow future trends."

The first three chapters of this volume, which deal with marine carotenoids (1), steroids (2), and diterpenoids (3), represent areas that are generating much current research activity. The topics of Chapters 1 and 2 have not previously been reviewed. While Volume I of this treatise contains a chapter on steroids that emphasizes uncommon molecular structures, the steroid chapter in this volume is oriented toward biosynthesis and comparative biochemistry. Thus it is our first departure from distinct emphasis on structural organic chemistry.

Chapters 4 and 5 underscore this trend. In Chapter 4, the authors scrutinize a single phylum, the Coelenterata, and its metabolites. This is an almost exclusively marine phylum of some 9000 described living species. Research to date would predict that the coelenterates will yield a rich harvest of organic metabolites. Finally, Chapter 5, in which the authors focus on the latest tool for structural elucidation, ^{13}C nmr spectroscopy, reveals the power of this instrumental method especially when applied to the difficult problems of polyhalogenated marine metabolites.

Again, I am grateful to the contributors to this volume for their cooperation and to the marine research community at large for their generosity in making research results available to the authors prior to publication.

PAUL J. SCHEUER

Preface to Volume I

"Chemistry of Marine Natural Products" (Academic Press, 1973), the progenitor of the present volume, covered the early literature of a budding research area through December 1971. Although barely six years have elapsed since then, the field of marine natural products has flowered beyond expectation. Research has grown geometrically; it has spread geographically; and it has begun to explore in earnest some fascinating phenomena at the interface between biology and chemistry. Since March 1973, when "Chemistry of Marine Natural Products" was published, it has become increasingly apparent to me that a review of the entire field by one person was no longer feasible; hence the present effort in which I have asked some of my colleagues to share the task of providing critical reviews and new perspectives for the marine research community. I am grateful for the enthusiastic and prompt response by the contributors to this as well as to subsequent volumes.

Another facet of the 1973 book also needed reexamination. In 1970–1971, when I planned and wrote the earlier book, the organizational choices were essentially between a phyletic and a biogenetic approach. I chose a broad structural biogenetic outline, a concept with which I was comfortable and which, in my opinion, filled a need at that time. Such a unidimensional design seems no longer satisfactory. It has now become desirable to highlight and review topics even though they may bear little lateral relationship to one another. It may be desirable to focus on an intensive research effort in a particular phylum, or on biosynthetic studies dealing with a single species, or on research that concentrates perhaps on a particular class of compounds, or on a given biological activity. The present volume and its successors, therefore, will not adhere to any overall plan. I will attempt to bring together, at convenient intervals, timely and critical reviews that are representative of major current researches and that, hopefully, will also foreshadow future trends. In this way the treatise should remain responsive to the needs of the marine research community. I will be grateful for comments and suggestions that deal with the present or future volumes.

It is indeed a pleasure to acknowledge the cooperation of all workers in the field who have responded so generously and have provided to the individual authors new results prior to publication.

PAUL J. SCHEUER

Chapter 1

Marine Carotenoids

SYNNØVE LIAAEN-JENSEN

MARINE NATURAL PRODUCTS

ISBN 0-12-624002-7

I. INTRODUCTION

Carotenoids are usually yellow to red isoprenoid polyene pigments formally derived from lycopene (**1**, Scheme 1) (IUPAC–IUB, 1974). In the marine environment carotenoids are encountered in bacteria, yeasts, and fungi. They are present in all algae and in several vertebrates and invertebrates. Carotenoids are produced *de novo* by all photosynthetic organisms, including algae and photosynthetic bacteria, and some other microorganisms, whereas some invertebrates and vertebrates have the ability to structurally modify carotenoids taken through the diet. The annual production of phytoplankton in the oceans has been estimated to be 4×10^{10} metric tons of organic matter (dry weight), of which around 0.1%, or several million tons, constitute carotenoids (Fox, 1974). Peridinin (**2**, Scheme 1) and fucoxanthin (**3**, Scheme 1), both exclusively marine pigments, are the major carotenoids of phytoplankton and the carotenoids produced in largest quantity in nature (Strain *et al.*, 1976).

Of the approximately 400 carotenoids hitherto described (Straub, 1976) more than 100 have been isolated from marine sources. Around 40 of these have been isolated from marine organisms exclusively. Typical examples of carotenoids in the latter category are peridinin (**2**), fucoxanthin (**3**), actinioerythrin (**4**), and 7,8,7′,8′-tetradehydroastaxanthin (**5**)

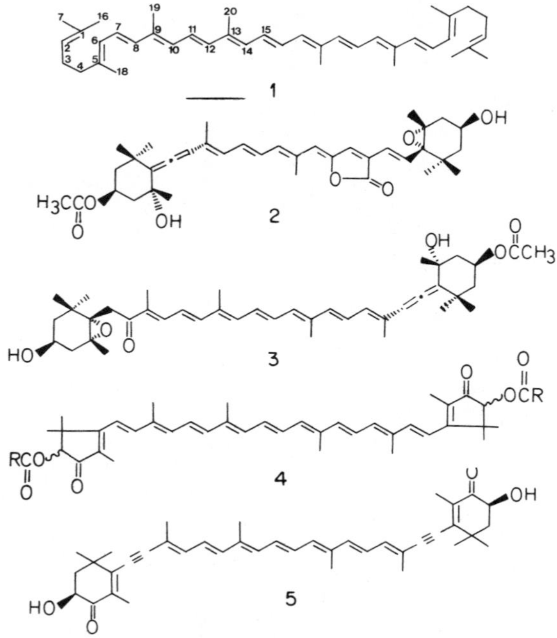

Scheme 1. Examples of marine carotenoids.

(Scheme 1). It is interesting that the most complex structures and the highest structural variation are found among marine carotenoids.

In this chapter an attempt is made to review our present knowledge of marine carotenoids with emphasis on structural aspects in a biochemical context. With the risk of generalizing too much, a chemosystematic approach is used. Key literature references are provided, and the literature treatment is selective rather than comprehensive. The outline follows evolutionary lines, starting with bacteria. Some work has been done on carotenoids of marine bacteria, but few typically marine species have been available. There have been very few studies of the carotenoids of marine yeasts (van Uden and Fell, 1968) and marine fungi (Barghoorn and Linder, 1976). However, yeasts of the *Rhodotorula* type are reported to have marine representatives (Litchfield, 1976). Several microscopic algae are now available through laboratory cultivation, and algal carotenoids have been studied rather systematically. Marine invertebrates and vertebrates have been investigated mainly according to the availability of pigmented species for analysis, but the literature in this area is rapidly increasing.

Why should we study marine carotenoids? The main reason, aside from the desire to satisfy our scientific curiosity, is that quantitative analysis of carotenoid composition, including structural elucidation, is required for biosynthetic considerations on a molecular level. Knowledge of absolute stereochemistry is a useful tool in biogenetic research. When sufficient evidence is compiled, chemosystematic considerations naturally emerge (Smith, 1975). Particularly for organisms capable of *de novo* carotenoid synthesis the potential use of carotenoids as chemosystematic markers, in addition to morphological, physiological, and biochemical criteria, is obvious. Knowledge of biosynthetic pathways and of end products is an even better criterion for chemosystematics. For functional biology it may prove useful to establish a specific locus of a particular carotenoid. In other biological contexts carotenoids may serve as indicators for monitoring the marine food chain. In the course of this research biologically active carotenoids may be discovered and definitive structures may serve as models for synthetic dyes or retinoids that have been shown to be capable of preventing cancer development (Sporn *et al.*, 1976).

The main body of this chapter is devoted to the discussion of marine algal, invertebrate, and vertebrate carotenoids (Sections VI–VIII). Short introductions are also given on the isolation, identification, structural elucidation, biosynthesis, and function of carotenoids (Sections II–IV). Available scattered information on carotenoids of marine nonphotosynthetic bacteria is discussed, and carotenoids of photosynthetic bacteria, treated more fully elsewhere, are briefly included for the sake of completeness (Section V). Finally, carotenoids of marine sediments are discussed in Section IX.

For carotenes the recent IUPAC–IUB (1974) nomenclature, denoting each end group by Greek letters (see Scheme 3) is partly used. Systematic names are otherwise used only when trivial names are lacking, and derivative names are used when convenient. The numbering of the carotenoid skeleton is shown for **1** in Scheme 1. Primed numbers are used for the righthand half of the molecule. In this chapter structural fromulas are numbered independently in each section or subsection. When the chirality of a particular carotenoid is known, absolute configuration is usually given for the structure, irrespective of the source of the carotenoid which chirality was proved. In only three cases have structurally identical carotenoids of different chirality been isolated. One case is that of astaxanthin from marine sources (**12**, Scheme 31) (Andrews *et al.*, 1974b) and its enantiomer from yeast (Andrewes and Starr, 1976). The second example is lutein (**4**, Scheme 7) (Buchecker *et al.*, 1974; Andrewes *et al.*, 1974a) and its C-3′ epimer calthaxanthin, (Dabbagh and Egger, 1974), and the third is that of the chiriquixanthins (Bingham *et al.*, 1977). Additional information about each carotenoid, is most easily obtained from the index by Straub (1976), in which complete references, physical and chemical properties, synonyms, etc., are compiled. Recommended background literature may be found in Isler's (1971) monograph on carotenoids.

II. ISOLATION, IDENTIFICATION, AND STRUCTURAL ELUCIDATION

A. Isolation

All carotenoids are unstable toward oxygen, light, heat, acids, and peroxides. Some carotenoids such as fucoxanthin and peridinin are also labile to alkali. Carotenol esters, of course, are hydrolyzed under alkaline conditions.

Common isolation procedures in which these factors are taken into account have been discussed elsewhere (Hager and Stransky, 1970a; Liaaen-Jensen, 1971; Liaaen-Jensen and Jensen, 1971; Davies, 1976). Isolation generally involves extraction and repeated column and thin-layer chromatography. For alkali-stable carotenoids saponification is usually included to remove any chlorophylls and lipids. Partition between petroleum ether and aqueous methanol represents an alternative for separating chlorophylls and carotenoids.

Several systems for separation of carotenoids by thin-layer chromatography (tlc) have been worked out, and (tlc) is the most common separation method used at present. Because of the color and high extinction coefficient of carotenoids in visible light, separation on a microgram scale is easily effected. High-pressure liquid chromatography is gradually being adapted to the carotenoid field (Stewart and Wheaton, 1971; Kramer, 1974; Watts *et al.*, 1977). Since carotenoids lack volatility, gas chromatography is used only for degradation products of carotenoids, including fatty acids from carotenol esters.

Removal of noncarotenoid contaminants can be effected by rechromatography in other systems, saponification, derivative formation (acetylation is frequently resorted to), and crystallization.

B. Identification

Criteria for the identification of carotenoids consist of chromatographic data, including cochromatography with authentic samples, partition coefficents in immiscible solvent systems, iodine-catalyzed stereomutation, spectroscopic evidence, and chemical derivative formation.

Carotenoids readily undergo trans–cis isomerization in solution (Zechmeister, 1962). This phenomenon complicates the isolation and identification of carotenoids and must be taken into account.

Spectroscopic data should include electronic spectra, revealing the chromophore; infrared spectra, providing information about functional groups; ^{1}H nuclear magnetic resonance (^{1}H nmr), ^{13}C nmr, circular dichroic (CD), and mass spectra. With Fourier transform techniques ^{1}H nmr spectra of microgram quantities can now be obtained. Information on ^{13}C nmr spectra of carotenoids is being obtained (Bremser and Paust, 1974; Englert, 1975; Moss, 1976), but currently sample sizes of 3–5 mg are required. Optical rotatory dispersion or CD spectra are required for chiral carotenoids (Bartlett *et al.*, 1969; Scopes, 1975; Liaaen-Jensen, 1976) and are indispensable in studies of absolute configuration. Mass spectra are essential for confirmation of molecular weight, and the fragmentation pattern on electron impact gives useful structural information. Use of field desorption mass spectrometry has recently been described (Watts and Maxwell, 1975). A comprehensive treatment of spectroscopic methods in the carotenoid field has been provided by Vetter *et al.* (1971). Use of chemical reactions for derivative formation has been reviewed by Liaaen-Jensen (1971).

A minimal requirement for the unequivocal identification of a carotenoid should be cochromatography with an authentic sample and verification by electronic and mass spectra. This requirement is frequently not fulfilled, even in the most recent studies.

C. Structural Elucidation

Spectroscopic data combined with chemical derivative formation are generally used for the elucidation of the structure of an unknown carotenoid, requiring about 1–30 mg of compound. In simpler cases spectroscopic data alone may define the structure. For more complex structures oxidative degradation to smaller molecules may also be required. The work on peridinin (Strain *et al.*, 1976; Kjøsen *et al.*, 1976; Johnansen and Liaaen-Jensen, 1977) may serve to illustrate the principles used. As in other areas of organic chemistry, total synthesis represents a definitive proof of structure.

D. Total Synthesis

The field of total synthesis of carotenoids has been expertly reviewed in detail by Mayer and Isler (1971). For later interesting developments see Weedon (1973, 1976), Fischli and Mayer (1975), Leuenberger *et al.* (1976), Liu (1976), and Paust (1976). Several of the carotenoids encountered in the marine environment have been prepared by total synthesis via one or more routes [see Straub's (1976) index in conjunction with the present review]. These include among others, several routes to β,β-carotene, echinenone, canthaxanthin, zeaxanthin, and aryl carotenoids such as renierapurpurin, as well as acetylenic carotenoids such as crocoxanthin and alloxanthin (Weedon, 1970), astaxanthin (Leftwick and Weedon, 1967; Englert *et al.*, 1977), and actinioerythrin (Kienzle and Minder, 1976). So far only a few optically active carotenoids have been prepared by total synthesis from chiral synthons, e.g., zeaxanthin (Mayer *et al.*, 1975), bacterioruberin derivatives (Johansen and Liaaen-Jensen, 1976), and astaxanthin (Englert *et al.*, 1977).

III. BIOSYNTHESIS

A. General Pathway

For several fairly recent reviews on the biosynthesis of carotenoids the reader should consult Straub's (1976) compilation of monographs and reviews. These include the selective treatment of bacteria (Schmidt, 1978; McDermott *et al.*, 1973; Weeks *et al.*, 1973; Liaaen-Jensen and Andrewes, 1972; Weeks, 1971; Liaaen-Jensen, 1965) and of yeasts and fungi (Davies, 1973; Goodwin, 1972; Simpson, 1972) and more general reviews on biosynthesis (Britton, 1971, 1976b; Goodwin, 1969, 1971b) and metabolism (Thommen, 1971). Data are most abundant for nonphotosynthetic microorganisms, photosynthetic bacteria, and tomatoes.

The early steps of carotenoid biosynthesis, which appear to be common to all carotenogenic systems, have been discussed by Davies and Taylor (1976), and Britton (1976a) has reviewed later reactions of carotenoid biosynthesis. Photoregulation of carotenoid biosynthesis in plants has also been discussed (Rau, 1976). For details the reader is referred to these reviews.

A brief summary of the present state of knowledge is given in the following paragraphs. The evidence is based on various techniques including analysis of mutants with a specific block in their carotenoid synthesis, inhibitors, endogenous carotenoid synthesis, stereospecifically labeled substrates, and cell-free preparations. Two major problems should be mentioned. Since all terpenoids and steroids have common low molecular weight precursors, use of labelled substrates resulting in the incorporation of label into many cell components requires extensive purification of the carotenoids before the measurement of radioactivity. This requirement is not always fulfilled in studies on carotenoid biosynthesis. A second problem is the hydrophobic nature of carotenoids, which precludes their ready use as substrates for aqueous enzyme preparations.

The primary events in carotenoid biosynthesis are depicted in Scheme 2. Mevalonic acid (MVA) is converted via its 5-phosphate (MVAP) and 5-pyrophosphate (MVAPP) by loss of C-1 as carbon dioxide to the C_5 unit Δ^3-isopentenyl pyrophosphate (IPP). Isomerization of IPP to dimethylallyl pyrophosphate (DMAPP) and condensation of these two units provide the C_{10} unit geranyl pyrophosphate (GPP). Subsequent additions of C_5 units (IPP) result in the formation of the C_{15} unit farnesyl pyrophosphate (FPP) and the C_{20} unit geranylgeranyl pyrophosphate (GGPP). Dimerization of GGPP via the cyclopropyl derivative prephytoene pyrophosphate (PPPP) provides the first C_{40} carotenoid phytoene (**1**) with a triene chromophore.

By stepwise dehydrogenation phytoene (**1**) is converted to colored carotenes via phytofluene (**2**) with a pentaene chromophore, ζ-carotene (**3**) or its unsymmetrical analog **4**, both with heptaene chromophores, neurosporene (**5**) with a nonaene chromophore, and lycopene (**6**) with an undecaene chromophore (Scheme 2). In photosynthetic bacteria the route via **4** is operative, whereas in higher plants ζ-carotene (**3**) appears to be the preferred intermediate. In other bacteria both alternative routes via **3** and **4** seem to operate. 15-*cis*-Phytoene (**1**) is considered to be the common biosynthetic intermediate in most organisms, except for bacteria producing C_{50} carotenoids from which only all-*trans*-phytoene (**1**) has been isolated. The pathway given in Scheme 2 has been demonstrated by incorporation experiments with labeled precursors and partial purified enzyme preparations.

Cyclization of an aliphatic carotenoid precursor to β, ϵ, or γ rings is thought to require the preformed Δ^7 double bond. If this is true, compounds **13** and **14** in Scheme 32 must be metabolic hydrogenation products, and an aliphatic precursor must be assumed in Scheme 20. The concepts that "half-molecules" rather than individual compounds are the substrates recognized by the enzymes has been introduced (Britton, 1976a)

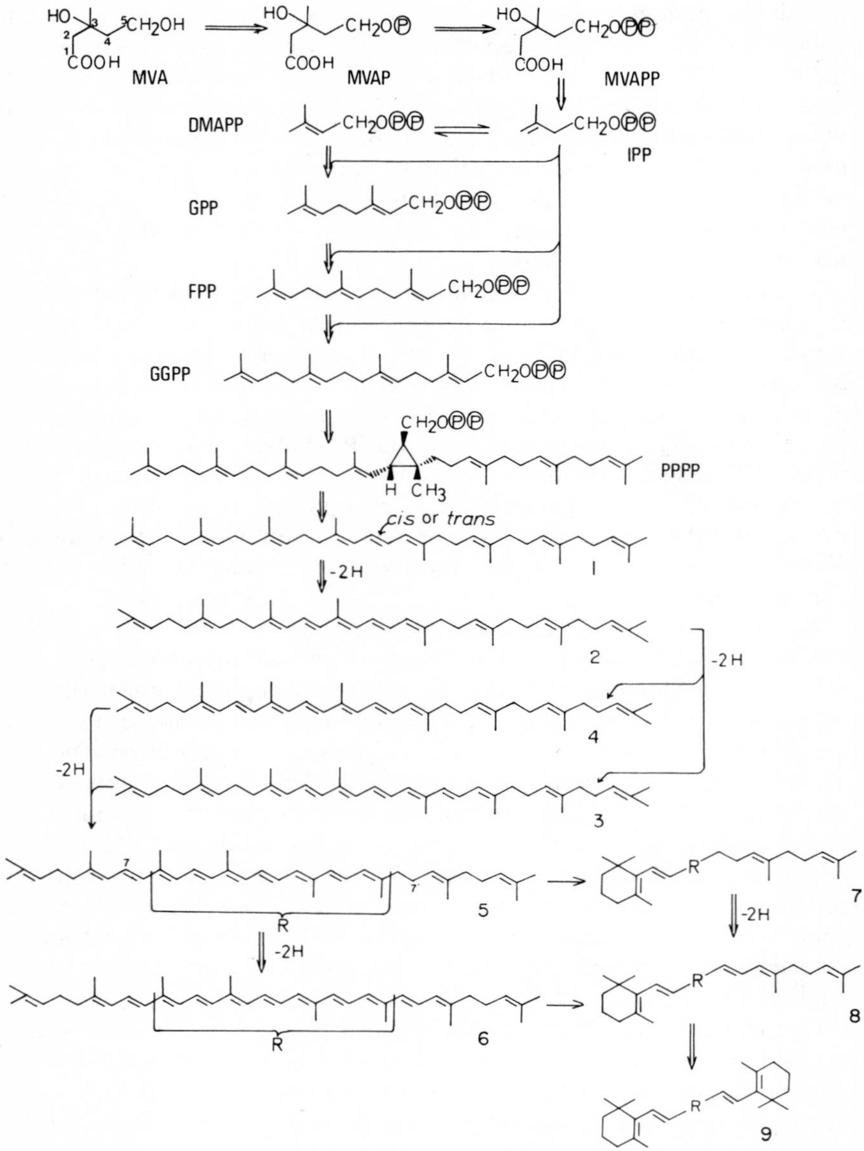

Scheme 2. General pathway of carotenoid biosynthesis.

and serves to simplify the overall picture of carotenoid biosynthesis. Thus, cyclization of neurosporene (**5**) via β-zeacarotene (**7**) and of lycopene (**6**) via β, ψ-carotene (**8**) to β, β-carotene (**9**) has been documented (Scheme 2). Different enzymes and different folding of the acyclic precursor appear to be involved in the formation of the β, ϵ, and γ rings (Scheme 3A). Details of the cyclization reactions have not been substantiated by work with cell-free preparations. However, some clues have been obtained from stereochemical considerations of the cyclic products.

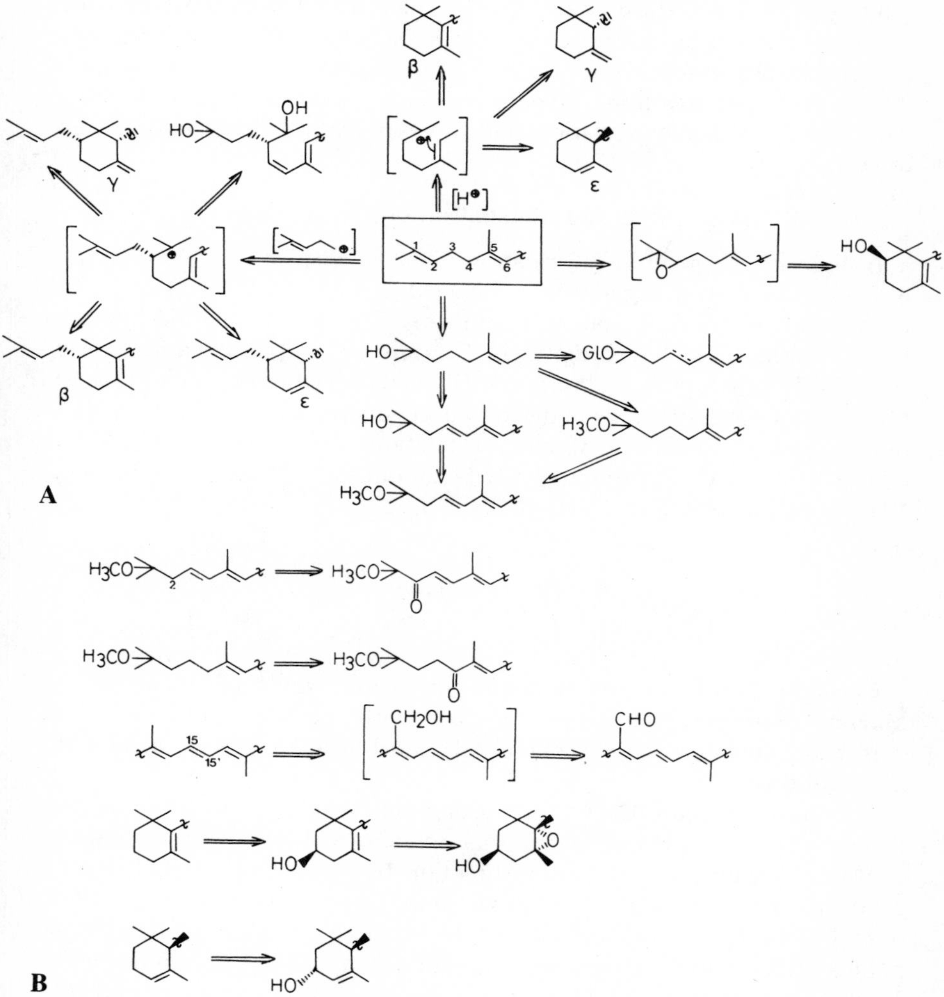

A

B

Scheme 3. Some later steps in carotenoid biosynthesis.

The steps described in Scheme 2 appear to be general. Much less evidence is available for the later steps in carotenoid biosynthesis including insertion of oxygen functions, introduction of allenic and acetylenic bonds, and skeletal modifications. In Scheme 3-A are listed various reactions ascribed to initial attack of the Δ^1 bond. They comprise, besides cyclization, anaerobic hydration, followed by *O*-methylation (photosynthetic bacteria) or *O*-glycosidation (bacteria) and isopentenylation with or without accompanying cyclization (certain bacteria). The 2-hydroxy β-ring feature could possibly be derived from a 1,2-epoxide.

Introduction of other oxygen functions (Scheme 3B) may be anaerobic or aerobic processes. Thus, insertion of conjugated 2-oxo, 4-oxo, and 20-oxo carbonyl functions in photosynthetic bacteria occurs anaerobically, whereas 3-hydroxylation and 5,6-epoxidation in β rings require oxygen.

B. Biosynthesis and Metabolism in Marine Organisms

In light of the general information given in Section III, A, a further discussion of marine organisms is included in subsequent sections on the carotenoids in various types of organisms. The reader will find information on carotenoid biosynthesis in Sections V,A (photosynthetic bacteria), V,B,1 (halophilic bacteria), V,B,2 (other bacteria), and VI (algae). Metabolism of carotenoids is discussed in general terms in Section VI,C (food chain considerations) and in Sections VII,A (sponges), VII,B (starfishes), VII,C (sea urchins), VII,D (mollusks), VII,E (sea anemones), VII,G (crustaceans), and VIII,H (fishes).

IV. FUNCTION

A. General Functions

Carotenoids have many functions in nature, some of them not yet known. For comprehensive reviews with detailed references on carotenoid function see Krinsky (1971) and Burnett (1976). A brief summary follows.

The most common function of carotenoids is photoprotection, as first demonstrated by Stanier and his associates (Stanier, 1960). In photosynthetic organisms carotenoids function by quenching singlet oxygen produced from excited chlorophyll and triplet oxygen (Foote, 1968; Krinsky, 1976). Carotenoids have the same function in various nonphotosynthetic organisms, where another sensitizer replaces chlorophyll. To fulfill this function a chromophore length of at least nine conjugated double bonds is required for the carotenoid. Treatment of patients suffering from light sensitivity with large doses of β,β-carotene is an interesting application of this carotenoid property (Mathews-Roth *et al.*, 1970).

The ability of carotenoids to function as accessory pigments of photosynthesis in certain organisms has long been recognized. Light energy harvested by carotenoids may be transferred to chlorophyll, thereby allowing the organism to utilize a broader part of the spectrum for photosynthesis.

Phototropism and phototaxis are initiated by carotenoids. Here carotenoids function as photoreceptors for light-directed movements of plants and bacteria. Visual stimulation caused by colored carotenoids in the flowers and fruits of many higher plants, which attracts or repells insects, serves an important function in nature.

Aside from the photofunctions of the carotenoids a theory has been proposed that epoxidic carotenoids participate in oxygen transport (Karrer, 1948). A possibility that carotenoids stabilize the proteins with which they are associated has been reviewed (Cheesman *et al.*, 1967). Although indirect evidence suggests that carotenoids may play a role in reproduction, such a function has been proved only for trisporic acid, which is a metabolite of β,β-carotene. Other metabolic products of carotenoids such as abscisic acid, the allenic grasshopper ketone, and vitamin A (retinol) have well-established biological effects. Although the role of retinal in the mechanism of vision was established long ago, a more recent interesting development is the cancer-preventing effect of certain retinoids (Sporn *et al.*, 1976). The question whether sporopollenin of spores and pollen grains is indeed a kind of carotenoid polymer has not yet been settled.

B. Marine Aspects

Although the photoprotective role of marine carotenoids is rather difficult to demonstrate for algae, it appears to be valid for both algae and photosynthetic bacteria (Burnett, 1976). It was also demonstrated that the wild type of a red halophilic bacterium better survived strong light than did a mutant devoid of colored carotenoids (Dundas and Larsen, 1963). One can only speculate why marine invertebrates and fishes of tropical waters are frequently so brilliantly colored in contrast to those of temperate waters. A photoprotective function is a possible explanation.

The role of carotenoids as accessory pigments in photosynthesis has been demonstrated for various algae, including diatoms (Dutton and Manning, 1941), green algae (Emerson and Lewis, 1943), green, red, and blue-green algae (Goedheer, 1965), and dinoflagellates (Song *et al.*, 1976). It has also been demonstrated in certain photosynthetic bacteria (Duysens, 1951).

The role of carotenoids in algal phototaxis is uncertain (Burnett, 1976).

A flavin or a flavanoid is stated to be the photoreceptor pigment (Tollin and Robinson, 1969).

As reviewed by Hager (1975), the violaxanthin–antheraxanthin–zeaxanthin cycle or its monoacetylenic analog is operative in all algae, except the most primitive classes lacking epoxy carotenoids. However, the role of the cycle in oxygen transport is disputed. The tolerance of marine mollusks to environmental pollution has been correlated with carotenoid concentration and has been taken as evidence that oxygenated carotenoids provide an intracellular reserve of oxygen (Karnaũkhov *et al.*, 1977).

Echinenone may play a functional role in the juvenile stage of sea urchins is discussed in Section VII,C. The facts that male and female gametes of brown algae display selective accumulations of β,β-carotene and fucoxanthin, respectively (Carter *et al.*, 1948), and that astaxanthin concentrates in lobster roe are suggestive of some function in reproduction. Carotenoproteins are particularly abundant in Crustacea, where control of protein configuration, coloration, or reproductional aspects have been considered as possible functions (Lee, 1977). The protective color of *Idotea* is commented on in Section VII,G.

V. BACTERIAL CAROTENOIDS

A. Photosynthetic Bacteria

Photosynthetic bacteria thrive in the anaerobic parts of all kinds of aquatic environments including marine habitats (Pfennig, 1967; Bergey, 1974; Litchfield, 1976).

The carotenoids of all photosynthetic bacteria that are available in pure culture have been studied, as reviewed by Pfennig and Trüper (1974) and Liaaen-Jensen (1978). The carotenoid composition of purple sulfur and nonsulfur bacteria, as well as that of photosynthetic green bacteria, falls into a pattern suitable for chemosystematic considerations (Liaaen-Jensen and Andrewes, 1972; Schmidt, 1978).

The structural features of the approximately 84 different carotenoids synthesized by photosynthetic bacteria are compiled in Scheme 4. Carotenoids of various degrees of dehydrogenation are encountered, encompassing variations in the central polyene chain A–H, to which two end groups a–n are attached in several combinations. Aliphatic end groups containing tertiary hydroxy or methoxy groups are common, as are trimethyl aryl end groups. Tertiary glycosides are rare. The chromophore frequently has conjugated keto groups in the 2 or 4 position. Cross conjugated aldehyde functions at C-20 are also encountered. It should be noted that methoxylated carotenoids as well as carotenoids

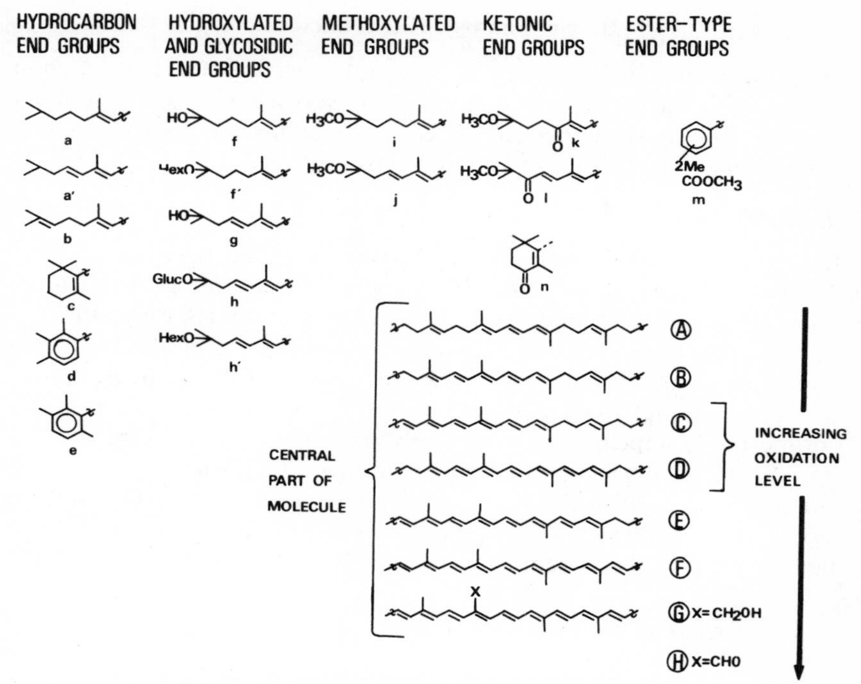

Scheme 4. Carotenoids of photosynthetic bacteria.

substituted at C-20 are restricted to photosynthetic bacteria. Further characteristic features of photosynthetic bacteria include their inability to cyclize their carotenoids, except for the aryl derivatives, or to introduce chirality (molecular asymmetry) into their carotenoids. Structural elements such as allenic or acetylenic bonds, norcarotenoids with abbreviated carbon skeletons, or C_{45} or C_{50} skeletons are not encountered.

Recent reports (Britton *et al.*, 1975, 1976a,b) on new minor carotenoids from photosynthetic bacteria add no new structural features to those previously reported. Several carotenoids from photosynthetic bacteria have been prepared by total synthesis (see compilations by Straub, 1976; Liaaen-Jensen, 1978).

From studies of the carotenoid composition of mutants possessing a block in their carotenoid synthesis, of oxygen effects, of effects of inhibitors such as diphenylamine, and of endogenous carotenoid synthesis in light, much information has been gained on the final steps of carotenoid synthesis in photosynthetic bacteria, which occur via chemically simple step reactions (Liaaen-Jensen *et al.*, 1961; Eimhjellen and Liaaen-Jensen, 1964; Schmidt, 1977). In fact, most of our evidence for carotenoid biosynthesis at the C_{40} level has been obtained with photosynthetic bacteria.

B. Nonphotosynthetic Bacteria

Only scattered information is available on carotenoids of non-photosynthetic bacteria. How many of these species may be encountered in marine habitats is uncertain. Few typically marine species, except true halophilic bacteria, have been investigated.

1. Halophilic Bacteria

In bacterial classification the distinction is made between extreme halophiles requiring more than 15% sodium chloride for growth (family Halobacteriaceae) and moderately halophilic bacteria (Bergey, 1974).

The main carotenoid of red, extremely halophic bacteria is the C_{50} tetrol bacterioruberin (**1**) (see Scheme 5) (Kelley *et al.*, 1970; Kushwaha *et al.*, 1974; Johansen and Liaaen-Jensen, 1976). A *cis* isomer (neo-U) of bacterioruberin (**1**) appears to be naturally occurring, even taking into account the facile cis isomerization of all-*trans*-**1** (Liaaen-Jensen, 1962; Kushwaha *et al.*, 1975). From the high *cis* peak of the neo-U isomer, a 13-*cis* or 15-*cis* configuration may be tentatively assigned, although 15-*cis* bonds have not yet been demonstrated in naturally occurring carotenoids.

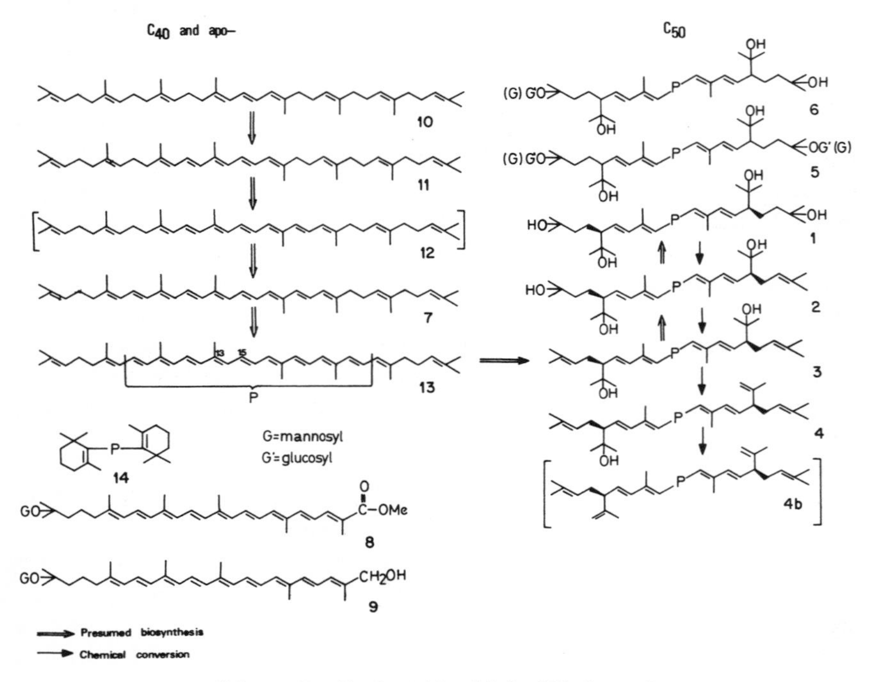

Scheme 5. Carotenoids of halophilic bacteria.

Monoanhydrobacterioruberin (**2**) and bisanhydrobacterioruberin (**3**) are also naturally occurring (Kelly *et al.*, 1970; Kushwaha *et al.*, 1975). Both may be prepared from **1** by dehydration with phosphorus oxychloride in pyridine. Further dehydration provides the trisanhydro derivative **4** and tetraanhydrobacterioruberin (**4b**) (Kelly *et al.*, 1970). Kushwaha *et al.* (1974) isolated a carotenoid thought to be trisanhydrobacterioruberin (**4**), but confirmatory evidence is required. Tetraanhydrobacterioruberin (**4b**), which is not naturally occurring, was prepared by total synthesis in the optically active form (Johansen and Liaaen-Jensen, 1976). Bacterioruberin (**1**) in the optically inactive form was prepared synthetically in low yield by the same workers. The structure of a $C_{50}H_{24}O_4$ diol with a dodecaene chromophore (Norgård, 1972; Straub, 1976) has not yet been established.

Kushwaha *et al.* (1974) state that C_{50} carotenoids are found only in extreme halophiles. However, bacterioruberin (**1**), its monoglycoside (**5**), and its diglycoside (**6**) were obtained from a moderately halophilic bacterium isolated from glacial mud (Arpin *et al.*, 1972). Glucose (80%) and mannose (20%) were identified after hydrolysis of **5** and **6**, revealing that **1** occurs both as the glucoside and the mannoside.

Aasen *et al.* (1969b) isolated neurosporene as a major carotenoid in one yellow halophilic coccus and the mannoside of a methyl apocarotenoate (**8**) as the main carotenoid in another. The alcohol **9**, corresponding to **8**, was also present in the two cocci.

Regarding the presence of other C_{40} carotenes in halophilic bacteria Kushwaha *et al.* (1974, 1975) claim the presence of *cis*- and *trans*-phytoene (**10**) and *cis*- and *trans*-phytofluene (**11**). ζ-Carotene (**12**) has been reported to be a biosynthetic intermediate (Kushwaha *et al.*, 1976). Lycopene (**13**) and β,β-carotene (**14**) represent minor carotenoids (Kelly et al., 1970; Kushwaha *et al.*, 1974, 1975).

Results of studies with cell-free preparations of *Halobacterium salinarium* (*cuticurubrum*) and labeled mevalonate and isopentenyl pyrophosphate by Kushwaha *et al.* (1976) were taken to support the biosynthetic sequence isopentenyl pyrophosphate → → → *trans*-phytoene (**10**) → *trans*-phytofluene (**11**) → ζ-carotene (**12**) → neurosporene (**7**) → lycopene (**13**) → β,ψ-carotene → β,β-carotene (**14**). The trans isomer of phytoene (**10**) was reported to be the main intermediate. The main isomer of phytoene from photosynthetic bacteria and higher plants is known to be 15-*cis*, whereas that of the bicyclic C_{50} carotenoid producer *Flavobacterium dehydrogenans* is 15-*trans* (Davies *et al.*, 1966; Khatoon *et al.*, 1972). Kushwaha and Kates (1976) reported the accumulation of lycopene (**13**) and bisanhydrobacterioruberin (**3**) in *H. salinarium* (*cuticurubrum*) grown in the presence of nicotine. Removal of nicotine resulted in the formation of monoanhydrobacterioruberin (**2**) and bacterioruberin (**1**) at the expense of **13** and **3**. These results are consistent with the early postulate that C_{50} carotenoids are probably formed by isopentenylation of a traditional C_{40} skeleton with isopropylidene end groups. Interestingly, the absolute configuration at C-2 is the same for all aliphatic and cyclic C_{50} carotenoids hitherto characterized (Johansen and Liaaen-Jensen, 1976).

2. *Other Bacteria*

Britton *et al.* (1977) studied the carotenoids of a *Flavobacterium* species produced by mutation from a marine *Flavobacterium* species. The main carotenoid was (3*R*,3′*R*)-zeaxanthin (β,β-carotene-3,3′-diol). Also present (see Scheme 2) were 15-*cis*-phytoene (**1**), phytofluene (**2**), ζ-carotene (**3**), neurosporene (**5**), lycopene (**6**), β-zeacarotene (**7**), β,ψ-carotene, β,β-carotene (**9**), β-cryptoxanthin (β,β-caroten-3-ol), rubixanthin (β,ψ-caroten-3-ol), 3-hydroxy-β-zeacarotene, and several apocarotenoids. The biosynthesis was discussed in terms of "half-molecule" sequence reactions. In the nonmarine *Flavobacterium dehydrogenans* belonging to another pigment group of *Flavobacterium* (McMeekin *et al.*, 1971), the bicyclic C_{50} carotenoid decaprenoxanthin is the major carotenoid (Liaaen-Jensen *et al.*, 1968).

The author is not aware of other true marine bacteria the carotenoids of which have been investigated from a structural standpoint. However, with increasing knowledge of isolation and cultivation of marine bacteria (Litchfield, 1976) progress in this field is anticipated.

VI. ALGAL CAROTENOIDS

A. Distribution and Structures

Since algae are photosynthetic organisms, they invariably contain carotenoids. Table 1, which describes the carotenoids encountered in 15 algal classes, follows essentially the classification of Christensen (1962), starting with the more primitive and finishing with the more highly developed algae. Marine representatives are found in all algal classes. No attempt has been made to exclude freshwater representatives. For previous reviews on algal carotenoids see Goodwin (1971c, 1976) and Liaaen-Jensen (1978).

1. *Cyanophyceae*

Although the carotenoid composition of unicellular blue-green algae has not yet been studied in detail, much evidence is available concerning the

TABLE 1

Classification of Algae[a]

Division	Class	Subclass	No. of orders
Prokaryotes			
Cyanophyta	Cyanophyceae		5
Eukaryotes			
Rhodophyta	Rhodophyceae	Bangiophyceae	5
		Florideophycideae	6
Chromophyta	Cryptophyceae		2
	Dinophyceae		6
	Rhapidophyceae		1
	Chrysophyceae		11
	Haptophyceae		2
	Bacillariophyceae		2
	Xanthophyceae		5
	Eustigmatophyceae		1
	Phaeophyceae		10
Chlorophyta	Euglenophyceae		2
	Loxophyceae		1
	Prasinophyceae		1
	Chlorophyceae		11

[a] Mainly according to Christensen (1962).

carotenoids of filamentous blue-green algae. Particularly, studies by Stransky and Hager (1970b) and by the Trondheim group (Hertzberg and Liaaen-Jensen, 1966a,b, 1967, 1969a,b, 1971; Francis *et al.*, 1970a; Halfen and Francis, 1972; Hertzberg *et al.*, 1971; Buchecker *et al.*, 1976a) define the characteristic distribution pattern in structural terms (see Scheme 6).

Characteristic carotenoids are the monocyclic rhamnoside myxoxanthophyll (**12**), the aliphatic dirhamnoside oscillaxanthin (**15**), less general, related rhamnosides such as 4-ketomyxoxanthophyll (**13**) and aphanizophyll (**14**), and *O*-methylmethylpentoside **16**, **17**, and **18**.

Also present are bicyclic carotenoids with hydroxy groups at C-3(3′), cryptoxanthin (**4**) and zeaxanthin (**5**) without and with conjugate keto groups at C-4(4′), echinenone (**6**), canthaxanthin (**7**), and 3′-hydroxyechinenone (**8**). β,β-Carotene (**2**) is always present in algae. From *Oscillatoria princeps* has been isolated β,ψ-carotene (**3**) and lycopene (**1**) as minor carotenes (Francis and Halfen, 1972). A trace constituent, mutatochrome (**11**), is the only epoxidic carotenoid so far encountered (Hertzberg and Liaaen-Jensen, 1967). Two carotenoids previously assigned allenic structures, caloxanthin and nostoxanthin (Stransky and Hager, 1970b), have been shown to be (2*R*,3*R*,3′*R*)-β,β-carotene-2,3,3′-triol (**9**) and (2*R*,3*R*,2′*R*,3′*R*)-β,β-carotene-2,3,2′,3′-tetrol (**10**), respectively (Buchecker *et al.*, 1976a; Smallidge and Quackenbush, 1973).

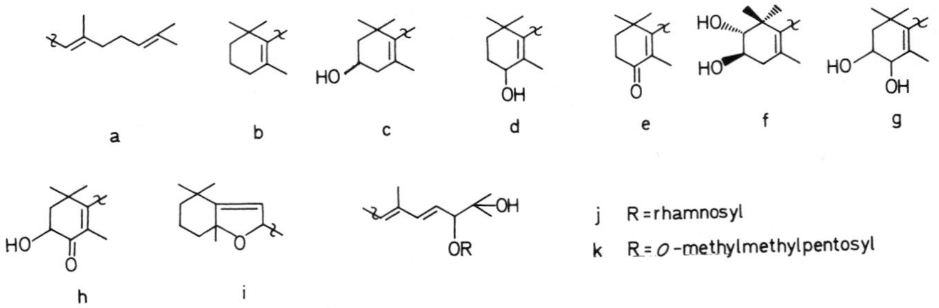

a b c d e f g

j R = rhamnosyl

k R = O-methylmethylpentosyl

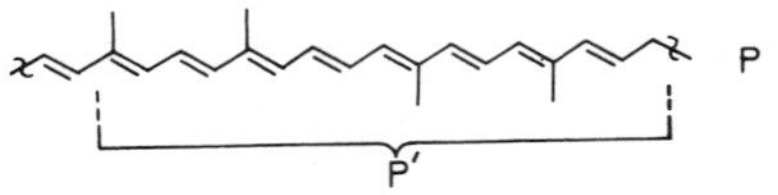

h i

P

P′

Lycopene (**1**) a–P–a
β,β-Carotene (**2**) b–P–b
β,ψ-Carotene (**3**) b–P–a
Cryptoxanthin (**4**) c–P–b
Zeaxanthin (**5**) c–P–c
Echinenone (**6**) e–P–b
Canthaxanthin (**7**) e–P–e
3′-Hydroxyechinenone (**8**) e–P–c
Caloxanthin (**9**) f–P–c
Nostoxanthin (**10**) f–P–f
Mutatochrome (**11**) i–P–b

Glycosides
Myxoxanthophyll (**12**) c–P–y
4-Ketomyxoxanthophyll (**13**) h–P–j
Aphanizophyll (**14**) g–P–j
Oscillaxanthin (**15**) j–P–j
Myxol-2′-*O*-methyl-methylpentoside (**16**) c–P–k
4-Ketomyxol-2′-*O*-methylmethylpentoside (**17**) h–P–k
Oscillol-2,2′-di(*O*-methyl)methylpentoside (**18**) k–P–k

Scheme 6. Carotenoids of blue-green algae (Cyanophyceae).

Noteworthy features of carotenoids of blue-green algae are (a) monocyclic and aliphatic carotenoids, (b) glycosidic carotenoids (rhamnosides and *O*-methylmethylpentoside), and (c) the absence of epoxidic carotenoids. No exception to the last feature has so far been encountered. A previous claim (Parsons, 1961) that the epoxidic antheraxanthin was a major carotenoid in *Agmenellum quadruplicatum* could not be confirmed (Antia and Cheng, 1977). The characteristic distribution pattern of carotenoids in blue-green algae clearly differentiates them from other algal classes.

Compilations of information on the species that have been examined and of quantitative data for individual carotenoids have been published (Stransky and Hager, 1970b; Hertzberg *et al.*, 1971; Goodwin, 1971c). Spectroscopic and chemical properties of glycosidic carotenoids are treated elsewhere (Hertzberg and Liaaen-Jensen, 1969a,b; Francis *et al.*, 1970a; Liaaen-Jensen, 1971).

2. *Rhodophyceae*

Data on the large number of red algae examined for carotenoids have been compiled by Goodwin (1971c) and Bjørnland and Aguilar-Martinez

(1976). Most studies were carried out without the use of mass or ^{1}H nmr spectroscopy. The rather simple distribution pattern encountered (Scheme 7) may therefore be questioned. However, a recent investigation of eight Rhodophyceae species by Bjørnland and Aguilar-Martinez (1976) confirms the earlier view. The four main carotenoids have repeatedly been identified as bicyclic carotenes with β (a) and ϵ (b) end groups and their 3,3′-dihydroxy derivatives: β,β-carotene (**1**), β,ϵ-carotene (**2**), zeaxanthin (**3**), and lutein (**4**). The identification of nine other carotenoids is somewhat questionable, and they occur only occasionally. These are 3(3′)-monohydroxy derivatives of **1** and **2**, α-cryptoxanthin (**5**), β-crytoxanthin (**6**), and various epoxidic carotenoids such as antheraxanthin (**7**), violaxanthin (**8**), auroxanthin (**9**) ≡ violaxanthin furan oxide), aurochrome (**10**, a difuran oxide), neoxanthin (**11**), fucoxanthin (**12**), and taraxanthin (**13**). The original literature is cited by Bjørnland and Aguilar-Martinez (1976).

The nearly general occurrence of carotenoids containing ϵ rings (b, d) is noteworthy and demonstrates the development of the ϵ-cyclase system in the Rhodophyceae.

Epoxidic carotenoids are generally not present. Positive identifications (see Scheme 7 and Table 1) always refer to the subclass Florideophyceae. Taraxanthin, now known to be identical with lutein epoxide (**13**) (Buchecker *et al.*, 1976b), is claimed to be a major carotenoid of *Grateloupia filicina* (de Nicola and Furnari, 1957).

The isolation of fucoxanthin (**12**) (Carter *et al.*, 1939; Kylin, 1939) from

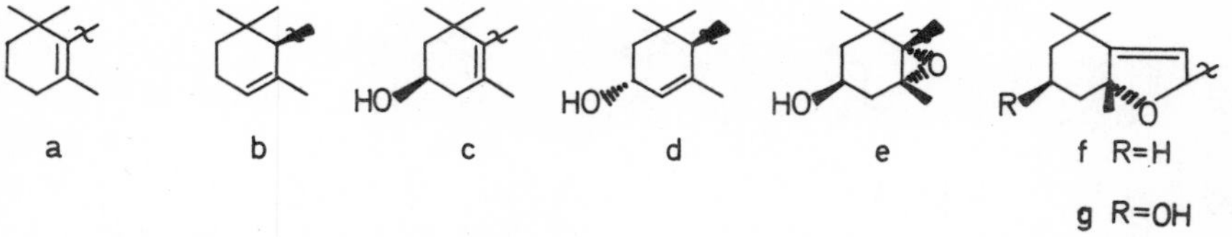

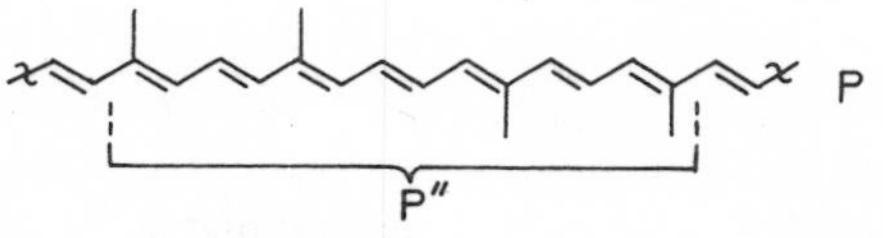

a b c d e f R=H g R=OH

P P″

Common carotenoids

β,β-Carotene (**1**) a–P–a
β,ϵ-Carotene (**2**) a–P–b
Zeaxanthin (**3**) c–P–c
Lutein (**4**) c–P–d

Questionable and occasional carotenoids

α-Cryptoxanthin (**5**) c–P–b
β-Cryptoxanthin (**6**) c–P–a
Antheraxanthin (**7**) e–P–c
Violaxanthin (**8**) e–P–e
Auroxanthin (**9**) g–P″–g
Aurochrome (**10**) f–P″–f
Neoxanthin (**11**) (**8** Scheme 11)
Fucoxanthin (**12**) (**2** Scheme 11)
Taraxanthin (**13**) e–P–d

Scheme 7. Carotenoids of red algae (Rhodophyceae).

red algae could be due to epiphytes such as diatoms. Thus, fucoxanthin was detected in naturally occurring material of *Bangia fusco-purpurea, Nemalion helminthoides, Ceramium rubrum*, and *Polysiphonia brodiaei*, but could not be found in cultured material of the first three species. However, the high content of fucoxanthin in wild-type *N. helminthoides* could not be correlated with the amounts of diatoms present at harvest (Bjørnland and Aguilar-Martinez, 1976).

Because of the capacity particularly of *Laurencia* to synthesize halogenated terpenoids (e.g., Howard and Fenical, 1975) an investigation, including mass spectroscopy, of the carotenoids of *Laurencia* was undertaken (Arnesen and Liaaen-Jensen, 1977). However, no halogenated carotenoids could be demonstrated in red algae. The partial synthesis and characterization of halogenated carotenoids have been reported (Johansen and Liaaen-Jensen, 1974; Buchecker and Liaaen-Jensen, 1975; Pfander and Leuenberger, 1976).

3. Cryptophyceae

The Cryptophyceae possess as do the Rhodophyceae, the ϵ-cyclase system and are furthermore characterized by their ability to synthesize acetylenic carotenoids (see Scheme 8). Thus, β,ϵ-carotene (**2**) may predominate over β,β-carotene (**1**) (Chapman and Haxo, 1963; Allen *et al.*, 1964:, Norgård *et al.*, 1974b), and ϵ,ϵ-carotene (**3**) is reported to occur in *Cryptomonas ovata* (Chapman and Haxo, 1963).

Acetylenic carotenoids were first detected in this class after the introduction of mass spectrometry. Acetylenic bonds are always restricted to

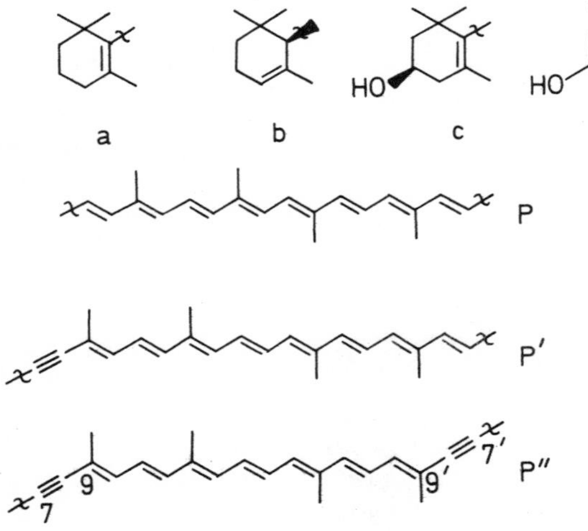

ϵ,ϵ-Carotene (**3**) b–P–b
Crocoxanthin (**4**) c–P′–b
Monadoxanthin (**5**) c–P′–d
Alloxanthin (**6**) c–P″–c
Manixanthin (**7**) = 9,9′-di-*cis*-**6**
β,β-Carotene (**1**) a–P–a
β,ϵ-Carotene (**2**) a–P–b

Scheme 8. Carotenoids of yellow algae (Cryptophyceae).

the 7(7′) position. Crocoxanthin (**4**) and monadoxanthin (**5**) are a monoacetylenic monool and diol, respectively, with one ϵ ring (Chapman, 1966; Mallams *et al.*, 1967). The diacetylenic alloxanthin (**6**) with two β rings (Chapman, 1966; Mallams *et al.*, 1967) is a characteristic major constituent. Epoxidic carotenoids have not been encountered in this class.

Species examined by recent methods include *Cryptomonas ovato* var. *palustris*, a *Rhodomonas* species, *Hemiselmis virescens* (Chapman and Haxo, 1963; Chapman, 1966), a *Cryptomonas* species (Hager and Stransky, 1970c), and *Chroomonas salina* (Norgård *et al.*, 1974b; Cheng *et al.*, 1974). Under particular growth and conditions the latter is stated to produce manixanthin (**7**), which is considered to be 9,9′-di-*cis*-alloxanthin.

4. Dinophyceae

The literature covering both the freeliving and the symbiotic dinoflagellates (zooxanthellae) has been reviewed up to 1971 by Goodwin (1971c).

The chemistry of dinoflagellate carotenoids has been extensively studied in recent years. The Dinophyceae produce structurally more complex carotenoids than do the three classes so far discussed (see Scheme 9). Their major carotenoid, peridinin (**8**), seems to be synthesized only by the Dinophyceae. Structural studies have revealed a unique C_{37} skeletal structure for peridinin (**8**) (Strain *et al.*, 1971b, 1976; Kjøsen *et al.*, 1976), bearing butenolide, allenic, epoxy, acetoxy, and alcohol functions.

In some dinoflagellates peridinin is replaced by fucoxanthin (**9**) (Jeffrey et al., 1975). A survey of the chloroplast ultrastructure in Dinophyceae (Dodge, 1975) demonstrated the presence of two nuclei in fucoxanthin-producing dinoflagellates such as *Glenodinium foliaceum* and *Peridinium balticum*, tentatively ascribed to an endosymbiont of the Crysophyta. However, other fucoxanthin-producing dinoflagellates have no second nucleus. Dodge advanced the interesting hypothesis that dinoflagellates were originally heterotrophic organisms and that the fucoxanthin-containing species represent an early stage of the photosynthetic apparatus before the development of peridinin. Recent studies on the absolute configuration of fucoxanthin (**9**) (Bernhard *et al.*, 1976) and peridinin (**8**) (Johansen and Liaaen-Jensen, 1977) support a close biosynthetic relationship between these carotenoids. Loeblich (1976), however, points out the combination of prokaryotic and eukaryotic features in dinoflagellates and suggests that species possessing fucoxanthin obtained their plastids by capture of photosynthetic eukaryotes.

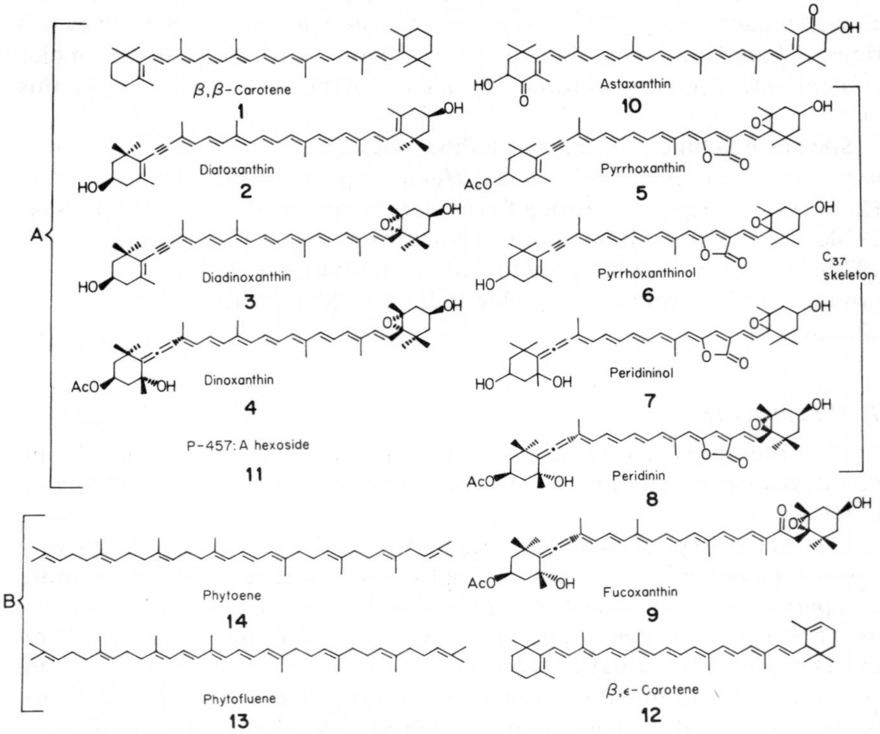

A
β,β-Carotene
1
Diatoxanthin
2
Diadinoxanthin
3
Dinoxanthin
4
P-457: A hexoside
11
Astaxanthin
10
Pyrrhoxanthin
5
Pyrrhoxanthinol
6
Peridininol
7
Peridinin
8
C_{37} skeleton
B
Phytoene
14
Phytofluene
13
Fucoxanthin
9
β,ε-Carotene
12

Scheme 9. Carotenoids of dinoflagellates (Dinophyceae).

Closer examination of the carotenoids of *Peridinium foliaceum* by Withers and Haxo (1975) revealed in addition to fucoxanthin (**9**) the presence of β,β-carotene (**1**), β,ε-carotene (**12**), as well as the colorless carotenoid precursors phytofluene (**13**) and phytoene (**14**) (Scheme 9B). The two latter carotenes, not previously isolated from wild algae, accumulate in extraplastidic oil globules.

A compilation of dinoflagellate species studied and carotenoids encountered before 1974 was presented by Johansen *et al.* (1974), together with a quantitative study of six Dinophyceae by modern spectrometric methods. A similar study of 22 species was carried out by classical methods only (Jeffrey *et al.*, 1975).

β,β-Carotene (**1**) is generally present as a minor constituent. The monoacetylenic diol diatoxanthin (**2**) and the corresponding 5,6-epoxide diadinoxanthin (**3**) seem to be of general occurrence. Dinoxanthin (Strain et al., 1944), shown to be the allenic neoxanthin 3-acetate (**4**) (Johansen *et al.*, 1974), is encountered. Pyrrhoxanthin (Loeblich and Smith, 1968) with a C_{37} skeleton is closely related to peridinin and has been assigned the acetylenic structure **5** (Johansen *et al.*, 1974). Pyrrhoxanthinol (**6**) was isolated from *Gyrodinium dorsum* as a minor carotenoid. Peridininol (**7**) is found in a number of species and is the alcohol corresponding to peridinin (**8**), which is a natural acetate. Astaxanthin (**10**) obtained from a *Glenodinium sp.* species bears structural resemblance only to **1** and **2**. An unidentified minor carotenoid, P-457 (**11**), present in several species, is a glycoside.

Inspection of the structures compiled in Scheme 9A,B demonstrates a capacity of Dinophyceae to produce carotenoids that possess the following structural features: (a) β rings, (b) triple bonds, (c) epoxides (high proportion), (d) allenic bonds, (e) acetates, and (f) C_{37} skeletons with a butenolide moiety, in addition to the capacity of all algae to produce carotenoid alcohols. Only the capability to synthesize carotenoids with functions a–c is found within the three algal classes so far treated here. Possible biogenetic interrelations are discussed in Section VI,B.

The molecular topology of the light-harvesting pigment complex of dinoflagellates consisting of peridinin–chlorophyll a–protein has been elucidated from CD and fluorescence polarization data (Song *et al.*, 1976).

5. *Rhapidophyceae (Chloromonadophyceae)*

The class Rhapidophyceae consists of a single order, Rhapidomonadales, with a single family, Rhapidomonadaceae. Two species, *Gonyostomum semen* and *Vacuolari virescens*, were investigated (Chapman and Haxo, 1966) in a preliminary manner (see Scheme 10). β,β-Carotene (**1**) was the only carotene isolated. The monoepoxide antheraxanthin (**2**) was the major carotenoid, accompanied by lesser amounts of lutein epoxide (**3**). A third xanthophyll, different from the allenic neoxanthin **6** (Scheme 14), was characterized as being trollixanthin-like. However, since trollixanthin was recently shown to be identical with neoxanthin

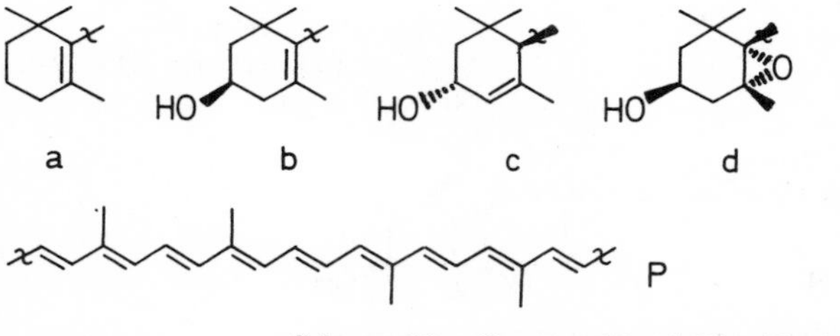

HO HO HO

a b c d

P

β,β-Carotene (**1**) a–P–a
Antheraxanthin (**2**) d–P–b
Lutein epoxide (**3**) d–P–c
Unknown (**4**)

Scheme 10. Carotenoids of Rhapidophyceae.

(Buchecker *et al.*, 1975), identification of the most polar carotenoid (heteroxanthin?) requires further study.

In conclusion, available evidence indicates a simple carotenoid pattern in the Rhapidophyceae requiring the presence of β- and ϵ-cyclase and a vigorous epoxidizing enzyme.

6. *Chrysophyceae*

The yellow algae, according to Christensen (1962), are divided into 11 orders comprising nearly 30 families. In spite of this abundance, modern investigations of their carotenoids are lacking. In his survey Goodwin (1971c) reviews earlier studies. However, most of the species dealt with are now classified as Haptophyceae. From work on *Ochromonas danica* and *Sphaleromantis sp.* (Allen *et al.*, 1960), *Dietateria inomata* and *Pseudopedinella sp.* (Dales, 1960), and *Ochromonas stipitata* (Hager and Stransky, 1970c) the tentative conclusion may be drawn that fucoxanthin (**2**) (Scheme 11) is the major carotenoid and that β,β-carotene (**1**) is generally present. From *O. stipitata* zeaxanthin (**3**), the epoxides antheraxanthin (**4**), violaxanthin (**5**), and trace amounts of cryptoxanthin (**6**), a cryptoxanthin monoepoxide (**7**), and neoxanthin (**8**) were reported. Further studies on the carotenoid distribution in Chrysophyceae are required for additional generalizations.

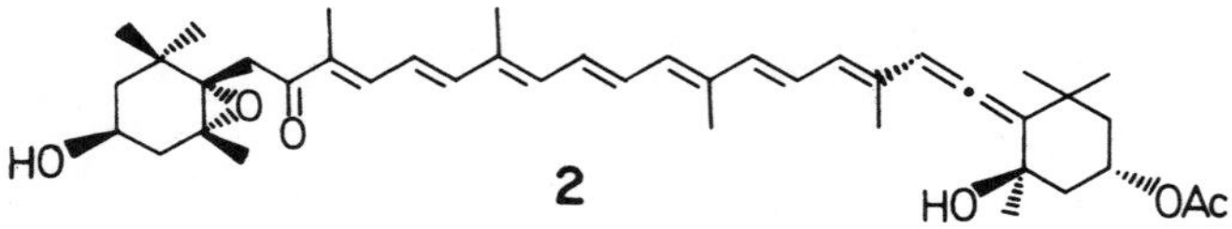

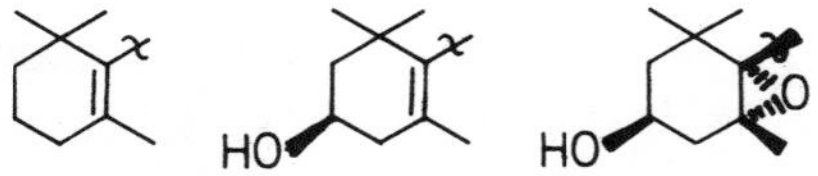

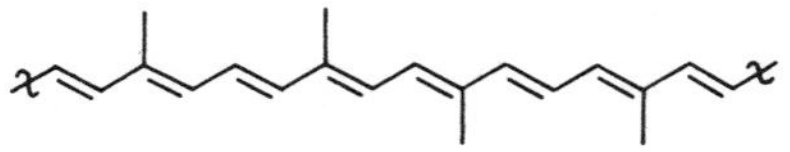

Characteristic
β,β-Carotene (**1**) a–P–a
Fucoxanthin (**2**)
Other
Zeaxanthin (**3**) b–P–b
Antheraxanthin (**4**) c–P–b
Violaxanthin (**5**) c–P–c
Cryptoxanthin (**6**) b–P–a
Cryptoxanthin epoxide (**7**) c–P–a?
Neoxanthin (**8**)

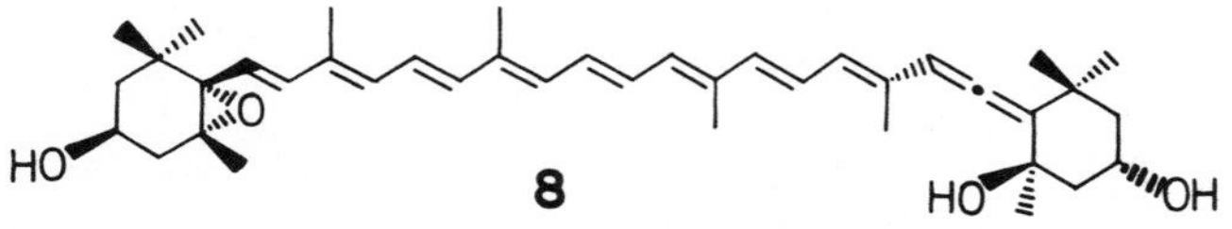

Scheme 11. Carotenoids of Chrysophyceae.

7. *Haptophyceae*

The Haptophyceae originally belonged to the Chrysophyceae. Recent work on some selected species, *Isochrysis galbana, Cricosphaera (Syracophaera) carterae, Prymnesium parvum, Monochrysis lutheri,* and a *Pavlova* species in our laboratory (Berger *et al.*, 1977b) has shown that all produce β,β-carotene (**1**), the acetylenic diatoxanthin (**2**), diadinoxanthin (**3**), and fucoxanthin (**4**), which is the major carotenoid, in addition to several minor carotenoids characteristic of each alga (see Scheme 12). These minor carotenoids include β,ϵ-carotene (**5**), β-carotene 5,6-epoxide (**6**), cryptoxanthin 5,6,5′,6′-diepoxide (**7**, tentative), cryptoxanthin (**8**), echinenone (**9**), canthaxanthin (**10**), fucoxanthinol (**11**), and deepoxyneoxanthin (**12**, tentative). The presence of fucoxanthin (**4**) as the major carotenoid, the presence of acetylenic carotenoids, and a high proportion of epoxidic carotenoids appear to be typical of the Haptohyceae.

Coccolithus huxleyi produces β,β-carotene (**1**), diatoxanthin (**2**), and diadinoxanthin (**3**) (Norgård *et al.*, 1974b; Hertzberg *et al.*, 1977). Fucoxanthin (**4**) could not be detected, and the major carotenoid is replaced by the previously unknown 19′-hexanoyloxyfucoxanthin (**13**) (Arpin *et al.*, 1976; Hertzberg *et al.*, 1977). The latter carotenoid was absent in the first-mentioned species. From one batch of *C. huxleyi* the apocarotenoid 19-hexanoyloxyparacentrone 3-acetate (**14**) was characterized (Arpin *et al.*, 1976). It is interesting that *C. huxleyi* differs from the other haptophytes in the sense that the haptonema never have been demonstrated.

8. *Bacillariophyceae*

Diatoms examined for carotenoids before 1970 have been listed by Goodwin (1971c). Early studies suggest that characteristic carotenoids are β,β-carotene (**1**), the acetylenic diatoxanthin (**2**), and diadinoxanthin (**3**) with fucoxanthin (**4**) as the major carotenoid (see Scheme 13). ϵ,ϵ-Carotene (**2**) was first discovered in *Nitzchia closterium* (Strain *et al.*, 1944).

Results on three kiesel algae, *Melosira sp., Phaeodactylum tricornutum (Navicula closterium)*, and *Navicula pelliculosa*, reported by Hager and Stransky (1970c) confirm this view. Neoxanthin (**6**) was detected in *P. tricornutum*.

9. *Xanthophyceae (Heterokontae)*

Xanthophyceae examined for carotenoids up to 1970 have been listed by Goodwin (1971c), who discusses the earlier work, including a quantitative study of nine species by Stransky and Hager (1970a). One species was

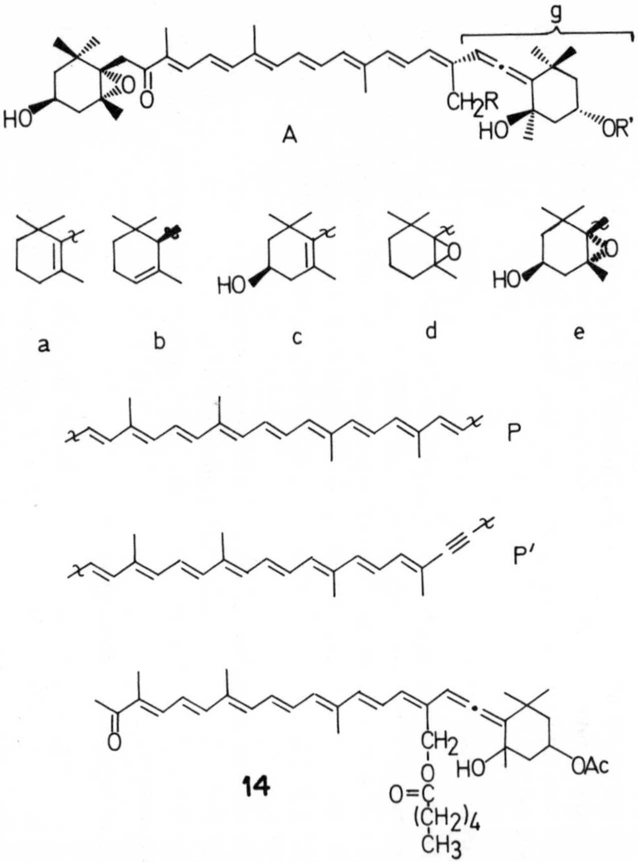

General

β,β-Carotene (**1**) a–P–a
Diatoxanthin (**2**) c–P′–c
Diadinoxanthin (**3**) e–P′–c
Fucoxanthin (**4**) A, R = H, R′ = Ac

Other

β,ε-Carotene (**5**) a–P–b
β-Carotene epoxide (**6**) d–P–a
Cryptoxanthin diepoxide (**7**) e–P–d
Cryptoxanthin (**8**) c–P–a
Echinenone (**9**) f–P–a
Canthaxanthin (**10**) f–P–f
Fucoxanthinol (**11**) A, R = H, R′ = H
Deepoxyneoxanthin (**12**) c–P–g, R = R′ = H
19′-Hexanoyloxyfucoxanthin (**13**) A, R = $OCO(CH_2)_4CH_3$
19′-Hexanoyloxyparacentrone 3- acetate (**14**)

Scheme 12. Carotenoids of Haptophyceae.

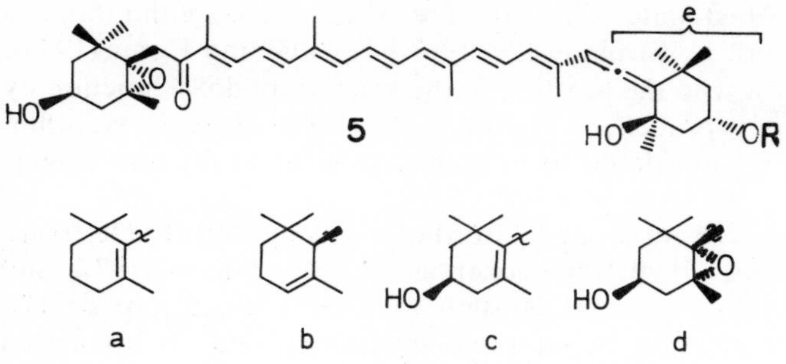

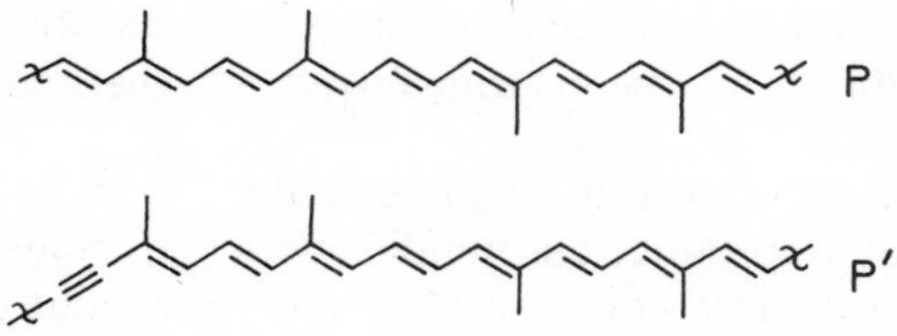

β,β-Carotene (1) a–P–a
ε,ε-Carotene (2) b–P–b
Diatoxanthin (3) c–P′–c
Diadinoxanthin (4) c–P′–d
Fucoxanthin (5) R = Ac
Neoxanthin (6) a–P–e, R = H

Scheme 13. Carotenoids of diatoms (Bacillariophyceae).

examined by Norgård *et al.* (1974b), and three were examined by Whittle and Casselton (1975b).

Major carotenoids (Scheme 14) appear to be the acetylenic epoxide diadinoxanthin (3) associated with smaller amounts of diatoxanthin (2). Also present in large amounts is vaucheriaxanthin (5), which occurs

General
β,β-Carotene (1) a–P–a
Diatoxanthin (2) b–P′–b
Diadinoxanthin (3) b–P′–d
Heteroxanthin (4) b–P′–e
Vaucheriaxanthin (5) A, R = CH_2OH

Other
Neoxanthin (6) A, R = CH_3*
β-Carotene diepoxide (7) c–P–c
Cryptoxanthin 5′,6′-epoxide (8) b–P–c
Cryptoxanthin diepoxide (9) d–P–c

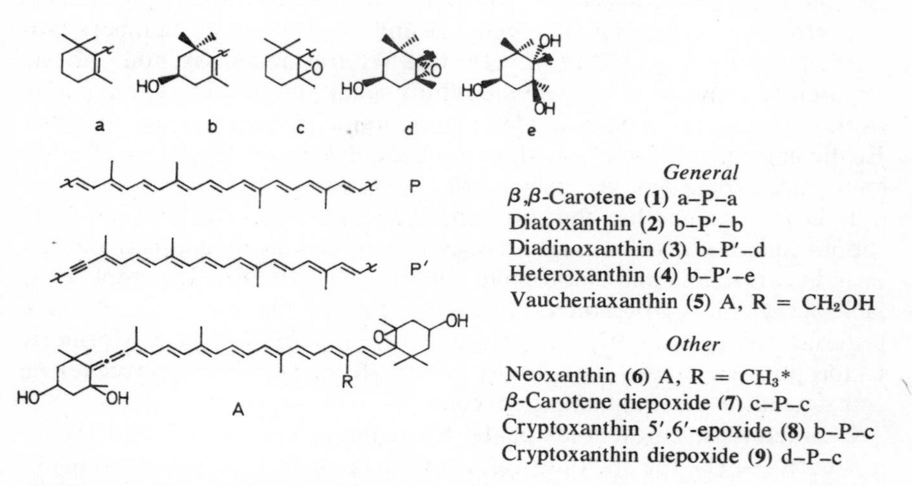

Scheme 14. Carotenoids of Xanthophyceae. *See absolute configuration for **6**, Scheme 13.

mostly in the esterified state. The structure of vaucheriaxanthin (**5**) was elucidated from work by Strain *et al.* (1968), Nitsche and Egger (1970), and Nitsche (1973a), who the location of the primary hydoxy function by ^{1}H nmr spectroscopy. Vaucheriaxanthin is thus 19′-hydroxyneoxanthin. The natural ester is considered to be a diacetate (Stransky and Hager, 1970a).

Other characteristic xanthophylls are the acetylenic tetrol heteroxanthin (**4**) (Strain *et al.*, 1970; Buchecker and Liaaen-Jensen, 1977) and neoxanthin (**6**). Minor carotenoids such as β-carotene diepoxide (**7**), cryptoxanthin 5′,6′-epoxide (**8**, structure uncertain), and cryptoxanthin diepoxide (**9**) are encountered. β,β-Carotene is the common carotene. *Pleurochloris commutata,* which produces no acetylenic carotenoids (Stransky and Hager, 1970b), has been transferred to the Eustigmatophyceae (Hibberd and Leedale, 1972).

On this basis the following generalizations can be made. Typical of the Xanthophyceae is their ability to produce carotenoids with (a) β rings, (b) epoxides (high proportion), (c) triple bonds, (d) allenes, (e) acetates, (f) 5,6-glycols, and (g) in-chain (19-) hydroxy groups. The last two properties (f and g) have not been noted in any of the algal classes so far discussed.

10. Eustigmatophyceae

On the basis of ultrastructural differences, Hibberd and Leedale (1971, 1972) transferred a number of species from the Xanthophyceae to their newly created class, the Eustigmatophyceae.

A study including mass spectrometry of the photosynthetic pigments in one prototype (*Pleurochloris magna*) and two potential members was carried out by Norgård *et al.* (1974b). Whittle and Casselton (1975a), apparently unaware of the previous study, later published an examination of the carotenoids of *Pleurochloris commutata, P. magna*, and six other Eustigmatophyceae species, drawing conclusions on the basis of chromatographic data and visible spectra.

It is interesting that the differences in ultrastructure between Xanthophyceae *sensu strictu* and Eustigmatophyceae nicely parallell a difference in carotenoid and chlorophyll composition. The Eustigmatophyceae lack acetylenic carotenoids and chlorophyll b. On the basis of these criteria, Antia *et al.* (1975) suggested that *Nannochloris oculata* (formerly Chlorophyceae) and *Monallantus salina* (formerly Xanthophyceae) be transferred to the Eustigmatophyceae.

Characteristic carotenoids of the Eustigmatophyceae (Scheme 15) are β,β-carotene (**1**), the epoxides violaxanthin (**2**, major), antheraxanthin (**3**), occasionally the parent diol zeaxanthin (**4**), the allenic vaucheriaxanthin

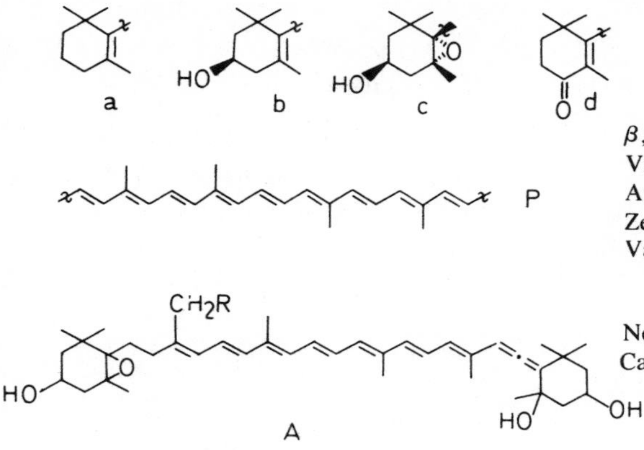

General
β,β-Carotene (**1**) a–P–a
Violaxanthin (**2**) c–P–c
Antheraxanthin (**3**) c–P–b
Zeaxanthin (**4**) b–P–b
Vaucheriaxanthin ester (**5**) A, R = OH

Occasional
Neoxanthin (**6**) A, R = H
Canthaxanthin (**7**) d–P–a

Scheme 15. Carotenoids of Eustigmatophyceae.

(**5**) ester, occasionally small amounts of the structurally related neoxanthin (**6**), and small amounts of the ketocarotenoid canthaxanthin (**7**). The proportion of epoxidic carotenoids is very high. In the algal classes discussed above the diepoxide violaxanthin (**2**) was rarely encountered.

11. Phaeophyceae

We shall now consider the last class in the division Chromophyta (Table 1), the brown algae.

Phaeophyceae examined for carotenoids up to 1970 have been reviewed by Goodwin (1971c), who also cites quantitative data. β,β-Carotene (**1**) is the major and usually sole carotene (Jensen, 1966a; Strain, 1966). Fucoxanthin (**3**) (Jensen, 1966a; Bonnett *et al.*, 1969; Bernhard *et al.*, 1976) is always the major carotenoid, and the diepoxide violaxanthin (**2**) (Bartlett *et al.*, 1969) is the second major xanthophyll (Scheme 16). Bernhard *et al.* (1974) have claimed that the 6′*S* isomer, which is formed by rotation of the allenic bond of fucoxanthin (**3**), occurs naturally in low amounts.

Hager and Stransky (1970c) reported small amounts of ε,ε-carotene (**4**) from a *Laminaria sp.* species. The acetylenic diatoxanthin (**7**) and diadinoxanthin (**8**) may occur as trace constituents and are believed not to originate from contaminating diatoms (Strain *et al.*, 1944; Jensen, 1966a). The isolation of small amounts of zeaxanthin (**5**) (Jensen, 1966a; Hager and Stransky, 1970c) could be due to post mortem changes of violaxanthin (**2**) (Heilbron and Phipers, 1935; Liaaen and Sørensen, 1956). Using modern techniques, Nitsche (1974a) demonstrated that fucoxanthinol (**9**)

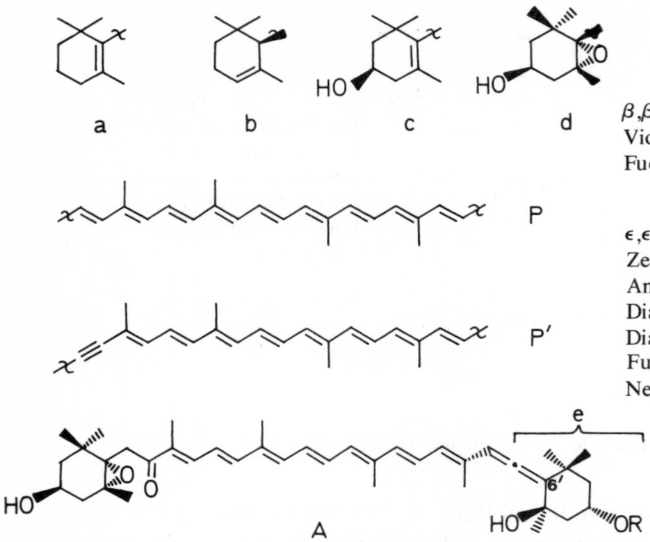

Major

β,β-Carotene (**1**) a–P–a
Violaxanthin (**2**) d–P–d
Fucoxanthin (**3**) A, R = Ac

Minor

ϵ,ϵ-Carotene (**4**) b–P–b
Zeaxanthin (**5**) c–P–c
Antheraxanthin (**6**) d–P–c
Diatoxanthin (**7**) c–P–c
Diadinoxanthin (**8**) c–P–d
Fucoxanthinol (**9**) A, R = H
Neoxanthin (**10**) d–P–e, R = H

Scheme 16. Carotenoids of brown algae (Phaeophyceae).

and neoxanthin (**10**) are minor constituents in *Fucus vesiculosus*.

In conclusion, we see that brown algae have a characteristic carotenoid composition with fucoxanthin (**3**) as the major xanthophyll, accompanied by violaxanthin (**2**) and β,β-carotene (**1**) as major carotenoids. Other carotenoids (Scheme 16) may be present in trace amounts.

12. Euglenophyceae

Earlier studies on the carotenoids of *Euglena* have been summarized briefly by Goodwin (1971c, 1976).

Euglena can grow heterotrophically in the dark without the production of chloroplasts and therefore offers the opportunity of studying carotenoid biosynthesis during the development of the photosynthetic apparatus in light (Krinsky *et al.*, 1964).

The major carotenoid of *Euglena* is now, after some earlier controversy, recognized as the acetylenic epoxide diadinoxanthin (**3**) (Aitzetmüller *et al.*, 1968) (see Scheme 17). The acetylenic diatoxanthin (**2**) (Johannes *et al.*, 1971), heteroxanthin (**4**) (Buchecker and Liaaen-Jensen, 1977; Nitsche, 1973b), and the allenic neoxanthin (**5**) (Krinsky and Goldsmidt, 1960; Nitsche, 1973b) are usually also present. β,β-Carotene (**1**) is the major carotene.

General
β,β-Carotene (**1**) a–P–a
Diatoxanthin (**2**) c–P–c
Diadinoxanthin (**3**) c–P–d
Heteroxanthin (**4**) c–P–e
Neoxanthin (**5**) f–Q–d

Eye spot
Echinenone (**6**) g–P–a
3-Hydroxyechinenone (**7**) h–P–a
Canthaxanthin (**8**) g–P–g
Astaxanthin (**9**) h–P–h

Other
Cryptoxanthin (**10**) c–P–a
Cryptoxanthin 5′,6′-epoxide (**11**) c–P–b

Scheme 17. Carotenoids of Euglenophyceae.

Hager and Stransky (1970c) reported small amounts of cryptoxanthin (**10**) and its 5′,6′-epoxide (**11**) in *Euglena gracilis*. Small amounts of ketocarotenoids such as echinenone (**6**), 3-hydroxyechinenone (**7**), euglenanone (= canthaxanthin) (**8**) (Goodwin and Gross, 1958; Krinsky and Goldsmidt, 1960), and astaxanthin (**9**) (Tischer, 1941) are likely to be associated with the eye spot.

13. Prasinophyceae and Loxophyceae

The carotenoids of these green flagellates have been studied by Ricketts (1966, 1967a,b, 1970, 1971a,b). The situation is not clear, but carotenoids that have been encountered (Scheme 18) include lycopene (**1**, minor), β,ψ-carotene (**2**, minor), β,β-carotene (**3**), β,ϵ-carotene (**4**), zeaxanthin (**5**) lutein (**6**), violaxanthin (**7**), and neoxanthin (**8**). Some species produce xanthophylls that were eventually identified as siphonaxanthin (**9**) and siphonein (**10**). Siphonein (**10**) constitutes a mixed 19-ester, mainly with C_{16} esterifying acids (Ricketts, 1971a) of siphonaxanthin (**9**) (Kleinig et al., 1969; Ricketts, 1971b). The structure of micronone (**11**) (Ricketts, 1966, 1967b) is not yet known.

The above information suggests that carotenoid synthesis in these green flagellates is characterized by an efficient ϵ-cyclase system. The proportion of epoxidic carotenoids is moderate, and acetylenic carotenoids have not been found.

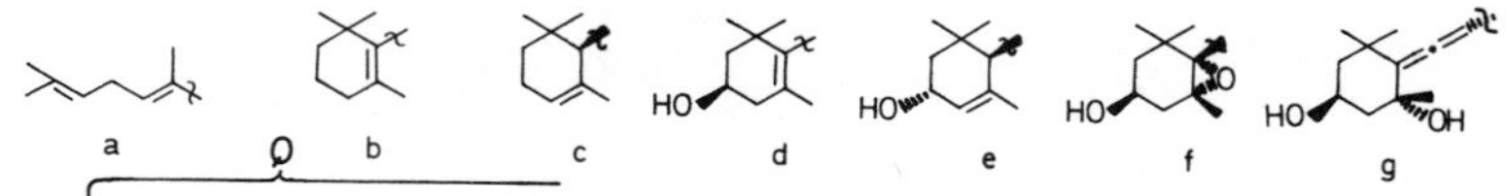

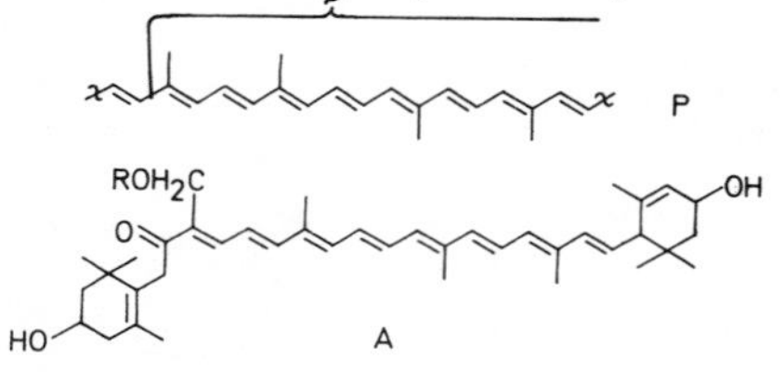

Lycopene (**1**) a–P–a
β,ψ-Carotene (**2**) b–P–a
β,β-Carotene (**3**) b–P–b
β,ϵ-Carotene (**4**) b–P–c
Zeaxanthin (**5**) d–P–d
Lutein (**6**) d–P–e
Violaxanthin (**7**) f–P–f
Neoxanthin (**8**) g–Q–f
Siphonaxanthin (**9**) A, R = H
Siphonein (**10**) A, R = acyl
Micronone (**11**)

Scheme 18. Carotenoids of green flagellates (Prasinophyceae and Loxophyceae).

14. Chlorophyceae

Green algae from all 11 orders examined for carotenoids have been listed by Goodwin (1971c). A quantitative study by Hager and Stransky (1970b) covering 16 Chlorophyceae species in several orders is particularly instructive.

Most green algae appear to have a carotenoid composition resembling that of higher plants (see Scheme 19). Characteristic carotenes are β,β-carotene (**1**) as well as β,ϵ-carotene (**2**). The dominating xanthophyll is generally lutein (**3**) with one ϵ ring. Also abundant is zeaxanthin (**4**) and the corresponding epoxides antheraxanthin (**5**) and violaxanthin (**6**). The allenic neoxanthin (**7**) is usually present.

Some carotenoids are occasionally encountered as minor constituents. These include β,ψ-carotene (**8**) ϵ,ϵ-carotene (**9**), cryptoxanthin (**10**), cryptoxanthin 5′,6′-epoxide (**11**, structural evidence insufficient) and lutein epoxide (**12**).

Loroxanthin (19-hydroxylutein) is encountered in *Chlorella* and *Scenedesmus* species (Aitzetmüller *et al.*, 1969) and in *Chlamydomonas reinhardii* (Francis *et al.*, 1973). Pyrenoxanthin from *Chlorella pyrenoidosa*, considered to be 20-hydroxylutein (Yamamoto *et al.*, 1969), has been found to be identical with loroxanthin (Nitsche, 1974b). It therefore appears that the unidentified β,ϵ-carotenetriol isolated by Hager and Stransky (1970b) from several species, including *Chlorella pyrenoidosa*, is loroxanthin (**13**). From the nonmarine *Trentepohlia iolithus*, (2*R*)-β,β-caroten-2-ol (**14**), (2*R*,2′*R*)-β,β-carotene-2,2′-diol (**15**), and (2*R*,6′*S*)-β,ϵ-caroten-2-ol (**16**) have been characterized (Kjøsen *et al.*, 1972; Buchecker *et al.*, 1974).

General

β,β-Carotene (**1**) b–P–b
β,ε-Carotene (**2**) b–P–c
Lutein (**3**) d–P–g
Zeaxanthin (**4**) d–P–d
Antheraxanthin (**5**) d–P–i
Violaxanthin (**6**) i–P–i
Neoxanthin (**7**) **6**, Scheme 14

Minor occasional

β,ψ-Carotene (**8**) b–P–a
ε,ε-Carotene (**9**) c–P–c
Cryptoxanthin (**10**) d–P–b
Cryptoxanthin 5′,6′-epoxide (**11**) d–P–h
Lutein epoxide (**12**) i–P–g

Special occasional

Loroxanthin (**13**)
β,β-Caroten-2-ol (**14**) e–P–b
β,β-Carotene-2,2′-diol (**15**) e–P–e
β,ε-Caroten-2-ol (**16**) e–P–c
Siphonaxanthin (**17**) A, R = H
Siphonein (**18**) A, R = lauroyl

Ketonic carotenoids produced during nitrogen starvation

Echinenone (**19**) j–P–b
4-Hydroxyechinenone (**20**) f–P–j
Canthaxanthin (**21**) j–P–j
3-Hydroxycanthaxanthin (**22**) k–P–j
Astaxanthin (**23**) k–P–k as ester

Scheme 19. Carotenoids of green algae (Chlorophyceae).

Species of the order Siphonales frequently also contain the ketone siphonaxanthin (**17**) and the ester siphonein (**18**) (Strain, 1965; Kleinig, 1969). Siphonaxanthin was eventually formulated as **16** (Kleinig and Egger, 1967; Kleinig *et al.*, 1969; Walton *et al.*, 1970; Strain *et al.*, 1971a; Ricketts, 1971a,b) with a hydrogen-bonded keto group at C-8. Siphonein is considered by the same workers to be the 19-laurate (**17**).

Other ketonic carotenoids with keto group(s) at C-4 accumulate in various green algae (Kleinig, 1966; Czygan, 1968a), particularly under conditions of nitrogen starvation. These include echinenone (**19**) 4-hydroxyechinenone (**20**), canthaxanthin (**21**), 3-hydroxycanthaxanthin (**22**), and astaxanthin (**23**) esters (Czygan and Kessler, 1967; Czygan, 1968a; Kleinig and Czygan, 1969). Caprylic, capric, lauric, myristic, palmitic, and stearic esters occur (Kleinig and Czygan, 1969). Related diosphenols with end group m (Scheme 19), which have been claimed to be present (Kleinig and Czygan, 1969; Hager and Stransky, 1970b), are considered to be artifacts.

Location of extraplastidic carotenoids synthesized under unfavorable growth conditions has been discussed by Goodwin (1976). The carotenoid content may be exceptionally high, e.g., 14% β,β-carotene (**1**) of the dry matter in *Dunaliella salina (Aasen et al.*, 1969a).

We may conclude, therefore, that the carotenoids of green algae closely resemble those of higher plants, that they reveal the presence of an active ϵ-cyclase, and that they constitute a fairly high proportion of epoxidic carotenoids. Allenic but no acetylenic carotenoids have been found. In-chain hydroxylation may occur (siphonaxanthin, **17**, and loroxanthin, **13**). In the Siphonales, siphonaxanthin (**17**) is encountered. 4(4′)-Ketocarotenoids are produced under conditions of nitrogen starvation and stored outside the chloroplasts. Carotenol esters include no acetates, only longer-chain acid moieties.

B. Biosynthetic, Chemosystematic, and Evolutionary Aspects

Little work toward defining the pathways of carotenoid biosynthesis in algae has been carried out. With our present knowledge of growing microscopic algae, work in this direction is anticipated.

Whereas bacteria frequently leave some of their biosynthetic precursors behind, algae carry out carotenoid biosynthesis more efficiently to structurally more complex products. Factors controlling carotenoid synthesis in algae have been reviewed by Goodwin (1971c) and Britton

(1976b). Research on biosynthetic routes by Claes (1954, 1957, 1959), who utilized mutants of *Chlorella vulgaris*, has demonstrated the presence of the more saturated precursors such as phytoene (**1**, Scheme 2), phytofluene (**2**, Scheme 2), and ξ-carotene (**3**, Scheme 2) in mutants having a block in their synthesis of colored carotenoids, and 3-hydroxylation of β rings of carotenes in the presence of oxygen and light. 5,6-Epoxidation of xanthophylls in chloroplasts in the presence of oxygen and light is a well-known phenomenon (Krinsky, 1966; Hager, 1967a,b; Sapozhnikov, 1973). Experiments with oxygen-18 have confirmed that the epoxidic oxygen is derived from molecular oxygen (Yamamoto and Chichester, 1965). The reverse, deepoxidation, occurs in darkness. A cell-free preparation of *Euglena gracilis* converted the epoxidic diadinoxanthin to diatoxanthin in the presence of a hydrogen donor (Bamji and Krinsky, 1965). Post-mortem deepoxidation of violaxanthin to zeaxanthin in brown algae has been recorded (Liaaen and Sørensen, 1956).

Evidence for the steps in the biosynthetic pathway that leads to more complex algal carotenoids, including allenic, acetylenic, and highly oxygenated carotenoids, is lacking (Britton, 1976a,b). However, when we consider the structural variations encountered in algal carotenoids (Section VI.A) and our knowledge of carotenogenesis in algal and other systems (Section III), it is possible to formulate a minimal number of steps that are required for the synthesis of the terminal carotenoid products at the C_{40} level (see Scheme 20 and Table 2).

Some of the reactions listed must clearly be multistep transformations, e.g., A, O, and P, but plausible intermediates have so far not been isolated. In some cases the immediate precursor is completely unknown, e.g., for reaction J or N. For the acetylenic carotenoids an allenic precursor has been considered (Weedon, 1970), a reaction that has an *in vitro* counterpart (Johansen and Liaaen-Jensen, 1974). However, in order to account for the abundance of acetylenic carotenoids in the lower algal classes, one may consider a structurally simpler precursor (see Scheme 21, alternative b), which involves direct cis dehydrogenation of a sterically hindered 7-cis double bond.

Yet even if we include the above reservations in our thinking, Scheme 20 gives an idea of the minimal number of enzymes required to effect the final transformations in the biosynthesis of algal carotenoids. Recent knowledge of the absolute configuration of algal carotenoids (Bartlett *et al.*, 1969; Buchecker *et al.*, 1973, 1974, 1976a; Bernhard *et al.*, 1976; Hertzberg *et al.*, 1977; Johansen and Liaaen-Jensen, 1977; Buchecker and

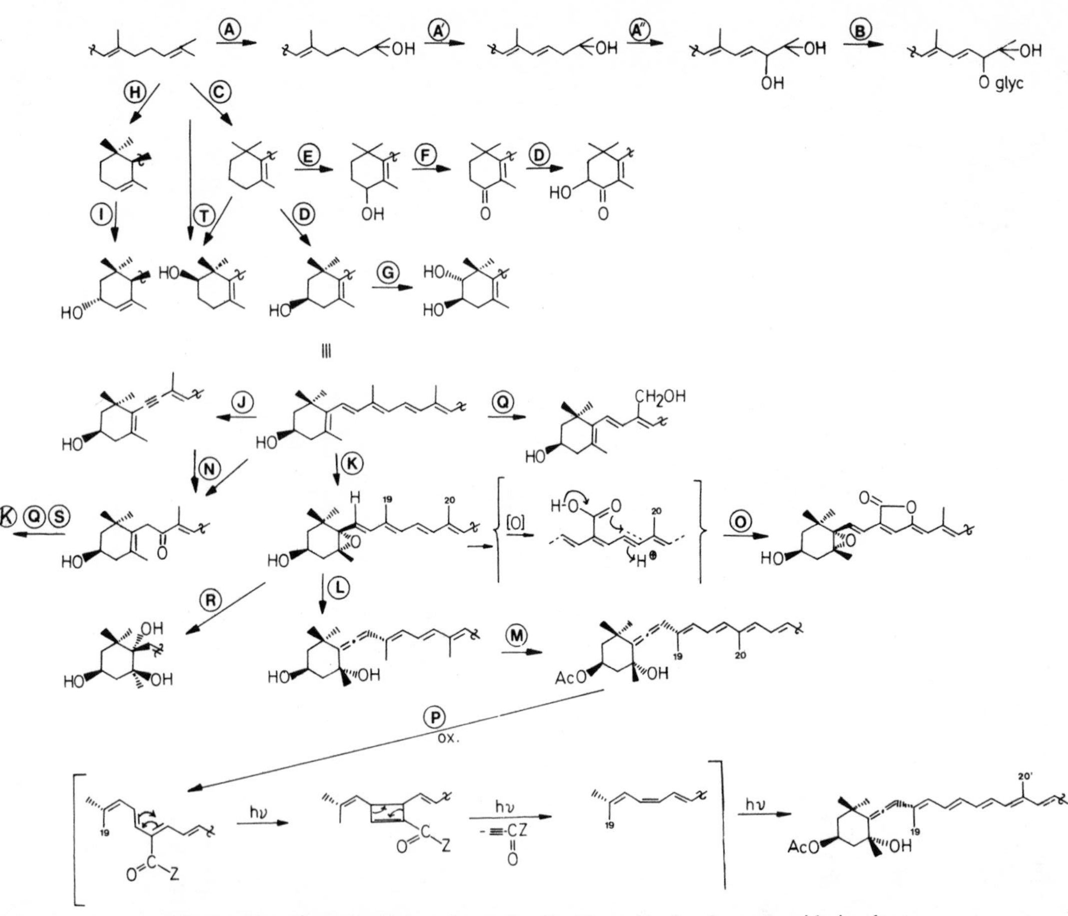

Scheme 20. Hypothetical pathway for the biosynthesis of carotenoids in algae.

TABLE 2

Required Steps for the Synthesis of Algal Carotenoids

	Reaction	Step
A.	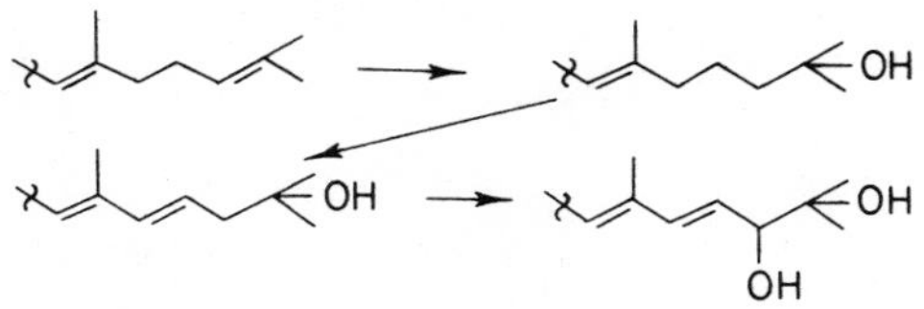 OH; OH; OH, OH	
B.	ROH → ROglyc.	Glycosidation
C.	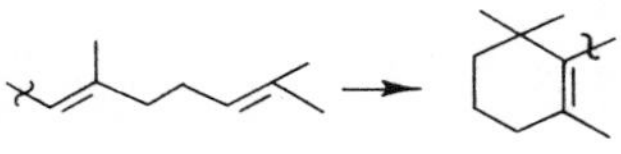	β-Cyclization
D.	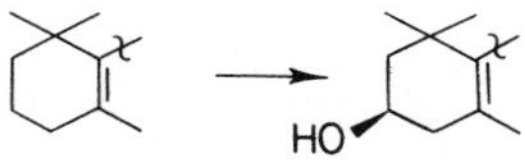 HO	3-Hydroxylation
E.	 OH	Allylic 4-hydroxylation
F.	 (OH); O	Allylic 4-keto formation
G.	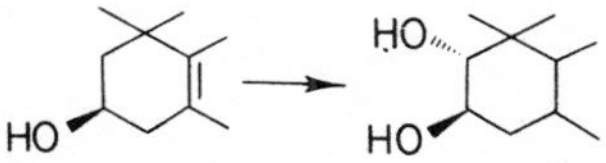 HO; HO, HO	Trans 2-hydroxylation
H.	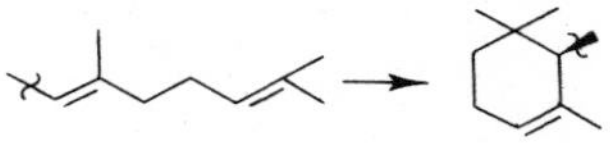	ϵ-Cyclization
I.	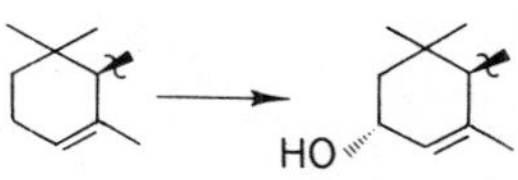 HO	Allylic 3-hydroxylation
J.	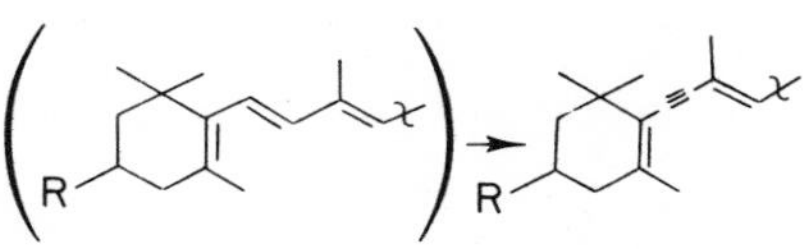 R; R	Triple-bond formation
K.	 R; R, O; R = H, OH	5,6-Epoxidation

(*Continued*)

TABLE 2 (*continued*)

	Reaction	Transformation
L.	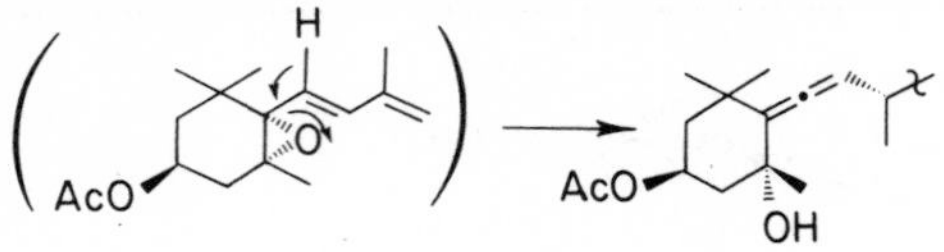 AcO → AcO, OH	Allene formation
M.	ROH → ROAc	Acetylation
N.	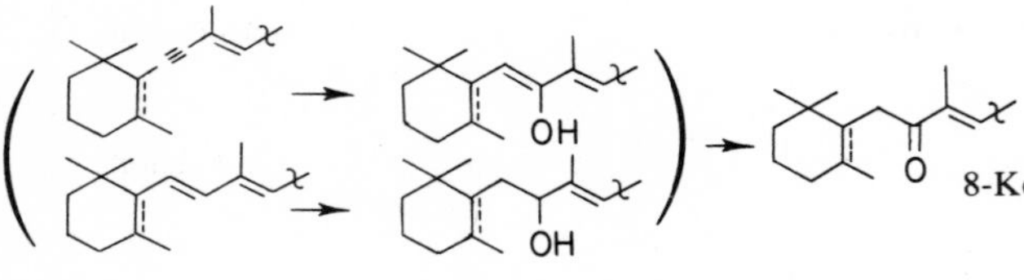 OH, OH → O	8-Keto formation
O.	 → → → O, O	Butenolide formation
P.	 → → →	C_3 Expulsion
Q.	 CH_3 → CH_2OH	19-Hydroxylation
R.	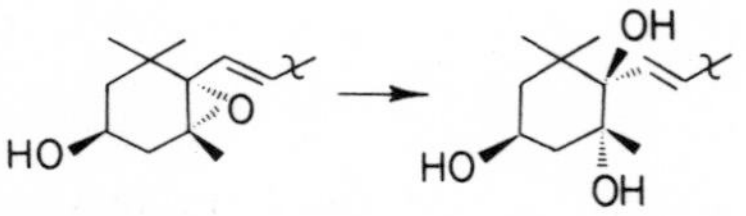 HO, O → OH, HO, OH	5,6-Glycol formation
S.	ROH → ROacyl	Acylation (higher fatty acids)
T.	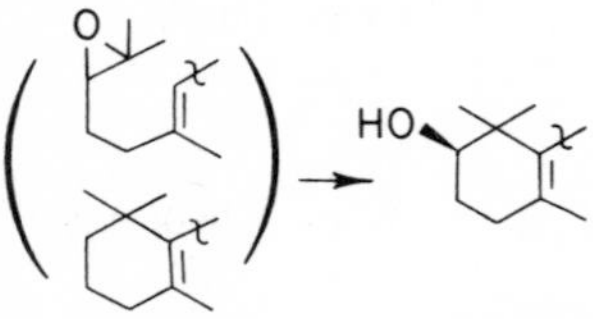 O → HO	2-Hydroxylation

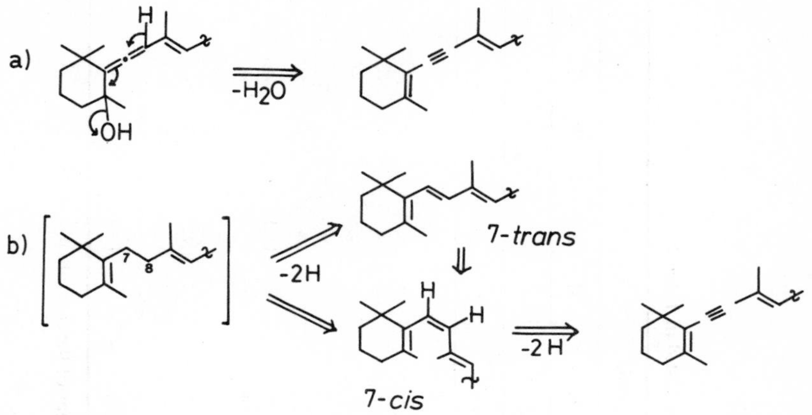

Scheme 21. Some alternatives for the biosynthesis of acetylenic carotenoids.

Liaaen-Jensen, 1977; Liaaen-Jensen, 1976) supports the enzymatic character of the transformations involved and indicates probable biosynthetic relationships between various chiral end groups (see Scheme 20).

When one considers the chemotaxonomic value of carotenoids in algal systematics, the presence or absence of enzymes capable of carrying out the various transformations is believed to be a better criterion than end-product identification alone. In order to emphasize this point Table 3 compares the biosynthetic capability of various algal classes, descending from the most primitive to the most highly developed, according to Christensen (1962). However, some changes in Christensen's original ordering have been made in order to demonstrate the logical progression in carotenoid synthesis.

It is evident that the carotenoid complement of blue-green algae is so specific that the class can easily be defined on this basis. A high proportion of monocyclic and aliphatic carotenoids and the presence of carotenoid glycosides are unique to this class (reactions A and B).

The following statements summarize our present knowledge of carotenoid biosynthesis in various algal classes. Letters in parantheses refer to reaction types or necessary enzymes that lead to a particular structural element (see Scheme 20 and Table 2).

Red algae contain simple carotenoids characterized by a strong

TABLE 3

Capability of Algal Classes to Effect the Reactions Listed in Table 2

Class[a]	Step reaction																			
	A	B	C	D	E	F	G	H	I	J	K	L	M	N	O	P	Q	R	S	T
1. Cyanophyceae	+	+	+	+	+	+	+	−	−	−	(−)[b]	−	−	−	−	−	−	−	−	−
2. Rhodophyceae	−	−	+	+	−	−	−	+	+	−	(−)	−	−	−	−	−	−	−	−	−
3. Cryptophyceae	−	−	+	+	−	−	−	+	+	+	−	−	−	−	−	−	−	−	−	−
5. Rhapidophyceae	−	−	+	+	−	−	−	+	+	−	+	−	−	−	−	−	−	−	−	−
6. Chrysophyceae	−	−	+	+	−	−	−	−	−	−	+	+	+	+	−	−	−	−	−	−
11. Phaeophyceae	−	−	+	+	−	−	−	(−)	−	−	+	+	+	+	−	−	−	−	−	−
7. Haptophyceae	−	−	+	+	−	−	−	+	−	+	+	+	+	+	−	−	(−)	−	(−)	−
8. Bacillariophyceae	−	−	+	+	−	−	−	+	−	+	+	+	+	+	−	−	−	−	−	−
4. Dinophyceae	−	(−)	+	+	(−)	(−)	−	−	−	+	+	+	+	(−)	+	+	−	−	−	−
10. Eustigmatophyceae	−	−	+	+	−	−	−	(−)	−	−	+	+	+	+	−	−	+	−	+	−
9. Xanthophyceae	−	−	+	+	−	−	−	−	−	+	+	+	−	−	−	−	+	+	+	−
12. Euglenophyceae	−	−	+	+	(+)	+	−	−	−	+	+	+	−	−	−	−	−	+	−	−
13. Prasinophyceae/ Loxophyceae	−	−	+	+	−	−	−	+	+	−	+	+	−	−	−	−	+	−	+	−
14. Chlorophyceae	−	−	+	+	+	+	−	+	+	−	+	+	−	(+)	−	−	+	−	−	+

[a] Numbers refer to the ranking used by Christensen (1962) and in the present treatment of the individual classes.
[b] ( ) Indicates minor reservations.

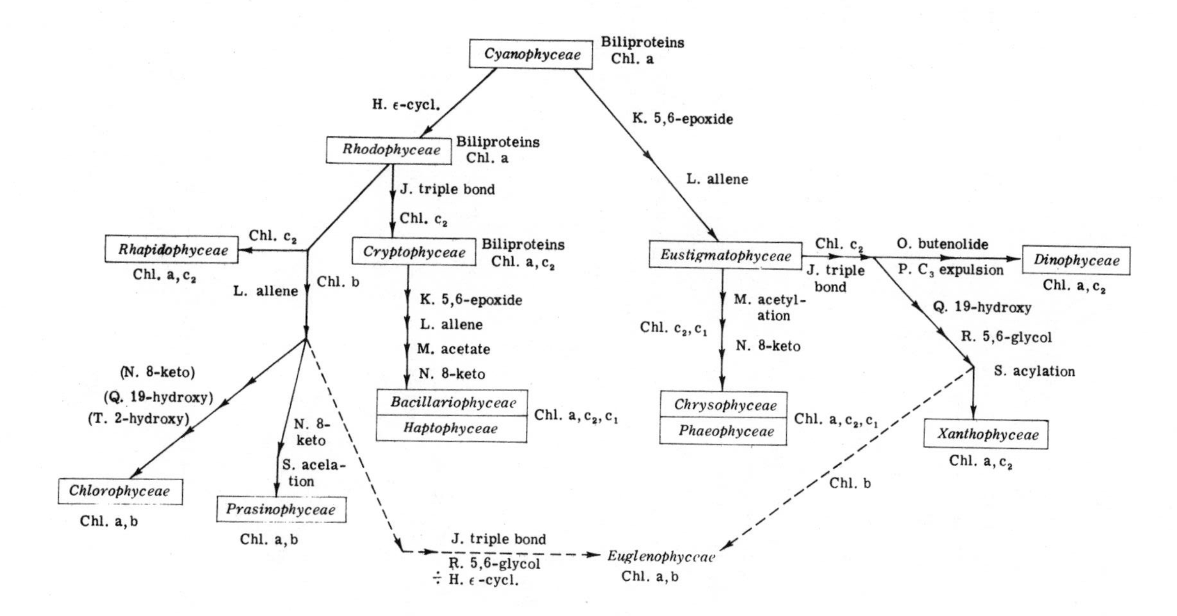

Scheme 22. Possible events in the evolution of algal chloroplast pigments. Capital letters refer to enzymes required to effect carotenoid transformations in Scheme 20 and Table 2.

ϵ-cyclase requirement (H). Cryptophyceae also contain simple carotenoids but a dominance of acetylenic derivatives (J). Epoxidic carotenoids (K) are rarely encountered in these three lowest algal classes but are present in all the other algal classes. Cyclic carotenoids with 3-hydroxylated β rings (C, D) are common to all algae and are therefore without chemotaxonomic significance.

Rhapidophyceae, Chrysophyceae, and Phaeophyceae contain hardly any acetylenic carotenoids (J) but synthesize epoxides (K). The two latter classes also elaborate allenic end groups with acetoxy function (L, M) and 8-keto groups (N), but they lack the ability to produce ϵ rings (H).

Haptophyceae and Bacillariophyceae have similar carotenoids including structural features such as ϵ ring (H), triple bonds (J), epoxides (K), acetylated allenic end groups (L, M), and 8-keto groups (N).

Dinophyceae have a unique carotenoid pattern, including C_{37} skeletal carotenoids of the peridinin type (O, P). Of the Eustigmatophyceae and Xanthophyceae only the Xanthophyceae have the ability to produce acetylenic carotenoids (J) and 5,6-glycols (R). Both classes synthesize 19-ols (Q) and higher fatty acid esters (S).

Euglenophyceae differ from the other Chlorophyta by the absence of ϵ-type carotenoids (H) and the presence of acetylenic carotenoids (J). A 5,6-glycol (R) is also encountered here. Acetates (M) and 8-ketocarotenoids (N) are not found in the Chlorophyta, except in the Siphonales (N). Some of the green flagellates and green algae contain 19-ols (Q) and epoxidic (K), allenic (L), and ϵ-type (I) xanthophylls.

We may conclude from these summary statements that the differences observed in carotenoid composition among most of the algal classes demonstrate the possible use of carotenoids as a taxonomic tool in algal classification. Modern reinvestigation and studies of additional species may somewhat alter the details of Table 3, but our present state of knowledge is sufficiently solid for chemosystematic evaluation.

The obvious diagonal trend of Table 3, which indicates the presence of certain structural features and the lack of others, makes it tempting to speculate what kind of evolutionary pattern among the algal classes would emerge from the concepts of Scheme 20, assuming that evolution is triggered by the development of new enzymes. Scheme 22 incorporates these concepts. Repression of originally developed enzymes has been kept to a minimum, but the parallel evolution of new enzymes is required. Consideration is also given to the presence of biliproteins (O'Carra and OhEocha, 1976) and to the distribution of chlorophylls (Bogorad, 1976; Jackson, 1976).

Cyanophyceae, which elaborate biliproteins, chlorophyll a only, and aliphatic and monocyclic carotenoids, are recognized as the most primitive algae.

Rhodophyceae, which similarly possess biliproteins and chlorophyll a only, could have evolved from the Cyanophyceae, including the development of ϵ-cyclase to produce ϵ-type carotenoids.

Rhapidophyceae appear to be further developed since carotenoid epoxides (enzyme K) and chlorophyll c_2 are present. The development of chlorophyll b and allenic carotenoid synthesis leads to a missing link from which Prasinophyceae/Loxophyceae could have evolved after acquiring enzyme N (8-keto introduction) and S (acylation), followed by Chlorophyceae after developing enzymes N (8-keto), Q (19-ol), and T (2-ol), which are present in some species.

Cryptophyceae, which still contain some biliproteins and chlorophyll a and c_2, could have evolved from the Rhodophyceae after evolving enzyme J (acetylenic carotenoids). Bacillariophyceae and Haptophyceae might follow by the introduction of chlorophyll c_1 synthesis and the synthesis of epoxides (K), allenic acetates (L, M), and 8-keto carotenoids (N).

In Scheme 22 algal classes without ϵ-type carotenoids are presumed to have evolved by a different route from the Cyanophyceae.

The Eustigmatophyceae, which possess chlorophyll a, may occupy only an intermediate position after the introduction of epoxidic (K) and allenic (L) carotenoids. To reach the Chrysophyceae and Phaeophyceae, chlorophyll c_2 and c_1 and enzymes M (acetate) and N (8-keto) are required.

Alternatively, the introduction of enzyme J (acetylenic carotenoids) and chlorophyll c_2 could lead to a missing link from which Dinophyceae evolved after the evolution of enzymes O (butenolide) and P (C_3 expulsion), which are unique to this class, and from which Xanthophyceae evolved after the development of enzymes Q (19-ol), R (5,6-glycol), and S (acetylation).

The Euglenophyceae pose a particular dilemma. The left dotted line avoids duplicate evolution of chlorophyll b but requires total repression of the ϵ-cyclase.

An important lesson of Scheme 22 is the realization that the total picture is not easy to rationalize. Several branching points are likely to have occurred in algal evolution. This complicates an arrangement of classes on an evolutionary scale (see Table 3). Missing links pose additional problems. Similar attempts with different results have been made by Goodwin (1971c). However, as our knowledge increases, there is a reasonable chance that the pigments of the photosynthetic apparatus, considered together with other important morphological and particularly physiological (life cycle) parameters, may throw additional light on these fascinating problems.

C. Algal Carotenoids as Food Chain Indicators

It is assumed that animals cannot carry out *de novo* carotenoid synthesis. It is further recognized that planktonic algae serve as food for various marine animals. Provided that the animal retains the carotenoids intact or metabolizes them in such a way that the original structure can be recognized, our knowledge of the distribution of algal carotenoids (Section VI,A) provides a tool for tracing the food chain.

Studies along this line were carried out by Jensen and Sakshaug (1970a,b), who demonstrated the correlation between total carotenoids in the edible sea mussel *Mytilus edulis* and the occurrence of phytoplankton in the Trondheimsfjord. Increased carotenoid content in the mussels was observed in two consecutive years during blooms of dinoflagellates such as *Ceratium fusus*. Other workers (Khare *et al.*, 1973) have demonstrated that the acetylenic mytiloxanthin (**1**) and isomytiloxanthin (**2**) are characteristic constituents of *M. edulis* and have suggested that end groups a and b (Scheme 23) may be biogenetically related to the epoxidic end groups d of fucoxanthin (**9**, Scheme 9). Of the C_{40} carotenoids encountered in dinoflagellates, diadinoxanthin (**3**, Scheme 9) is a possible precursor.

Lederer (1938) observed some 40 years ago that the carotenoid content of the scallop *Pecten maximus* varied greatly with the season. Pectenolone (**3**, Scheme 23) (Campbell *et al.*, 1967) from *P. maximus* is a likely metabolic product of diatoxanthin (**2**, Scheme 9) with one end group c. Alloxanthin (= pectenoxanthin = cynthiaxanthin) (**4**, Scheme 23) is another constituent of *P. maximus* (Campbell *et al.*, 1967), thereby suggesting cryptophycean (Scheme 8) origin.

Brightly colored nudibranchs that contain triophaxanthin (**5**) and hopkinsiaxanthin (**6**) are reported to be sponge, bryozoan, or hydroid feeders; the carotenoids are believed to be resorbed unchanged from the food source (McBeth, 1972a,b). Alloxanthin (**4**) and diatoxanthin, therefore, are possible precursors of **5** and **6** (Scheme 23).

The occurrence of β,β-caroten-2-ol with end group e in *Idotea* (Lee and Gilchrist, 1975) indicates that this carotenoid may also occur in marine algae (see Section VI, A, 14). 4-Hydroxy- and 4-ketocarotenoids (end group f) obtained from *Idotea granulosa*, which is known to feed on red and brown algae (Lee, 1966a), are metabolic products of β,β-carotene (see Scheme 31). The best-known metabolic product of β,β-carotene is, of course, vitamin A, which is stored in fish liver.

Fucoxanthinol (**7**) and paracentrone (**8**) have been obtained from the sea

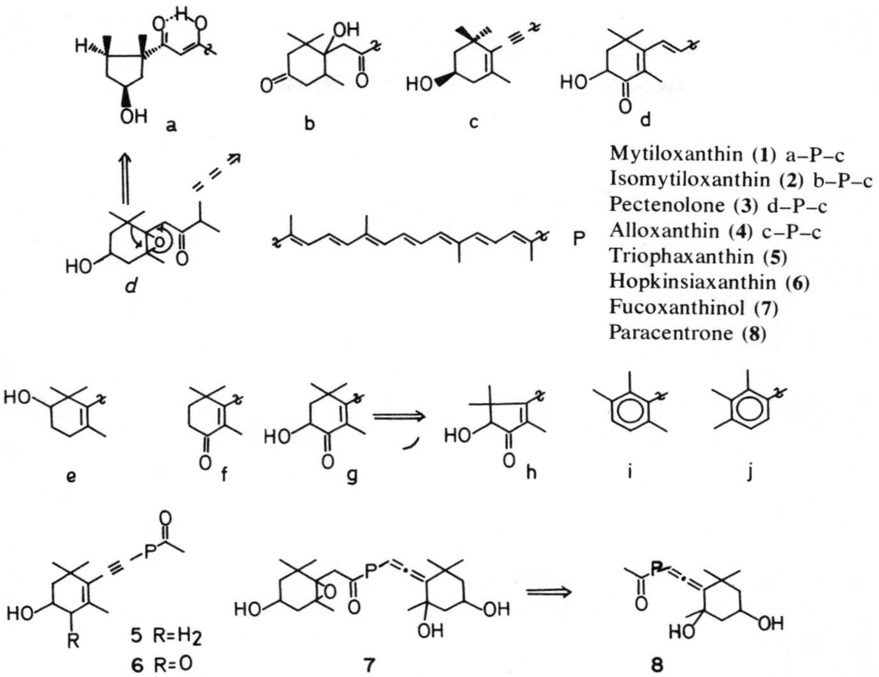

Mytiloxanthin (**1**) a–P–c
Isomytiloxanthin (**2**) b–P–c
Pectenolone (**3**) d–P–c
Alloxanthin (**4**) c–P–c
Triophaxanthin (**5**)
Hopkinsiaxanthin (**6**)
Fucoxanthinol (**7**)
Paracentrone (**8**)

Scheme 23. Examples of metabolic carotenoids of presumed algal origin.

urchin *Paracentrus lividus* (Scheme 23), and the plausible suggestion has been made that paracentrone (**8**) is a metabolic product of fucoxanthinol (**7**) which is discussed in Section VII,C. Fucoxanthin is a major carotenoid in diatoms (see Scheme 13) and brown algae (Scheme 16).

Tunaxanthin (**20**, Scheme 32, structure uncertain) (Crozier 1967) from fish is probably the 3,3′-diol corresponding to the rare algal carotene ϵ,ϵ-carotene (**2**, Scheme 13). Chiriquixanthins A and B, which are reported to have the same structural features as **9**, have been characterized from a terrestrial source (Bingham *et al.*, 1977).

Marine invertebrate and/or vertebrate carotenoids such as idoxanthin (**9**, Scheme 31), astaxanthin (**12**, Scheme 31), asterinic acid (**2** and **3**, Scheme 25), and doradexanthin (**18**, Scheme 32), all with α-ketol end group g, are examples of reasonably well defined algal carotenoid precursors of the β,β-carotene type. This is also probably true for the aromatic sponge carotenoids with end groups i and j; algae are apparently incapable of carrying out the aromatization reaction. Actinioerythrin (**1**, Scheme 28) with end group h, isolated from the sea anemone *Actinia equina,* is presumably an example of a metabolic product of astaxanthin, which gives rise to an α-ketol (g) end group further down the food chain (see Scheme 28).

These examples illustrate the potential use of algal carotenoids for tracing the food chain in the marine environment.

VII. CAROTENOIDS OF MARINE INVERTEBRATES: DISTRIBUTION, STRUCTURES, AND POSSIBLE ORIGIN

Available evidence supports the view that animals are not able to carry out *de novo* carotenoid synthesis. Hence, invertebrate carotenoids originate from indiscriminate or selected storage of dietary carotenoids or are metabolic modifications of dietary carotenoids. Since neither the marine food chain, nor the detailed carotenoid composition of a particular organisms' diet are well established, carotenoids that are considered typical metabolic products of that organism may in some cases indeed originate from the food (see McBeth, 1972a).

Previous compilations on carotenoids of marine invertebrates have been made by Lederer (1935), (Goodwin 1952, 1968) Cheesman *et al.* (1967) and Fox (1974). Much progress in identification and structural elucidation has since been made.

To date no comprehensive, systematic studies on the carotenoids of marine invertebrates are available. Species examined have frequently been dictated by the availability of materials and preference for organisms with strongly pigmented appearance. Although scattered examples must be resorted to, available results seem to justify the semisystematic approach attempted below.

A. Sponges (Porifera)

Early work has been summarized by Goodwin (1952, 1968) listing species examined. Modern re-investigations are needed in many cases.

Sponges are often brilliantly coloured from carotenoids. Sponge carotenoids are frequently strongly dehydrogenated with aryl end groups and acetylenic bonds and have none or few oxygen functions. Characteristic sponge carotenoids are compiled in Scheme 24. Included are aryl carotenes with 1,2,5 and 1,2,3-trimethylphenyl end groups of well-established structures such as isorenieratene (**1**), renieratene (**2**), and renierapurpurin (**3**), of Yamaguchi (for complete references including total syntheses, see Straub, 1976), and the acetylenic hydrocarbons 7,8-didehydroisorenieratene (**4**) and 7,8-didehydrorenieratene (**5**) (Hamasaki *et al.*, 1973). The location of the triple bond was based on ^1H nmr

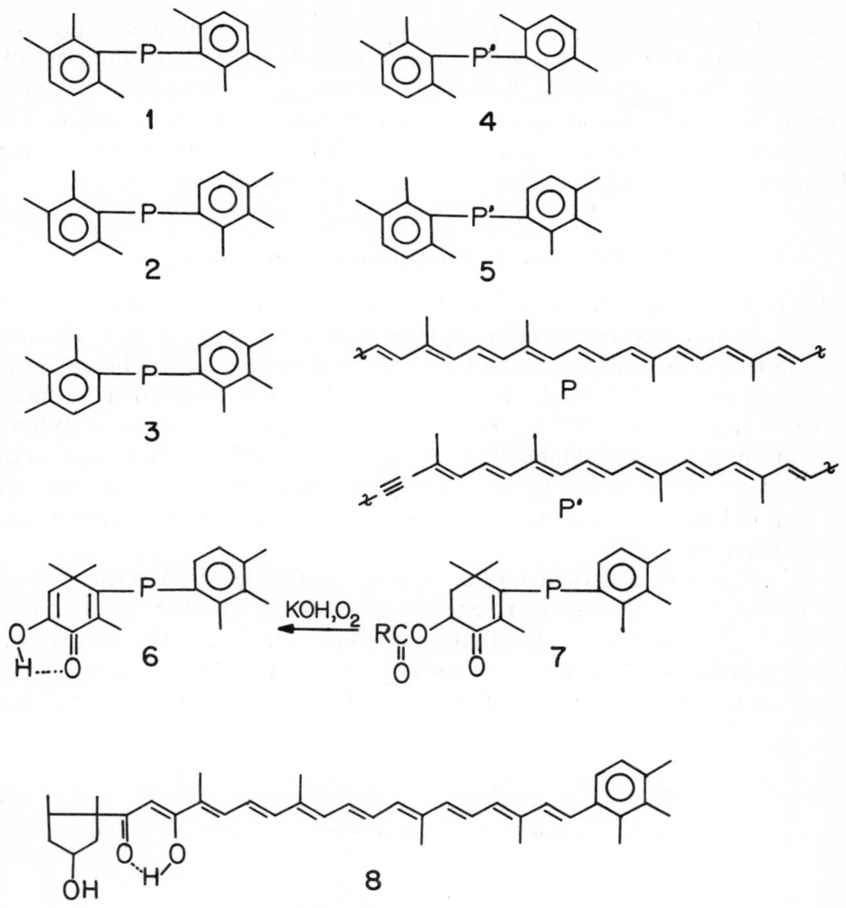

Scheme 24. Carotenoids from sponges.

arguments. The diosphenol tedanin **6** from *Tedania digitata* (Okukado, 1975) was isolated after saponification. The native carotenoid may therefore be identical with the esterified α-ketol clathriaxanthin (**7**) obtained from *Clathria frondifera* (Tanaka and Katayama, 1976) and also from *Microciona prolifera* (Litchfield and Liaaen-Jensen, 1977). Trikentriorhodin (**8**) from *Trikentrion helium* (Aguilar-Martinez and Liaaen-Jensen, 1974) is an enolized β-diketone; this structural feature was first encountered in mytiloxanthin (Khare *et al.*, 1973).

Our current knowledge indicates that aryl carotenoids are restricted to sponges and certain photosynthetic and nonphotosynthetic bacteria. In view of the wide taxonomic gap between bacteria and sponges I suggested (Liaaen-Jensen, 1967) that aryl carotenoids in sponges might possibly originate from inhabiting bacteria. However, *Halichondria panicea* Eimhjellen (1967) found no carotenoids of typical microbial origin. The aromatization reaction, therefore, is probably a capacity of the sponge; dietary β-type 3-hydroxycarotenoids are possible precursors.

B. Starfishes (Echinodermata, Asteroideae)

Early work was summarized by Goodwin (1952). Astaxanthin (**1**) or its esters and free and esterified acetylenic derivatives (**2** and **3**) of astaxanthin appear to be typical (see Scheme 25). Asterinic acid has been shown to be a mixture of 7,8-didehydro- and 7,8,7′,8′-tetradehydroastaxanthin (**2** and **3**) (Francis *et al.*, 1970b). The absolute configuration of the acetylenic derivatives has been established (Berger *et al.*, 1977a). Metridine (Fox and Scheer, 1941) from *Pisaster giganteus* is probably the diosphenol corresponding to 3-hydroxy-β,β-carotene-4,4′-dione (**4**) (Upadhyay and Liaaen-Jensen, 1970).

The occasional dominance of acetylenic derivatives in *Asterias rubens* (Sørensen *et al.*, 1968) gives rise to the question of whether the starfish is capable of effecting 7,8-dehydrogenation, although the acetylenic diatoxanthin and alloxanthin are possible precursors. Mass spectrometric evidence for the presence of alloxanthin (**5**) in *Asterias rubens* has been

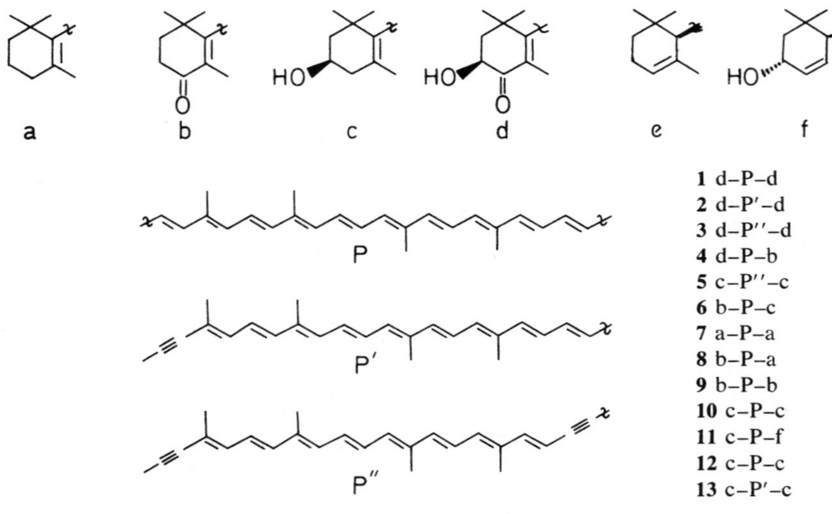

Scheme 25. Carotenoids from starfishes.

published (Francis *et al.*, 1970b). De Nicola (1959) reported the occurrence of ketocarotenoids related to the metabolism of carotenoids in starfish, namely, asteroidenone (**6**) and hydroxyasteroidenone (**4**).

In addition to the carotenoids mentioned above, Tanaka and Katayama (1976) reported the presence of β,β-carotene (**7**), echinenone (**8**), canthaxanthin (**9**), α-cryptoxanthin (**10**), lutein (**11**), zeaxanthin (**12**), and diatoxanthin (**13**) in some Japanese starfishes.

The skin of starfishes contains red, green, or blue carotenoid–protein complexes (see the compilation by Cheesman *et al.*, 1967). As discussed in Section VIII,A, the 4-keto group is considered essential for complex formation with the protein (Cheesman *et al.*, 1967).

C. Sea Urchins (Echinodermata, Echinodea)

The carotenoids of sea urchins have received much attention. Again, the early work was discussed by Goodwin (1952). In a recent evaluation Hallenstvet *et al.* (1978) presented an overall picture (Scheme 26), taking into particular consideration results by Lederer (1935b, 1938), de Nicola and Goodwin (1954), Galasko *et al.* (1969), and Griffiths and Perrott (1976).

Carotenoids of presumed dietary origin (red and brown algae) include β,β-carotene (**1**), β,ϵ-carotene (**2**), zeaxanthin (**3**), lutein (**4**), and fucoxanthin (**5**). Fucoxanthinol (**6**) is probably formed in the gut by nonenzymatic, weak, alkaline hydrolysis. Pentaxanthin, assumed to be identical with isofucoxanthinol (**7**) (Galasko *et al.*, 1969), is another likely metabolite of fucoxanthin under alkaline conditions (see also Jensen, 1966b, for the formation of isofucoxanthin by laying hens). Paracentrone (**8**), which represents a more sophisticated rearrangement product of fucoxanthinol (**6**), is possibly formed by retro-aldol fission of a 6-hydroxy-8-keto intermediate (Galasko *et al.*, 1969) (isofucoxanthinol, **7**) or by oxidation of the 3-hydroxy group to give an activated methylene at C-4, followed by fragmentation. The latter alternative has an *in vitro* counterpart in the preparation of paracentrone (**8**) from fucoxanthin (**5**) on treatment with aluminum isopropoxide in cyclohexanone (Toube and Weedon, 1970). Evidence for the metabolic transformation of dietary β,β-carotene (**1**) via isocryptoxanthin (**9**) to echinenone (**10**) in the ovary of *Strongylocentrotus* (*Paracentrotus*) *droebachiensis* was presented by Griffiths and Perrott (1976). Preferential incorporation of echinenone (**10**) into eggs of *S. droebachiensis* and also into unfed larvae of *Psammechinus militaris* (Hallenstvet *et al.*, 1978) supports a functional role of echinenone (**10**) in the juvenile stage of sea urchins. Again, a carotenoprotein may be involved (Monroy and de Nicola, 1952).

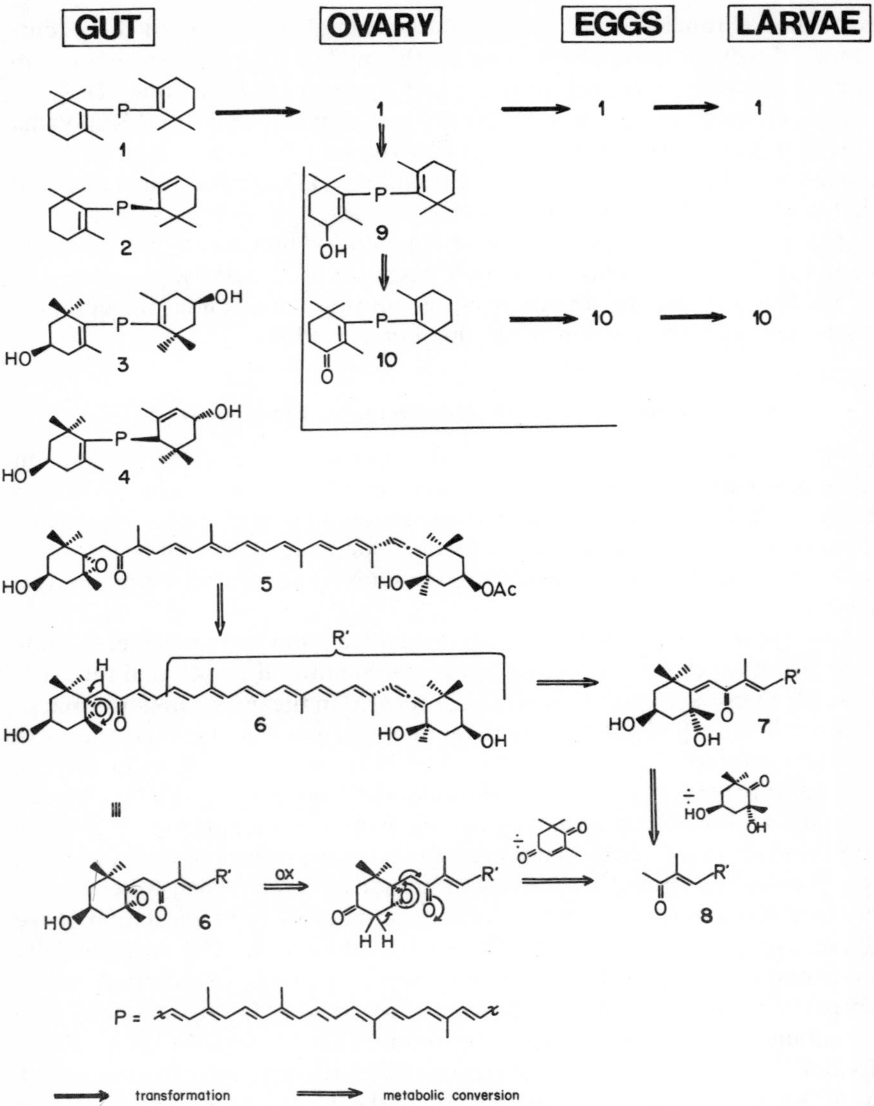

GUT

OVARY

EGGS

LARVAE

transformation

metabolic conversion

Scheme 26. Carotenoids from sea urchins.

D. Mollusks (Mollusca)

In his survey of the early work on carotenoids in the phylum Mollusca, Goodwin (1952) states that β,β-carotene (**1**) and lutein (**2**) occur generally together with some specific (at that time structurally unidentified) carotenoids (see Scheme 27).

The characteristic carotenoids of the mussel *Mytilus edulis*, mytiloxanthin (**3**) and isomytiloxanthin (**4**), have been assigned interesting structures (Khare *et al.*, 1973) with a hypothetical carotenoid such as **5** as a possible biological precursor. Pinakolic rearrangement of the epoxide would give mytiloxanthin (**3**), whereas hydride reduction of the epoxide and oxidation at C-3′ could give isomytiloxanthin (**4**), as suggested by Khare *et al.* (1970). Pectenoxanthin from *Pecten maximus* is now known

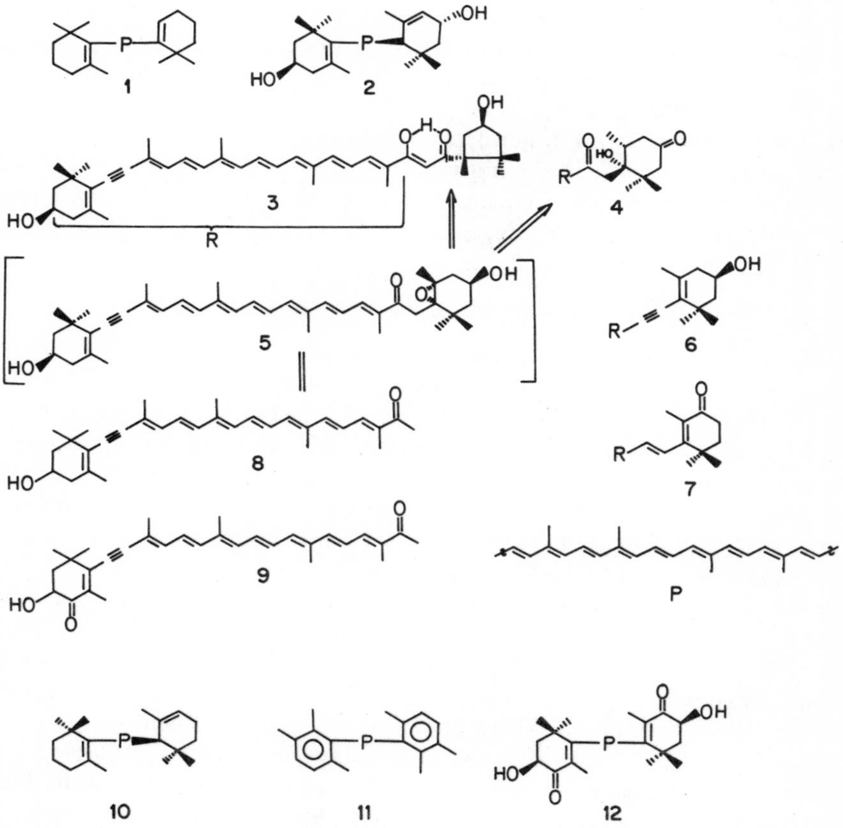

Scheme 27. Carotenoids from mollusks.

to be identical with alloxanthin (**6**), and pectenolone is therefore 4′-ketodiatoxanthin (**7**) (Campbell *et al.*, 1967).

Isolated from nudibranchs and studied by McBeth (1972a,b), triophaxanthin (**8**, Scheme 27) and hopkinsiaxanthin (**9**, Scheme 27) have been formulated as methyl ketones analogs to paracentrone (**8**, Scheme 26). Provided that they were formed *in vivo* by the process suggested for paracentrone formation (Scheme 26), the hypothetical carotenoid **5** (Scheme 27) could be predicted to occur further down the food chain. Methyl ketones also may be artifacts generated during isolation from the corresponding carotenals by aldol condensation (Schmidt *et al.*, 1971; Stewart and Wheaton, 1973). However, acetone was apparently not involved in the isolation of **8** and **9** (Scheme 27). Other carotenoids identified from nudibranchs were β,β-carotene (**1**), β,ϵ-carotene (**10**), isorenieratene (**11**), and astaxanthin (**12**) (Scheme 27) (McBeth, 1972a). McBeth's (1972a,b) carotenoid investigations also included food studies of sponge and algal origin. For instance, *Hopkinsia rosacea* was reported to feed on the bryozoan *Eurystomella bilabiata*, from which its carotenoids were obtained. Fox (1974) states that nudibranchs in general merely assimilate carotenoids from their prey.

From six species of tridacnid clams Jeffrey and Haxo (1968) isolated the symbiotic zooxanthellae, which were shown to have carotenoids of the dinoflagellate type (see Scheme 9).

E. Sea Anemones (Coelenterata, Anthozoa)

Early studies were summarized by Goodwin (1952, 1968). As far as we now know the structurally interesting 2,2′-dinorcarotenoid actinioerythrin (**1**) is restricted to sea anemones (see Scheme 28). First isolated by Lederer (1933), its structure was elucidated in our laboratory (Hertzberg *et al.*, 1969). The corresponding 2-norastaxanthin diester (**2**) was subsequently reported (Francis *et al.*, 1972). Upon careful alkali treatment in the presence of oxygen, **1** and **2** afford the corresponding cyclopentenedione derivatives violerythrin (**3**) and roserythrin (**4**). Partial syntheses of the dark blue violerythrin (**3**) and of roserythrin (**4**) have been achieved by oxidation of astacene (**5**) with manganese dioxide (Holzel *et al.*, 1969; Francis *et al.*, 1972), a reaction that mimics the biosynthetic route previously postulated from astaxanthin (**6**) to actinioerythrin (**1**) via a hypothetical triketone (**7**) by benzilic acid rearrangement.

Conversion of violerythrin (**3**) to actinioerythrol (the free diol corresponding to **1**) had previously been effected with sodium borohydride (Hertzberg and Liaaen-Jensen, 1968), and a total synthesis of violerythrin (**3**) was achieved by Kienzle and Minder (1976). The middle portion of

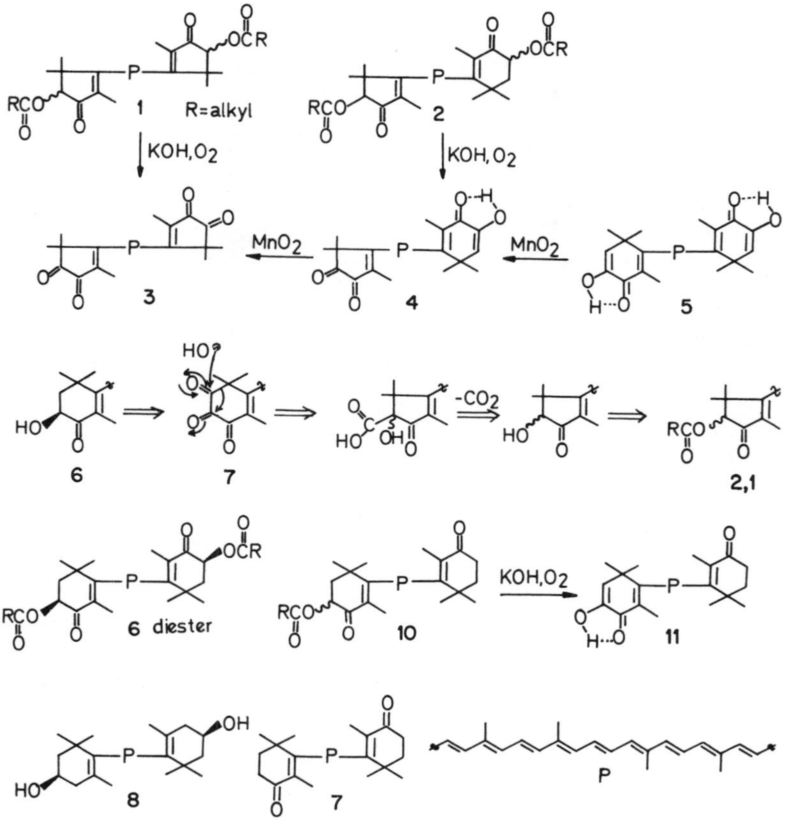

Scheme 28. Carotenoids from sea anemones.

their synthesis, outlined with two alternatives in Scheme 29, is particularly instructive because it demonstrates modern principles of carotenoid synthesis, including alkylation of sulfone carbanions followed by elimination (Julia reaction) as well as phosphorus ylids (Wittig reaction) for carbon–carbon double-bond formation when building up the carotenoid skeleton (see Mayer and Isler, 1971).

Czygan and Seefrid (1970) claim that the ability of *Actinia equina* to produce actinioerythrin (**1**) (see Scheme 28) from astaxanthin (**6**) is genetically controlled and characteristic of strains from southern Europe, since astaxanthin (**6**) esters were dominant in Atlantic strains. However, in *A. equina* from Norwegian waters, **1** is dominant. As stated by Goodwin (1952) and Upadhyay and Liaaen-Jensen (1970), actinioerythrin (**1**) is encountered in various sea anemones. The plumose anemone *Metridium*

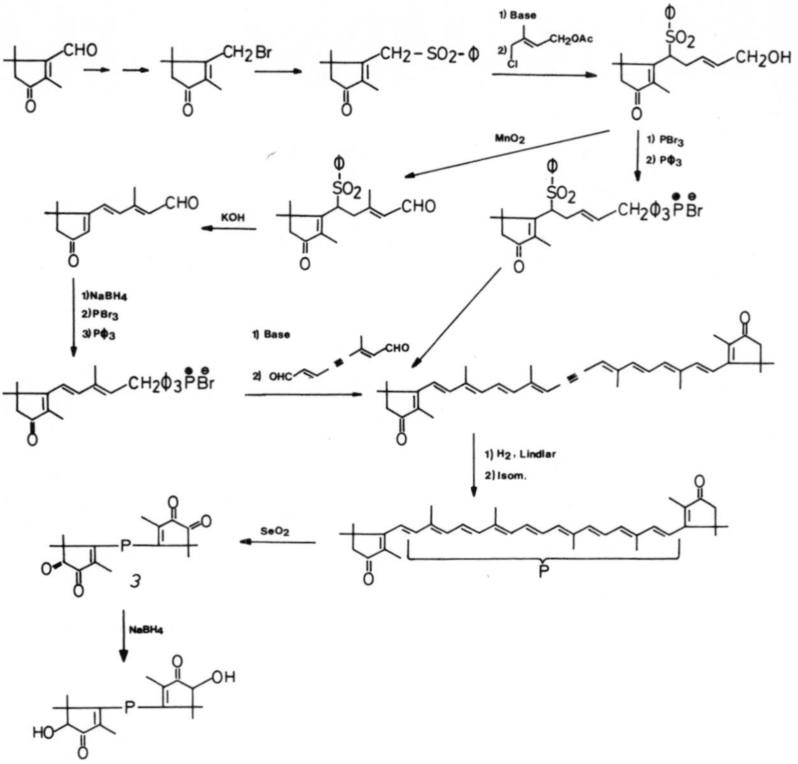

Scheme 29. Total synthesis of actinioerythrol and violerythrin.

senile (Fox and Pantin, 1941; Fox *et al.*, 1967; Upadhyay and Liaaen-Jensen, 1970) is reported to contain zeaxanthin (**8**) esters, canthaxanthin (**9**), astaxanthin (**6**) esters, and a natural ester (probably **10**), which on alkali treatment produces metridin (**11**, structure needs further proof).

In summary, it appears that a rough picture of carotenoid metabolism in sea anemones may be obtained from Scheme 28, in which actinioerythrin (**1**) is regarded as a terminal product and the *in vitro* products are disregarded.

The absolute configuration of actinioerythrin (**1**) is not yet known, but chiroptical properties have been reported (Andrewes *et al.*, 1974b). The esterifying acids of actinioerythrin (**1**) are a mixture containing hexadecanoic, octadecanoic, hexadecenoic, octadecenoic, and hexadecadienoic acids (Czygan and Seefrid, 1970). It is interesting that actinioerythrin (**1**) occurs as a protein complex (Lederer, 1933; de Nicola and Goodwin, 1954; Cheesman *et al.*, 1967).

Sulcatoxanthin from *Anemonia sulcata* (Heilbron *et al.*, 1935) is now known to be identical with peridinin (**8**, Scheme 9) (Strain *et al.*, 1976). Its presence may be ascribed to zooxanthellae (dinoflagellate symbionts) (see studies by Taylor, 1969).

F. Corals (Coelenterata)

The carotenoids of soft, horny, and stony corals have been investigated. Several corals have a dinoflagellate carotenoid pattern, and the carotenoids may be ascribed to zooxanthellae symbionts, as demonstrated by Jeffrey and Haxo (1968). They isolated the zooxanthellae from nine soft and hard corals of the orders Scleractinia, Coenthecalia, Alcyonaceae, and Milleporina and analyzed the carotenoids by chromatography and visible spectroscopy in a comparative study with dinoflagellate extracts. Later work on various Australian soft corals in our laboratory, employing mass spectrometry, confirmed the dinoflagellate carotenoid pattern (see Scheme 9) (Hallenstvet and Liaaen-Jensen, 1977). In the gorgonian *Isis hippuris*, the presence of peridinin has been documented by all spectroscopic criteria (Murphy *et al.*, 1978).

Studies on five other soft corals from Norwegian waters (Upadhyay and Liaaen-Jensen, 1970) suggested that astaxanthin (**1**) (see Scheme 30)

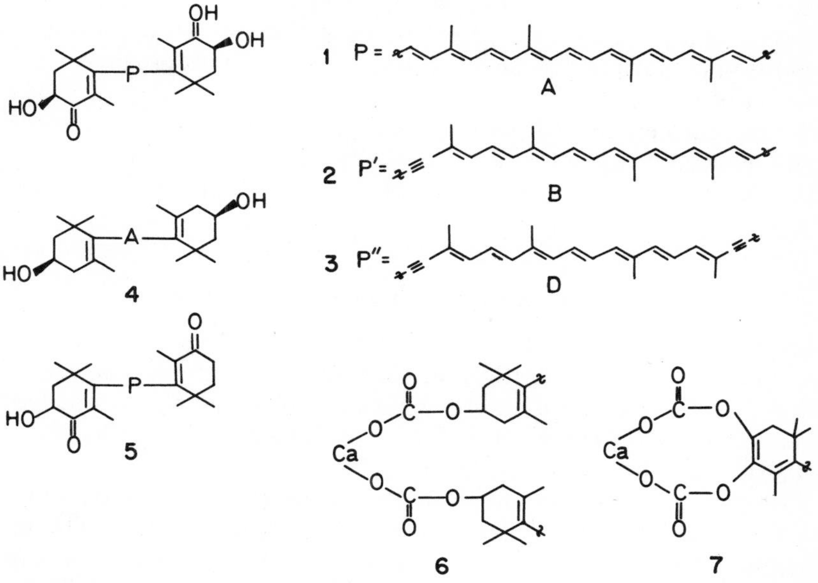

Scheme 30. Carotenoids from corals. In addition carotenoids due to zooxanthellae symbionts may be encountered (Scheme 9).

is a rather general carotenoid. The 7,8-didehydro (**2**) and 7,8,7′,8′-tetradehydro (**3**) derivatives were encountered in *Alcyonium digitatum*.

The pigments of the calcareous spicules of horny corals have attracted the attention of Fox and his collaborators (Fox *et al.*, 1969; Fox and Wilkie, 1970; Fox, 1972). For these pigments, which are not extractable with hot alkali or pyridine, three alternative isolation procedures were worked out: (1) extraction with acidified methanol; (2) treatment with potassium oxalate to provide insoluble calcium and magnesium oxalate, followed by extraction of the carotenoids with pyridine; and (3) treatment with aqueous disodium EDTA, followed by solvent extraction. Although the first method gave higher pigment yield, the other two seem preferable in view of the acid lability of carotenoids. The question of the authenticity and structure of eugorgianoic acid (Fox *et al.*, 1969) remains open. Most interesting is the successful extraction of astaxanthin (**1**) (see Scheme 30) from the calcerous skeletons of several corals of the orders Gorgonacea and Hydrocorallina by Fox and Wilkie (1970) and Fox (1972) following the above procedures. *Stylaster elegans* contained a bound carotenoid with properties of zeaxanthin (**4**). Fox suggested that the 3-hydroxyxanthophyll is bound as a carbonate ester as in **6** and astaxanthin (enol form) as in **7**. Circular dichroic studies of the released astaxanthin, which should be optically inactive, would be rewarding. From the soft tissues of *Allopora californica* Fox and Wilkie (1970) reported the presence of astaxanthin (**1**, free and esterified) together with smaller amounts of 3-hydroxy-β,β-carotene-4,4′-dione (**5**), in accordance with the isolation of bound astaxanthin from its skeleton.

Besides carotenoid fatty acid esters, glycosides, and carotenoproteins, the inorganically bound carotenoids, if further documented of horny corals, represent a fascinating extension of the state in which naturally occurring carotenoids are encountered.

G. Crustaceans (Arthropoda, Crustaceae)

This class of arthropods has five subclasses. Three of these (branchiopods, ostracods, and copepods) are microscopic or small animals, and the copepods comprise major ingredients of zooplankton. The fourth subclass consists of barnacles, and the fifth (Malacostraca) includes krills, shrimps, crayfish, crabs, and lobsters.

Various aspects of carotenoids in Crustaceae were reviewed by Goodwin (1952, 1971a), Cheesman *et al.* (1967), and Fox (1974). The early generalization of Goodwin (1952) that β,β-carotene (**1**) and astaxanthin (**12**) (see Scheme 31) are the most abundant carotenoids still holds with certain modifications.

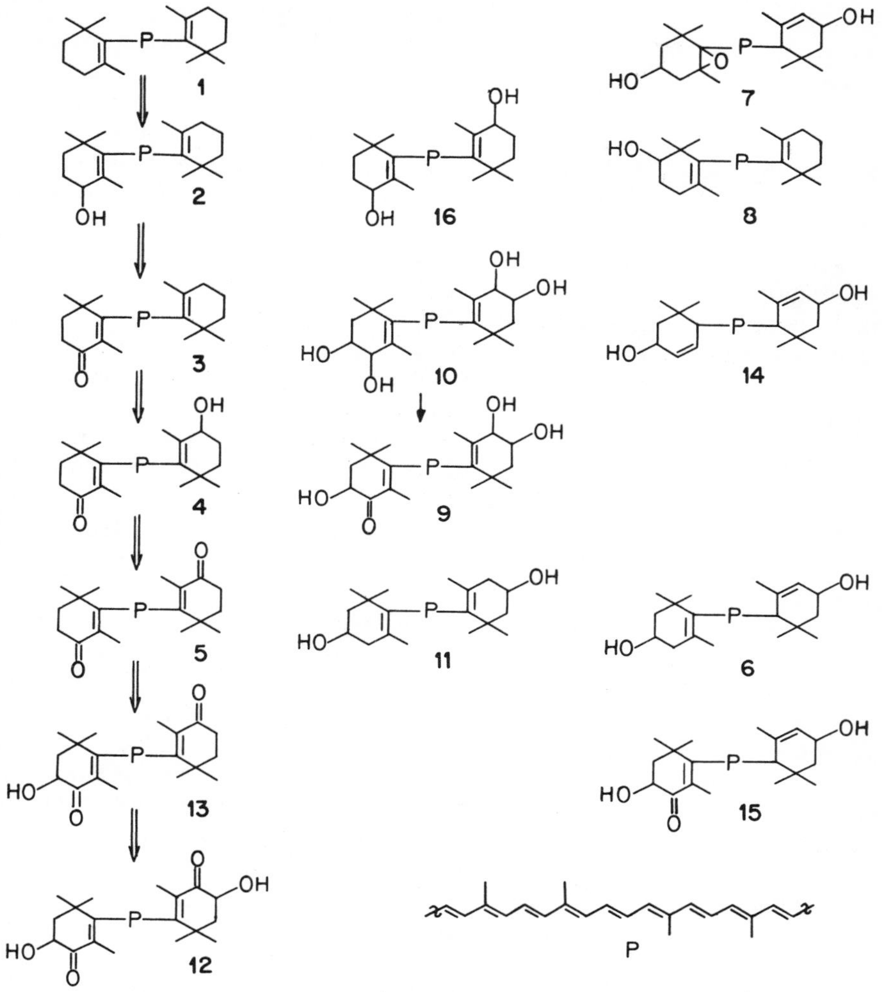

Scheme 31. Carotenoids from Crustaceae.

Lee (1966a,b,c), Lee and Gilchrist (1975), and Gilchrist and Lee (1976) have diverse interests within the Isopoda. *Idotea monthereyensis* contains β,β-carotene (**1**), isocryptoxanthin (**2**), echinenone (**3**), 4′-hydroxy-β,β-caroten-4-one (**4**), canthaxanthin (**5**), lutein (**6**), and lutein epoxide (**7**) in different proportions in the green, brown, and red varieties, thus matching their environment. To achieve this, admixture with a blue canthaxanthin–protein complex is involved. Surprisingly, β,β-caroten-

2-ol (**8**), which was previously encountered only in terrestrial green algae and insects (Kjøsen *et al.*, 1972; Kayser, 1976, was encountered in three species of *Idotea* (Lee and Gilchrist, 1975). Previous identifications of monohydroxycarotenes in Crustaceae have been critically discussed (Lee and Gilchrist, 1975). Results of biosynthetic experiments by Gilchrist and Lee (1976) using ^{14}C-labeled β,β-carotene were taken as supporting hydroxy intermediates in the conversion of β,β-carotene (**1**) to cantaxanthin (**5**). A new carotenoid, idoxanthin (**9**), has been isolated from *Idotea metallica* (Herring, 1969). Idoxanthin (**9**) was later prepared by partial oxidation of crustaxanthin (**10**) (Hodler *et al.*, 1974). The tetrol crustaxanthin (**10**) was first obtained from the copepod *Arctodiaptomus salinus* (Bodea *et al.*, 1966) and claimed to have *trans*-glycol end groups (Nicoara *et al.*, 1967). The absolute configuration is not yet settled. Esterification studies of synthetic, optically inactive crustaxanthin have been reported (Kienzle and Hodler, 1975). Carotenoids similar to those encountered in *Idotea montereyensis* were reported in *Carcinus maenas* (Decapoda) and included zeaxanthin (**11**) (Gilchrist and Lee, 1967).

The carotenoids of the brine shrimp *Artemia salina* have received much attention (Krinsky, 1965; Czygan, 1968b; Davies *et al.*, 1965, 1970; Hsu *et al.*, 1970). Canthaxanthin (**5**) is the terminal product of carotenoid synthesis, and the question of undetected hydroxy intermediates remains open.

Carotenoids of the prawn (Katayama *et al.*, 1971a, 1972a,b), crab (Katayama *et al.*, 1973a,b; Czeczuga, 1974; Harashima *et al.*, 1976), and lobster (Katayama *et al.*, 1973c) have also attracted much interest in recent years, and biosynthetic routes to astaxanthin (**12**) have been studied. The long-known astaxanthin (**12**) (Kuhn and Sørensen, 1938) is a major carotenoid in many Crustaceae (Kuhn *et al.*, 1939). The absolute configuration of astaxanthin (**1**) has been established (Andrewes *et al.*, 1974b). Partial synthesis of optically inactive astaxanthin was carried out by oxidation of canthaxanthin (**5**) with potassium *tert*-butoxide to astacene, followed by sodium borohydride reduction to astaxanthin (Cooper et al., 1975).

Ovoverdin and crustacyanin and other protein complexes of astaxanthin (**12**) have been further characterized (Jencks and Buten, 1964; Cheesman and Prebble, 1966; Cheesman *et al.*, 1967; Lee and Zagalsky, 1966; Kuhn and Kühn, 1967; Buchwald and Jencks, 1968; Zagalsky, 1976). α-Crustacyanin has a molecular weight of about 320,000; it reversibly dissociates into eight subunits with a molecular weight of 38,000 in neutral solution of low ionic strength. Only astaxanthin (**12**), its esters, and canthaxanthin (**5**) have been unequivocally established as prosthetic groups of true carotenoproteins. Recombination experiments between various carotenoids and the apoprotein of crustacyanin have indicated the requirement of both 4- and 4′-keto groups for interaction with the protein. Aside from that point, the linkage between carotenoid and protein remains an open question. Covalent bonding must be disregarded for reasons of instability. Bonding to enolized astaxanthin, as suggested by Kuhn and Sørensen (1938), is not possible since astaxanthin liberated from the protein complex is optically active (Andrewes *et al.*, 1974b). In order to account for the spectral characteristics of crustacyanin, Buchwald and Jencks (1968) developed a model with distortion of the polyene chain around the double bonds. Zagalsky (1976) further suggested a mode of binding involving hydrogen bonding of the 4-keto groups of the carotenoid to imide groups of the apoprotein.

Fox (1973) reported that 4-ketocarotenoids such as echinenone (**3**), canthaxanthin (**5**), 3-hydroxy-β,β-carotene-4,4-dione (**13**), and astaxanthin (**12**) are chitin bound in a crustacean carapace, which would imply still another type of carotenoid complex in nature.

The biosynthesis of astaxanthin in the lobster was studied by Katayama *et al.* (1973c) using tritium-labeled β,β-carotene (**1**). The results were taken as evidence for the left vertical sequence of Scheme 31.

Other carotenoids reported in Crustacea include tunaxanthin (**14**) (Tanaka *et al.*, 1976b), the structure of which has not been satisfactorily proved, α-doradexanthin (**15**) (Harashima *et al.*, 1976), and isozeaxanthin (**16**) (Czeczuga, 1974). The carotenoids encountered in Crustacea are compiled in Scheme 31 in a manner that emphasizes their chemical relationship. Crustaceae in general seem to have the ability to introduce keto groups in the 4,4′-positions of dietary carotenoids. Whether a single common pathway is operating for the biosynthesis of canthaxanthin (**5**) and astaxanthin (**12**) is not yet clear. From structural considerations carotenoids such as lutein epoxide (**7**), β,β-caroten-2-ol (**8**), and tunaxanthin (**14**) appear not to be metabolized.

H. Miscellaneous Taxa

In addition to the marine invertebrates discussed above, some reports on the carotenoids belonging to other phyla or classes are available. Many of these appeared before unequivocal identification criteria existed (Goodwin, 1952). Since only scattered reports are available, no generalizations are possible or desirable. A few selected data are reviewed here.

From the bryozoan *Eurystomella bilabiata* the apocarotenoid hopkinsiaxanthin (**9**, Scheme 27) has been isolated (McBeth, 1972a,b). Within the Annelida a polychaete worm (*Sabella penicillus*) contained zeaxanthin, lutein, and a series of carotenoids that may represent intermediates in the formation of canthaxanthin. These were, besides canthaxanthin, β,β-carotene, isocryptoxanthin, echinenone, and 4′-hydroxyechinenone (Lee *et al.*, 1967). In two other worms Lee *et al.* (1967) reported astaxanthin as the major carotenoid. Echinenone was also present.

VIII. CAROTENOIDS OF MARINE VERTEBRATES: DISTRIBUTION, STRUCTURES, AND METABOLISM

A. Tunicates (Protochordata)

Tunicates, comprising ascidians and sea squirts, are classified as primitive chordates. They have a diet rich in phytoplankton (Fox, 1974). Both pectenoxanthin from *Botryllus schlosseri* and cynthiaxanthin from *Halocynthia papillosa* (Lederer, 1938) are now known to be identical with the diacetylenic diol alloxanthin (Campbell *et al.*, 1967). Astaxanthin, as its ester, was obtained from the latter source and from *Dondroda grossularia* (Lederer, 1938). Other early work (Goodwin, 1952) should be reinvestigated by modern methods.

B. Fishes (Pisces)

An early general treatment of piscine carotenoids, including localization of the carotenoids in different organs of the fish, was given by Goodwin (1952). Many fishes are brightly colored as a result of carotenoids in the skin. Goodwin (1952) lists astaxanthin (**7**), lutein (**17**), and taraxanthin (**16**) as common skin carotenoids (see Scheme 32). Taraxanthin is now known to be identical with lutein epoxide (**16**) (Buchecker *et al.*, 1976b). Lutein epoxide (**16**), however, has not been mentioned in more recent studies on carotenoids from fishes and may have been misidentified.

Carotenoids reported to occur in fishes are arranged in Scheme 32 in a way that illustrates their chemical relationship, using consecutive numbering of formulas. The identifications are only partly supported by satisfactory spectroscopic criteria (mass spectroscopy, ^{1}H nmr).

In recent years much interest has been displayed in fish carotenoids, particularly in Japan. The goldfish *Carassius auratus* contains esters of astaxanthin (**7**), lutein (**17**), β-doradexanthin (**21**, previously described by different names), and α-doradexanthin (**18**) (Katayama *et al.*, 1970a,b,c). Compound **18** was characterized by infrared and ^{1}H nmr data of the disophenol doradecin and metal hydride reduction. The fancy red carp *Cyprinus carpio* also contains astaxanthin (**7**) and α-doradexanthin (**18**) (Katayama *et al.*, 1971b). From a teleost (*Oryzia latipes*) Hirao *et al.* (1969) identified β,β-carotene (**1**), lutein (**17**), and tunaxanthin (**20**), among

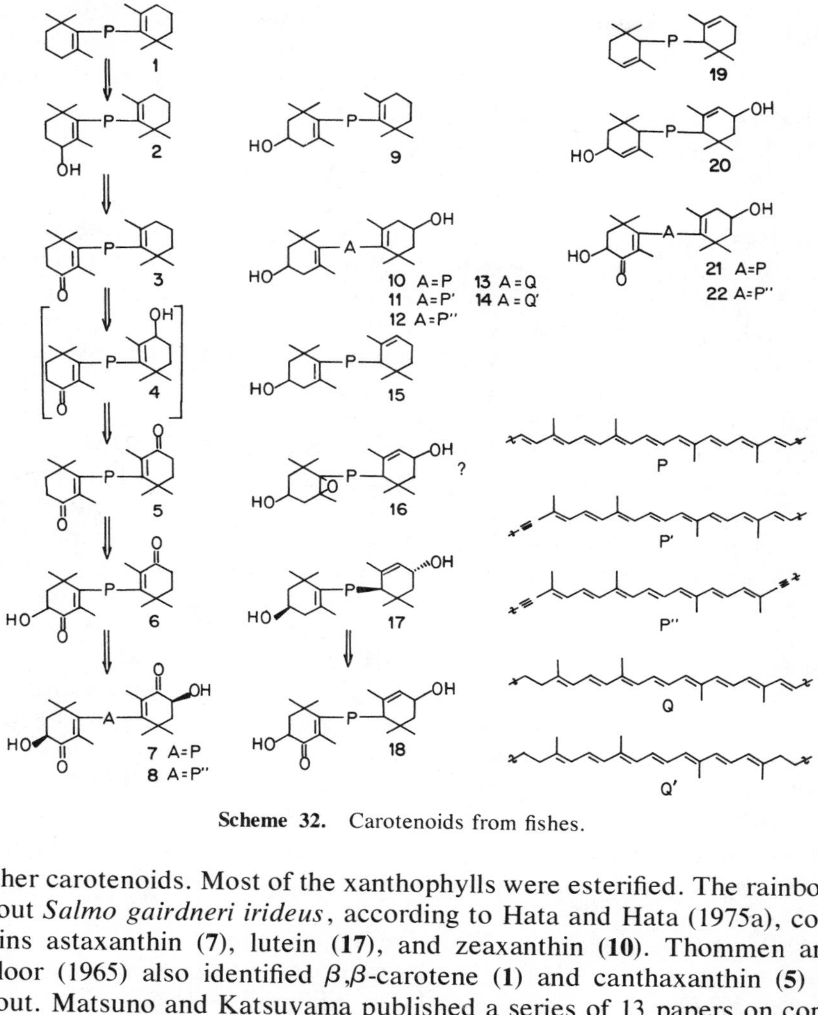

Scheme 32. Carotenoids from fishes.

other carotenoids. Most of the xanthophylls were esterified. The rainbow trout *Salmo gairdneri irideus*, according to Hata and Hata (1975a), contains astaxanthin (**7**), lutein (**17**), and zeaxanthin (**10**). Thommen and Gloor (1965) also identified β,β-carotene (**1**) and canthaxanthin (**5**) in trout. Matsuno and Katsuyama published a series of 13 papers on comparative biochemical studies on carotenoids of a large variety of fishes. Matsuno and Katsuyama (1976a,b,c,d) and Matsuno *et al.* (1976a) considered tunaxanthin (**20**) to be a chemical indicator for Percichthyes. Full characterization of **20** is still lacking. Additional identified carotenoids were β,β-carotene (**1**), astaxanthin (**7**), cryptoxanthin (**9**), zeaxanthin (**10**), diatoxanthin (**11**), alloxanthin (**12**), α-cryptoxanthin (**15**), lutein (**17**), and α-doradexanthin (**18**, as diosphenol). The previously unknown carotenoids prasilioxanthin (**13**) and 7,8-dihydroprasilioxanthin (**14**) were isolated from catfish (*Paragilurus asotus*) by Matsuno *et al.* (1976b) in a modern investigation that included ^{1}H nmr and mass spectroscopic evidence. These compounds represent 7,8-dihydro and 7,8,7′,8′-tetrahydro derivatives of zeaxanthin (**10**). ε,ε-Carotene (**19**) is stated to occur in sea breams, together with tunaxanthin (**20**), α-doradexanthin (**18**), and astaxanthin (**7**) (Katayama *et al.*, 1970a).

An early account of the distribution of astaxanthin (**7**) in fishes was made by Kuhn *et al.* (1939). Chichester, Katayama, and their co-workers have reported the biosynthesis of astaxanthin (**7**) in fishes in a series of papers. A biosynthetic pathway from lutein (**17**) to astaxanthin (**7**) via α-doradexanthin (**18**) in goldfish has been suggested on the basis of studies including ^{14}C-labeling experiments (Katayama *et al.*, 1970a,b; Hsu *et al.*, 1971). This proposal is not compatible with the absolute configuration of astaxanthin (**7**) (Andrewes *et al.*, 1974b) and lutein (**17**) (Buchecker *et al.*, 1974; Andrewes *et al.*, 1974a) that was later established, since an epimerization at C-3′ would be required and was not involved in the proposed scheme. Also, biosynthetic findings by Hata and Hata (1975b) for red carp contradict a conversation of lutein (**17**) to astaxanthin (**7**). A biosynthetic route from β,β-carotene (**1**) to astaxanthin (**7**) via isocryptoxanthin (**2**), echinenone (**3**), the hypothetical intermediates 4-hydroxyechinenone (**4**) and canthaxanthin (**5**), and finally via 3-hydroxycanthaxanthin (**6**) has been suggested (Rodriguez *et al.*, 1973; Hata and Hata, 1975b). Support for this pathway was obtained by feeding the intermediates isocryptoxanthin (**2**), echinenone (**3**), and canthaxanthin (**5**) to depleted goldfish, which resulted in increased astaxanthin content. Isozeaxanthin (β,β-carotene-4,4′-diol) was not metabolized (Rodriguez *et al.*, 1974). Sea breams, however, were not able to metabolize canthaxanthin (**5**) and zeaxanthin (**10**) (Tanaka *et al.*, 1976a). Hata and Hata (1970) reported the conversion of alloxanthin (**12**) to 4-ketoalloxanthin (**22**) and 7,8,7′,8′-tetradehydroastaxanthin (**8**) in goldfish. The infrared spectra of the diosphenols corresponding to **8** and **22** are not convincing (Francis *et al.*, 1970b).

The pink-colored flesh of cultured salmon (*Salmo salar*) and trout is of commercial interest. Synthetic canthaxanthin (**5**) is successfully used as a feed additive to trout for this purpose (Bauernfeind *et al.*, 1971).

Other interesting biological aspects such as the localization and function of carotenoids in fishes, including the role of carotenoids containing unsubstituted β rings as vitamin A precursors, will not be discussed here.

IX. CAROTENOIDS IN MARINE SEDIMENTS

The ultimate fate of carotenoid molecules after structural modifications have been effected through the marine food chain may be their deposition on the ocean floor and consolidation in marine sediments. In such surroundings structural changes, microbial or chemical, may still occur. Identification of carotenoid derivatives in sediments represents a difficult analytical problem, requiring modern instrumentation (Hajibrahim *et al.*, 1978).

Various aspects of carotenoids in marine sediments have been discussed by Watts *et al.* (1977), Maxwell *et al.* (1971), Schwendinger (1969), Vallentyne (1957), Dunning (1963), Goodwin (1952), and Fox *et al.* (1944).

Recent sediments contain intact carotenoids, which from structural considerations may be traced back to algal or bacterial origin (Watts *et al.*, 1977) (see Scheme 33). The presence of β,β-carotene (**1**), echinenone (**2**), canthaxanthin (**3**), lutein (**4**), zeaxanthin (**5**), and spheroidenone (**6**) in marine sediments has been demonstrated by modern techniques (Peake *et al.*, 1974; Ikan *et al.*, 1975; Watts and Maxwell, 1977). An intact carotenoid assigned as isorenieratene (**7**) or renieratene (**8**) has also been isolated from a 30-million-year-old deep-sea sediment (Watts, 1977). The surprising stability of carotenoids in sediments is reflected by the total amount of carotenoids reported in the 5000-year-old Cariaco Trench sediments as compared to present-day algal mats (Watts *et al.*, 1977). However, a decrease in carotenoid concentration with increasing age of sediment (5000–340,000 years) has been demonstrated (Watts and Maxwell, 1977).

Older marine sediments contain saturated hydrocarbons such as perhydro-β,β-carotene (**9**), which represents a fully hydrogenated deriva-

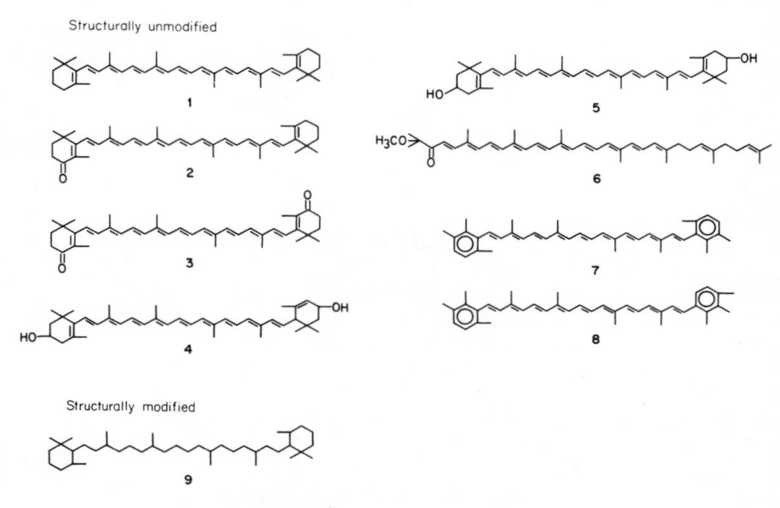

Scheme 33. Carotenoids from marine sediments.

tive of the parent carotene (Simoneit and Burlingame, 1971; Watts and Maxwell, 1977). References is made here to the Cariaco Trench formations, which are at least 340 thousand years old. Evidence indicating the occurrence of partly hydrogenated lutein and canthaxanthin with saturation of up to five double bonds (see structures **3** and **4**) in the Cariaco Trench sediments has recently been presented (Watts and Maxwell, 1977). So far the nature of such a progressive reduction of carotenoids in the geological column is not known.

ACKNOWLEDGMENTS

The author wishes to thank her many collaborators who have contributed ideas and results on marine carotenoids in our laboratory. Without financial support over many years from the Norwegian Research Council of Science and the Humanities and from Hoffmann-La Roche, Basel, our research in the field could not have been maintained.

REFERENCES

Aasen, A. J., Eimhjellen, K. E., and Liaaen-Jensen, S. (1969a). *Acta Chem. Scand.* **23**, 2544.

Aasen, A. J., Francis, G. W., and Liaaen-Jensen, S. (1969b). *Acta Chem. Scand.* **23**, 2605.

Aguilar-Martinez, M., and Liaaen-Jensen, S. (1974). *Acta Chem. Scand., Ser. B* **28**, 1247.

Aitzetmüller, K., Svec, W. A., Katz, J. J., and Strain, H. H. (1968). *J. Chem. Soc. Chem. Commun.* 32.

Aitzetmüller, K., Strain, H. H., Svec, W. A., Grandolfo, M., and Katz, J. J. (1969). *Phytochemistry* **8**, 1761.

Allen, M. B., Goodwin, T. W., and Phagpolngarm, S. (1960). *J. Gen. Microbiol.* **23**, 93.

Allen, M. B., Fries, L., Goodwin, T. W., and Thomas, D. M. (1964). *J. Gen. Microbiol.* **34**, 259.

Andrewes, A. G., and Starr, M. (1976). *Phytochemistry* **15**, 1009.

Andrewes, A. G., Borch, G., and Liaaen-Jensen, S. (1974a). *Acta Chem. Scand., Ser. B* **28**, 139.

Andrewes, A. G., Borch, G., Liaaen-Jensen, S., and Snatzke, G. (1974b). *Acta Chem. Scand., Ser. B* **28**, 730.

Antia, N. J., and Cheng, J. Y. (1977). *J. Fish. Res. Board Can.* **34**, 659.

Antia, N. J., Bisalputra, T., Cheng, J. Y., and Kalley, J. P. (1975). *J. Phycol.* **11**, 339.

Arnesen, U., and Liaaen-Jensen, S. (1977). Unpublished observations.

Arpin, N., Fiasson, J.-L., and Liaaen-Jensen, S. (1972). *Acta Chem. Scand.* **26**, 2526.

Arpin, N., Svec, W. A., and Liaaen-Jensen, S. (1976). *Phytochemistry* **15**, 529.

Bamji, M. S., and Krinsky, N. I. (1965). *J. Biol. Chem.* **240**, 467.

Barghoorn, E. S., and Linder, D. H. (1976). *In* "Marine Microbiology" (C. D. Litchfield, ed.), pp. 431–446. Halstead Press, Stroudsbourg, Pennsylvania.

Bartlett, L., Klyne, W., Mose, W. P., Scopes, P. M., Galasko, G., Mallams, A. K., Weedon, B. C. L., Szabolcs, J., and Tóth, G. (1969). *J. Chem. Soc.* (series C) 2527.

Bauernfeind, J. C., Brubacher, G. B., Kläui, H. M., and Marusich, W. L. (1971). *In* "Carotenoids" (O. Isler, ed.), pp. 743–770. Birkhaeuser, Basel.

Berger, R., Borch, G., and Liaaen-Jensen, S. (1977a). *Acta Chem. Scand.* **31**, 243.
Berger, R., Liaaen-Jensen, S., McAllister, V., and Guillard, R. R. L. (1977b). *Biochem. Syst. Ecol.* **5**, 71.
"Bergey's Manual of Determinative Bacteriology" (1974). 8th Ed. Williams & Wilkins, Baltimore, Maryland.
Bernhard, K., Moss, G. P., Tóth, G., and Weedon, B. C. L. (1974). *Tetrahedron Lett.* 3899.
Bernhard, K., Moss, G. P., Tóth, G., and Weedon, B. C. L. (1976). *Tetrahedron Lett.* 115.
Bingham, A., Mosher, H. S., and Andrewes, A. G. (1977). *J. Chem. Soc., Chem. Commun.* 96.
Bjørnland, T., and Aguilar-Martinez, M. (1976). *Phytochemistry* **15**, 291.
Bodea, C., Nicoara, E., Illyes, G., and Suteu, M. (1966). *Rev. Roum. Chim.* **24**, 153.
Bogorad, L. (1976). *In* "Chemistry and Biochemistry of Plant Pigments" (T. W. Goodwin, ed.), Vol. 1, pp. 64–148. Academic Press, New York.
Bonnett, R., Mallams, A. K., Spark, A. A., Tee, J. L., Weedon, B. C. L., and McCormick, A. (1969). *J. Chem. Soc. C.* p. 429.
Bremser, W., and Paust, J. (1974). *Org. Magn. Reson.* **6**, 433.
Britton, G. (1971). *In* "Aspects of Terpenoid Chemistry and Biochemistry" (T. W. Goodwin, ed.), pp. 255–289. Academic Press, New York.
Britton, G. (1976a). *Pure Appl. Chem.* **47**, 223.
Britton, G. (1976b). *In* "Chemistry and Biochemistry of Plant Pigments (T. W. Goodwin, ed.), Vol. 1, pp. 262–327. Academic Press, New York.
Britton, G., and Goodwin, T. W. (1971). *In* "Vitamins and Coenzymes," Part C (D. B. McCormick and L. D. Wright, eds.), "Methods in Enzymology," Vol. 18, pp. 654–701. Academic Press, New York.
Britton, G., Singh, R. K., Goodwin, T. W., and Ben-Aziz, A. (1975). *Phytochemistry* **14**, 2427.
Britton, G., Malhotra, H. C., Singh, R. K., Goodwin, T. W., and Ben-Aziz, A. (1976a). *Phytochemistry* **15**, 1749.
Britton, G., Malhotra, H. C., Singh, R. K., Taylor, S., Goodwin, T. W., and Ben-Aziz, A. (1976b). *Phytochemistry* **15**, 1971.
Britton, G., Brown, D. J., Goodwin, T. W., Leuenberger, F. J., and Schochter, A. J. (1977). *Arch. Mikrobiol.* **113**, 33.
Buchecker, R., and Liaaen-Jensen, S. (1975). *Helv. Chim. Acta* **58**, 89.
Buchecker, R., and Liaaen-Jensen, S. (1977). *Phytochemistry* **16**, 729.
Buchecker, R., Hamm, P., and Eugster, C. H. (1972). *Chimia* **26**, 134.
Buchecker, R., Eugster, C. H., Kjøsen, H., and Liaaen-Jensen, S. (1973). *Helv. Chim. Acta* **56**, 2899.
Buchecker, R., Hamm, P., and Eugster, C. H. (1974). *Helv. Chim. Acta* **57**, 631.
Buchecker, R., Liaaen-Jensen, S., and Eugster, C. H. (1975). *Phytochemistry* **14**, 797.
Buchecker, R., Liaaen-Jensen, S., Borch, G., and Siegelman, H. W. (1976a). *Phytochemistry* **15**, 1015.
Buchecker, R., Liaaen-Jensen, S., and Eugster, C. H. (1976b). *Helv. Chim. Acta* **59**, 1360.
Buchwald, M., and Jencks, W. P. (1968). *Biochemistry* **7**, 834.
Burnett, J. H. (1976). *In* "Chemistry and Biochemistry of Plant Pigments" (T. W. Goodwin, ed.), Vol. 1, pp. 655–680. Academic Press, New York.
Campbell, S. A., Mallams, A. K., Waight, E. S., Weedon, B. C. L., Barbier, M., Lederer, E., and Salaque, A. (1967). *J. Chem. Soc. Chem. Commun.* 941.
Carter, P. W., Heilbron, I. M., and Lythgoe, B. (1939). *Proc. Roy. Soc., Ser. B* **128,** 82.

Carter, P. W., Heilbron, I. M., and Lythgoe, B. (1939). *Proc. Roy. Soc., Ser. B* **128**, 82.
Chapman, D. J. (1966). *Phytochemistry* **5**, 1331.
Chapman, D. J., and Haxo, F. T. (1963). *Plant Cell Physiol.* **4**, 57.
Chapman, D. J., and Haxo, F. T. (1966). *J. Phycol.* **2**, 89.
Cheesman, D. F., and Prebble, J. (1966). *Comp. Biochem. Physiol.* **17**, 929.
Cheesman, D. F., Lee, W. L., and Zagalsky, P. F. (1967). *Biol. Rev. Cambridge Philos. Soc.* **42**, 131.
Cheng, J. Y., Don-Paul, M., and Antia, N. J. (1974). *J. Protozool.* **21**, 761.
Christensen, T. (1962). "Systematisk Botanik, Alger." Munksgaard, Copenhagen.
Claes, H. (1954). *Z. Naturforsch., B* **9**, 462.
Claes, H. (1957). *Z. Naturforsch., B* **12**, 401.
Claes, H. (1959). *Z. Naturforsch.,* **14**, 4.
Cooper, R. D. G., Davies, J. B., Leftwick, A. P., Price, C., and Weedon, B. C. L. (1975). *J. Chem. Soc., Perkin Trans. 1* p. 2195.
Crozier, G. F. (1967). *Comp. Biochem. Physiol.* **23**, 179.
Czeczuga, B. (1974). *Int. Rev. Gesamten Hydrobiol.* **59**, 87.
Czygan, F.-C. (1968a). *Arch. Mikrobiol.* **61**, 81.
Czygan, F.-C. (1968b). *Z. Naturforsch., Teil B* **23**, 1367.
Czygan, F.-C., and Kessler, E. (1967). *Z. Naturforsch., Teil B* **22**, 1085.
Czygan, F.-C., and Seefrid, H. (1970). *Z. Naturforsch., Teil B* **25**, 761.
Dabbagh, A. G., and Egger, K. (1974). *Z. Pflanzenphysiol.* **72**, 177.
Dales, R. R. (1960). *J. Mar. Biol. Assoc. U.K.* **39**, 693.
Davies, B. H. (1973). *Pure Appl. Chem.* **35**, 1.
Davies, B. H. (1976). *In* "Chemistry and Biochemistry of Plant Pigments" (T. W. Goodwin, ed.), 2nd Ed., Vol. 2, pp. 38–155. Academic Press, New York.
Davies, B. H., and Taylor, R. F. (1976). *Pure Appl. Chem.* **47**, 211.
Davies, B. H., Hsu, W.-J., and Chichester, C. O. (1965). *Biochem. J.* **92**, 26P.
Davies, B. H., Hsu, W.-J., and Chichester, C. O. (1970). *Comp. Biochem. Physiol.* **33**, 601.
Davies, J. B., Jackman, L. M., Siddons, P. T., and Weedon, B. C. L. (1966). *J. Chem. Soc.* (Series C) 2154.
de Nicola, M. G. (1959). *Boll. Sedute Accad. Gioenia Sci. Nat. Catania* **5**, 201.
de Nicola, M. G., and Furnari, F. (1957). *Boll. Ist. Bot. Univ. Catania* **1**, 180.
de Nicola, M. G., and Goodwin, T. W. (1954a). *Exp. Cell Res.* **7**, 23.
de Nicola, M. G., and Goodwin, T. W. (1954b). *Pubbl. Staz. Zool. Napoli* **25,** 145.
Dodge, J. D. (1975). *Phycologia* **14**, 253.
Dundas, I. D., and Larsen, H. (1963). *Arch. Mikrobiol.* **46**, 19.
Dunning, H. N. (1963). *In* "Organic Geochemistry" (I. A. Breger, ed.), pp. 367–430. MacMillan, New York.
Dutton, H. J., and Manning, W. M. (1941). *Am. J. Bot.* **28**, 516.
Duysens, L. N. M. (1951). *Nature (London)* **168**, 548.
Eimhjellan, K. E. (1967). *Acta Chem. Scand.* **21**, 2280.
Eimhjellen, K. E., and Liaaen-Jensen, S. (1964). *Biochim. Biophys. Acta* **82**, 21.
Emerson, R., and Lewis, C. M. (1943). *Am. J. Bot.* **30**, 165.
Englert, G. (1975). *Helv. Chim. Acta* **58**, 2367.
Englert, G., Kienzle, F., and Noack, F. (1977). *Helv. Chim. Acta* **60**, 1209.
Fischli, A., and Mayer, H. (1975). *Helv. Chim. Acta* **58**, 1492.
Foote, C. S. (1968). *Acc. Chem. Res.* **1**, 104.
Fox, D. L. (1972). *Comp. Biochem. Physiol. B* **43**, 919.
Fox. D. L. (1973). *Comp. Biochem. Physiol. B* **44**, 953.
Fox, D. L. (1974). *In* "Biochemical and Biophysical Perspectives in Marine Biology" (D. C. Malins and J. R. Sargent, eds.), pp. 169–211. Academic Press, New York.

Fox, D. L., and Pantin, C. F. A. (1941). *Philos. Trans. R. Soc. London, Ser. B* **230**, 415.
Fox, D. L., and Scheer, B. T. (1941). *Biol. Bull. (Woods Hole, Mass.)* **80**, 441.
Fox, D. L., and Wilkie, D. W. (1970). *Comp. Biochem. Physiol.* **36**, 49.
Fox, D. L., Updegraff, D. M., and Novelli, D. G. (1944). *Arch. Biochem. Biophys.* **5**, 1.
Fox, D. L., Crozier, G. F., and Smith, V. E. (1967). *Comp. Biochem. Physiol.* **22**, 177.
Fox, D. L., Smith, V. E., Grigg, R. W., and Mac Leod, W. D. (1969). *Comp. Biochem. Physiol.* **28**, 1103.
Francis, G. W., and Halfen, L. N. (1972). *Phytochemistry* **11**, 2347.
Francis, G. W., Hertzberg, S., Andersen, K., and Liaaen-Jensen, S. (1970a). *Phytochemistry* **9**, 629.
Francis, G. W., Upadhyay, R. R., and Liaaen-Jensen, S. (1970b). *Acta Chem. Scand.* **24**, 3050.
Francis, G. W., Hertzberg, S., Upadhyay, R. R., and Liaaen-Jensen, S. (1972). *Acta Chem. Scand.* **26**, 1097.
Francis, G. W., Knutsen, G., and Lien, I. (1973). *Acta Chem. Scand.* **27**, 3599.
Galasko, G., Hora, J., Toube, T. P., Weedon, B. C. L., André, D., Barbier, M., Lederer, E., and Villanueva, V. R. (1969). *J. Chem. Soc. C* p. 1264.
Gilchrist, B. M., and Lee, W. L. (1967). *J. Zool.* **151**, 171.
Gilchrist, B. M., and Lee, W. L. (1976). *Comp. Biochem. Physiol. B* **54**, 343.
Goedheer, J. C. (1965). *Biochim. Biophys. Acta* **102**, 73.
Goodwin, T. W. (1952). "The Comparative Biochemistry of the Carotenoids." Chapman & Hall, London.
Goodwin, T. W. (1968). *In* "Chemical Zoology" (M. Florkin and B. T. Scheer, eds.), Vol. 2, Chs. 2 and 3. Academic Press, New York.
Goodwin, T. W. (1969). *Pure Appl. Chem.* **20**, 483.
Goodwin, T. W. (1971a). *In* "Chemical Zoology" (M. Florkin and B. T. Scheer, eds.), Vol. 6, pp. 290–306. Academic Press, New York.
Goodwin, T. W. (1971b). *In* "Carotenoids" (O. Isler, ed.), pp. 577–636, Birkhaeuser, Basel.
Goodwin, T. W. (1971c). *In* "Aspects of Terpenoid Chemistry and Biochemistry" (T. W. Goodwin, ed.), pp. 315–356. Academic Press, New York.
Goodwin, T. W. (1972). *Prog. Ind. Microbiol.* **11**, 29.
Goodwin, T. W. (1976). *In* "Chemistry and Biochemistry of Plant Pigments" (T. W. Goodwin, ed.), 2nd Ed., Vol. 1, pp. 230–245. Academic Press, New York.
Goodwin, T. W., and Gross, J. A. (1958). *J. Protozool.* **5**, 292.
Griffiths, M., and Perrott, P. (1976). *Comp. Biochem. Physiol. B* **55**, 435.
Hager, A. (1967a). *Planta* **74**, 148.
Hager, A. (1967b). *Planta* **76**, 138.
Hager, A. (1975). *Ber. Dtsch. Bot. Ges.* **88**, 27.
Hager, A., and Stransky, H. (1970a). *Arch. Mikrobiol.* **71**, 132.
Hager, A., and Stransky, H. (1970b). *Arch. Mikrobiol.* **72**, 68.
Hager, A., and Stransky, H. (1970c). *Arch. Mikrobiol.* **73**, 77.
Hajibrahim, S., Tibbetts, T. C., Watts, C. D., Maxwell, J. R., Eglinton, G., Colin, H., and Guiochon, G. (1978). *Anal. Chem.* (in press).
Halfen, L. N., and Francis, G. W. (1972). *Arch. Mikrobiol.* **81**, 25.
Hallenstvet, M., and Liaaen-Jensen, S. (1977). Unpublished observations.
Hallenstvet, M., Ryberg, E., and Liaaen-Jensen, S. (1978). *Comp. Biochem. Physiol.* **60B**, 173.
Hamasaki, T., Okukado, N., and Yamaguchi, M. (1973). *Bull. Chem. Soc. Jpn.* **46**, 1884.
Harashima, K., Nakahara, J., and Kato, G. (1976). *Agric. Biol. Chem.* **40**, 711.
Hata, M., and Hata, M. (1970). *Tohoku J. Agric. Res.* **21**, 183.
Hata, M., and Hata, M. (1975a). *Tohoku J. Agric. Res.* **26**, 35.

Hata, M., and Hata, M. (1975b). *Nippon Suisan Gakkaishi* **41**, 653.
Heilbron, I. M., and Phipers, R. F. (1935). *Biochem. J.* **29**, 1369.
Heilbron, I. M., Jackson, H., and Jones, R. N. (1935). *Biochem. J.* **29**, 1384.
Herring, P. J. (1969). *J. Mar. Biol. Assoc. U.K.* **49**, 766.
Hertzberg, S., and Liaaen-Jensen, S. (1966a). *Phytochemistry* **5**, 557.
Hertzberg, S., and Liaaen-Jensen, S. (1966b). *Phytochemistry* **5**, 565.
Hertzberg, S., and Liaaen-Jensen, S. (1967). *Phytochemistry* **6**, 1119.
Hertzberg, S., and Liaaen-Jensen, S. (1968). *Acta Chem. Scand.* **22**, 1714.
Hertzberg, S., and Liaaen-Jensen, S. (1969a). *Phytochemistry* **8**, 1259.
Hertzberg, S., and Liaaen-Jensen, S. (1969b). *Phytochemistry* **8**, 1281.
Hertzberg, S., and Liaaen-Jensen, S. (1971). *Phytochemistry* **12**, 3251.
Hertzberg, S., Liaaen-Jensen, S., Enzell, C. R., and Francis, G. W. (1969). *Acta Chem. Scand.* **23**, 3290.
Hertzberg, S., Liaaen-Jensen, S., and Siegelman, H. W. (1971). *Phytochemistry* **12**, 3121.
Hertzberg, S., Mortensen, T., Borch, G., Siegelman, H. W., and Liaaen-Jensen, S. (1977). *Phytochemistry* **16**, 587.
Hibberd, D. J., and Leedale, G. F. (1971). *Taxon* **20**, 523.
Hibberd, D. J., and Leedale, G. F. (1972). *Ann. Bot. (London)* **36**, 49.
Hirao, S., Kikichi, R., and Hama, T. (1969). *Nippon Suisan Gakkaishi* **35**, 187.
Hodler, M., Thommen, H., and Mayer, H. (1974). *Chimia* **28**, 723.
Holzel; R., Leftwick, A. P., and Weedon, B. C. L. (1969). *J. Chem. Soc. Chem. Commun.* 128.
Howard, B. M., and Fenical, W. (1975). *Tetrahedron Lett.* p. 1687.
Hsu, W.-J., Chichester, C. O., and Davies, B. H. (1970). *Comp. Biochem. Physiol.* **32**, 69.
Hsu, W.-J., Rodriguez, D. B., and Chichester, C. O. (1971). *Int. J. Biochem.* **3**, 333.
Ikan, R., Aizenshtat, Z., Baedecker, M. J., and Kaplan, I. R. (1975). *Geochim. Cosmochim. Acta* **39**, 173.
Isler, O. (1971). "Carotenoids." Birkhaeuser, Basel.
IUPAC–IUB (1974). "Nomenclature of Carotenoids" (Rules Approved 1974). Butterworth, London.
Jackson, A. H. (1976). *In* "Chemistry and Biochemistry of Plant Pigments" (T. W. Goodwin, ed.), Vol. 1, pp. 1–62. Academic Press, New York.
Jeffrey, S., and Haxo, F. T. (1968). *Biol. Bull. (Woods Hole, Mass.)* **135**, 149.
Jeffrey, S., Sielicki, M., and Haxo, F. T. (1975). *J. Phycol.* **11**, 374.
Jencks, W. P., and Buten, B. (1964). *Arch. Biochem.* **107**, 511.
Jensen, A. (1966a). "Carotenoids of Norwegian Brown Seaweeds and of Seaweed Meals," Rep. No. 31. Norw. Inst. Seaweed Res., Trondheim.
Jensen, A. (1966b). *Acta Chem. Scand.* **20**, 1728.
Jensen, A., and Sakshaug, E. (1970a). *J. Exp. Mar. Biol. Ecol.* **5**, 180.
Jensen, A., and Sakshaug, E. (1970b). *J. Exp. Mar. Biol. Ecol.* **5**, 246.
Johannes, B., Brzezinka, H., and Budzikiewics, H. (1971). *Z. Naturforsch., B* **26,** 377.
Johansen, J. E., and Liaaen-Jensen, S. (1974). *Acta Chem. Scand., Ser. B* **28**, 949.
Johansen, J. E., and Liaaen-Jensen, S. (1976). *Tetrahedron Lett.* 955.
Johansen, J. E., and Liaaen-Jensen, S. (1977). *In* "Marine Natural Products Chemistry" (D. J. Faulkner and W. H. Fenical, eds.), pp. 225–237. Plenum, New York.
Johansen, J. E., Svec, W. A., Liaaen-Jensen, S., and Haxo, F. T. (1974). *Phytochemistry* **13**, 2261.
Karnaŭkhov, V. N., Melovidova, N. Y., and Kargopolova, I. N. (1977). *Comp. Biochem. Physiol. A* **56**, 189.
Karrer, P. (1948). *Fortschr. Chem. Org. Naturst.* **5**, 1.

Katayama, T., Hirata, K., Yokoyama, H., and Chichester, C. O. (1970a). *Nippon Suisan Gakkaishi* **36**, 709.
Katayama, T., Yokoyama, H., and Chichester, C. O. (1970b). *Int. J. Biochem.* **1**, 438.
Katayama, T., Yokoyama, H., and Chichester, C. O. (1970c). *Nippon Suisan Gakkaishi* **36**, 702.
Katayama, T., Hirata, K., and Chichester, C. O. (1971a). *Nippon Suisan Gakkaishi* **37**, 614.
Katayama, T., Tsuchiya, H., and Chichester, C. O. (1971b). *Kagoshima Daigaku Suisan Gakubu Kiyo* **20**, 173.
Katayama, T., Kamata, T., and Chichester, C. O. (1972a). *Int. J. Biochem.* **3**, 363.
Katayama, T., Kamata, T., Shimaya, M., Deshimaru, O., and Chichester, C. O. (1972b). *Nippon Suisan Gakkaishi* **38**, 1171.
Katayama, T., Kunisaki, Y., Shimaya, M., Sameshima, M., and Chichester, C. O. (1973a). *Nippon Suisan Gakkaishi* **39**, 283.
Katayama, T., Shiemaya, M., Sameshima, M., and Chichester, C. O. (1973b). *Nippon Suisan Gakkaishi* **39**, 215.
Katayama, T., Shimaya, M., Sameshima, M., and Chichester, C. O. (1973c). *Int. J. Biochem.* **4**, 223.
Kayser, H. (1976). *Z. Naturforsch., Teil C* **31**, 121.
Kelly, M., Norgård, S., and Liaaen-Jensen, S. (1970). *Acta Chem. Scand.* **24**, 2169.
Khare, H., Moss, G. P., and Weedon, B. C. L. (1973). *Tetrahedron Lett.* p. 3921.
Khatoon, N., Loebe, D. E., Toube, T. P., and Weedon, B. C. L. (1972). *J. Chem. Soc. Chem. Commun.*, 996.
Kienzle, F., and Hodler, M. (1975). *Helv. Chim. Acta* **58**, 198.
Kienzle, F., and Minder, R. E. (1976). *Helv. Chim. Acta* **59**, 439.
Kjøsen, H., Arpin, N., and Liaaen-Jensen, S. (1972). *Acta Chem. Scand.* **26**, 3053.
Kjøsen, H., Norgård, S., Liaaen-Jensen, S., Svec, W. A., Strain, H. H., Wegfahrt, P., Rapoport, H., and Haxo, F. T. (1976). *Acta Chem. Scand., Ser. B* **30**, 157.
Kleinig, H. (1966). *Ber. Dtsch. Bot. Ges.* **79**, 126.
Kleinig, H. (1969). *J. Phycol.* **5**, 281.
Kleinig, H., and Czygan, F.-C. (1969). *Z. Naturforsch., Teil B* **24**, 927.
Kleinig, H., and Egger, K. (1967). *Phytochemistry* **6**, 1681.
Kleinig, H., Nitsche, H., and Egger, H. (1969). *Tetrahedron Lett.* 5139.
Kramer, U. (1974). Thesis, Univ. of Bern, Bern.
Krinsky, N. I. (1965). Comp. Biochem. Physiol. **16**, 181.
Krinsky, N. I. (1966). *In* "Biochemistry of Chloroplasts" (T. W. Goodwin, ed.), Vol. 1, pp. 423–430. Academic Press, New York.
Krinsky, N. I. (1971). *In* "Carotenoids" (O. Isler, ed.), pp. 669–716. Birkhaeuser, Basel.
Krinsky, N. I. (1976). *In* "The Survival of Vegetative Microbes" (T. G. R. Gray and J. R. Postgate, eds.), pp. 209–239. Cambridge Univ. Press, London and New York.
Krinsky, N. I., and Goldsmidt, T. H. (1960). *Arch. Biochem. Biophys.* **91**, 271.
Krinsky, N. I., Gordon, A., and Stern, A. (1964). *Plant Physiol.* **39**, 441.
Kuhn, R., and Kühn, H. (1967). *Eur. J. Biochem.* **2**, 349.
Kuhn, R., and Sørensen, N. A. (1938). *Ber. Dtsch. Chem. Ges.* **71**, 1879.
Kuhn, R., Stene, J., and Sørensen, N. A. (1939). *Ber. Dtsch. Chem. Ges.* **72**, 1688.
Kushwaha, S. C., and Kates, M. (1976). *Can. J. Biochem.* **54**, 824.
Kushwaha, S. C., Gochnauer, M. B., Kushner, D. J., and Kates, M. (1974). *Can. J. Microbiol.* **20**, 241.
Kushwaha, S. C., Kramer, J. K. G., and Kates, M. (1975). *Biochim. Biophys. Acta* **398**, 303.
Kushwaha, S. C., Kates, M., and Porter, J. (1976). *Can. J. Biochem.* **54**, 816.

Kylin, H. (1939). *K. Fysiogr. Saellsk. Lund, Foerh.* **9**, 213.
Lederer, E. (1933). *C. R. Soc. Biol.* **113**, 1391.
Lederer, E. (1935a). "Les Caroténoïdes des Animaux." Hermann, Paris.
Lederer, E. (1935b). *C. R. Acad. Sci.* **201**, 300.
Lederer, E. (1938). *Bull. Soc. Chim. Biol.* **20**, 567.
Lee, W. L. (1966a). *Comp. Biochem. Physiol.* **18**, 17.
Lee, W. L. (1966b). *Ecology* **47**, 930.
Lee, W. L. (1966c). *Comp. Biochem. Physiol.* **19**, 13.
Lee, W. L. (1977). "Carotenoproteins in Animal Colorations" (W. L. Lee, ed.), Wiley, New York.
Lee, W. L., and Gilchrist, B. M. (1975). *Comp. Biochem. Physiol. B* **51**, 247.
Lee, W. L., and Zagalsky. P. F. (1966). *Biochem. J.* **101**, 9c.
Lee, W. L., Gilchrist, B. M., and Dale, R. P. (1967). *J. Mar. Biol. Assoc. U.K.* **47**, 33.
Leftwick, A. P., and Weedon, B. C. L. (1967). *J. Chem. Soc. Chem. Commun.*, 49.
Leuenberger, H. G. W., Boguth, W., Widmer, E., and Zell, R. (1976). *Helv. Chim. Acta* **59**, 1832.
Liaaen, S., and Sørensen, N. A. (1956). *Proc. Int. Seaweed Symp., 2nd* p. 25.
Liaaen-Jensen, S. (1962). *K. Nor. Vidensk. Selsk. Skr.* No. 8.
Liaaen-Jensen, S. (1965). *Annu. Rev. Microbiol.* **19**, 163.
Liaaen-Jensen, S. (1967). *Pure Appl. Chem.* **14**, 227.
Liaaen-Jensen, S. (1971). *In* "Carotenoids" (O. Isler, ed.), pp. 61–188. Birkhaeuser, Basel.
Liaaen-Jensen, S. (1976). *Pure Appl. Chem.* **47**, 129.
Liaaen-Jensen, S. (1977). *In* "Marine Natural Products" (D. J. Faulkner and W. H. Fenical, eds.), pp. 239–259. Plenum, New York.
Liaaen-Jensen, S. (1978). *In* "Photosynthetic Bacteria" (R. K. Clayton and W. R. Sistrom, eds.), Plenum, New York.
Liaaen-Jensen, S., and Andrewes, A. G. (1972). *Annu. Rev. Microbiol.* **26**, 225.
Liaaen-Jensen, S., and Jensen, A. (1971). *In* "Photosynthesis and Nitrogen Fixation" (A. San Pietro, ed.), Methods in Enzymology, Vol. 23, pp. 586–602. Academic Press, New York.
Liaaen-Jensen, S., Cohen-Bazire, G., and Stanier, R. Y. (1961). *Nature (London)* **192**, 1168.
Liaaen-Jensen, S., Hertzberg, S., Weeks, O. B., and Schwieter, U. (1968). *Acta Chem. Scand.* **22**, 1171.
Litchfield, C., and Liaaen-Jensen, S. (1977). Unpublished observations.
Litchfield, C. D. (1976). *In* "Marine Microbiology" (C. D. Litchfield, ed.), pp. 413–418. Halstead Press, Stroudsbourg, Pennsylvania.
Liu, R. S. H. (1976). *Abstr. Int. Symp. Carotenoids, 4th, Berne*, p. 37.
Loeblich, A. F. (1976). *J. Protozool.* **23**, 13.
Loeblich, A. R., and Smith, V. E. (1968). *Lipids* **3**, 5.
McBeth, J. W. (1972a). *Comp. Biochem. Physiol. B* **41**, 55.
McBeth, J. W. (1972b). *Comp. Biochem. Physiol. B* **41**, 69.
McDermott, J. C. B., Ben-Aziz, A., Singh, R. K., Britton, G., and Goodwin, T. W. (1973). *Pure Appl. Chem.* **35**, 47.
McMeekin, T. A., Patterson, J. T., and Murray, J. G. (1971). *J. Appl. Bacteriol.* **34**, 699.
Mallams, A. K., Waight, E. S., Weedon, B. C. L., Chapman, D. J., Haxo, F. T., Goodwin, T. W., and Thomas, D. M. (1967). *Proc. Chem. Soc.* 301.
Mathews-Roth, M. M., Pathak, M. A., Fitzpatrick, T. B., Harber, L. L., and Kass, E. H. (1970). *New Engl. J. Med.* **282**, 1231.
Matsuno, T., and Katsuyama, M. (1976a). *Nippon Suisan Gakkaishi* **42**, 645.
Matsuno, T., and Katsuyama, M. (1976b). *Nippon Suisan Gakkaishi* **42**, 761.

Matsuno, T., and Katsuyama, M. (1976c). *Nippon Suisan Gakkaishi* **42**, 765.
Matsuno, T., and Katsuyama, M. (1976d). *Nippon Suisan Gakkaishi* **42**, 847.
Matsuno, T., Katsuyama, M., and Takashi, A. (1976a). *Nippon Suisan Gakkaishi* **42**, 651.
Matsuno, T., Nagata, S., and Kitamara, K. (1976b). *Tetrahedron Lett.*, 4601.
Maxwell, J. R., Pillinger, G. T., and Eglinton, G. (1971). *Q. Rev., Chem. Soc.* **25**, 571.
Mayer, H., and Isler, O. (1971). *In* "Carotenoids" (O. Isler, ed.), pp. 325–576. Birkhaeuser, Basel.
Mayer, H., Bogurth, W., Leuenberger, H. G. W., Widmer, E., and Zell, R. (1975). *Abstr. Int. Symp. Carotenoids, 4th, Berne*, p. 43.
Monroy, A., and de Nicola, M. (1952). *Experientia* **8**, 29.
Moss, J. (1976). *Pure Appl. Chem.* **47**, 97.
Murphy, P. T., Wells, R. J., Johansen, J. E., and Liaaen-Jensen, S. (1977). *Biochem. Syst. Ecol.* **60B**, 173.
Nicoara, E., Illyes, G., Suteu, M., and Bodea, C. (1967). *Rev. Roum. Chim.* **12**, 547.
Nitsche, H. (1973a). *Z. Naturforsch., Teil C* **28**, 641.
Nitsche, H. (1973b). *Arch. Mikrobiol.* **90**, 151.
Nitsche, H. (1974a). *Biochim. Biophys. Acta* **338**, 572.
Nitsche, H. (1974b). *Arch. Mikrobiol.* **95**, 79.
Nitsche, H., and Egger, K. (1970). *Tetrahedron Lett.* 1435.
Norgård, S. (1972). Thesis, Univ. of Trondheim, Trondheim.
Norgård, S., Svec, W. A., Liaaen-Jensen, S., Jensen, A., and Guillard, R. R. L. (1974a). *Biochem. Syst. Ecol.* **2**, 3.
Norgård, S., Svec, W. A., Liaaen-Jensen, S., Jensen, A., and Guillard, R. R. L. (1974b). *Biochem. Syst. Ecol.* **2**, 7.
O'Carra, P., and OhEocha, C. (1976). *In* "Chemistry and Biochemistry of Plant Pigments" (T. W. Goodwin, ed.), Vol. 1, pp. 328–376. Academic Press, New York.
Okukado, N. (1975). *Bull. Chem. Soc. Jpn.* **48**, 1061.
Parsons, T. R. (1961). *J. Fish. Res. Board Can.* **18**, 1017.
Paust, J. (1976). *Abstr. Int. Symp. Carotenoids, 4th Berne*, p. 46.
Peake, E., Casagrande, D. J., and Hodgson, G. W. (1974). *Mem., Am. Assoc. Pet. Geol.* **20**, 505.
Pfander, H., and Leuenberger, U. (1976). *Chimia* **30**, 71.
Pfennig, N. (1967). *Annu. Rev. Microbiol.* **21**, 285.
Pfennig, N., and Trüper, H. G. (1974). *In* "Bergeys Manual of Determinative Bacteriology," 8th Ed., pp. 24–75. Williams & Wilkins, Baltimore, Maryland.
Rau, W. (1976). *Pure Appl. Chem.* **47**, 237.
Ricketts, T. R. (1966). *Phytochemistry* **5**, 571.
Ricketts, T. R. (1967a). *Phytochemistry* **6**, 19.
Ricketts, T. R. (1967b). *Phytochemistry* **6**, 669.
Ricketts, T. R. (1970). *Phytochemistry* **9**, 1835.
Ricketts, T. R. (1971a). *Phytochemistry* **10**, 155.
Ricketts, T. R. (1971b). *Phytochemistry* **10**, 161.
Rodriguez, D. B., Simpson, K. L., and Chichester, C. O. (1973). *Int. J. Biochem.* **4**, 213.
Rodriguez, D. B., Simpson, K. L., and Chichester, C. O. (1974). *Int. J. Biochem.* **5**, 157.
Sapozhnikov, D. I. (1973). *Pure Appl. Chem.* **35**, 47.
Schmidt, K. (1978). *In* "Photosynthetic Bacteria" (R. K. Clayton and W. R. Sistrom, eds.), (in press), Plenum; New York.
Schmidt, K., Francis, G. W., and Liaaen-Jensen, S. (1971). *Acta Chem. Scand.* **25**, 2476.
Schwendinger, R. B. (1969). *In* "Organic Geochemistry" (G. Eglinton and M. T. I. Murphy, eds.), Ch. 17, Springer-Verlag, Berlin and New York.

Scopes, P. M. (1975). *Fortschr. Chem. Org. Naturst.* **32**, 167.
Simoneit, B. R., Howells, W. G., and Burlingame, A. L. (1973). *Deep Sea Drilling Proj. Leg. 15. Initial Rep. Deep Sea Drilling Proj.* No. 20, p. 907.
Simpson, K. L. (1972). *In* "Advances in the Chemistry of Plant Pigments" (C. O. Chichester, ed.), Advances in Food Research, Suppl. 3, pp. 9–22. Academic Press, New York.
Smallidge, R. L., and Quackenbush, F. W. (1973). *Phytochemistry* **12**, 2481.
Smith, P. M. (1975). "The Chemotaxonomy of Plants." Arnold, London.
Sørensen, N. A., Liaaen-Jensen, S., Børdalen, B., Haug, A., Enzell, C., and Francis, G. (1968). *Acta Chem. Scand.* **22**, 344.
Song, P. S., Koka, P., Prézelin, B. B., and Haxo, F. T. (1976). *Biochemistry* **15**, 4422.
Sporn, M. B., Dunlop, N. M., Newton, D. L., and Henderson, W. R. (1976). *Nature (London)* **263**, 110.
Stanier, R. Y. (1960). *Harvey Lect.* **54**, 219.
Stewart, I., and Wheaton, T. A. (1971). *J. Chromatogr.* **55**, 325.
Stewart, I., and Wheaton, T. A. (1973). *Phytochemistry* **12**, 2947.
Strain, H. H. (1965). *Biol. Bull. (Woods Hole, Mass.)* **129**, 366.
Strain, H. H. (1966). *In* "Biochemistry of Chloroplasts" (T. W. Goodwin, ed.), Vol. 1, pp. 387–406. Academic Press, New York.
Strain, H. H., Manning, W. M., and Hardin, G. S. (1944). *Biol. Bull. (Woods Hole, Mass.)* **86**, 169.
Strain, H. H., Svec, W. A., Aitzetmüller, K., Grandolfo, M., and Katz, J. J. (1968). *Phytochemistry* **7**, 1417.
Strain, H. H., Aitzetmüller, K., Svec, W. A., and Katz, J. J. (1970). *J. Chem. Soc. Chem. Commun.*, 876.
Strain, H. H., Svec, W. A., Aitzetmüller, K., Cope, B. T., Harkness, A. L., and Katz, J. J. (1971a). *Org. Mass Spectrom.* **5**, 565.
Strain, H. H., Svec, W. A., Aitzetmüller, K., Grandolfo, M. C., Katz, J. J., Kjøsen, H., Norgård, S., Liaaen-Jensen, S., Haxo, F. T., Wegfahrt, P., and Rapoport, H. (1971b). *J. Am. Chem. Soc.* **93**, 1823.
Strain, H. H., Svec, W. A., Wegfahrt, P., Rapoport, H., Haxo, F. T., Norgård, S., Kjøsen, H., and Liaaen-Jensen, S. (1976). *Acta Chem. Scand., Ser. B* **30**, 109.
Stransky, H., and Hager, A. (1970a). *Arch. Mikrobiol.* **71**, 164.
Stransky, H., and Hager, A. (1970b). *Arch. Mikrobiol.* **72**, 84.
Straub, O. (1976). "Key to Carotenoids—Lists of Natural Carotenoids." Birkhaeuser, Basel.
Tanaka, Y., and Katayama, T. (1976). *Nippon Suisan Gakkaishi* **42**, 801.
Tanaka, Y., Katayama, T., Simpson, K. L., and Chichester, C. O. (1976a). *Nippon Suisan Gakkaishi* **42**, 1187.
Tanaka, Y., Matsuguchi, H., Katayama, T., Simpson, K. L., and Chichester, C. O. (1976b). *Comp. Biochem. Physiol. B* **54**, 391.
Taylor, D. L. (1969). *J. Mar. Biol. Assoc. U.K.* **49**, 1057.
Thommen, H. (1971). *In* "Carotenoids" (O. Isler, ed.), pp. 637–668. Birkhaeuser, Basel.
Thommen, H., and Gloor, U. (1965). *Naturwissenschaften* **52**, 161.
Tischer, J. (1941). *Moppe-Seylers Z. Physiol. Chem.* **267**, 281.
Tollin, G., and Robinson, M. I. (1969). *Photochem. Photobiol.* **9**, 411.
Toube, T., and Weedon, B. C. L. (1970). *J. Chem. Soc.* 241.
Upadhyay, R. R., and Liaaen-Jensen, S. (1970). *Acta Chem. Scand.* **24**, 3055.
Vallentyne, J. R. (1957). *J. Fish. Res. Board Can.* **14**, 33.
van Uden, N., and Fell, J. W. (1968). *Adv. Microbiol. Sea* **1**, 167.

Vetter, W., Englert, G., Rigassi, V., and Schwieter, U. (1971). *In* "Carotenoids" (O. Isler, ed.), pp. 189–266. Birkhaeuser, Basel.
Walton, T. J., Britton, G., Goodwin, T. W., Diner, B., and Moshier, S. (1970). *Phytochemistry* **9**, 2545.
Watts, C. D. (1977). Personal communication.
Watts, C. D., and Maxwell, J. R. (1975). *Org. Mass Spectrom.* **10**, 1102.
Watts, C. D., and Maxwell, J. R. (1977). *Geochim. Cosmochim. Acta* **41**, 493.
Watts, C. D., Maxwell, J. R., and Kjøsen, H. (1977). *Adv. Org. Geochem., Proc. Meet. Org. Geochem., 7th, Madrid, 1975* p. 391.
Weedon, B. C. L. (1970). *Pure Appl. Chem.* **20**, 51.
Weedon, B. C. L. (1973). *Pure Appl. Chem.* **35**, 113.
Weedon, B. C. L. (1976). *Pure Appl. Chem.* **47**, 161.
Weeks, O. B. (1971). *In* "Aspect of Terpenoid Chemistry and Biochemistry" (T. W. Goodwin, ed.), pp. 291–313. Academic Press, New York.
Weeks, O. B., Saleh, F. K., Wirahadikusumah, M., and Berry, R. A. (1973). *Pure Appl. Chem.* **35**, 63.
Whittle, S. J., and Casselton, P. J. (1975a). *Br. Phycol. J.* **10**, 179.
Whittle, S. J., and Casselton, P. J. (1975b). *Br. Phycol. J.* **10**, 192.
Withers, N., and Haxo, F. T. (1975). *Plant Sci. Lett.* **5**, 7.
Yamamoto, H., and Chichester, C. O. (1965). *Biochim. Biophys. Acta* **109**, 303.
Yamamoto, H. R., Yokoyama, H., and Boettger, H. (1969). *J. Org. Chem.* **34**, 4207.
Zagalsky, P. F. (1976). *Pure Appl. Chem.* **47**, 103.
Zechmeister, L. (1962). "*Cis-trans* Isomeric Carotenoids, Vitamins A and Arylpolyenes." Springer-Verlag, Berlin and New York.

Chapter 2

The Sterols of Marine Invertebrates: Composition, Biosynthesis, and Metabolites

L. J. GOAD

MARINE NATURAL PRODUCTS

ISBN 0-12-624002-7

I. INTRODUCTION

There has been a continuing interest in the sterols and steroids of marine organisms ever since the earliest studies of Henze (1904) and Dorée (1909) revealed that sponges contained sterols other than cholesterol, which, although its structure was not known at that time, was established as the major mammalian sterol. This interest was maintained by Heilbron with his work on marine algal sterols, which resulted in the characterization of fucosterol, and later by Tsuda's studies on seaweeds. However, it was the classic investigations of Bergmann (1962) that laid much of the foundations for present-day studies of marine invertebrate sterols.

An examination of the literature before about 1960 indicates that, for the most part, because of methodological limitations, only the major sterol of an invertebrate was isolated and identified with any confidence. Many animals were reported to contain "new" sterols, to which were given rather exotic trivial names such as magakesterol, shakosterol, meritristerol, and pteropodasterol. However, it is now apparent that the pioneer workers were dealing with complex mixtures of up to 20 or more related sterols, which are notorious for cocrystallization. Since crystallization was the main purification method available, it is hardly surprising that many of the intractable sterol mixtures remained unidentified. Indeed, one can only admire the perseverance and skill of some of the early chemists who did manage to isolate and identify a pure sterol.

The resurgence of interest in marine sterols since about 1960 stems in part from the search for new marine natural products with useful pharmacological properties and with challenging structures for elucidation by the chemist. Another stimulus to the resumption of interest has been the development of highly improved separation methods [for example, gas–liquid chromatography (glc) and argentation chromatography] and the great technical advances in analytical equipment and methods such as nuclear magnetic resonance (nmr) and mass spectrometers, X-ray crystallography, and computerized combined gas chromatography–mass spectrometry. These various techniques now make possible the identification of sterol components that constitute only 0.1% or less of a mixture. This is an important factor in the search for potential short-lived intermediates in a biosynthetic or metabolic sequence. The development of improved underwater breathing equipment has also been beneficial because of the increased ease with which many animals can now be collected and supplied to the laboratory in a fresh condition, a particularly important point if biosynthetic studies are envisaged.

The most exciting fruits of recent studies have been the characterization of a host of novel sterols, many with unique alkylation patterns in the side chain, but some with modified ring structures (Minale, 1977; Schmitz, 1977). However, equally important to an understanding of the origins of invertebrate sterols has been the unraveling of complex sterol mixtures and the identification and quantification of the various components, many of them previously known plant sterols. Also, the continuing studies on algal sterols and algal sterol biosynthesis are important since, as is discussed later, it now seems likely that many of the sterols found in invertebrates are either unchanged or modified dietary sterols of algal origin.

The earlier work on marine sterols was reviewed by Bergmann (1949, 1958, 1962), Faulkner and Anderson (1974), Austin (1970), Scheuer (1973), and Goad (1976). Many sterols have been given trivial names, and these are listed together with their systematic names according to the IUPAC–IUB 1971 Definitive Rules for Steroid Nomenclature in the appendix at the end of this chapter. The structures of the sterol side chains and ring systems are presented in Figs. 1 and 2, respectively.

In this chapter the nature and distribution of the sterols found in marine algae and invertebrate animals are surveyed, and then their biosynthetic origins are discussed. Two other areas of current interest are the biosynthesis and transformations of steroid hormones in invertebrates and of the ecdysones (molting hormones) in crustaceans. Finally, recent work on the steroidal saponins of echinoderms is reviewed.

II. STEROLS OF MARINE ALGAE AND FUNGI

Since algae are undoubtedly the prime source of many sterols found in the marine environment a summary of marine algal sterols is a prerequisite to the consideration of marine invertebrate sterols. Detailed reviews of the nature, distribution, and biosynthesis of algal sterols have appeared elsewhere (Patterson, 1971; Goodwin, 1974; Goad, 1976, 1977; Goad *et al.*, 1974). Only some of the pertinent recent publications on algal sterols are cited here.

The vast majority of the Rhodophyta contain cholesterol (**1b**) as the major constituent, although in a few species desmosterol (**2b**) is contained in quite substantial amounts and may even be the major sterol, as in *Rhodymenia palmata* (Gibbons *et al.*, 1967; Idler and Wiseman, 1970). However, Idler and Atkinson (1976) found that, although the cholesterol content stayed at a constant low value, the desmosterol (**2b**) content varied somewhat in samples of *R. palmata* collected over a period of one year. More seasonal studies of this kind would be useful with other algal species, and certainly possible seasonal changes in sterol composition

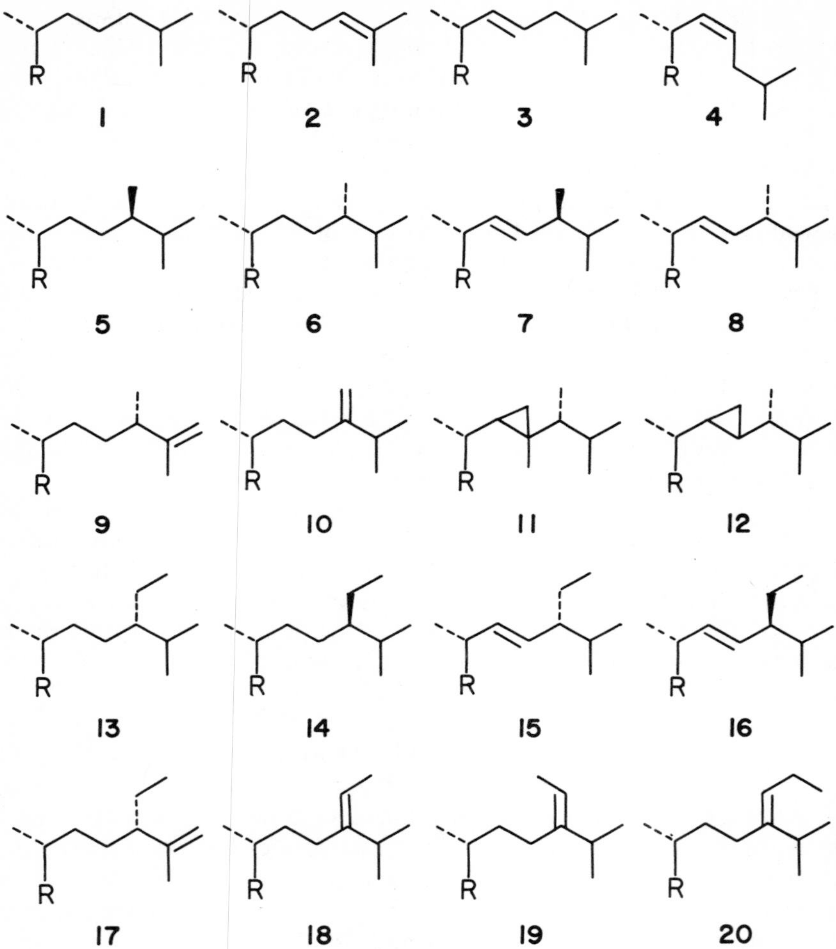

Fig. 1. Structures of sterol side chains (R represents one of the ring systems shown in Fig. 2).

should be considered when various sterol analyses of algae (and invertebrates?) are compared.

Cholesta-5,23*E*-diene-3β,25-diol (liagosterol, **39b**) was isolated from two species of red algae (*Liagora distenta* and *Scinaia furcenata*) by Fattorusso *et al.* (1975c) and identified together with cholesta-5,25-diene-3β,24-diol (**40b**) in *Rhodymenia palmata* (Morisaki *et al.*, 1976). 22-Dehydrocholesterol (**3b**) is the main sterol found in *Hypnea* (Tsuda *et al.*,

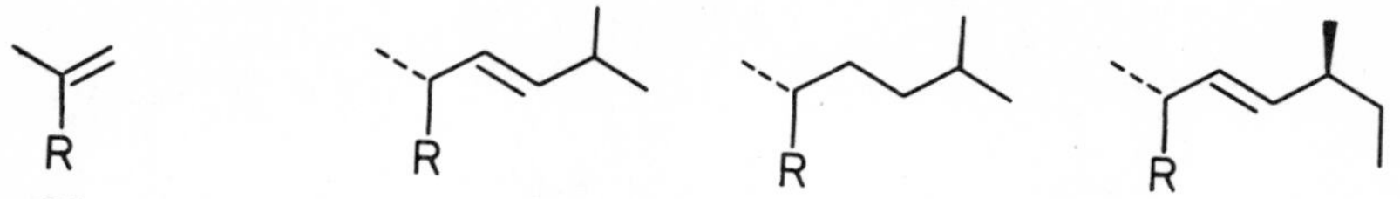

21 22 23 24

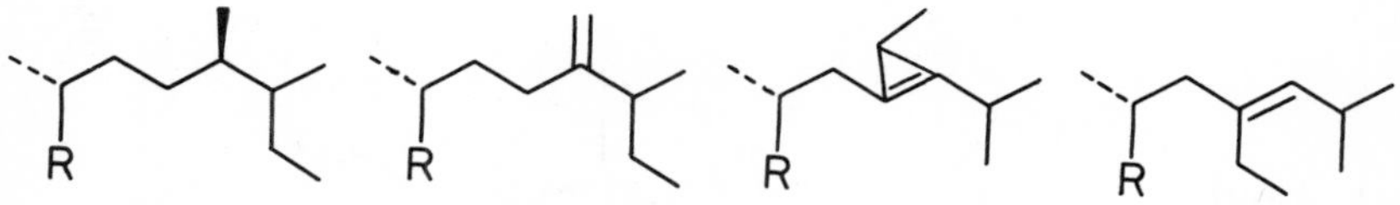

25 26 27 28

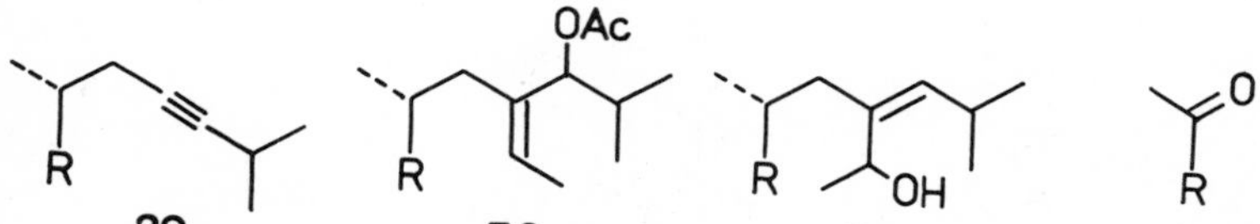

29 30 31 32

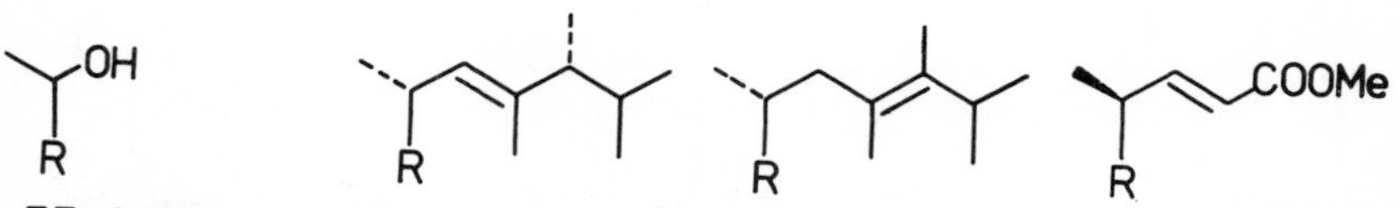

33 20α–
34 20β–
35 36 37

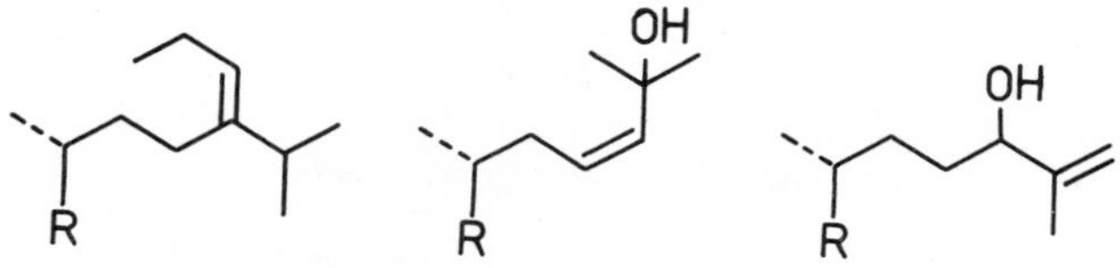

38 39 40

Fig. 1. (*Continued*)

1960; Fattorusso *et al.*, 1975c) and several species of the unicellular red alga *Phorphyridium* (Beastall *et al.*, 1971; Kanazawa *et al.*, 1972; Cargile *et al.*, 1975). Cholestanol (**1a**) and other ring-saturated sterols were reported for the first time in large amounts in red algae by Chardon-Loriaux *et al.* (1976), who showed that they were the major sterols in three species from the family Gelidiaceae and in a species from the family Callymeniaceae. It is noteworthy that Kanazawa and Yoshioka (1971a,b)

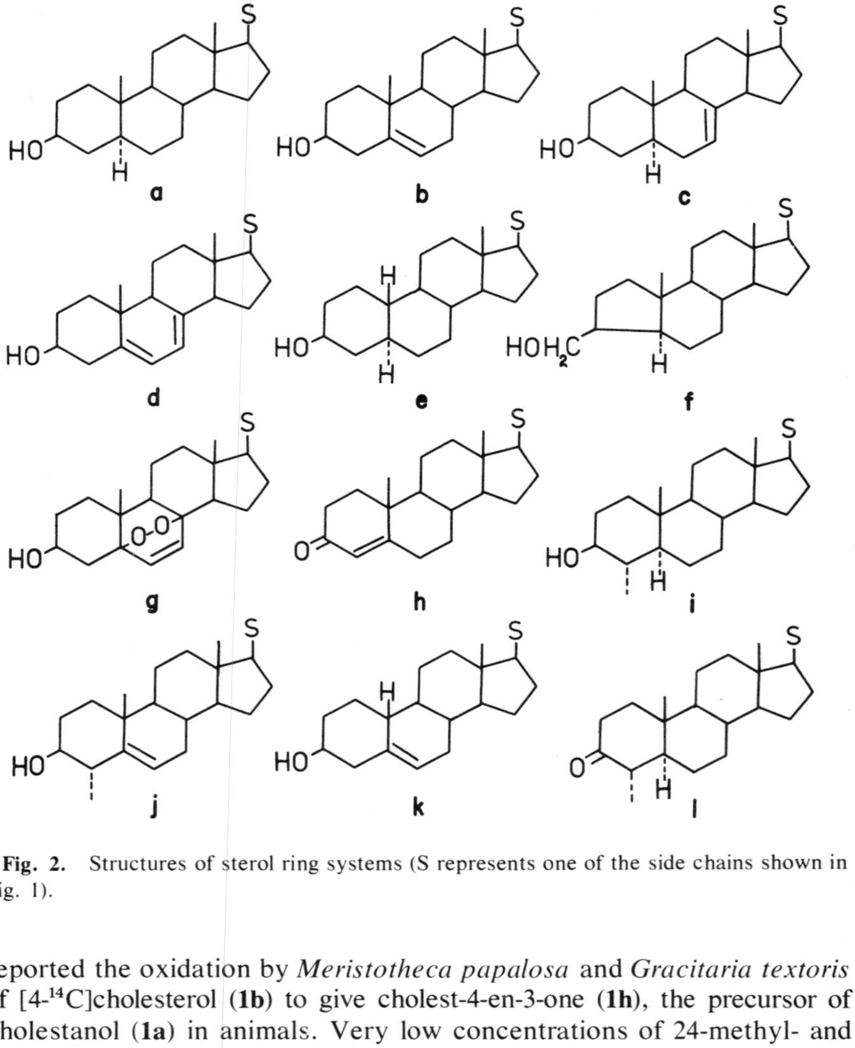

Fig. 2. Structures of sterol ring systems (S represents one of the side chains shown in Fig. 1).

reported the oxidation by *Meristotheca papalosa* and *Gracitaria textoris* of [4-^{14}C]cholesterol (**1b**) to give cholest-4-en-3-one (**1h**), the precursor of cholestanol (**1a**) in animals. Very low concentrations of 24-methyl- and 24-ethylsterols have been noted in some red algae (Idler *et al.*, 1968; Idler and Wiseman, 1970; Beastall *et al.*, 1971, 1974), but substantial amounts of C_{28} sterol (**5b** or **6b**) have been reported only in *Rytiphlea tinctoria* and *Vidalia volubilis* (Alcaide *et al.*, 1969; Fattorusso *et al.*, 1976).

The Phaeophyta all contain the 24*E*-ethylidene compound fucosterol (**19b**) with often smaller amounts of 24-methylenecholesterol (**10b**) and cholesterol (**1b**) (Goad, 1976; Amico *et al.*, 1976; Shimadate *et al.*, 1977). A number of oxidized derivatives of fucosterol have been identified in brown seaweeds (Goad, 1976), but it may be speculated that these are

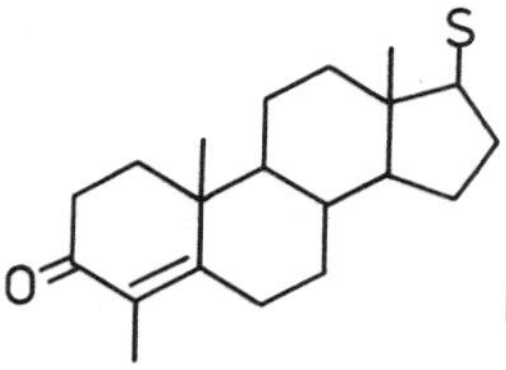

m 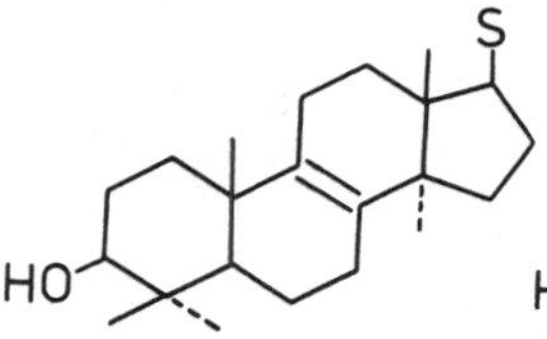 n 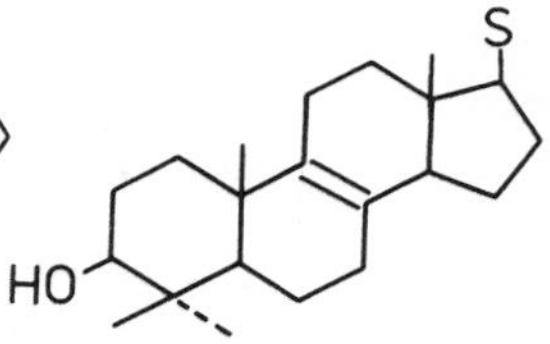 o

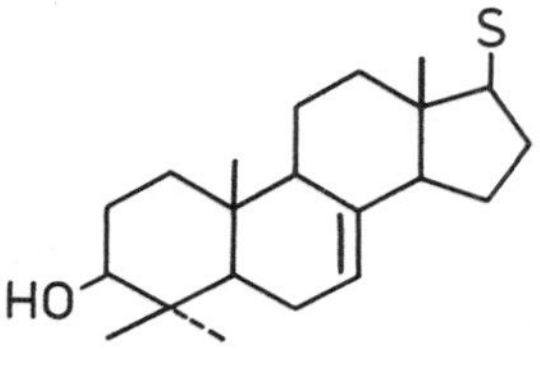

p 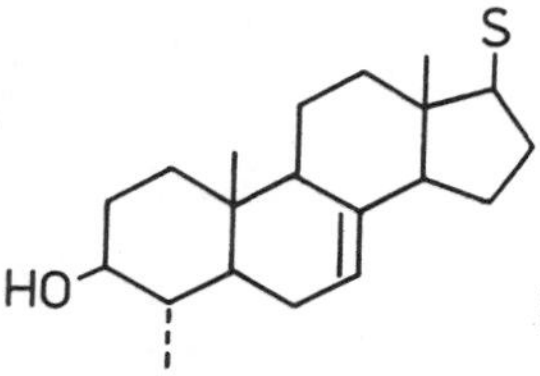 q 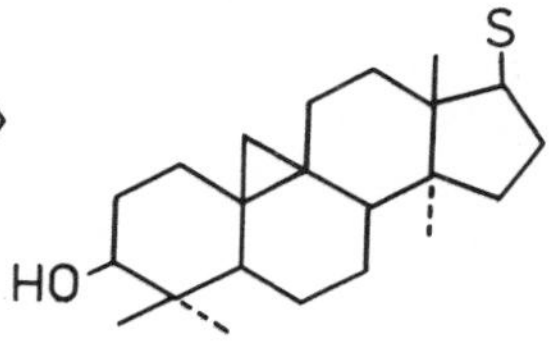 r

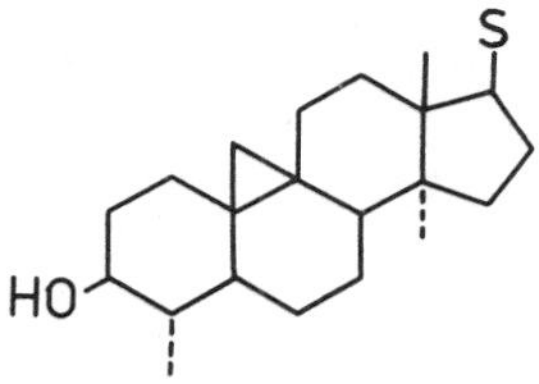

s 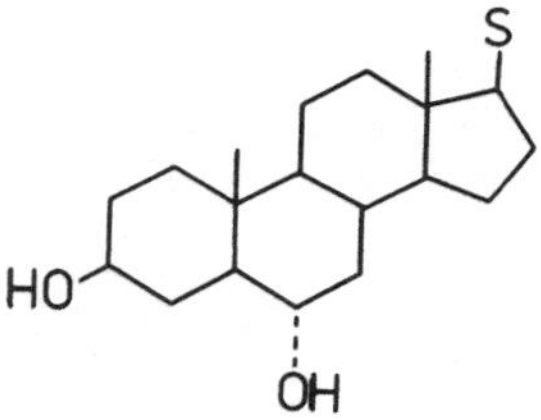 t

Fig. 2. (*Continued*)

oxidation artifacts resulting from the air drying of the seaweeds before extraction (Knights, 1970).

The Chlorophyta species so far examined (Goad, 1976) do not allow a clear phylogenetic pattern of sterol distribution to be deduced with any degree of certainty other than that the 24*Z*-ethylidenesterol isofucosterol (**18b**) has been observed as the major component in four species of Ulotrichales (Tsuda and Sakai, 1960; Gibbons *et al.*, 1968; Doyle and Patterson, 1972).

From the order Siphonales, *Codium fragile* contained two 25-methylenesterols—clerosterol (**17b**, 95%) and codisterol (**9b**, 5%)—both with the 24β configuration (Rubinstein and Goad, 1974a).* The only other

* The 24α and 24β system of assigning configurations is used for the most part in this chapter since it makes biosynthetic relationships more readily apparent than the 24*R* and 24*S* system recommended by IUPAC–IUB (Goad, 1976).

species from this order examined was *Halimedia incrassata* (Patterson, 1974), but this contained two sterols with a 24β-alkyl-saturated side chain: clionasterol (**13b**, 94%) and ergost-5-enol (**6b**, 4%). Patterson (1974) also examined *Cladophora flexuosa* from the order Cladophorales and found that it contained isofucosterol (**18b**, 45%), 24-methylenecholesterol (**10b**, 21%), and 2–6% of clionasterol (**13b**) and ergost-5-enol (**6b**). *Chaetomorpha crassa* from the same order contained larger amounts of 24-ethyl-(**13b** or **14b**, 40%) and 24-methylsterols (**5b** or **6b**, 10%; **7b** or **8b**, 6%) and 24-methylenecholesterol (**10b**, 12%) but no isofucosterol (**18b**).

Relevant to considerations of the origins of marine sterols are the observations of 24β-ethyl- and 24β-methyl-, Δ^5, Δ^7, and $\Delta^{5,7}$ sterols in unicellular algae of the order Chlorococcales (Patterson, 1971) and possibly ergosterol (**8d**) and 7-dehydroporiferasterol (**15d**) in *Chlamydomonas reinhardi* (Eichenberger, 1976) since related algae with similar sterol patterns presumably occur in the phytoplankton.

The diatoms have not received the attention that they merit in view of their importance as major algal constituents of the phytoplankton, and only recently have any detailed investigations been undertaken on their sterol content. Low (1955) tentatively identified chondrillasterol (**15c**) in *Navicula pelliculosa,* and Kanazawa *et al.* (1971c) identified the only sterol of *Nitzschia closterium* and *Cyclotella nana* as brassicasterol (**8b**). The latter sterol was assigned the 24β configuration on the basis of agreement of its melting point with that of the authentic sterol (**8b**). However, melting point data are not particularly reliable criteria for assignment of C-24 configuration since several 24α/24β epimeric sterol pairs have rather similar melting points, and obviously very pure sterol samples must be obtained in order to measure a meaningful melting point. It has now been shown that 100 MHz, or better 220 MHz, nmr spectroscopy is a much more reliable method of assigning C-24 configuration. Moreover, in this method the sterols do not have to be absolutely pure; indeed, mixtures of 24α/24β epimeric sterols can often be quantitated by nmr (Thompson *et al.*, 1972; Mulheirn, 1973; Rubinstein, 1973; Rubinstein et al., 1976; Nes *et al.*, 1976; Sucrow *et al.*, 1976).

When the sterol of the diatom *Phaeodactylum tricornutum* (formally *Nitzschia closterium* forma *minutissima*) was examined by nmr spectroscopy it was identified unambiguously as (24*S*)-24-methylcholesta-5,22*E*-dien-3β-ol (**7b**), i.e., the 24α epimer of brassicasterol (**8b**) (Rubinstein and Goad, 1974b). Since **7b** and its acetate had melting points close to the reported values for brassicasterol (**8b**) and brassicasteryl acetate, respectively, it seemed desirable that the *N. closterium* sterol isolated by Kanazawa *et al.* (1971c) be rechecked by nmr. This was done by Orcutt and Patterson (1975), who established that in fact it is the 24α epimer (**7b**) that occurs in *N. closterium*. These authors also found **7b** in four other species of diatom (*N. ovalis, N. frustulum, Navicula pelliculosa,* and *Biddulphia aurita*). In this study cholesterol (**1b**) was shown to be the principal sterol of *Nitzschia longissima,* whereas the 24α-ethylsterol, stigmasterol (**16b**), was tentatively identified on melting point evidence in *Amphora exigua* and a second *Amphora* species. Other species (*Thallasiosira pseudonana, Biddulphia aurita, and Fragilaria*) were found to contain more complex mixtures of 24-methylsterols; *Nitzschia frustulum* contained cholesterol (42%) in addition to **7b** (54%), and *N. ovalis* had a significant content (33%) of 22-dehydrocholesterol (**3b**), which is an important constituent in many marine invertebrates.

The diatom *Chaetoceros simplex calcitrans* is reported to contain cholesterol (**1b**, 40%), 24-methylenecholesterol (**10b**, 40%), and small amounts of other C_{28} and C_{29} sterols (Boutry and Barbier, 1974). It has also been shown that both the sterol composition and content vary with different light conditions (Boutry *et al.*, 1976a). *Dynamena pamila* contains an appreciable amount of cholesterol (**1b**, 37%) in its sterol mixture, which comprises several C_{27}, C_{28}, and C_{29} $\Delta^{5,22}$ sterols and stanols (Ando *et al.*, 1975). 24-Methylenecholesterol (**10b**), a common sterol in invertebrates, has been identified as the only sterol in the nonphotosynthetic diatom *Nitzschia alba* (Anderson *et al.*, 1978).

Cholesterol (**1b**) is the most abundant component in the sterol mixture of plankton (Baron and Boutry, 1963; Boutry and Baron, 1967; Boutry *et al.*, 1971; Kanazawa and Teshima, 1971c). Although much of this cholesterol (**1b**) might reasonably be expected to originate in the zooplankton, it has been shown that the phytoplankton also makes a significant cholesterol contribution, and in view of the above-mentioned results diatoms may well be the main donors. For example, Boutry and Jacques (1970) observed that the sterols from a sample of phytoplankton consisted of about 75% cholesterol (**1b**). Similarly, Boutry *et al.* (1976b) found that algae taken from surface waters contained sterol mixtures with 40–50% cholesterol (**1b**) in addition to **3b** (10–15%), **7b/8b** (15–22%), **13b/14b** (15–20%), and traces of other sterols. Gagosian (1976) also showed that cholesterol (**1b**) persists in water collected from depths down to 4500 m in the Sargasso sea. Low concentrations (2–2000 μg/liter) of sterol have been reported in true solution in seawater, and the literature on this topic and on the sterols of sediments has been reviewed (Goad, 1976).

Despite the extensive interest in the toxins produced by dinoflagellates (Shimizu *et al.*, 1977), these quantitatively important members of the phytoplankton escaped the attention of sterol chemists until the studies of *Gonyaulax tamarensis* by Shimizu *et al.* (1976) and *Pyrocystis lunula* by Ando and Barbier (1975). The *P. lunula* contained mainly cholesterol (**1b**) together with 22-dehydrocholesterol (**3b**), desmosterol (**2b**), and a number of Δ^5, $\Delta^{5,22}$, $\Delta^{5,24(28)}$, and saturated C_{28} and C_{29} sterols. However, examination of *G. tamarensis* (Shimizu *et al.*, 1976) was rewarded by the isolation of cholesterol (**1b**) and a unique sterol with exciting biosynthetic implications. The new sterol was named dinosterol (**35i**), and it has an unprecedented 4-monomethyl-saturated ring system and a side chain bearing in addition to the usual 24-methyl group a C-23 methyl group on a Δ^{22} bond. This compound has obvious affinities with gorgosterol (**11b**), previously identified in gorgonians (see page 89). The biosynthetic relation of dinosterol (**35i**) to gorgosterol (**11b**) has been reinforced by X-ray diffraction analyses, which have shown that both sterols have the 24*R* or 24β configuration (Hale *et al.*, 1970; Ling *et al.*, 1970; Finer *et al.*, 1977). The study of Finer *et al.* (1977) also established the Δ^{22} bond of dinosterol (**35i**) as trans. Examination of another dinoflagellate, *Crypthecodinium cohnii,* by Withers *et al.* (1978) revealed that this organism also contains dinosterol (**35i**) as the major sterol, but a number of other biosynthetically related sterols were also identified. These include compounds **6i**, **6j**, **35j**, and, significantly, from the viewpoint of ring saturation the 3-oxosteroids **35l** and **35m**; also present were small quantities of cholesterol (**1b**) and other usual 4-demethylsterols.

Finally, when considering the origins of invertebrate sterols, one must include the contribution made by fungi since marine fungi, mainly Ascomycetes and Fungi Imperfecti, are thought to have an important ecological role in the marine environment (Roth *et al.*, 1962; Johnson and Sparrow, 1961; Kohlmeyer and Kohlmeyer, 1971; Oppenheimer, 1963). To date all the marine fungi examined contain the 24β-methyl-$\Delta^{5,7,22}$-triene ergosterol (**8d**), which is typical of terrestial fungi (Kirk and Catalfomo, 1970; Teshima and Kanazawa, 1971a). The report of the presence of 24-methylcholest-5-en-3β-ol (**5b** or **6b**) in some species (Teshima and Kanazawa, 1971a) requires confirmation since Δ^5 sterols are very rare in fungi. The report is based on work in which the organisms were cultured in media containing molasses, which may have been the source of this sterol.

III. STEROL COMPOSITION OF INVERTEBRATES

A. Porifera

Since the first studies of the Henze (1904) and Dorée (1909), the sponges have continued to yield the most varied and biogenetically unprecedented array of sterols found among the invertebrate phyla. Much of the pioneering work on sponges was conducted by Bergmann and co-workers

(Bergmann, 1949, 1962). They characterized several sterols, such as clionasterol (**13b**) and poriferasterol (**15b**), which have subsequently been recognized as more typically algal products. With the recent resumption of interest in sponge sterols, Minale and his group have isolated and characterized a number of unique sterols. The structural studies on these sterols have been reviewed (Minale, 1976, 1977; Minale *et al.*, 1976; Minale and Sodano, 1977), and they include compounds with an unusual side chain alkylation pattern as in aplysterol (**25b**) from *Verongia* spp. species (de Luca *et al.*, 1972, 1973), a series of 19-norsterols (e.g., **1e**) from *Axinella polypoides* (Minale and Sodano, 1974a), and a novel group of 3β-hydroxymethyl-A-norstanols (e.g., **1f**) in *Axinella verucosa* (Minale and Sodano, 1974b).

Fattorusso and associates also identified unusual sterols in Mediterranean sponges. Thus, 5,8-epoxysterols such as **1g**, **3g**, and **8g** were obtained from *Axinella cannabina* (Fattorusso *et al.*, 1974), which also contains the parent $\Delta^{5,7}$ sterols (de Rosa *et al.*, 1973a; Cafieri *et al.*, 1975). Calysterol (**27b**) with an unprecedented cyclopropene group in the side chain was isolated from *Calyx nicaensis* (Fattorusso *et al.*, 1975a) together with small amounts of 23-ethylcholesta-5,23-dien-3β-ol (**28b**) and a remarkable cholesterol derivative (**29b**) with a 23,24 acetylenic linkage (cited in Minale and Sodano, 1977). Vigorous acetylation conditions cause rearrangement of calysterol (**27b**) to give a diacetate (**30b**, 3-acetate) (Minale and Sodano, 1977), and chromatography over alumina impregnated with silver nitrate also opens the cyclopropene ring to give diol **31b** (Fattorusso *et al.*, 1975b). These facts may explain why calysterol (**27b**) or related cyclopropene sterols have not been isolated previously since both acetylation and argentation chromatography are commonly employed in the resolution of complex sterol mixtures.

An examination of five sponges collected along the California coast led to the identification of complex sterol mixtures that included 5,8-epidioxysterols in *Tethya aurantia* and 3-oxo-Δ^4-steroids (e.g., **1h**, **10h**) in *Stelleta clarella* (Sheikh and Djerassi, 1974). The occurrence of the latter compounds (**1h**, **10h**, etc.) may be significant in the production of the stanols found in several sponges (Table 1). Ballantine *et al.* (1977) identified small amounts of the pregnane derivatives 3β-hydroxy-17β-pregn-5-en-20-one (**32b**), 17β-pregn-5-ene-3β,20α-diol (**33b**), and 17β-pregn-5-ene-3β,20β-diol (**34b**) together with the more usual sterols in the sponge *Haliclona rubens*.

Table 1 summarizes recent reports on the sterols of sponges. At this stage, although the sterol compositions of all sponges are very complex, a few broad generalizations can be made which may be helpful in discussions regarding the origins of sponge sterols. Thus, sterols with the

TABLE 1

Sterols of Sponges

Classification	Sterols[a]	Reference
Order Dictioceratida		
Family Spongidae		
Cacospongia mollior	C_{27}, C_{28}, C_{29}, Δ^5, $\Delta^{5,22}$, $\Delta^{5,7}$, $\Delta^{5,7,22}$	Cafieri and de Napoli (1976)
Euspongia officianalis	**1b**, **2b**, **3b**, **4b**, **6b**, **8b**, **10b**, **13b**, **15b**, **19b**, **22b**	Voogt (1976)
Ircinia muscarum, I. spinosula, Spongia nitens, S. officinalis	C_{27}, C_{28}, C_{29}, $\Delta^{5,7}$	de Rosa *et al.* (1973a)
Family Verongidae		
Verongia aerophoba, V. archeri, V. fistularis, V. thiona	**25b**, **26b**; other C_{27}, C_{28}, C_{29}, Δ sterols	de Rosa *et al.* (1973a)
Family Dysidaidae		
Ianthella ardia	C_{27}, C_{28}, C_{29}, Δ^5	de Rosa *et al.* (1973a)
Dysidae avara	C_{27}, C_{28}, C_{29}, Δ^5, $\Delta^{5,7}$	de Rosa *et al.* (1973a)
Order Haposclerida		
Family Haliclonidae		
Haliclona permollis,	**1b**, **8b**, **10b**, **15b**, **18b**, **22b**	Sheikh and Djerassi (1974)
Haliclona rubens	**1b**, **3b**, **8b**, **13a**, **13b**, **15b**, **22b**, **24b**, **32b**, **33b**, **34b**	Ballantine *et al.* (1977)
Family Callyspongiidae		
Callyspongia diffisa	**10b**, **18b**	Erdman and Scheuer (1975)
Order Poccilosclerida		
Family Agelasidae		
Agelasidae oroides	C_{27}, C_{28}, C_{29} stanols and Δ^7	de Rosa *et al.* (1973a)
Family Mixillidae		
Lissodendoryx noxiosa	**1a**, **1b**, **3b**, **8b**, **10b**, **13b**, **15b**, **18b**, **22b**	Sheikh and Djerassi (1974)
Tedania ignis	C_{27}, C_{28}, C_{29}, Δ^5	de Rosa *et al.* (1973a)

Order Axinellida		
Family Axinellidae		
Axinella damicornis	C_{27}, C_{28}, C_{29} stanols	de Rosa *et al.* (1973a)
	1a, **6a**, **8a**, **13a**, **15a**, **3b**, **22b**, **1c**	Voogt (1976)
A. polypoides	**8e** and other 19-norstanols	de Rosa *et al.* (1973a)
Axinella sp.	C_{27}, C_{28}, C_{29}, Δ^5	de Rosa *et al.* (1973a)
A. cannabini	C_{27}, C_{28}, C_{29}, $\Delta^{5,7}$, Δ^7	de Rosa *et al.* (1973a)
	1g and other sterol 5,8-epoxides	Fattorusso *et al.* (1974), Cafieri *et al.* (1975)
A. acuta	C_{27}, C_{28}, C_{29}, $\Delta^{5,7}$	de Rosa *et al.* (1973a)
A. verucosa	**1f**, **3f**, **5f**, **7f**, **14f**, **16f**	Minale and Sodano (1974b)
Family Halichondrilidae		
Halichondria sp. I	**1b**, **3b**, **4b**, **6b**, **8b**, **10b**, **13b**, **15b**, **19b**, **22b**	Voogt (1976)
Halichondria sp. II	**1a**, **6a**, **8a**, **13a**, **15a**, **3b**, **22b**, **1c**, **6c**, **13c**, **15c**	Voogt (1976)
Family Hymeniacidonidae		
Hymeniacidon perleve	**1a**, **3a**, **8a**, **10a**, **22a**, **8b**, **14b**, **19b**	Erdman and Thomson (1972), Edmonds *et al.* (1977)
Order Hadromerida		
Family Clionidae		
Cliona celata	**1b**, **3b**, **6b**, **8b**, **10b**, **13b**, **15b**, **22b**	Erdman and Thomson (1972)
C. viridis	C_{27}, C_{28}, C_{29}, Δ^5	de Rosa *et al.* (1973a)
Order Epipolasida		
Family Tethyidae		
Tethyia aurantium	C_{27}, C_{28}, C_{29}, Δ^5	de Rosa *et al.* (1973a)
T. auriantia	**1a**, **6a**, **1b**, **3b**, **8b**, **10b**, **15b**, **18b**, **20b**, **22b**	Sheikh and Djerassi (1974)
Order Choristidae		
Family Geodiidae		
Geodia cynodium	C_{27}, C_{28}, C_{29}, Δ^5	de Rosa *et al.* (1973a)
G. mülleri	**1b**, **2b**, **3b**, **4b**, **6b**, **8b**, **10b**, **13b**, **15b**, **18b**, **19b**	Voogt (1976)

(*Continued*)

TABLE 1 (*Continued*)

Classification	Sterols[a]	Reference
Family Stellettidae		
Stelleta clarella	**1a**, **6a**, **13a**, **1b**, **3b**, **6b**, **8b**, **10b**, **13b**, **18b**, **22b**, **1h**, **3h**, **8h**, **10h**, **19h**, **22h**	Sheikh and Djerassi (1974)
Order Carnosa		
Family Chondrillidae		
Chondrilla nucula	C_{27}, C_{28}, C_{29} stanols	de Rosa *et al.* (1973a)
Family Grantiidae		
Grantia compressa	**1a**, **3a**, **8a**, **10a**, **22a**, **1b**, **1c**	Edmonds *et al.* (1977)

[a] The 24-methyl- and 24-ethylsterols have been indicated as either the 24α or 24β epimers only for convenience; they should more correctly be regarded as 24ξ-methyl- or 24ξ-ethylsterols since the analytical methods employed did not permit the determination of the C-24 configurations.

side chains **25** and **26** are found only in species of the Verongidae; 19-norstanols and 3β-hydroxymethyl-A-norstanols are present in some Axinellidae species; stanols (**1a**, **3a**, **10a**) occur in substantial amounts in some representatives of the Axinellidae, Halichondridae, Haliclonidae, Chondrillidae, and Hymeniacidae; and sterol-5,7-dienes are found in sponges of the Spongidae and Axinellidae families.

B. Cnidaria

The sterols found in coelenterate animals are listed in Table 2. Most animals contain the diverse spectrum of sterols encountered in other invertebrate phyla. However, with the exception of some gorgonian species and members of the Zoanthidea, cholesterol (**1b**) is usually the predominant sterol.

The gorgonians are distinguished by their content of gorgosterol (**11b**) and related sterols (**12b**, **11i**, **11j**) with a cyclopropane group in the side chain (Schmitz, 1977). A reexamination of the sterols of *Pseudoplexaura porosa,* which has a high content of gorgosterol (**11b**), resulted in the identification of 19-norcholesterol (**1k**) and its higher homologs (**8k** and **15k** but with undetermined C-24 configuration) (Popov *et al.*, 1976a,b). This observation is significant in two respects. First, it reveals that 19-norsteroids are not unique to the sponges in which the 19-norstanols were first recorded by Minale and Sodano (1974a). This may indicate a biosynthetic origin in another organism rather than transformation of dietary sterol by the sponge or gorgonian. Second, the retention of the Δ^5 bond in **1k** apparently shows that ring saturation or double-bond migration to Δ^4 is not a prerequisite for the presumed C-19 demethylation of a normal sterol precursor (e.g., cholesterol, **1b**).

Two sterols, which are clearly related biosynthetically to gorgosterol (**11b**), have been isolated from the soft coral *Sarcophyta elegans*. These are 23,24-dimethylcholesta-5,22-dien-3β-ol (**35b**) and 23,24-dimethylcholesta-5,23-dien-3β-ol (**36b**), respectively (Kanazawa *et al.*, 1974c, 1977a). The only species of sea pen so far examined, *Ptilosarcus gurneyi*, has yielded a unique steroid, (20*S*)-3β-hydroxychol-5,22-dien-24-oic acid methyl ester (**37b**) (Vanderah and Djerassi, 1977). Not only is this the first steroid to be reported in an invertebrate animal with the degraded bile acid type of side chain but, more remarkably, its C-20 configuration is opposite to that found in virtually all other naturally occurring steroids, and it clearly raises intriguing conjectures regarding its biosynthetic origins.

Several soft corals have been reported to contain polyhydroxylated sterols among their major steroids (Braekman, 1977). An unidentified

TABLE 2

Sterols of Coelenterates

Classification	Major sterol	Other sterols[a]	Reference
Class Hydrozoa			
Order Siphonophora			
Physalia physalis	**1b**		Hamilton *et al.* (1964), Middlebrook and Lane (1968)
Alolla wyvillei	**1b**		Ballantine *et al.* (1976)
Pelagia noctiluca	**1b**		Ballantine *et al.* (1976)
Order Anthomedusae			
Spirocodon saltatrix	**1b**	**2b**, **3b**, **6b**, **10b**, **13b**, **19b**, **22b**	Yasuda (1974a)
Periphylla periphylla	**1b**	**1a**, **3a**, **8a**, **10a**, **13a**, **18a**, **22a**, **2b**, **3b**, **6b**, **13b**, **15b**, **18b**, **22b**	Ballantine *et al.* (1975a)
Class Scyphozoa			
Order Rhizostomae			
Rhizostoma sp.	**1b**		van Aarem *et al.* (1964)
Stomophus sp.	**1b**	**13b**, **19b**, **22b**	Yasuda (1974a)
Cassiopea xamachana[b]	C_{27}	C_{28}, C_{29}, C_{30}	Ciereszko *et al.* (1968)
Rhopilema esculentia	**1b**	**3b**, **8b**, **14b**, **22b**, **24b**	Kanazawa *et al.* (1977b)
Order Semascostomae			
Aurelia aurita	**1b**	**1a**, **2b**, **3b**, **8b**, **10b**, **13b**, **19b**, **22b**	Yasuda (1974a), Kanazawa *et al.* (1977b)
Class Anthozoa			
Order Gorgonaceae			
Briareum asbestinum[b]	C_{30}	C_{27}, C_{28}	Ciereszko *et al.* (1968)
Eunicea mamosa[b]	C_{27}	C_{28}, C_{29}, C_{30}	Ciereszko *et al.* (1968)
Eunicea succinea	**1b**		Flores and Rosas (1966)
Eunicella verrucosa	C_{27}, C_{28}	C_{29}	Ciereszko *et al.* (1968)
Eugorgia ampla	**1b**	**3b**, **6b**, **8b**, **10b**, **14b**, **16b**, **22b**	Block (1974)
Gorgonia flabellum, *Gorgonia ventilina*		Complex mixture, **11b**, **12b**	Schmitz and Pattabhiraman (1970)
Isis hippuris[b]	C_{28}, C_{30}	C_{27}	Ciereszko *et al.* (1968)
Melithaea flabillifera	**1b**	**2b**, **3b**, **8b**, **10b**, **22b**, **24b**	Kanazawa *et al.* (1977b)

Muricea appressa	**1b**	**3b, 6b, 8b, 14b, 22b**	Block (1974)
Muricea atlantica	**1b**	**1h**	Flores and Rosas (1966)
Plexaura sp.	**1b**	**6b, 8b, 11b, 14b, 16b, 19b**	Block (1974)
Plexaura flexuosa[b]	C_{27}, C_{28}	C_{28}, C_{29}, C_{30}	Ciereszko *et al.* (1968)
Plexaurealla nutans	**1b**	**1h**	Flores and Rosas (1966)
Pseudoplexaura porosa[b]	C_{28}, C_{30}	C_{27}, C_{29}, **1k**	Ciereszko *et al.* (1968), Popov *et al.* (1976a,b)
Pterogorgia anceps[b]	C_{27}	C_{28}, C_{29}, C_{30}	Ciereszko *et al.* (1968)
Rumphella antipathes[b]	C_{28}	C_{27}, C_{28}, C_{29}, C_{30}	Ciereszko *et al.* (1968)
Order Zoanthidea			
Zoanthus proteus	**10b**		Bergmann and Dusza (1957)
Zoanthus confertus	**8b**	**1b, 6b, 10b**	Gupta and Scheuer (1969)
Palythoa tuberculosa	**6b**	**1b, 8b, 11b, 14b**	Gupta and Scheuer (1969)
Palythoa sp.	**10b**		Gupta and Scheuer (1969)
Palythoa sp.	**6b**	**1b, 3b, 8b, 10b, 11b**	Kanazawa *et al.* (1977b)
Order Actinaria			
Anemonia sulcata	**1b**	**2b, 3b, 6b, 8b, 10b, 14b, 18b, 22b**	Voogt *et al.* (1974)
Anthopleura midori	**10b**	**1b, 2b, 8b**	Kanazawa *et al.* (1977b)
Anthopleura fuscoviridis	**1b**	**6b, 8b, 10b, 13b**	Yasuda (1974a)
Calliactis japonica	**1b**		Toyama and Tanaka (1956b)
Calliactis parasitica	**1b**	**1a, 3b, 8b, 10b**	Ferezou *et al.* (1972)
Dofleinia armata	**10b**	**1b, 3b, 8b**	Kanazawa *et al.* (1977b)
Metridium senile	**1b**	**2b, 3b, 6b, 8b, 10b, 14b, 18b, 22b**	Mason (1972), Voogt *et al.* (1974), Habermehl *et al.* (1976)
Phymactis clematis	**1b**	**2b, 3b, 4b, 6b, 8b, 10b, 13b, 18b, 19b, 22b**	Orlando Munoz *et al.* (1976)
Stoichactis kenti	**1b, 10b**		Kanazawa *et al.* (1977b)
Order Pennatulaceae			
Ptilosarcus gurneyi	**37b**		Vanderah and Djerassi (1977)
Order Scleractinia			
Acropora palmata[b]	C_{28}	C_{29}, C_{30}	Steudler *et al.* (1977)
Acropora studeri	**6b**	**1a, 1b, 3b, 8b, 10b, 11b, 12b**	Kanazawa *et al.* (1977b)
Pavana clecussata	**6b**	**8b, 10b, 11b**	Kanazawa *et al.* (1977b)
Porites porites var. *furcata*[b]	C_{28}	C_{27}, C_{29}, C_{30}	Steudler *et al.* (1977)

(*Continued*)

TABLE 2 (*Continued*)

Classification	Major sterol	Other sterols[a]	Reference
Order Ceriantharia			
Cerianthus membranaceus	**1b**	**2b**, **3b**, **6b**, **8b**, **10b**, **14b**, **18b**, **22b**,	Voogt *et al.* (1974)
Order Stolonifera			
Clavularia sp.	**6b**, **12b**	**1b**, **8b**, **10b**	Kanazawa *et al.* (1977b)
Order Alcyonaea			
Cladieria sp.	**6b**	**1b**, **8b**, **10b**, **11b**, **12b**	Kanazawa *et al.* (1977b)
Cladieria digitulata	**11b**	**1b**, **6b**, **8b**, **12b**, **35b**	Kanazawa *et al.* (1977b)
Cladieria sphaerophora	**6b**, **11b**	**1b**, **8b**, **10b**, **35b**, **12b**	Kanazawa *et al.* (1977b)
Lobophytum[b] sp.	C_{28}	C_{27}, C_{29}, C_{30}	Ciereszko *et al.* (1968)
Leioptilus guerneyi[b]	C_{27}	C_{28}, C_{29}	Ciereszko *et al.* (1968)
Manicina areolata[b]	C_{27}	C_{28}, C_{29}, C_{30}	Ciereszko *et al.* (1968)
Nephthea chabrolii	**1b**	**3b**, **6b**, **8b**, **10b**, **22b**, **24b**	Kanazawa *et al.* (1977b)
Nephtea sp.	**1b**	**5b**, **11b**, **27b**	Englebrecht *et al.* (1972)
Nephtea[b] sp.	C_{28}	C_{27}, C_{28}, C_{29}	Ciereszko *et al.* (1968)
Sarcophyta sp.	**11b**	**1b**, **6b**, **8b**, **10b**, **35b**	Kanazawa *et al.* (1977b)
Sarcophyta sp.	**6b**	**1b**, **8b**, **10b**, **11b**, **12b**	Kanazawa *et al.* (1977b)
Sarcophyta elegans	**6b**	**1b**, **8b**, **10b**, **11b**, **12b**, **35b**, **36b**	Kanazawa *et al.* (1977b)
Sarcophyta glaucum	**6b**	**1b**, **8b**, **10b**, **11b**, **35b**	Kanazawa *et al.* (1977b)
Sinularia sp.	**11b**	**1b**, **6b**, **8b**, **10b**, **12b**, **35b**	Kanazawa *et al.* (1977b)
Sinularia sp.	**11b**	**1b**, **6b**, **8b**, **10b**, **35b**, **36b**	Kanazawa *et al.* (1977b)
Sinularia polydactyla	**6b**	**1b**, **8b**, **10b**, **11b**, **35b**	Kanazawa *et al.* (1977b)
Siphonogorgia dispsacea	**1b**	**3b**, **6b**, **8b**, **10b**, **22b**, **24b**	Kanazawa *et al.* (1977b)
Xenia elongata	**11b**	C_{27}, C_{28}, C_{30}	Steudler *et al.* (1977)

[a] See footnote a of Table 1.

[b] The sterol mixtures were examined by mass spectrometry but not further characterized. Therefore, they are designated by carbon number only. The C_{30} compound was assigned as gorgosterol (**11b**) (Ciereszko *et al.*, 1968).

alcyonarian thought to be probably a *Nephtea* species by the authors (Englebrecht *et al.*, 1972) but cited as *Simalaria mayi* by Braekman (1977) contained 24ξ-methylcholest-5-ene-3β,25-diol. *Srarcophyton elegans* yielded 24ξ-methylcholestane-3β,5α,6β-25-tetrol 25-acetate (Moldowan *et al.*, 1974) and 24ξ-methylcholestane-3β,5α,6β,12β,25-pentol 25-acetate (Moldowan *et al.*, 1975). *Lobophytum pauciflorum* gave (24*S*)-24-methylcholestane-3β,4β,5β,25-tetrol-6-one 25-acetate (Tursch *et al.*, 1976); *Sinularia dissecta* gave 24ξ-methylcholestane-3β,5α,6β-triol, 24ξ-methylcholestane-3β,5α,6β-triol 6-acetate, and the corresponding 24-methylene derivatives (Bortolotto *et al.*, 1976a); and *Pseudopterogorgia elisabethae* produced cholestane-3β,5α,6β,9α-tetrol (Schmitz *et al.*, 1976). Finally, 24-methylenecholest-5-ene-3β,7β,19-triol and its 7-acetate have been isolated from *Litophyton viridis* (Bortolotto *et al.*, 1976b).

C. Nemertea and Annelida

Only one nemertean worm and five polychaete worms have been rigorously analyzed for their sterol composition, but in all cases cholesterol (**1b**) is the major component of the complex mixtures of C_{26}, C_{27}, C_{28}, and C_{29} Δ^5 and $\Delta^{5,22}$ sterols (Table 3). The occurrence of cholesterol (**1b**) in two polychaetes (*Marphysa mossambica, Terebella ehrenbergi*), an echiuroid (*Ochetostoma erthyrogrammon*), and a sipunculid (*Sipunculus rudus*) was briefly mentioned by Marsden (1976), but detailed sterol analyses were not performed.

In their first analysis of the sterols of the polychaete *Pseudopotamilla occelata* Kobayashi *et al.* (1973a) tentatively identified a minor component as the Δ^{22} -cis compound cholesta-5,22*Z*-dien-3β-ol (**4b**). However, isolation and reexamination of the sterol showed it to be a Δ^{22}-trans side chain, and the structure was found to be 27-nor-(24*S*)-24-methylcholesta-5,22*E*-dien-3β-ol (**24b**), to which was given the trivial name occelasterol (Kobayashi and Mitsuhashi, 1974a). The characterization of this sterol is important on two counts. First, a minor component has been observed in the sterol mixtures of several marine animals, most often mollusks, which on the basis of its glc retention time has been assigned as the cis compound **4b**. It now seems likely that these identifications may require revision to compound **24b**. Second, as is discussed on p. 128, it may provide an important clue to the biosynthetic origins of the ubiquitous C_{26} sterols in marine organisms.

D. Mollusca

More species of mollusks have been analyzed for their sterol content and composition than have those of any other phylum, in part because of

TABLE 3

Percent Composition of the Sterols of Nemertea and Annelida Species

Species	Sterol[a] (%)											Reference
	1b	**2b**	**3b**	**5b/6b**	**7b/8b**	**10b**	**13b/14b**	**15b/16b**	**18b/19b**	**22b/23b**	**24b**	
Phylum Nemertea												
Cerebratulus marginatus	89.5	—	1.8	1.8	1.8	1.8	1.3	0.6	0.6	0.5	0.3	Voogt (1973a)
Phylum Annelida												
Order Polychaeta												
Aphrodite aculeata	85.1	—	7.7	1.5	2.8	0.7	0.9	0.4	—	0.7	—	Voogt (1974)
Nephthys hombergii	89.7	—	4.3	1.4	1.3	1.0	0.6	—	—	0.6	—	Voogt (1974)
Nereis diversicolor	85.1	1.1	4.3	2.8	2.7	1.0	1.2	1.1	—	0.5	—	Voogt (1974)
Spirographis spallanzani	47.4	—	13.2	2.7	14.4	5.6	5.7	2.9	1.9	5.6	—	Voogt (1974)
Pseudopotamilla occelata	50.1	17.9	6.0	3.4	5.5[b]	9.3	2.0	0.5	—	4.4	0.5	Kobayashi *et al.* (1973a), Kobayashi and Mitsuhashi (1974a)

[a] See footnote a of Table 1 regarding the C-24 configuration of most of these sterols.

[b] Shown to have the 24α or $24S$ configuration.

their nutritional value. The results are summarized in Table 4; in most cases only sterols occurring in excess of 1% of the mixture are included. Earlier literature is discussed in the reviews of Toyama (1958), Bergmann (1962), Austin (1970), Idler and Wiseman (1972), Voogt (1972a), and Goad (1976).

The chitons (class Amphineura) stand apart by virtue of their high content of cholest-7-en-3β-ol (**1c**), whereas the other classes of Mollusca contain predominantly Δ^5 sterols. The Cephalopoda appear to have the simplest sterol composition, with cholesterol (**1b**) usually constituting 90% or more of the total mixture. In contrast the Gastropoda and Pelecypoda contain much more complicated sterol mixtures, with the various C_{28} and C_{29} Δ^5 and $\Delta^{5,22}$ compounds often comprising 40–50% of the total, although cholesterol (**1b**) is again the most abundant sterol in virtually all cases. A notable exception to this generality are the four clams *Tridacna crocea, T. roae, T. squamosa,* and *Hipponps hipponps,* which contained 34–65% of dihydrobrassicasterol (**6b**) compared to 12–29% of cholesterol (**1b**) (Teshima *et al.*, 1974a).

The sterol composition recorded in different laboratories for a particular species sometimes varies markedly. This is illustrated by the results obtained with four samples of *Mytilus edulis* collected from different localities (Table 5). The overall sterol patterns were the same in all four samples, and the two Japanese samples collected in 1971 and 1972 had very similar compositions. However, the animals collected in Halifax had a higher cholesterol (**1b**) content, a markedly lower amount of 24-methylcholesta-5,22-dien-3β-ol (**8b**), and apparently negligible 24-methylenecholesterol (**10b**) when compared to the samples from Zeeland and Kagoshima. On the other hand, the Kagoshima samples had a considerably higher level of 22-dehydrocholesterol (**3b**) and a smaller amount of desmosterol (**2b**) than did the other two samples. As discussed later, such variations undoubtedly result, at least in part, from the sterol contributions made by the various planktonic dietary sources to which the mollusk has access. These presumably vary with geographical location and season, and consequently the sterol compositions reported in Table 4 for mollusks, and indeed those for all invertebrates, should be regarded only as a guide to the sterols likely to be encountered in a given species.

Many species of mollusks have been observed to contain $\Delta^{5,7}$ sterols (Table 4), and these sometimes apparently constitute a fairly high proportion of the total sterol mixtures (Idler and Wiseman, 1972; Voogt, 1972a). For the most part, these $\Delta^{5,7}$ sterols have not been identified, but 7-dehydrocholesterol (**1d**), cholesta-5,7,22-trien-3β-ol (**3d**), and 7-dehydrostigmasterol (**16d**) have been reported (Table 4). The detailed procedures outlined recently for the extraction and isolation of marine

TABLE 4

Sterols of Mollusks

Classification	Sterols[a]	$\Delta^{5,7}$ Sterol detected	Reference
Class Amphineura			
Acanthochiton rubrolineatus	**1c**		Takagi and Toyama (1956b)
Chaetopleura apiculata	**1c**		Idler and Wiseman (1971c)
Chiton tuberculatus	**1c**		Kind and Meigs (1955)
Cryptoplax japonica	**1c**		Takagi and Toyama (1956b)
Lepidochitona cinerea[b]	**1c**, **1b**, **7c**		Voogt and Van Rheenen (1974)
Liolophura japonica[b]	**1c**, **1b**, **5b**		Toyama and Tanaka (1953), Teshima and Kanazawa (1972b, 1973c)
Onithochiton hirasei	**1c**		Kita and Toyama (1959)
Class Gastropoda			
Amacea testudinalis	**1b**, **2b**		Idler and Wiseman (1971c)
Aplysia depilans	**1b**		Voogt and Van Rheenen (1973a)
Aplysia kwodai	**1b**		Tanaka and Toyama (1959)
Archidoris tuberculata[b]	**1b**, **2b**, **3b**, **6b**, **8b**, **14b**, **16b**, **22b**		Voogt (1973c)
Arianta arbustorum	**1b**, **1d**, **2b**, **5b**, **8b**, **14b**, **22b**	+	Voogt (1972b), Voogt and Van der Horst (1972)
Arion empiricorum	**1b**, **8d**	+	Bock and Wetter (1938)
Arion rufus[b]	**1b**, **2b**, **3b**, **5b**, **8b**, **14b**, **16b**	+	Voogt (1967a, 1972b)
Astralium haematragum	**1b**, **5b**, **8b**, **14b**, **19b**		Teshima and Kanazawa (1972b)
Buccinum perryi	**1b**	+	Kind and Goldberg (1953), Toyama and Tanaka (1956a)
Buccinum undatum[b]	**1b**, **2b**, **3b**, **22b**		Idler and Wiseman (1971c), Voogt (1967b, 1972d)
Busycon canaliculatum	**1b**, **6b**, **14b**	+	Toyama and Tanaka (1954), Idler and Wiseman (1971c)
Cellana nigrisquamata	**1b**, **2b**, **5b**, **8b**		Teshima and Kanazawa (1972b)
Cellana nigrolineata	**1b**, **2b**		Teshima and Kanazawa (1972b)
Chlorostoma argyristina	**1b**		Hayashi and Yamada (1974)
Charonia [illegible]	**1b**, **2b**, **3b**, **8b**, **10b**, **14b**		Teshima and Kanazawa (1972b)

Conomurex luhuanus	**1b, 3b, 8b, 10b, 14b**		Teshima and Kanazawa (1972b)
Crepidula fornicata[b]	**1b, 3b, 3d, 6b, 8b, 10b**	+	Idler and Wiseman (1971c)
Cymatiidae sp.	**1b, 3b, 5b, 8b, 14b**		Teshima and Kanazawa (1972b)
Dendronotus frondosus[b]	**1b, 2b, 3b, 6b, 8b, 10b**	+	Voogt (1973c)
Euhadra herklotsi	**1b**	+	Takagi and Toyama (1958)
Haliotis gigantea	**1b**		Tsujimoto and Koyanagi (1934a)
Haliotis gurneri[b]	**1b, 3b, 22b**		Teshima and Kanazawa (1972b, 1974)
Hemifusus ternatanus	**1b**	+	Toyama *et al.* (1955b)
Incillaria confusa	**14b**(?)	+	Tanaka and Toyama (1957b)
Lamellidoris bilamellata[b]	**1b, 2b, 3b, 8b, 10b**		Voogt (1973c)
Littorina brevicula	**1b**	+	Tanaka and Toyama (1959)
Littorina littorea[b]	**1b, 2b, 3b, 5b, 6b, 8b, 10b, 14b, 18b, 19b, 22b**	+	Kind and Herman (1948), Idler and Wiseman (1971c), Voogt (1969, 1971a)
Lunatia groenlandica	**1b**		Idler and Wiseman (1971c)
Lunella coronata coreensis	**1b**	+	Toyama *et al.* (1955b)
Melampus lineatus	**1b, 3b, 6b, 10b, 14b, 22b**	+	Idler and Wiseman (1971c)
Monodonta labio	**1b**	+	Toyama *et al.* (1955a), Teshima and Kanazawa (1972b)
Monodonta turbinata[b]	**1b, 2b, 1d, 3d, 10b, 14b, 19b**	+	Voogt (1968, 1971a)
Murex asianus	**1b, 3b, 5b, 8b, 10b, 14b**		Teshima and Kanazawa (1972b)
Murex brandaris[b]	**1b, 1d, 2b, 5b, 8b, 10b, 14b, 16b**	+	Voogt (1972c)
Nassa obseleta	**1b**		Kind and Goldberg (1953)
Nassa[b] sp.	**1b**		Voogt and Van Rheenen (1973b)
Natica cataena[b]	**1b, 1d, 3b, 3d, 8b, 10b, 14b, 16b, 22b**	+	Voogt (1971b)
Neptunea antiqua[c]	**1b, 1d, 2b, 3b, 5b, 8b, 14b, 16b, 22b**	+	Voogt (1972d)
Neptunea decemcostrata	**1b, 3b, 6b, 10b**		Idler and Wiseman (1971c)
Nerita peleronta	**1b**		Kind *et al.* (1948)
Patella coerulea[b]	**1b, 2b, 10b, 14b, 19b**		Voogt (1968, 1971a)
Patella vulgata[b]	**1b, 2b, 19b**		Collignon-Thiennot *et al.* (1973)
Patelloida saccharina	**1b**		Teshima and Kanazawa (1972b)
Peltodoris atromaculata	**1b, 2b, 3b, 6b, 8b, 10b, 14b, 16b, 19b**	+	Voogt (1973c)
Peribolus arabica	**1b, 1d, 2b, 5b, 8b, 16b**		Teshima and Kanazawa (1972b)
Plearoploca trapezium	**1b, 3b, 5b, 8b, 14b**		Teshima and Kanazawa (1972b)

(*Continued*)

TABLE 4 (*Continued*)

Classification	Sterols[a]	$\Delta^{5,7}$ Sterol detected	Reference
Purpura bronni	**1b**, **3b**, **8b**		Teshima and Kanazawa (1972b)
Purpura lapillus[b]	**1b**, **1d**, **2b**, **5b**, **8b**, **16b**	+	Voogt (1972c)
Rapana thomasiana	**1b**		Tsujimoto and Koyanagi (1934c)
Ravitrona captserpentis	**1b**		Teshima and Kanazawa (1972b)
Spiratella helicina	**1b**, **2b**, **3b**, **6b**, **8b**, **10b**, **14b**, **18b**, **19b**, **22b**	+	Idler and Wiseman (1971c)
Succinea putris[b]	**1b**, **1d**, **2b**, **3b**, **5b**, **14b**, **22b**	+	Voogt and Van der Horst (1972)
Teculatus maximus	**1b**, **3b**, **5b**, **14b**		Voogt (1972b), Teshima and Kanazawa (1972b)
Tegula argyrostoma	**1b**	+	Toyama and Tanaka (1956a)
Tegula xanthostigma	**1b**		Tsujimoto and Koyanagi (1934b)
Thais clavigera	**1b**		Hayashi and Yamada (1974)
Tonna luteostoma	**1b**	+	Toyama *et al.* (1955b), Tanaka and Toyama (1957a)
Trochus maculatus	**1b**, **3b**, **5b**, **8b**, **14b**	+	Teshima and Kanazawa (1972b)
Turbo cornutus	**1b**	+	Toyama *et al.* (1955b)
Urosalpinx cinerea	**1b**, **6b**		Kind and Goldberg (1953), Idler and Wiseman (1971c)
Viviparus fasciatus[b]	**1b**, **1d**, **3b**, **3d**, **5b**, **8b**, **10b**, **14b**, **16b**, **19b**	+	Voogt (1969, 1971a)
Viviparus japonicus	**1b**	+	Matsumoto and Tamura (1955a)
Viviparus malleatus	**1b**	+	Matsumoto and Tamura (1955a)
Class Pelecypoda (Bivalvia)			
Anodonta cygnea[b]	**1b**, **3b**, **6b**, **8b**, **14b**, **15b**, **22b**, **24b**		Voogt (1975b)
Anomia simplex	**1b**, **8b**, or **10b**		Kind and Goldberg (1953)
Arctica islandica	**1b**, **2b**, **3b**, **6b**, **8b**, **10b**, **14b**, **18b**, **22b**	+	Idler and Wiseman (1971c), Grossert and Swee (1977)
Artrina fragilis[c]	**1b**, **2b**, **3b**, **6b**, **8b**, **10b**, **14b**, **15b**, **22b**, **24b**		Voogt (1975a)
Atrina pectinata	**1b**, **3b**, **10b**, **14b**, **16b**, **19b**, **22b**		Teshima and Kanazawa (1972b)
Cardium corbis	**10b**		Fagerlund and Idler (1956)
Cardium edule[b]	**1b**, **2b**, **3b**, **5b**, **8b**, **10b**, **14b**, **16b**, **18b**, **22b**, **24b**		Voogt and Van der Horst (1972), Voogt (1975b)
Cerastoderma edule	**1b**, **2b**, **3b**, **5b**, **8b**, **10b**, **14b**, **16b**, **18b**, **22b**		Ballantine *et al.* (1975b)

Chlamys nipponensis akazara	1b	+	Hayashi and Yamada (1973)
Corbicula leana	**1b, 8b, 13b, 15b, 16d**	+	Tamura *et al.* (1956), Matsumoto and Toyama (1944), Toyama and Tanaka (1954)
Crassostrea virginica	**1b, 3b, 6b, 8b, 10b, 14b, 15b, 18b, 22b**	+	Tamura *et al.* (1964a), Idler and Wiseman (1971c)
Cristaria spatiosa	**16d**	+	Matsumoto and Tamura (1955b)
Cyprina islandica[b]	**1b, 3b, 6b, 8b, 10b, 14b, 15b, 18b, 22b, 24b**		Voogt (1975b)
Donax trunculus	**1b, 3b, 8b, 10b, 13b**		Ando *et al.* (1976)
Hipponps hipponps	**1b, 6b, 8b, 10b**		Teshima *et al.* (1974a)
Modiolus demissus	**1b, 8b, 13d**?	+	Bergmann and Ottke (1949), Petering and Waddell (1951)
Mya arenaria[b]	**1b, 2b, 3b, 6b, 8b, 10b, 14b, 15b, 18b, 22b**	+	Idler and Wiseman (1971c), Voogt (1975b)
Mytilus edulis	**1d, 3d, 8d**	+	Van der Vliet (1948)
Mytilus edulis	**1b, 13b, 15b**		Toyama and Tanaka (1956a)
Mytilus edulis	**1b, 2b, 3b, 8b, 10b, 14b, 15b, 22b**		Idler and Wiseman (1971c), Teshima and Kanazawa (1972b)
Mytilus edulis[c]	**1b, 1d, 3b, 5b, 8b, 10b, 14b, 16b**	+	Voogt (1972a, 1975a)
Nucula sp.	**1a, 1b, 3b, 6b, 8b, 10b, 14b**	+	Idler and Wiseman (1971c)
Ostrea edulis[c]	**1b, 3b, 6b, 8b, 10b, 14b, 15b, 18b, 22b, 24b**		Voogt (1975a)
Ostrea gigas	**10b**	+	Idler and Fagerlund (1955, 1957a)
Ostrea gryphea	**1b, 3b, 10b, 14b, 16b, 19b**		Salaque *et al.* (1966)
Ostreal sp. (?)	**1b, 3b, 8b, 10b**		Kritchevsky *et al.* (1967)
Ostrea virginica	**10b**		Fagerlund and Idler (1956), Idler and Fagerlund (1957b)
Patinopecten yessoensis	**1a, 3a, 5a, 13a, 24a, 1b, 2b, 3b, 5b, 7b, 10b, 13b, 15b, 18b, 19b, 20b, 22b, 24b**		Kita and Toyoma (1960), Kobayashi and Mitsuhashi (1975)
Pecten caurinus	**10b**		Fagerlund and Idler (1956)
Pecten maximus	**1b**		Zagalsky *et al.* (1967)
Pecten yessoensis	**10b, 13b**		Kita and Toyama (1959)
Pinctada martensi	**7b**		Ashikaga (1957)
Pinna pectinata japonica	**13b**	+	Takagi *et al.* (1956)
Placopecten megellanicus	**1a, 1b, 3a, 3b, 4b, 8d, 10b, 14b, 16b, 19b, 20b, 22a, 22b, 23b, 38b**	+	Idler *et al.* (1964, 1970, 1971), Tamura *et al.* (1964b), Wainai *et al.* (1964), Idler and Wiseman (1971b), Idler *et al.* (1976)

(*Continued*)

TABLE 4 (*Continued*)

Classification	Sterols[a]	$\Delta^{5,7}$ Sterol detected	Reference
Placopecten sp. (?)	**1b**, **3b**, **8b**, **10b**		Kritchevsky *et al.* (1967)
Saxidomus giganteus	**10b**		Fagerlund and Idler (1956), Idler and Fagerlund (1955, 1957b)
Solemya velum	**1b**		Idler and Wiseman (1971c)
Spisula sachalinensis	**14b**	+	Toyama *et al.* (1955a)
Spisula solidissima	**1b**, **3b**, **6b**, **8b**, **10b**, **14b**, **18b**, **22b**	+	Idler and Wiseman (1971c)
Tapes japonica	**1b**, **3b**, **8b**, **10b**, **16d**, **18b**	+	Yasuda (1966, 1970)
Tapes philippinarum	**1b**, **2b**, **3b**, **8b**, **10b**, **14b**, **16b**, **19b**, **21b**, **22b**		Teshima *et al.* (1971, 1972, 1974b), Kanazawa and Teshima (1971b), Teshima and Kanazawa (1972b)
Tridacna crocea, *T. noae*, *T. squamosa*	**1b**, **6b**, **8b**, **10b**		Teshima *et al.* (1974a)
Venus sp. (?)	**1b**, **3b**, **8b**, **10b**		Kritchevsky *et al.* (1967)
Venus gallina	**1b**, **3b**, **8b**, **10b**, **13b**, **15b**, **22b**		Piretti and Viviani (1976)
Class Cephalopoda			
Eledone aldrovandi[c]	**1b**, **2b**, **3b**, **10b**		Voogt (1972a, 1973b)
Illex illecebrosus	**1b**, **1d**	+	Gouseeva *et al.* (1973)
Octopus dofleini	**1b**		Hatano (1961)
Octopus vulgaris[b]	**1b**, **2b**, **10b**		Henze (1908), Voogt (1972a, 1973b)
Ommastrefes sloani pacificus	**1b**		Marumo *et al.* (1955)
Sepia officinalis[b]	**1b**, **2b**, **10b**		Deffner (1943), Zandee (1967b), Voogt (1973b)
Todarus sagittatus	**1b**		Endre and Canal (1926)
Phylum Molluscoidae			
Class Brachiopoda			
Terebratalia transversa	**1b**, **3b**, **6b**, **8b**, **10b**, **18b**, **19b**, **22b**		Idler and Wiseman (1971c)

[a] See footnote a of Table 1 regarding the C-24 configuration of these sterols.
[b] [^{14}C]Acetate or [2-^{14}C]mevalonate incorporated into sterols.
[c] Incorporation of [^{14}C]acetate or [2-^{14}C]mevalonate into sterols inconclusive or not demonstrated.

TABLE 5

Sterol Composition of *Mytilus edulis* in Three Locations

Collection point	Sterol[a] (%)									
	1b	**2b**	**3b**	**6b**	**8b**	**10b**	**13b**	**15b**	**18b**	**22b**
Zeeland, Netherlands[b]	34.0	10.5	9.0	3.0	21.5	9.0	3.5	3.0	0.3	5.0
Halifax, Canada[c]	58.7	8.1	8.6	0.6	9.6	—	1.2	6.4	—	5.9
Kagoshima, Japan[d]	30.0	4.0	25.0	—	23.0	12.0	2.0	—	—	2.0
Kagoshima, Japan[e]	35.0	2.0	18.0	—	25.0	16.0	1.0	—	—	1.0

[a] See footnote a of Table 1 concerning C-24 configuration.
[b] Voogt (1975a).
[c] Idler and Wiseman (1971c).
[d] Teshima and Kanazawa (1972b).
[e] Teshima and Kanazawa (1973a).

sterols (Popov *et al.*, 1976a; Djerassi *et al.*, 1977) do not make specific provisions for the identification of $\Delta^{5,7}$ sterols. However, more attention to these compounds in future examinations of the sterols of mollusks, and indeed of other marine animals, could be rewarding in view of the importance of $\Delta^{5,7}$ dienes as precursors of Δ^5 sterols and the D vitamins, which may be involved in calcium metabolism and hence may play a role in shell production (Wilbur, 1972).

Among the unusual sterols found in mollusks can be cited (22*E*)-24-norcholesta-5,22-dien-3β-ol (**22b**), which was first characterized from a scallop (Idler *et al.*, 1970) but which, together with other related C_{26} sterols, is now recognized as being widespread in the marine environment. Other rare mollusk sterols are the isomeric pair of propylidenesterols (**20b** and **38b**) reported in the scallop *Placopecten magellanicus* (Idler *et al.*, 1971, 1976) and the C_{22} sterol (**21b**) tentatively identified in *Tapes philippinarum* (Kanazawa and Teshima, 1971b).

E. Arthropoda

The crustaceans have the simplest sterol compositions recorded in marine invertebrates (Table 6). In every species examined cholesterol (**1b**) is the major constituent, usually in excess of 90% of the total mixture. In those species with a cholesterol (**1b**) content of less than 90% the balance is usually made up by the related C_{27} sterol desmosterol (**2b**). The trace amounts of C_{26}, C_{28}, and C_{29} sterols detected by glc are the more typical 24-methyl- and 24-ethyl-Δ^5-sterols found in other phyla.

TABLE 6

Major Sterols in Crustaceans

Classification	Major sterol(s)[a]	Reference
Order Cirripedia		
Balanus balanoides	**1b**	Dawson and Barnes (1966)
Balanus glandula	**1b**(60), **2b**(34)	Fagerlund and Idler (1957)
Balanus nubilis	**2b**(35)	Whitney (1967)
Order Euphausiacea		
Euphausia pacifica	**1b**	Fagerlund (1962)
Nyctiphanes norvegica	**1b**(99)	Idler and Wiseman (1971a)
Order Copepoda		
Euchaeta japonica	**1b**(99), **2b**(1)	Lee *et al.* (1974)
Order Amphipoda		
Caprella sp.	**1b**(73), **2b**(22), **3b**(4)	Teshima and Kanazawa (1971b)
Order Isopoda		
Ligia exotica	**1b**	Toyama and Shibano (1943)
Order Mysidacea		
Neomysis intermedia	**1b**(78), **3b**(3), **8b**(8), **10b**(10)	Teshima and Kanazawa (1971b)
Order Decapoda		
Arctodiaptomus salinus	**1b**	Bodea and Ciurdaru (1968)
Artemia salina	**1b**	Bodea and Ciurdaru (1968)
Callinectes sapidus	**1b**	Whitney (1967), Thompson (1964)
Cancer borealis	**1b**(97), **2b**(3)	Idler and Wiseman (1971a)
Cancer irroratus	**1b**(96), **2b**(3)	Idler and Wiseman (1971a)
Cancer magister	**1b**	Kind and Fasolino (1945), Thompson (1964)
Cancer pagarus	**1b**	Van den Oord (1964)
Cancer pagarus (ovary)	**1b**	Zagalsky *et al.* (1967)
Cancer pagarus	**1b**(57), **8b**(37)	Kritchevsky *et al.* (1967)
Carcinides maenas	**1b**(97), **2b**(1)	Idler and Wiseman (1971a)
Carcinus maenas	**1b**	Dorée (1909), Bodea and Ciurdaru (1968)

Chionoecetes opilio	**1b**(94), **2b**(6)	Idler and Wiseman (1968, 1971a)
Chionoecetes opilio	**1b**(91), **2b**(8)	Yasuda (1973)
Erimacrus isenbeckii	**1b**(98), **2b**(2)	Idler and Wiseman (1971a)
Erimacrus isenbeckii	**1b**(97), **2b**(2)	Yasuda (1973)
Eriphia spinifrons	**1b**	Leulier and Policard (1930)
Geryon quinquedens	**1b**(100)	Idler and Wiseman (1971a)
Hemigrapsus sanguineus	**1b**(90), **2b**(8)	Yasuda (1973)
Homarus americanus	**1b**(42)	Bligh and Scott (1966)
Homarus americanus	**1b**(99)	Idler and Wiseman (1971a), Gagosian (1975)
Homarus gammarus	**1b**	Zandee (1967a)
Homarus vulgaris	**1b**	Zandee (1964)
Homarus sp.	**1b**(99)	Kritchevsky *et al.* (1967)
Hyas araneus	**1b**(88), **2b**(12)	Idler and Wiseman (1971a)
Ilyoplax pusillus	**1b**(92), **5b** or **6b**(2)	Yasuda (1973)
Metapenaeus joyneri	**1b**(100)	Yasuda (1973)
Nephthrops norvegicus	**1b**	Burkhardt *et al.* (1934)
Pagurus bernhardus	**1b**(92), **2b**(7)	Idler and Wiseman (1971a)
Pagurus longicarpus	**1b**(75), **2b**(24)	Idler and Wiseman (1971a)
Pagurus samuelis	**1b**(74), **2b**(25)	Yasuda (1973)
Palaemon nipponensis	**1b**	Takagi and Toyama (1956a)
Palaemon palaemon	**1b**	Bodea and Ciurdaru (1968)
Palaemon squitta	**1b**	Bodea and Ciurdaru (1968)
Pandalus borealis	**1b**(99), **2b**(1)	Idler and Wiseman (1971a)
Pandalus borealis	**1b**(94), **2b**(4)	Gagosian (1975)
Panulirus argus	**1b**(99)	Idler and Wiseman (1971a)
Panulirus japonica	**1b**(100)	Teshima and Kanazawa (1971b)
Paralithodes brevipes	**1b**(82), **2b**(17)	Yasuda (1973)
Paralithodes camtschatica	**1b**(62), **2b**(31)	Idler and Wiseman (1968), Yasuda (1973)
Paratya compressa compressa (f.w.)	**1b**(99)	Yasuda (1973)
Panaeus aztecus	**1b**	Thompson (1964)

(*Continued*)

TABLE 6 (*Continued*)

Classification	Major sterol(s)[a]	Reference
Penaeus japonicus	**1b**(90), **3b**(3), **10b**(7)	Teshima and Kanazawa (1971b)
Penaeus setiferus	**1b**	Thompson (1964)
Plesionika edwardsi (eggs)	**1b**	Zagalsky *et al.* (1967)
Portunus plicatus	**1b**	Leulier and Policard (1930)
Portunus tribuberculatus	**1b**(100)	Teshima and Kanazawa (1971b)
Order Stomatopoda		
Gonodactylus chiragra	**1b**(100)	Teshima and Kanazawa (1971b)
Gonodactylus falcatus	**1b**(100)	Teshima and Kanazawa (1971b)
Odontodactylus scayllarus	**1b**(100)	Teshima and Kanazawa (1971b)

[a] Number in parentheses indicates approximate percentage.

F. Echinodermata

Ever since the earlier studies on echinoderm sterols (see reviews in Bergmann, 1962; Austin, 1970; Goad *et al.*, 1972a; Goad, 1976) the most striking feature of this phylum has been the dichotomy between the crinoids, ophiuroids, and echinoids, which contain Δ^5 sterols, and the holothurians and asteroids, which contain Δ^7 sterols. More recent sterol analyses of echinoderms are listed in Table 7, and it can be seen that they contain mixtures of a complexity comparable to that of mixtures in most other invertebrate phyla. In the majority of animals the most abundant sterol is cholesterol (**1b**) in the crinoids, ophiuroids, and echinoids or cholest-7-en-3β-ol (**1c**) in the holothurians and asteroids, but occasionally, as in the asteroids *Acanthaster planci* (Gupta and Scheuer, 1968) and *Echinaster sepositus* (Voogt, 1973d), the C_{28} and C_{29} sterols may predominate.

A number of unique sterols have been encountered in echinoderms, for example, asterosterol (**22c**) (Kobayashi *et al.*, 1972, 1973b, 1974), 24-norcholest-5-en-3β-ol (**23b**) (Rubinstein, 1973), 24-norcholest-22-en-3β-ol (**22a**) (Goad *et al.*, 1972a), amurasterol (**24c**) (Kobayashi and Mitsuhashi, 1974b; Teshima *et al.*, 1976a), 24ξ-methylcholesta-7,22,25-trien-3β-ol (Teshima *et al.*, 1974c), acanthasterol (**11c**) (Gupta and Scheuer, 1968; Sheikh *et al.*, 1971), and gorgostanol (**11a**) (Kanazawa *et al.*, 1974a).

With the discovery of considerable amounts of sterol sulfates in echinoderms, the earlier view that holothurians and asteroids contain exclusively Δ^7 sterols must be abandoned. In an examination of lipids of the starfish *Asterias rubens* Björkman *et al.* (1972a,b) isolated a component which they identified on the basis of infrared and mass spectral evidence as cholesterol sulfate. This conflicted with the identification of predominantly cholest-7-en-3β-ol (**1c**) and other Δ^7 sterols in the free sterol mixture (Goad *et al.*, 1972a; Smith *et al.*, 1973). However, careful investigation (Goodfellow and Goad, 1973) of the sterol sulfates confirmed that cholesterol sulfate was indeed the major constituent of the mixture (Table 8), although cholest-7-en-3β-ol sulfate was also a significant component. An equally striking difference between the free and sulfated sterols was the considerably higher proportion of C_{27} sterols in the sulfated form (85%) than in the free form (46%).

Yoshizawa and Nagai (1974) reported the presence of cholesterol sulfate in the oocytes of a sea urchin (*Anthocidaris crassipina*), and Goodfellow (1974) demonstrated the occurrence of sterol sulfates in all five classes of echinoderms (Table 9) but, apart from traces in a crab, sterol sulfates were not detected in representatives of other invertebrate phyla. In all classes of echinoderms, cholesterol (**1b**) was the major sulfated

TABLE 7

Sterols of Echinoderms

Classification	Sterols[a]	Reference
Class Crinoidea		
Antedon bifida	**1b, 3b, 5b, 7b, 10b, 14b, 16b, 18b, 19b, 22b**	Rubinstein (1973)
Antedon sp.	**1b, 3b, 5b, 7b, 14b, 16b, 18b**	Gupta and Scheuer (1968)
Comatula sp.	**7b**	Bolker (1967a,b)
Class Echinoidea		
Anthocidaris crassipina	**1b, 19b**	Yasuda (1974b)
Astriclypeus manni	**1b, 3b, 8b, 10b, 14b**	Yasuda (1974b)
Echinocardium cordatum	**1b, 3b, 8b, 10b, 14b**	Yasuda (1974b)
Echinothrix diadema	**1b, 5b, 14b**	Gupta and Scheuer (1968)
Echinus acutus	**1b, 2b, 3b, 8b, 10b, 14b, 16b, 22b**	Voogt (1972e)
Echinus esculentus	**1b, 2b, 3b, 7b, 10b, 14b, 18b, 22b**	Smith and Goad (1974)
Paracentrotus lividus	**1b, 10b, 14b, 16b**	Salaque *et al*. (1966), Voogt (1972e)
Psammechinus miliaris	**1b, 3b, 5b, 7b, 10b, 22b**	Voogt (1972e), Rubinstein (1973)
Scaphechinus mirabilis	**1b, 3b, 8b, 10b, 14b**	Yasuda (1974b)
Temnopleurus toreumatus	**1b, 2b**	Yasuda (1974b)
Class Ophiuroidea		
Ophiocoma insularia	**1b, 5b, 7b, 14b, 16b, 18b**	Gupta and Scheuer (1968)
Ophiocomina nigra	**1b, 3b, 5b, 7b, 10b, 14b, 16b, 18b, 19b, 22b**	Rubinstein (1973)
Ophioderma longicauda	**1b, 3b, 4b, 5b, 8b, 14b, 16b, 18b, 22b**	Voogt (1973e)
Ophiura albida	**1b, 3b, 4b, 5b, 7b, 14b, 16b, 18b, 22b**	Goad *et al*. (1972a), Voogt (1973c)
Class Holothuroidea		
Cucumaria elongata	**1c, 3c, 5c, 7c, 14c, 16c, 22c**	Goad *et al*. (1972a)
Cucumaria hydriani	**1c, 3c, 5c, 7c, 14c, 16c, 22c**	Goad *et al*. (1972a)
Cucumaria lactea	**1c, 3c, 5c, 7c, 10c, 16c, 18c, 22c**	Rubinstein (1973)
Cucumaria planci	**1c, 3c, 5c, 7c, 14c, 16c, 18c, 1a, 3a, 7a**	Voogt and Over (1973)
Holothuria tubulosa	**1c, 3c, 5c, 7c, 14c, 16c, 1a, 3a, 7a, 1b**	Nomura *et al*. (1969a), Voogt and Over (1973)
Stichopus japonicus	**1c, C_{28} Δ^7, C_{29} Δ^7, 1b**	Nomura *et al*. (1969a), Toyama and Tanaka (1956b)
Stichopus regalis	**1c, 3c, 5c, 7c, 14c, 16c, 1a, 3a, 7a, 14a**	Voogt and Over (1973)

Class Asteroidea		
Acanthaster planci	**1c, 3c, 6c, 7c, 11c, 14c, 1a**	Gupta and Scheuer (1968), Sheikh *et al.* (1973), Kanazawa *et al.* (1974a)
Asterias amurensis	**1c, 3c, 5c, 7c, 10c, 14c, 18c, 22c, 24c, 1a, 1b**	Matsuno *et al.* (1972a,b), Kobayashi *et al.* (1973b), Kobayashi & Mitsuhashi (1974b)
Asterias rubens	**1c, 3c, 5c, 7c, 10c, 14c, 16c, 18c, 19c, 22c, 1a, 1b**	Goad *et al.* (1972a), Smith *et al.* (1973), Voogt (1973d)
Asterina pectinifera	**1c, 3c, 5c, 7c, 10c, 14c, 18c, 22c, 24c, 1a, 1b**	Matsuno *et al.* (1972a,b), Kobayashi *et al.* (1972), Kobayashi and Mitsuhashi (1974b), Masada *et al.* (1973)
Astropecten aurantiacus	**1c, 3c, 5c, 7c, 10c, 14c, 16c, 19c, 1a, 1b, 10b**	Voogt (1973d)
Astropecten irregularis	**1c, 3c, 5c, 7c, 10c, 14c, 16c, 18c, 19c, 22c, 1a, 1b**	Goad *et al.* (1972a)
Astropecten polyacanthus	**1c, 3c, 5c, 7c, 14c, 1a**	Matsuno *et al.* (1972b)
Astropecten scoparius	**1c, 3c, 5c, 7c, 14c, 16c, 1a**	Matsuno *et al.* (1972a,b),
Certonardoa semiregularis	**1c, 3c, 5c, 7c, 10c, 14c, 18c, 22c, 24c, 1a, 1b**	Kobayashi *et al.* (1972), Kobayashi and Mitsuhashi (1974b)
Coscinasterias acutispina	**1c, 3c, 5c, 7c, 10c, 14c, 16c, 27c, 1a, 1b**	Matsuno *et al.* (1972a,b), Kanazawa *et al.* (1973)
Ctenodiscus crispatus	**1c, 5c, 10c, 14c, 19c, 20c, 1a**	Grossert *et al.* (1973)
Culcita schmideliana	**1c, 5c, 7c, 11c, 14c, 16c, 1a**	Sheikh *et al.* (1973)
Distolasterias sticantha	**1c, 3c, 5c, 7c, 10c, 14c, 18c, 22c, 24c, 1a, 1b**	Kobayashi *et al.* (1972), Kobayashi and Mitsuhashi (1974b)
Echinaster sepositus	**1c, 3c, 5c, 7c, 14c, 16c, 19c, 1a, 1b, 19b, 3a, 7a, 14a, 16a**	Voogt (1973d)
Henricia sanguinolenta	**1c, 3c, 5c, 7c, 10c, 14c, 16c, 18c, 19c, 22c, 1a, 3a, 14a, 18a, 22a, 1b, 14b, 18b**	Goad *et al.* (1972a)
Laiaster leachii	**1c, 3c, 5c, 7c, 10c, 14c, 16c, 18c, 22c, 1a, 1b**	Kanazawa *et al.* (1973), Teshima *et al.* (1974c)
Linkia multiflora	**1c, 2c, 5c, 11c, 16c, 1a**	Sheikh *et al.* (1973)
Luidia ciliaris	**1c, 3c, 5c, 7c, 10c, 14c, 16c, 18c, 19c, 22c, 1a, 1b**	Goad *et al.* (1972a)

(Continued)

TABLE 7 (*Continued*)

Classification	Sterols[a]	Reference
Luidia quinaria	**1c, 3c, 5c, 7c, 14c, 16c, 1a**	Matsuno *et al.* (1972a,b)
Lysastrosoma anthostictha	**1c, 3c, 5c, 7c, 10c, 14c, 18c, 22c, 24c, 1a, 1b**	Kobayashi *et al.* (1972), Kobayashi and Mitsuhashi (1974b)
Marthasterias glacialis	**1c, 3c, 5c, 7c, 10c, 14c, 16c, 18c, 19c, 22c, 1a, 1b**	Voogt (1973d)
Nasdoa variolata	**1c, 5c, 7c, 14c, 16c, 1a**	Sheikh *et al.* (1973)
Pisaster ochraceus	**10c**	Fagerlund and Idler (1959)
Porania pulvillus	**1c, 3c, 5c, 7c, 10c, 14c, 16c, 18c, 19c, 22c, 1a, 1b**	Goad *et al.* (1972a)
Protoreaster lincki	**1c, 5c, 7c, 11c, 14c, 15c, 1a**	Sheikh *et al.* (1973)
Protoreaster nodosus	**1c, 5c, 7c, 11c, 14c, 15c, 19c, 1a**	Sheikh *et al.* (1973)
Solaster papposus	**1c, 3c, 5c, 7c, 10c, 14c, 16c, 18c, 19c, 22c, 1a, 1b**	Goad *et al.* (1972a)
Solaster paxillatas	**1c, 3c, 5c, 7c, 10c, 14c, 18c, 22c, 24c, 1a, 1b**	Kobayashi *et al.* (1972)
Stichastrella rosea	**1c, 3c, 5c, 7c, 10c, 14c, 16c, 18c, 19c, 22c, 1a, 1b**	Goad *et al.* (1972a)

[a] See footnote a of Table 1 concerning the C-24 configuration of these sterols.

TABLE 8

Percent Composition of Free Sterols and Sulfated Sterols of *Asterias rubens*

Sterol	Free sterols (%)	Sterol sulfates (%)
Cholesta-5,22-dien-3β-ol (**3b**)	—	3.3
Cholesta-7,22-dien-3β-ol (**3c**)	5.0	3.5
5α-Cholestan-3β-ol (**1a**)	Trace	6.1
Cholesterol (**1b**)	Trace	48.1
Cholest-7-en-3β-ol (**1c**)	41.0	25.1
24-Methylcholesta-7,22-dien-3β-ol (**7c**, **8c**)	16.0	4.2
24-Methylenecholest-7-en-3β-ol (**10c**)	26.0	2.5
24-Ethylcholesta-7,22-dien-3β-ol (**15c**, **16c**)	2.0	Trace
24-Ethylidenecholest-7-en-3β-ol (**18c**, **19c**)	9.0	1.8
Other	1.0	5.5

compound, including the sea cucumber, in which the free sterols were Δ^7 compounds, as in other holothurians. Björkman *et al.* (1972b) reported that in their analysis the sterol sulfates were present at a concentration of 1.3 mg/gm dry weight of tissue. Similar high concentrations of sterol sulfate, comparable to the amounts of free sterol, were encountered (Table 9) in all asteroids, the ophiuroid, and the crinoid analyzed by Goodfellow (1974). However, the echinoids contained relatively low levels of sterol sufate and, moreover, these animals contained about the same proportion of C_{27} sterol in both the free and sulfated forms. This contrasted with the other four classes of echinoderms, in which C_{27} sterols constituted a much larger proportion of the sterol sulfate mixture than found in the free sterol fraction.

G. Urochordata

Several species of ascidians have been analyzed, and the recent results are presented in Table 10. These animals contain the typical complex complement of sterols found in other phyla. The cholesterol (**1b**) content in the mixtures ranges from 65% in the sample of *Styela plicata* analyzed by Voogt and Van Rheenen (1975) to about 25% in the samples of the same species examined by Yasuda (1975), who, in addition, found about 10% cholestanol (**1a**). These observations again emphasize the variability in sterol composition that can be expected in animals collected from different localities. It seems that caution may be required before too much is read into the physiological implications of the sterol composition of an animal if only one analysis has been performed.

TABLE 9

Occurrence of Sterol Sulfates in Echinoderms and Other Marine Invertebrates

Classification	Sterol sulfate	Free sterol (mg/gm dry wt)	Sterol sulfate (mg/gm dry wt)	C_{27} Sterols (%)	
				Free	Sulfate
Phylum Echinodermata					
Class Asteroidea					
Asterias rubens	+	4.51	2.20	47	86
Marthasterias glacialis	+	3.40	1.15	66	94
Porania pulvillis	+	—	—	50	99
Henricia sanguinolenta	+	3.20	3.20	54	77
Class Holothuroidea					
Cucumaria lactea	+	—	—	54	67
Class Ophiuroidea					
Ophiocomina nigra	+	3.20	3.00	37	60
Class Echinoidea					
Echinus esculentus	+	2.10	0.22	88	89
Echinocardium cordatum	+	—	0.09	64	59
Class Crinoidea					
Antedon bifida	+	3.90	2.00	32	70
Phylum Arthropoda					
Carcinus maenus	Trace				
Phylum Mollusca					
Chlamys opercularis	—				
Buccinum undatum	—				
Phylum Porifera					
Hymeniacidon perleve	—				
Phylum Cnidaria					
Metridium senile	—				
Phylum Annelida					
Arenicola marina	—				
Phylum Urochordata					
Ciona intestinalis	—				

IV. ORIGINS OF INVERTEBRATE STEROLS

A. General Considerations

From the foregoing account of the distribution of sterols in the various invertebrate phyla, one is first impressed by the large number of sterols involved and by the complexity of the sterol mixtures encountered. However, on closer examination some generalities and patterns emerge which may give vital clues to the origins of the sterols found in various animals.

TABLE 10

Sterols in Ascidians

Species	Sterols[a]	References
Amaroucium pliciferum	**1b, 3b, 6b, 8b, 10b, 13b, 18b, 20b, 22b**	Yasuda (1976)
Ciona intestinalis	**1b, 1c, 3b, 4b, 6b, 8b, 13b, 15b, 18b, 19b, 22b**	Voogt and Van Rheenen (1975)
Microcosmus sulcatus	**1b, 1c, 3b, 4b, 6b, 8b, 10b, 13b, 15b, 18b, 19b, 22b**	Voogt and Van Rheenen (1975)
Halocynthia papillosa	**1b, 3b, 4b, 6b, 8b, 10b, 13b, 15b, 18b, 19b, 22b**	Voogt and Van Rheenen (1975)
Halocynthia roretzi	**1a, 1b, 1c, 10b, 13c, 20b, 22a, 22b, 22c**	Alcaide *et al.* (1971), Nomura *et al.* (1972), Viala *et al.* (1972), Yasuda (1976)
Styela clava	**1a, 1b, 2b, 3b, 6b, 8b, 10b, 13b, 18b, 22b**	Yasuda (1975)
Styela plicata	**1a, 1b, 3b, 4b, 6b, 8b, 10b, 13b, 15b, 18b, 19b, 22b**	Voogt and Van Rheenen (1975), Yasuda (1975)

[a] See footnote a of Table 1 regarding C-24 configuration of these sterols.

Thus, with certain notable exceptions, such as the unique sterols of the sponges and gorgosterol (**1b**) in the soft corals, basically the same range of sterols occurs in all phyla. The differences among phyla and individual species are more a question of actual concentrations than of the overall spectrum of sterols present. Indeed, as the trace sterols are being identified by the application of highly sensitive and refined analytical techniques, this similarity in sterol compositions is becoming more apparent. The C_{26} sterol **22b** is a case in point. When this sterol was first isolated and characterized in a scallop (Idler *et al.*, 1970) in which it occurs in relatively high concentration, it might reasonably have been thought of as a mollusk sterol. However, it is now quite apparent that it is present in small amounts in animals of all phyla. Moreover, it can no longer be thought of even as a unique marine compound since C_{26} sterols have been detected in the contemporary lacustrine sediments of a freshwater lake in Cheshire, England, by Gaskell and Eglinton (1976).

In general, C_{27} sterols are most often predominant. However, in filter feeders, such as sponges, bivalve mollusks, and ascidians, and in detritus feeders, such as some annelids and echinoderms, more complex mixtures with larger amounts of C_{26}, C_{28}, C_{29}, and C_{30} sterols are found than is the case with many carnivorous species, such as some coelenterates, cephalopods, and starfish, and particularly in the crustaceans, in which C_{27} sterols usually exceed 90% of the mixture. Those species with symbiotic zooxanthellae, such as clams, alcyonarians, and gorgonians, often have amounts of C_{28} and C_{30} sterols in excess of the amount of C_{27} sterols, indicating that the zooxanthellae may be making a very significant contribution to the overall sterol pattern.

To understand how the final sterol complement of an animal originates one must consider the following four contributory sources: (1) the *de novo* biosynthesis of sterols from acetyl-CoA via mevalonic acid and squalene, (2) the absorption and assimilation of dietary sterols, (3) the modifications of dietary sterols, and (4) the passage of sterols from symbiotic algae or other associated organisms (e.g., fungi or bacteria) to the host animal. The unraveling of the relative importance of these various routes calls not only for ingenuity on the part of the natural products chemists and biochemists tackling the biosynthetic problems, but also for a close collaboration with marine biologists in order to understand the subtleties of marine food chains and symbiotic relationships. A challenge to the biologist is presented by the need to isolate and culture some of the individual diatoms, dinoflagellates, and zooxanthellae, which the present author considers to be one of the more immediate problems in the search for the ultimate origins of many of the unique sterol skeletons encountered in the marine environment.

Several general reviews on sterol biosynthesis in animals and plants have appeared, which give the background for the following discussions on invertebrate sterol production (Frantz and Schroepfer, 1967; Goad, 1970, 1975, 1977; Goad and Goodwin, 1972; Goad *et al.*, 1974; Mulheirn and Ramm, 1972; Dempsey, 1974; Bean, 1973; Grundwald, 1975).

Mixed results have been obtained from efforts to demonstrate sterol synthesis in invertebrates, and they illustrate some of the problems to be considered in attempting to evaluate sterol-synthesizing ability and the extent of the contribution of *de novo* sterol synthesis to the overall sterol content of the organism. Feeding a water-soluble radioactive substrate such as acetate or mevalonate to a marine animal presents an immediate problem because great dilution results when a compound of high specific radioactivity is added to the relatively large volume of seawater in which it is necessary to maintain healthy marine invertebrates, particularly for long incubation periods. Even if the labelled compound is injected into the body of many invertebrates, there is the likelihood that it will pass through the animal to equilibrate with the aquarium water more rapidly than it is utilized by the animal. Consequently, even with the known ability of many invertebrates to concentrate compounds from dilute solution, only a small proportion of the added substrate may be absorbed, which must be assumed to enter an intracellular unlabeled pool. In the case of acetate such a pool may well be large in relation to the amount absorbed. This dilution, coupled perhaps with rapid metabolism of the substrate for other purposes, e.g., fatty acid synthesis, may result in such small incorporation of radioactivity into the sterols that the results either may be inconclusive or may even be interpreted as a failure of the animal to utilize the substrate for sterol synthesis.

[^{14}C]Acetate may not be a good substrate with which to demonstrate sterol synthesis for two other reasons. First, it is acetyl-CoA that is produced in the cell from pyruvate or fatty acid catabolism, for example, whilst added acetate required conversion first to acetyl-CoA before it can be converted to mevalonic acid or fatty acids. Thus, if acetyl kinase is rate limiting, sterol biosynthesis may not be observed. This point can be monitored by observing the incorporation of the added [^{14}C]acetate into fatty acids. In several studies this was done, and very active fatty acid synthesis was observed, indicating that this may not be a serious problem. The second point to consider in using [^{14}C]acetate concerns the enzyme hydroxymethylglutaryl-CoA (HMG-CoA) reductase, which is now well established as the regulatory enzyme in cholesterol biosynthesis in mammals (Rodwell *et al.*, 1973). In mammals HMG-CoA reductase shows diurnal variations in activity, and its activity is reduced by high-sterol diets. A similar regulatory role of HMG-CoA reductase in invertebrates could well prevent the detection of sterol synthesis when [^{14}C]-acetate is used. This problem can be circumvented by the use of [2-^{14}C]-mevalonate as substrate, but secondary controls may be operative at the squalene synthesis step or at squalene cyclization (Pennock, 1977).

At this point it is worth drawing a distinction between the ability of an invertebrate to synthesize sterol and its need to do so. The use of [2-^{14}C]-mevalonate may well prove that an animal has the enzyme systems to synthesize sterols but, in fact, all the sterol requirements may be adequately met from dietary sources or symbiotic organisms. This situation may then be truly reflected by the absence or very low incorporation of [^{14}C]acetate into its sterols. Finally, if invertebrates, in contrast to mammals, have rather slow sterol turnover rates, incubations of a few hours, or even days, may be insufficient for incorporation of detectable levels of [^{14}C]acetate or [2-^{14}C]mevalonate, and more rapid competing pathways such as fatty acid and polyisoprenoid production will remove the trace amounts of substrates added, thus rendering protracted incubation periods of little benefit unless fresh ^{14}C-labeled substrate is added periodically.

Clearly, only healthy animals are used for biosynthetic studies, but attention should perhaps be paid to the length of time and manner in which animals are maintained, both before and during experiments, since MacDonald and Wilber (1955) reported variations in sterol content in marine annelids that were dependent on the aquarium temperature and time. Age, state of sexual maturity, and season may also be important. This is illustrated by the results presented in Fig. 3, which show that during the winter, late summer, and autumn a brittlestar synthesized mainly squalene from [2-^{14}C]mevalonate with only low incorporation into 4-demethylsterols. However, in the spring and early summer, possibly coinciding with gonad development, there was a decline in the radioactivity in squalene since this was utilized to produce labeled 4-demethylsterols, which increased (Rubinstein and Goad, 1973).

An evaluation of the contribution of dietary sterols to the sterol content of an invertebrate presents some problems, especially in the case of cholesterol (**1b**). Many studies demonstrating sterol synthesis in invertebrates have simply recorded the incorporation of [^{14}C]acetate or [2-^{14}C]-mevalonate into the total sterol fraction, often after several crystallizations. Unfortunately, this gives no indication as to which sterol or type of sterol (C_{27}, C_{28}, or C_{29}) is labeled and hence a clue as to which sterols are dietary. Moreover, it may even result in an underestimation of sterol synthesis since by fractional crystallization a minor labeled constituent may be lost, leaving only unlabeled dietary sterols. This phenomenon was observed in the case of *Asterias rubens* sterols synthesized from [2-^{14}C]-

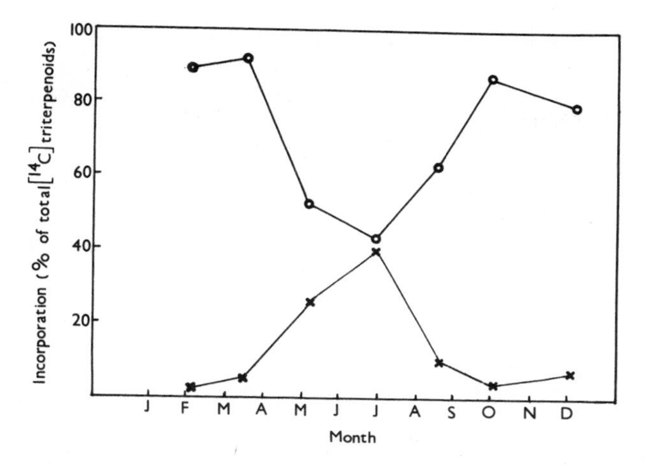

Fig. 3. Seasonal variations in the incorporation of [2-^{14}C]mevalonic acid into squalene (—o—o—) and 4-demethylsterols (—x—x—) by the brittlestar *Ophiocomina nigra*.

mevalonate (Goad *et al.*, 1972a; Smith and Goad, 1975a). The initial sterol mixture had a specific radioactivity of 137 dpm/mg and contained 60% of cholest-7-en-3β-ol (**1c**) and 24% of 24-methylcholesta-7,22-dien-3β-ol (**7c**). After nine crystallizations the specific radioactivity fell progressively to a value of 11 dpm/mg, and this could have been erroneously interpreted as a failure of the starfish to synthesize sterols. However, analysis of the final sterol mixture showed that it had become enriched with the C_{28} sterol **7c** (68%) whereas the content of cholest-7-en-3β-ol (**1c**), which was shown by preparative glc to be radioactive, had declined to only 5% of the mixture.

If *de novo* sterol synthesis can be shown to be restricted to C_{27} sterols as in the case cited above, the reasonable assumption can be made that the C_{26}, C_{28}, C_{29}, and C_{30} sterols present are dietary and either unchanged or perhaps modified by the animal. However, a proportion of the C_{27} sterol may also be derived from the diet since cholesterol (**1b**), desmosterol (**2b**), and 22-dehydrocholesterol (**3b**) are dominant in red algae and cholesterol (**1b**) is the major component of the plankton and some diatoms (Section II). Also, as discussed on p. 123, some animals, most notably the crustaceans but also probably some coelenterates and mollusks, have the ability to dealkylate C_{28} and C_{29} sterols to produce cholesterol (**1b**).

An interesting approach to the dietary sources of invertebrate sterols is that adopted by Steudler *et al.* (1977) and Kanazawa *et al.* (1976b). These groups have analyzed the sterol compositions of animals living on a coral reef and have shown that the predator animals have sterol patterns resembling those of their prey, thus illustrating how a sterol mixture can be handed down the food chain. Tracing back such food chains should help to locate the prime source of particular sterols, which can then be used for biosynthetic studies.

The C-24 configuration of 24-alkylsterols may also prove to be a valuable clue to their origin, particularly when considered in conjunction with the phylogenetic relationships that have emerged from a study of algal sterol side chain alkylation mechanisms (Goad *et al.*, 1974). In virtually none of the studies of invertebrate sterols has the C-24 configuration been determined, and it is only with the use of X-ray diffraction or high-resolution nmr spectroscopy that the configurations of a few sterols have been assigned with any confidence. Particularly from the studies on echinoderms (Rubinstein, 1973; Smith *et al.*, 1973; Kobayashi and Mitsuhashi, 1974b) and polychaetes (Kobayashi *et al.*, 1973a; Kobayashi and Mitsuhashi, 1974a) it is becoming apparent that in some phyla C_{28} sterols with the 24α configuration may predominate over the 24β epimers. This points to a diatom origin of these sterols since the diatoms are now becoming recognized as producing a 24α-methylsterol (**7b**) (see Section II), which contrasts with all other groups of algae, which produce 24β-sterols (Patterson, 1971; Goad *et al.*, 1974). Conversely in coelenterates, compounds with the 24β configuration seem more common, which indicates their derivation from chrysophyte, chlorophyte, or perhaps dinoflagellate sources.

B. Porifera

de Rosa *et al.* (1973b) were unable to demonstrate any incorporation of [^{14}C]acetate, [2-^{14}C]mevalonate, or [^{14}C]methionine into the sterols of *Verongia aerophoba*. Similarly, Walton and Pennock (1972) observed no incorporation of [2-^{14}C]mevalonic acid into the sterols of *Suberites domuncula*. With *Axinella verrucosa* (de Rosa *et al.*, 1975a) and *Axinella polypoides* (de Rosa *et al.*, 1975b) [1-^{14}C]acetate was incorporated efficiently into fatty acids, but the labeling of sterols (types **1f** and **1e**, respectively) was very low, and the authors concluded that there was little or no *de novo* sterol biosynthesis. In contrast low incorporation of [^{14}C]-acetate or [2-^{14}C]mevalonate into the sterols of *Grantia compressa* (Walton and Pennock, 1972), *Euspongia officinalis, Axinella damicornis, Geodia mülleri*, and two *Halichondria* species (Voogt, 1976) has been reported in which cholesterol (**1b**) ranges from 20 to 60% of the sterol mixtures.

It is noticeable that in some sponges in which incorporation results were negative or inconclusive cholesterol (**1b**) comprises a negligible or only small proportion of the total sterol mixture. In the two *Axinella* species there is no cholesterol (**1b**), and the major sterols are the 3β-hydroxymethylsterane (**1f**) or 19-norsterane (**1e**) types. In *Verongia aerophoba* there is 70% aplysterol (**25b**) and 24,28-didehydroaplysterol (**26b**), about 18% of C_{28}, C_{29}, or C_{30} sterols, and only 12% C_{27} sterols (de Rosa *et al.*, 1973a). If these sponges are actually unable to synthesize sterols, clearly their sterol requirements must be met from external sources and the above sterols (**1e**, **1f**, **25b**, **26b**, etc.) must then be either assimilated from the diet, acquired from associated organisms (symbiotic algae or bacteria?), or produced by the sponge from ingested precursor sterols. Conversely, it can be argued that an abundant and readily absorbed supply of sterols (**1e**, **1f**, **25b**, **26b**, etc.) may easily satisfy the sterol needs of sponges, especially if there is a slow turnover rate, and hence sterol biosynthesis will be regulated to the extent that it is undetectable by the methods employed.

The 26-methylsterols aplysterol (**25b**) and 24,28-didehydroaplysterol (**26b**) are apparently found only in sponges of the *Verongia* family, since other sponge species from the same habitat as *V. aerophoba* did not contain **25b** or **26b** (de Rosa *et al.*, 1973b). This led to the suggestion that these sterols may be synthesized by *Verongia* species, possibly by transmethylation of a suitable exogenous dietary sterol (de Rosa *et al.*, 1973b). However, incubation of *V. aerophoba* with [*methyl*-^{14}C]methionine (de Rosa *et al.*, 1973b) gave no incorporation of radioactivity into aplysterol (**25b**), and it has been suggested more recently that the 26-methylsterols may arise by a methyl rearrangement of a C_{29} sterol such as furcosterol (**19b**) (Minale and Sodano, 1977). Notwithstanding the failure to detect aplysterol (**25b**) in other sponge species, an external algal source of the 26-methylsterols is still conceivable since Reiswig (1971) observed that *Verongia gigantea* shows some particle size selectivity in removing suspended material from seawater. *Verongia* species may thus be more efficient than other sponges in the capture of the organisms that are the primary source of aplysterol (**25b**). From present knowledge on the formation of algal sterol side chains (Goad *et al.*, 1974) it can be suggested that the transmethylation sequence outlined in Fig. 4 may operate to produce side chains **25** and **26**. So far, among algae, 24β-methyl-25-methylenesterols have been reported in only one member of the Siphonales (Rubinstein and Goad, 1974a) and as precursors of 24β-methylsterols in members of the order Chlorococcales (Goad *et al.*, 1972b, 1974). This suggests that the source of aplysterol should perhaps be sought among the marine representatives of these orders of algae.

Regarding the biogenesis of calysterol (**27b**) Minale and Sodano (1977) reported that *Calyx nicaensis* did not incorporate either [^{14}C]acetate or [*methyl*-^{14}C]methionine into the sterols, which implies modification of a

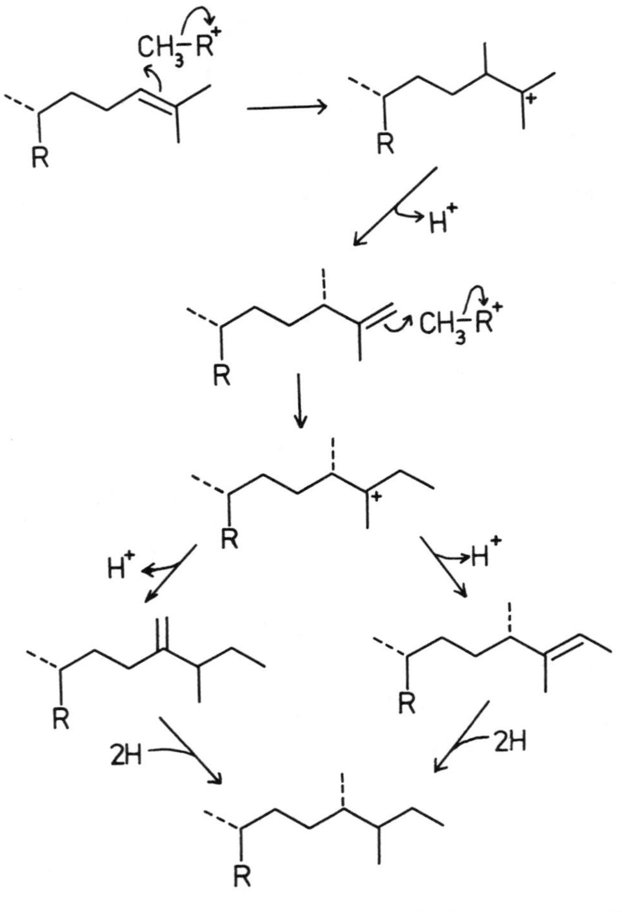

Fig. 4. A suggested mechanism for the biosynthesis of the aplysterol side chain.

dietary sterol. Administration of tritium-labeled sitosterol (**14b**) and stigmasterol (**16b**) to *C. nicaensis* gave no label in calysterol (**27b**); however, 2.7% conversion of [7-$^3H_2$]fucosterol (**19b**) into calysterol (**27b**) was observed, indicating the ability of the sponge to oxidize and rearrange a nutritional sterol with a 24-ethylidene side chain (Minale and Sodano, 1977).

Minale and co-workers reported investigations on the production of the sterols with a modified nucleus found in *Axinella* species. [26-^{14}C]-Cholesterol added as an ethanolic solution to seawater was absorbed by *Axinella polypoides* and metabolized to give 78% of the recovered radioactivity in 19-norcholestanol (**1e**) after 290 hr of incubation (de Rosa et al., 1975b). A Δ^5 double bond in the sterol substrate is essential for C-19 demethylation since feeding a mixture of [4-^{14}C]cholesterol and [7α-^{3}H]-cholestanol (**1a**) to *A. polypoides* resulted in the isolation of only ^{14}C-labeled 19-norcholestanol (**1e**) (Minale and Sodano, 1977).

Feeding [4-^{14}C]cholesterol to *A. verrucosa* resulted in efficient conversion to 3β-hydroxymethyl-A-nor-5α-cholestane (**1f**), and degradation showed that the ^{14}C was located at C-3 of compound **1f** (de Rosa *et al.*, 1975a, 1976). Administration of a mixture of [4-^{14}C]cholesterol (**1b**) and [7α-^{3}H]cholestanol (**1a**) to *A. verrucosa* gave preferential incorporation of ^{14}C into the 3β-hydromethyl-A-norsterane (**1f**), suggesting that cholestanol (**1a**) is not an obligatory intermediate. However, some incorporation of tritium into **1f** occurred, and this does indicate that cholestanol (**1a**) can be metabolized and therefore presumably a Δ^5 double bond is not essential, although it may facilitate the conversion (de Rosa *et al.*, 1976).

Although there have been no biosynthetic studies on the origin of the 5α-stanols (**1a**, **6a**, **13a**) reported in some sponges, the isolation of 3-oxo-4-ene steroids by Sheikh and Djerassi (1974) implies stanol production from Δ^5 sterols by the sequence involving Δ^5,3β-hydroxysterol dehydrogenase, steroid Δ^5–Δ^4-isomerase, and steroid Δ^4-reductase systems. This is the well-established route for the production of stanols in mammalian tissues (Makin, 1975), echinoderms (see Section IV, G), and some bacteria. The latter fact suggests that more consideration should be given to the possible importance of bacterial metabolism in the production of some of the sponge sterols, since sponges have a marked ability to capture bacteria during seawater filtration feeding (Reiswig, 1971) and some harbor large populations of symbiotic bacteria (Lévi and Lévi, 1965; Madri *et al.*, 1967; Vacelet, 1967).

C. Cnidaria

From the evidence so far available, it is uncertain whether coelenterates are able to synthesize sterols. No incorporation of [1-^{14}C]acetate into sterols was observed with a jellyfish, *Rhizostoma* sp. (van Aarem *et al.*, 1964), or an anemone, *Calliactis parasitica* (Ferezou *et al.*, 1972), although fatty acids were labeled. Similarly, although some terpenoids were labeled after injection of [2-^{14}C]mevalonic acid, no radioactivity was found in the sterols obtained from *Metridium senile* (Walton and Pennock, 1972), *Actinea equina*, or *Tealia felina* (Pennock, 1977).

Voogt *et al.* (1974) found that the free sterols of *Cerianthus membranaceus* and *Metridium senile* were unlabeled after incubation with [1-^{14}C]acetate, although the sterols obtained from the ester fraction of *M.*

senile were apparently radioactively labeled (about 20 dpm/mg after incubation of 10 animals with 125 μCi of [1-^{14}C]acetate for 24 hr). Using [2-^{14}C]mevalonate (50 μCi injected into 40 animals), Voogt *et al.* (1974) recorded specific radioactivities of about 9 dpm/mg in both the free and the esterified sterols recovered from *Anemonia sulcata*. These results were interpreted by Voogt *et al.* as evidence for a very slow rate of sterol synthesis by the animals, possibly the synthesis being inhibited by the intake of dietary sterol. They suggested that the labeling of the sterol esters was the consequence of the need to transport newly synthesized sterol as the ester. Kanazawa *et al.* (1974b) tentatively concluded on the basis of thin-layer chromatography (tlc) and radioautographic evidence that [2-^{14}C]mevalonate was incorporated into squalene and/or sterols by the alcyonarians *Sarcophyta* sp. and *Stereonephtya japonica* and the actinarians *Dofleinia armata* and *Parasicyonis actinostoloides,* but in no case could the sterols be crystallized to constant specific radioactivity. Two gorgonians, *Acalycigorgia inermis* and *Ellisella rubra,* were considered incapable of sterol synthesis.

In order to interpret the significance of a low incorporation of radioactivity into the sterol fraction, Pennock (1977) has emphasized the need to consider two points. First, and most important, the sterols must be rigorously purified to ensure the elimination of possible high specific activity contaminants such as isoprenoid alcohols (farnesol, geranylgeraniol, etc.), which Pennock and co-workers found to be labeled during incubation of several invertebrates with [2-^{14}C]mevalonate and which have chromatographic properties similar to those of sterols (see Section IV,A) for further examples and discussion of this problem). Second, the role of other organisms, particularly of symbiotic algae (Muscatine, 1974) but also of parasitic organisms, the digestive tract flora and fauna, or possibly organisms in the aquarium water, must be taken into account since these may be responsible for any observed sterol synthesis or sterol transformation. Identification of the labeled sterol may give some indication on this point if one accepts the generality that animals produce only C_{27} sterols (see Section IV,A) whereas most algae and fungi synthesize C_{28} and C_{29} sterols. The coelenterate *Anemonia sulcata* is reported to have a dinoflagellate symbiont (Taylor, 1969), and it is therefore open to question whether this alga was responsible for the low [2-^{14}C]mevalonate incorporation observed by Voogt *et al.* (1974).

The fact that C-24 alkylated sterols predominate in the sterol mixtures obtained from members of the Zoanthidae and that these sterols together with gorgosterol (**11b**) are abundant in gorgonians reveals the probable importance of the sterol contribution of the zooxanthellae (see Muscatine, 1974, for a discussion of symbiotic algae). This was further emphasized by

Ciereszko *et al.* (1968), who found that the zooxanthellae obtained from the gorgonians *Eunicea mammoma* and *Pseudoplexaura porosa* contained C_{28} and C_{29} sterols and, most significantly, gorgosterol (**11b**). The movement of compounds between symbiotic algae and their coelenterate hosts has been established (Trench, 1971a,b,c; Muscatine, 1974), and this could possibly include sterols to fulfill the requirements of the animals for membranes.

There have been no biosynthetic studies on gorgosterol (**11b**), but the isolation of dinosterol (**35i**) from dinoflagellates (Shimizu *et al.*, 1976; Withers *et al.*, 1978) and related sterols (**35b**, **36b**) from soft corals (Kanazawa *et al.*, 1974c, 1977a) points to a dinoflagellate origin for **11b** and indicates the possible route of side chain elaboration shown in Fig. 5.

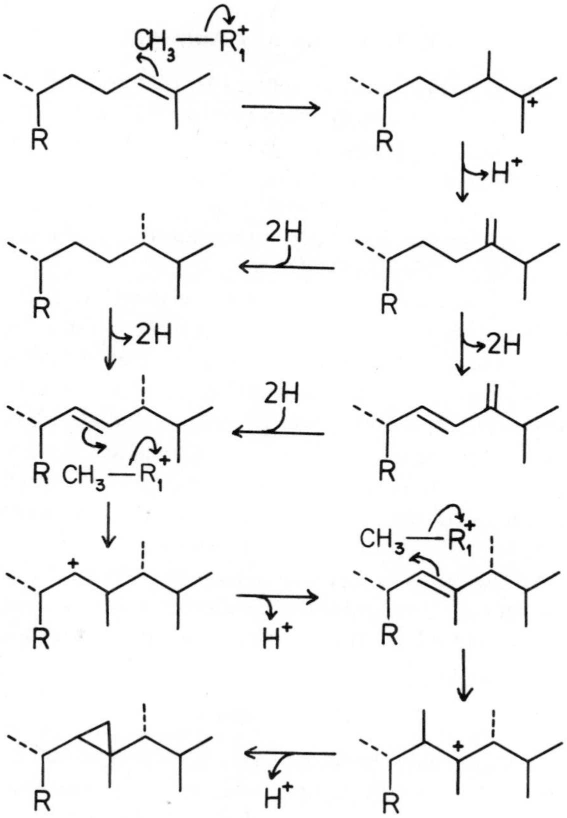

Fig. 5. Hypothetical mechanisms for the production of the gorgosterol side chain.

The mechanisms shown for the first two transmethylations to give side chain **35** have been verified by Withers *et al.* (1978) who used $[CD_3]$ methionine fed to the dinoflagellate *Crypthecodinium cohnii.*

There has been one report on sterol metabolism in a coelenterate. Saliot and Barbier (1973) demonstrated the transformation of $[^3H]$fucosterol (**19b**) into cholesterol (**1b**) in the anemone *Calliactis parasitica.* The relatively high proportions of cholesterol (**1b**) in the sterol mixtures of Hydrozoa, Scyphozoa, and most Actinaria species may possibly be explained by the ability of these animals to dealkylate C_{28} and C_{29} sterols. However, the alternative possibility must be considered that they are able to absorb selectively C_{27} sterol from the food, which may well comprise largely plankton and small crustacean species that are in any case rich in cholesterol (**1b**).

D. Nemertini and Annelida

Following incorporation studies using $[1\text{-}^{14}C]$acetate it was concluded that the nemertean *Cerebratulus marginatus* is capable of limited sterol synthesis (Voogt, 1973a). Interestingly, the specific activity of the esterified sterol (51 dpm/mg) was higher than that of the free sterol (9 dpm/mg), which is a similar situation to that observed in coelenterate studies (Voogt et al., 1974). *Linus ruber* has been reported to synthesize sterol from $[2\text{-}^{14}C]$mevalonic acid (Pennock, 1977).

That several marine polychaete worms are capable of sterol synthesis has been well established by the studies on *Arenicola marina, Nereis diversicolor,* and *Amphitrite ornata* by Wooton and Wright (1962), *Nereis pelagica* by Walton and Pennock (1972), and *Nereis diversicolor, Nephthys hombergii*, and *Spirographis spallonzani* by Voogt (1974).

None of the above-mentioned studies established the identities of the labeled sterols. This important facet of the problem should receive increased attention in future studies on all invertebrate phyla since it will permit more meaningful assessment of dietary sterol contribution and biosynthetic regulation. At present one can only surmise that the C_{26}, C_{28}, and C_{29} sterols recorded in annelids are dietary in origin but the extent to which they are modified by the animal, if at all, is unknown.

E. Arthropoda

It is now certain that crustaceans are incapable of *de novo* sterol synthesis from acetate or mevalonic acid and are entirely dependent on a dietary sterol source, as are insects (Rees, 1971). Sterol synthesis could not be demonstrated in *Astacus astacus* (Zandee, 1962, 1966), *Homarus vulgaris* (Zandee, 1964), *H. gammarus* (Zandee, 1967a), *Cancer apgurus* (Van den Oord, 1964), *Astacus fluviatilis* (Gosselin, 1965), *Carcinus maenus, Eupagurus bernhardus* (Walton and Pennock, 1972), *Rhithropanopeus harrisii, Libinia euarginata* (Whitney, 1969), *Artemia salina, Penaeus japonicus, Panulirus japonica, Portunus tritaberculatus* (Teshima and Kanazawa, 1971c,d), *Sesarma dehaani,* or *Helice tridens* (Teshima *et al.*, 1976b).

The prawn *Penaeus japonicus* had a poor growth and survival rate when fed a diet free of cholesterol (**1b**), but it grew normally on a diet supplemented with cholesterol (Kanazawa *et al.*, 1970, 1971a,b). Good survival rates were also observed with diets containing phytosterols (ergosterol, **8d**; stigmasterol, **16b**; sitosterol, **14b**), but growth was inferior to that observed with cholesterol feeding. Teshima *et al.* (1974d) showed that dietary cholesterol and several C-24 alkylated sterols are very efficiently absorbed by *P. japonicus*. The apparent percent absorption of the sterols was estimated by an indirect method using chromium oxide as an indicator, and the values obtained ranged from 61 to 99%; with cholesteryl esters the values recorded were 67–87% absorption.

When *P. japonicus* and the brine shrimp *Artemia salina* were fed diets containing various phytosterols (ergosterol, **8d**; campesterol, **6b**; sitosterol, **14b**; stigmasterol, **16b**) as the only sterol source, the sterol isolated from the animals was predominantly cholesterol (**1b**) thus indicating that the crustaceans can perform the C-24 dealkylation reaction previously documented in the Insecta (Kanazawa *et al.*, 1970, 1971a,b; Teshima and Kanazawa, 1971c; Teshima, 1972). The ability of crustaceans to dealkylate phytosterols was substantiated by Teshima and Kanazawa (1971e), who grew *Euglena gracilis* with added [1-^{14}C]acetate to produce labeled ergosterol (**8d**) and then fed the alga to *Artemia salina*. After 3 days the sterols of the shrimp contained radioactive cholesterol (**1b**). The conversion of labeled 24-methylcholest-5-en-3β-ol (**5b** or **6b**, produced by the yeast *Cryptoccus albidus*) and 24α-methylcholesta-5,22-dien-3β-ol (**7b**, produced by *Cyclotella nana;* see Section II concerning its C-24 configuration) to cholesterol (**1b**) by *A. salina* was also reported by Teshima (1971a) and Teshima and Kanazawa (1972a). The production of cholesterol (**1b**) from phytosterols was also demonstrated with a prawn (*Penaeus japonicus*) that was fed [4-^{14}C]sitosterol (**14b**) and a crab (*Purtunus trituberculatus*) that was fed [^{14}C]ergosterol (**8d**) (Teshima, 1971a,b; Kanazawa *et al.*, 1976a).

The mechanisms of sitosterol (**14b**) C-24 dealkylation operating in insects have been studied in detail (Fig. 6). As a result, fucosterol (**19b**), fucosterol 24,28-epoxide, and desmosterol (**2b**) have been identified as intermediates (Rees, 1971; Svoboda *et al.*, 1969; Pettler *et al.*, 1974;

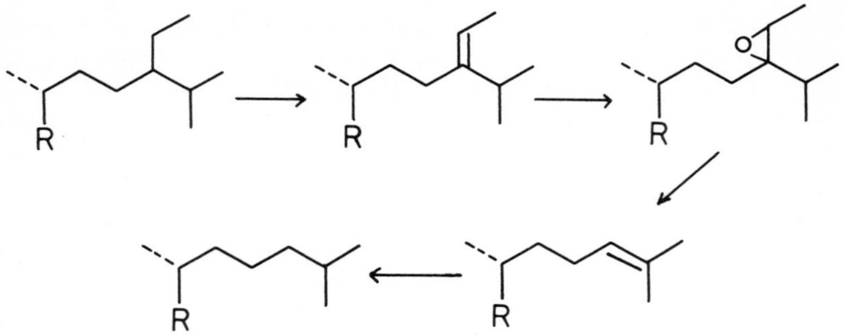

Fig. 6. An outline of the mechanism of sterol C-24 alkyl group removal operative in insects and possibly crustaceans.

Morisaki *et al.*, 1972; Allais *et al.*, 1973). The probability that this dealkylation route also operates in crustaceans is provided by the demonstration that a prawn (*Penaeus japonicus*), a shrimp (*Palaemon serratus*), and a crab (*Sesarma dehanni*) can reduce desmosterol (**2b**) to cholesterol (Teshima and Kanazawa, 1973b; Teshima *et al.*, 1975, 1976b). The conversion of desmosterol (**2b**) to cholesterol (**1b**) was examined at various stages in the molting cycle of *Pajaemon serratus* (Teshima *et al.*, 1975) and reported to be maximal at the postmolt stage B_1. It will be interesting to investigate the other individual steps in the dealkylation sequence and ascertain whether they also show any activity correlation with the molting process.

The production of desmosterol (**2b**) as an intermediate in dietary phytosterol dealkylation provides an attractive explanation for the occurrence of significant amounts of this sterol in many crustaceans (Table 6). Certainly it cannot be produced as an intermediate in *de novo* cholesterol (**1b**) biosynthesis (Frantz and Schroepfer, 1967) since crustaceans are incapable of sterol synthesis. However, it may perhaps arise, at least in part, from a dietary source since desmosterol (**2b**) is found in some red algae and is a constituent of some mollusk sterol mixtures; also, it is absorbed efficiently from the diet by two crabs and a prawn (Teshima *et al.*, 1975, 1976b).

F. Mollusca

The balance of evidence suggests that many mollusks are capable of sterol synthesis but that the rate of sterol production is slow. Species investigated for sterol biosynthetic ability by injection of [^{14}C]acetate and/or [2-^{14}C]mevalonate and the results obtained are indicated in Table 4, and they have also been tabulated in the review by Pennock (1977).

Cholest-7-en-3β-ol (**1c**), the major sterol of the class Amphineura, was synthesized from [1-^{14}C]acetate or [2-^{14}C]mevalonate by the chitons *Liolophura japonica* (Teshima and Kanazawa, 1973c) and *Lipodochitona cinera* (Voogt and Van Rheenen, 1974). Cholesterol (**1b**) was present in both animals (3 and 10% of the sterol mixture, respectively) but, although it contained negligible radioactivity in the former study, it did appear to be significantly labeled in the latter incubation. Voogt and Van Rheenen (1974) consider this to be evidence for an endogenous rather than an exogenous origin of the cholesterol (**1b**). Studies of a similar kind with other chitons seem warranted to determine whether the accumulation of cholest-7-en-3β-ol (**1c**) is due to a complete or only a partial block in the normal biosynthetic sequence to cholesterol (**1b**).

There appears to be general agreement that gastropods can synthesize sterol. An early study by Voogt (1967b) failed to demonstrate [^{14}C]acetate incorporation by the whelk *Buccinum undatum*. However, independent reinvestigations by Voogt (1972d) and Walton and Pennock (1972) using [2-^{14}C]mevalonate as substrate and incubation times of 69–140 and 24 hr, respectively, established conclusively that the whelk is able to produce sterols. This is substantiated by further studies on *Buccinum undatum* which show that it can convert lanosterol (**2n**) to cholesterol (**1b**) (Khalil and Idler, 1976). The failure of the first experiment of Voogt (1967b) to yield labeled sterol was probably due in part to the short incubation period (6.5 hr) and possibly also to the use of acetate as the substrate in view of the likely role of HMG-CoA reductase as the regulatory enzyme in isoprenoid production (see Section IV,A).

Four cephalopods have been investigated for sterol biosynthetic capability. An early study of *Sepia officinalis* with [^{14}C]acetate (Zandee, 1967b) and a more recent incubation of *Eledone cirrhosa* with [2-^{14}C]-mevalonate (Pennock, 1977) were both negative, and Pennock (1977) has expressed the view that cephalopods do not produce sterols. However, a reexamination of the situation with *Sepia officinalis* resulted in incorporation of [2-^{14}C]mevalonate into material that crystallized to constant specific activity with the sterols (Voogt, 1973b). This study also resulted in a very low apparent incorporation of [1-^{14}C]acetate into the sterols of *Octopus vulgaris* but not those of *Eledone aldrovandi*. It thus seems possible that cephalopods are capable of limited sterol synthesis, but it is suggested that the rate may be strictly controlled by HMG-CoA reductase and is perhaps dependent on the dietary sterol intake (Voogt, 1973b).

Investigations on sterol biosynthesis and interconversions in bivalves have given the most contradictory results of any group of invertebrates, and this has led to conflicting views concerning their ability to produce sterols *de novo* (Pennock, 1977). The earliest studies on bivalve sterol biosynthesis were by Fagerlund and Idler (1960, 1961a,b), who obtained evidence using *Mytilus californianus* and *Saxidomus giganteus* for the labeling of cholesterol (**1b**) from [2-^{14}C]acetate and a low transformation (0.05–0.3%) of [4-^{14}C]cholesterol (**1b**) into 24-methylenecholesterol (**10b**). In contrast, Salaque *et al.* (1966) could find no incorporation of either [2-^{14}C]-mevalonate or [*methyl*-^{14}C]methionine into the sterols of the oyster *Ostrea gryphea*. Likewise, Walton and Pennock (1972 and unpublished observations) failed to observe any labeling of the sterols of *Mytilus edulis, Cardium edule, Cyprina islandica, Glycymeris glycymeris,* or *Pecten maximus* after injection of [2-^{14}C]mevalonate. However, radioactivity was found in ubiquinone and dolichol and, most significantly, in other compounds such as farnesol and geranylgeraniol with chromatographic properties similar to those of the sterols. Therefore, in these animals it was concluded that the sterols were not synthesized, (Pennock, 1977) although the substrate was clearly available for isoprenoid biosynthesis. Voogt (1975a) also failed to prove sterol biosynthesis in *Mytilus edulis, Atrina fragilis,* or *Ostrea edulis* but expressed the view that their ability to synthesize sterols remains an open question since the synthetic rate may be extremely slow and below the detection level of the experimental conditions. The results reported by Voogt (1975a) with *Ostrea edulis* highlight the difficulty of obtaining conclusive proof of sterol synthesis. Two of the three experiments performed gave good incorporation of [^{14}C]acetate into the crude sterol fractions, which were then crystallized several times with a gradual increase in specific radioactivity (e.g., 264–348 dpm/mg), showing enrichment of the radioactive component. If the specific radioactivities had been lower, this change would not have been so apparent and the results may then have led to the conclusion that the sterols were labeled. However, in the case of *O. edulis,* Voogt (1975a) subjected the sterols to three purifications by tlc, which led to a progressive loss of the radioactivity (289, 107, 75, 63 dpm/mg). Moreover, preparative glc of the sterols showed that the radioactivity emerged from the column ahead of any of the sterols. These results are in accord with the labeling of nonsterol components as indicated by Pennock (1977).

The investigations of Teshima and Kanazawa (1974) on *Mytilus edulis* are at variance with the above-mentioned studies on this mollusk. These authors reported incorporation of [2-^{14}C]mevalonic acid into both squalene and the sterols, which were crystallized to constant specific radioactivity. Argentation chromatography of the steryl acetates revealed that radioactivity was associated with the fractions containing cholesterol (**1b**), 22-dehydrocholesterol (**3b**), desmosterol (**2b**), and 24-

methylenecholesterol (**10b**). The association of radioactivity with **10b** implies that sterol C-24 alkylation occurred. The ability of *M. edulis* to alkylate the sterol side chain was also suggested by incubation with [26-^{14}C]desmosterol. Chromatography of the recovered radioactive sterols on silica gel–silver nitrate indicated three products with chromatographic properties corresponding to cholesterol (**1b**), 22-dehydrocholesterol (**3b**), and, most significantly, 24-methylenecholesterol (**10b**), respectively (Teshima and Kanazawa, 1973a). These results, taken in conjunction with the earlier report of Fagerlund and Idler (1961a,b) describing **10b** synthesis in a clam, indicate that bivalve mollusks may be the first recorded animals capable of 24-alkylsterol production, which has hitherto been regarded as the preserve of fungi, algae, and higher plants. The phylogenetic importance of this point and its implications regarding the origins of at least some of the 24-alkylsterols found in mollusks are apparent, and confirmation of this work should be sought with other bivalves. However, the possibility must be borne in mind that the 24-alkylsterols may be produced not by the mollusk, but by associated organisms such as symbiotic algae (Yonge, 1953; Taylor, 1969; Trench, 1969, 1971a) or fungi in the digestive tract.

The possibility that sterol transmethylation can occur in bivalves has also been advanced by Voogt (1975b) to explain the apparently anomalous higher incorporation of [2-^{14}C]acetate compared to [1-^{14}C]acetate into the sterols of *Cardium edule*. Preparative glc of the *C. edule* sterols resulted in the trapping of some radioactivity in the C_{27} sterol, but the samples containing C_{28} plus C_{29} sterols were four to five times more radioactive. Voogt (1975b) speculates that the methyl group of acetate might be utilized preferentially over the carboxyl carbon for sterol synthesis, presumably via conversion to methionine followed by transmethylation to give C_{28} and C_{29} sterols.

Pennock (1977) expressed the view, based on many experiments from his laboratory, that bivalves do not synthesize sterols. Certainly caution is advisable, for the reasons already discussed, in interpreting the available evidence. Mollusks can survive anoxia for extended periods, and there have been extensive studies on the anaerobic metabolism of bivalves (de Zwaan and Wijsman, 1976). Since molecular oxygen is required for sterol synthesis (formation of squalene 2,3-oxide, C-4 demethylation, and Δ^5 bond introduction) it can be speculated that anaerobiosis may be the explanation for the difficulty in obtaining convincing evidence for sterol synthesis. To resolve the problem of bivalve sterol synthesis future studies should perhaps take more into account the time of year, possible associated algae or fungi, age, sexual maturity, and physiological condition of the animals used.

Phytosterol dealkylation has been demonstrated in mollusks. Cholesterol (**1b**) and desmosterol (**2b**) were identified as the metabolites of [3-^{3}H]fucosterol (**19b**) in *Ostrea grypha* (Saliot and Barbier, 1973). The same sterols also resulted when [3-^{3}H]sitosterol (**14b**) and [3-^{3}H]fucosterol (**19b**) were fed to *Patella vulgata* (Collignon-Thiennot *et al.*, 1973). *Buccinum undatum* converted labeled 24-methylenecholesterol (**10b**) and sitosterol (**14b**) to cholesterol (**1b**) (Khalil and Idler, 1976). Chondrillasterol (**15c**) was not transformed into cholesterol (**1b**), but the explanation that the animal may lack the enzyme for $\Delta^7 \rightarrow \Delta^5$ rearrangement seems unlikely in view of the observed conversion of lanosterol (**2n**) to cholesterol (**1b**), which also presumably requires this step (Khalil and Idler, 1976).

The identification of desmosterol (**2b**) as a demethylation product implicates the dealkylation pathway that has been established for the Insecta and that is probably operative in crustaceans (Fig. 6). Desmosterol (**2b**) has been identified in several mollusks (Table 4), and this provides circumstantial evidence for the speculation that dealkylation of dietary phytosterols may be of some importance in providing cholesterol (**1b**) in the absence of active *de novo* sterol synthesis. The possibility that various mollusks may have evolved both C-24 alkylation and C-24 demethylation enzyme systems presents a phylogenetic conundrum that will be interesting to resolve.

The identification of 24α-methylcholesta-5,22-dien-3β-ol (**7g**) in a scallop (*Patinopecten yessoensis*) indicates a dietary origin with diatoms as the primary source (Section II) of this and possibly other C_{28} and C_{29} sterols. The elucidation of the 24α configuration of occelasterol (**24b**), another *P. yessoensis* sterol, permitted Kobayashi and Mitsuhashi (1975) to propose the demethylation route outlined in Fig. 7 for the elaboration of the 24α-methyl-27-norsterol (**24**) and C_{26} sterol (**22**) types of sterol side chain. Since sterols with these side chains are widely distributed in invertebrates of several phyla it seems likely that the primary producer may be an organism early in the marine food chains.

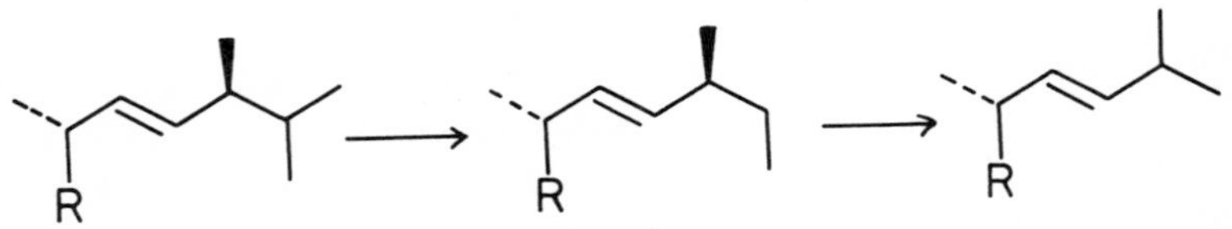

Fig. 7. Proposed route of sterol side chain dealkylation to give the 26-nor and 26,27-dinor side chains found in some marine sterols. (After Kobayashi and Mitsuhashi, 1975.)

G. Echinodermata

Representative species from all five classes of echinoderms have been examined and found to synthesize squalene and C_{27} sterols. Although Salaque *et al.* (1966) obtained no incoporation of [1,2-^{14}C]acetate into squalene or sterols by the echinoid *Paracentrotus lividus*, a subsequent study by Voogt (1972f) showed that this species as well as the urchins *Echinus acutus* and *Psammechinus miliaris* produced radioactive but unidentified sterols after injection of [1-^{14}C]acetate. Similarly, incorporation of [2-^{14}C]mevalonate into squalene and sterols by *Psammechinus miliaris* was observed by Walton and Pennock (1972) and by Rubinstein (1973) as well as by Kanazawa *et al.* (1974b) in the sea urchin *Clypeaster japonicus*. In the study by Rubinstein (1973) one of the labeled 4,4-dimethylsterols was identified as lanosterol (**2n**). Preparative glc of the demethylsterols showed cholesterol (**1b**) to be labeled, and this was confirmed by purification of **1b**, as the acetate, by silver nitrate–silica gel tlc, formation of and regeneration from the dibromide, and crystallization to constant specific radioactivity. There was no evidence for labeling of C_{28} and C_{29} sterols, but a compound with chromatographic properties indicating it to be desmosterol (**2b**) was significantly labeled.

Seasonal incubations of *P. miliaris* with [2-^{14}C]mevalonate showed that incorporation into squalene, expressed as percent total incorporation into nonsaponifiable lipid, was maximal in February (60%) but then declined to its lowest value (~10%) in May–July before gradually increasing again until December (Rubinstein, 1973). Conversely, incorporation into 4-demethylsterols was very low (~5%) in February but increased to its maximum (~40%) in May–July before again declining to around 5% in December. Similar results with a brittlestar have already been referred to (Fig. 3). Although this work warrants a statistically more meaningful reexamination using larger numbers of animals, it does illustrate the probable need to consider seasonal fluctuations in biosynthetic studies.

A similar seasonal study on [1-^{14}C]acetate incorporation by *P. miliaris* (Rubinstein, 1973) was inconclusive and did not yield any changes comparable to those obtained with [2-^{14}C]mevalonate. The former results indicate that there may be a secondary regulatory step in sterol biosynthesis that is operating between squalene and 4-demethylsterols (see also Pennock, 1977) and which can be observed only when [2-^{14}C]mevalonate is the precursor. However, with [1-^{14}C]acetate as substrate the presumed major regulatory step of HMG-CoA reductase may be sufficient to control the production of squalene to the point where the secondary control is not required, and thus a buildup of label in squalene is not observed.

In a detailed study of sterol biosynthesis in the sea urchin *Echinus esculentus* Smith and Goad (1974) found that [2-^{14}C]mevalonic acid was readily incorporated into squalene, lanosterol (**2n**), and desmosterol (**2b**), but cholesterol (**1b**) was only poorly labeled. Radioactive dihydrolanosterol (**1n**) was converted to cholesterol (**1b**) but [26-^{14}C]-desmosterol (**2b**) was not, thus revealing that all enzymes required for the **2n** to **1b** transformation, with the exception of the sterol Δ^{24}-reductase, were active in the urchin. Although desmosterol (**2b**) is present at about 8.6% of the sterol mixture of *E. esculentus* (Smith and Goad, 1974), its excessive accumulation, which might be anticipated in the absence of significant sterol Δ^{24}-reductase activity, may be prevented by a regulatory role of either HMG-CoA reductase or squalene cyclization, as discussed above. The possibility that desmosterol (**2b**) may arise by dealkylation of phytosterols, as seems possible in crustaceans (Section IV,E), was not supported since Smith and Goad (1974) failed to detect any metabolism of injected [4-^{14}C]sitosterol (**14b**) by *Echinus esculentus*.

Voogt (1973c) reported that the brittlestars *Ophiura albida* and *Ophioderma longicauda* produced labeled but unidentified sterols from injected [1-^{14}C]acetate. [2-^{14}C]Mevalonate was incorporated into squalene and sterols by *Ophiocomina nigra* (Walton and Pennock, 1972; Rubinstein, 1973), and the sterol contained labeled cholesterol (**1b**) and a second component tentatively identified by tlc and glc as desmosterol (**2b**) (Rubinstein, 1973). The C_{28} and C_{29} sterols were not radioactive. Seasonal fluctuations (Fig. 3) in the labeling of squalene and of sterols of *O. nigra* have already been discussed. Rubinstein (1973) also showed low incorporation of [2-^{14}C]mevalonate into squalene and sterol fractions by the crinoid *Antedon bifida*.

The ability of holothurians to synthesize sterols was first examined by Nomura *et al.* (1969b) using *Stichopus japonicus* injected with [1,2-^{14}C]-acetate; squalene was labeled but the Δ^7 sterols and lanosterol (**2n**) were not. Voogt and Over (1973) later observed [^{14}C]acetate labeling of the sterols of *Cucumaria planci, C. elongata, Holothuria tubulosa,* and *Stichopus regalis*. Similarly, Rubinstein (1973) obtained low incorporation of [2-^{14}C]mevalonate into squalene and sterols of *Cucumaria elongata* and *C. lactea*. Preparative glc indicated that the cholest-7-en-3β-ol (**1c**), one of the major sterols, was labeled, but there was also considerable radioactivity in other unidentified compounds.

De novo synthesis of sterols from mevalonate by starfish was first demonstrated by Smith and Goad (1971a), who found in *Asterias rubens* and *Henricia sanguinolenta* good incorporation into squalene and 4,4-dimethylsterols but rather low labeling of the 4-demethylsterols. Careful examination of the 4-demethylsterol fraction showed that 5α-cholest-7-en-3β-ol (**1c**) was the major radioactive constituent; labeled 5α-cholesta-

7,24-dien-3β-ol (**2c**) was also tentatively identified on tlc evidence, but there was no measurable incorporation into C_{28} or C_{29} sterols (Goad *et al.*, 1972a; Smith and Goad, 1975a).

Several 4,4-dimethyl- and 4-monomethylsterols were identified as minor sterol components of *Asterias rubens* (Goad *et al.*, 1972a; Smith *et al.*, 1973), and a number of these compounds were labeled by this animal after injection of [2-^{14}C]mevalonate (Goad *et al.*, 1972a; Smith and Goad, 1975a). These results led to the proposed biosynthetic routes to 5α-cholest-7-en-3β-ol (**1c**) shown in Fig. 8 (Smith and Goad, 1975a). The fact that dihydrolanosterol (**1n**) and 4,4-dimethylcholest-8-en-3β-ol (**1o**) have not been identified in the starfish, coupled with the high labeling of the Δ^{24} sterols, suggests that Δ^{24} double bond reduction may be delayed until at least the 4-monomethyl stage or even the last step in the sequence. The 9β,19-cyclopropane sterols cycloartanol (**1r**), cycloartenol (**2r**), and 31-norcycloartanol (**1s**) identified in *A. rubens* (Smith *et al.*, 1973) were not labeled from [2-^{14}C]mevalonate, indicating that they are dietary and probably have been passed down the food chain from algae in which cyclopropane sterols are established as phytosterol precursors (Goad and Goodwin, 1972; Goad, 1977). However, it would be interesting to explore the possibility that *A. rubens*, or other starfish, can modify ingested cycloartenol (**2r**) to give sterols **1r** and **1s**.

Synthesis of sterols has been reported by other authors using various starfish species. Voogt (1973d) observed incorporation of [^{14}C]acetate into the sterols of *Astropecten aurantiacus, Echinaster sepositus,* and *Asterias rubens*. In a more detailed study with *Asterias rubens* Voogt and Van Rheenen (1976b) confirmed that C_{28} and C_{29} sterol synthesis does not occur and that cholest-7-en-3β-ol (**1c**) is the major labeled compound after [2-^{14}C]mevalonate injection. They also reported low incorporation by *A. rubens* gonad and perivisceral fluid into cholesterol (**1b**), which comprised 1.4% of the sterol mixture, whereas hitherto it had been assumed that Δ^5 bond introduction into starfish sterol was blocked, thus accounting for the accumulation of cholest-7-en-3β-ol (**1c**) (Goad, 1976). In an examination of this problem in our laboratory (Teshima and Goad, 1975), even prolonged incubations of several weeks failed to reveal incorporation of [2-^{14}C]mevalonate into cholesterol (**1b**) either in the free form or as cholesterol sulfate (see below) by *A. rubens* or *Porania pulvillus,* although cholest-7-en-3β-ol (**1c**) was always labeled. In view of the high concentration of cholesterol sulfate in starfish tissues (Goodfellow and Goad, 1973) it is important to resolve this point. Possibly the present discrepancy can be explained by the fact that Voogt and Van Rheenen (1976b) analyzed only the sterols synthesized in the gonads rather than in the whole animal and that the two studies may have been conducted at different times of the

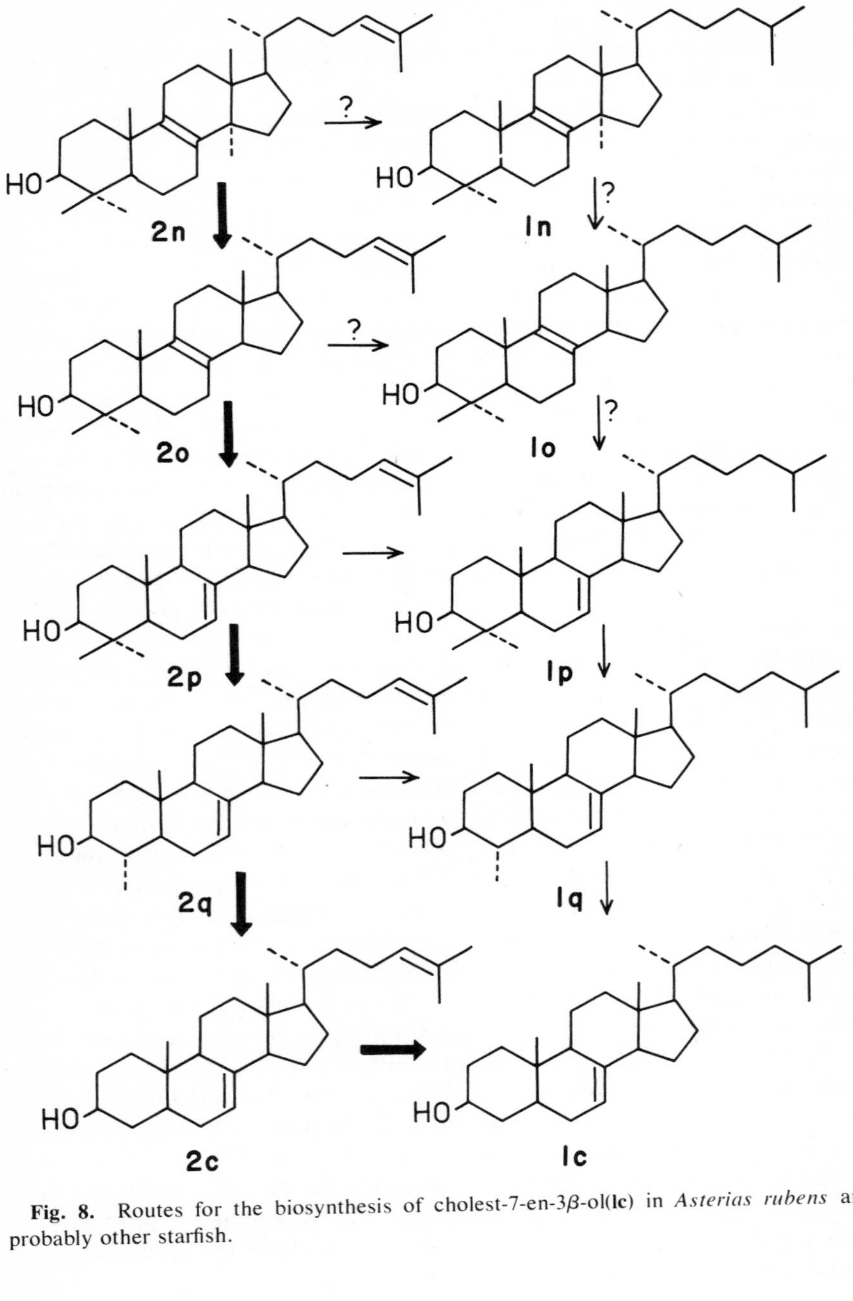

HO ? HO
2n ?
In
HO ? HO
2o ?
Io
HO HO
2p
Ip
HO HO
2q
Iq
HO HO
2c
Ic

Fig. 8. Routes for the biosynthesis of cholest-7-en-3β-ol(**lc**) in *Asterias rubens* and probably other starfish.

year, when the gonads may have been at different stages of development. Kanazawa *et al.* (1974b) reported incorporation of [2-^{14}C]mevalonate into the sterols of several other starfish. Incorporation of [2-^{14}C]mevalonate into 5α-cholest-7-en-3β-ol (**1c**) and probably 5α-cholesta-7,24-dien-3β-ol (**2c**) but not cholesterol (**1g**) or C_{28} and C_{29} sterols by *Laiaster leachii* (Teshima and Kanazawa, 1975) and *Coscinasterias acutispina* (Teshima and Kanazawa, 1976) has also been reported.

The many species of starfish analyzed all contain appreciable amounts of C_{28} and C_{29} sterols (Table 7), but biosynthetic studies have revealed the inability of these animals to produce 24-alkylsterols, which must therefore be regarded as dietary. However, this presents a problem since the free C_{28} and C_{29} sterols of asteroids are Δ^7 compounds, whereas predominantly Δ^5 sterols are found in all the presumed food sources of these animals. The possibility that asteroids can convert Δ^5 sterols to Δ^7 sterols was first tested by Fagerlund and Idler (1960), who proved that *Pisaster ochraceus* could convert [4-^{14}C]cholesterol to labeled 5α-cholest-7-en-3β-ol (**1c**). This transformation was confirmed in other starfish (*Asterias rubens, Solaster papposus, Porania pulvillus*) in studies that also established 5α-cholestanol (**1a**) as an important metabolite of cholesterol (**1b**) (Goad *et al.*, 1972a; Smith and Goad, 1971b, 1975b; Smith *et al.*, 1972). Cholest-4-en-3-one (**1h**) was shown to be an intermediate in the production of 5α-cholestanol (**1a**), and the conversion of labeled 5α-cholestanol (**1a**) to 5α-cholest-7-en-3β-ol (**1c**) by starfish was also demonstrated, thus establishing the scheme shown in Fig. 9 as a likely route for the production of Δ^7 sterols from dietary Δ^5 sterol. Support for this scheme is provided by the observation of small amounts of C_{27}, C_{28}, and C_{29} stanols together with traces of the Δ^5 sterols in the free sterols of starfish (Table 7) and the presence of larger amounts of stanols in the sulfated sterols (Goodfellow, 1974). The origin of the Δ^7 C_{28} and C_{29} sterols in starfish by this route was substantiated by the conversion by *Asterias rubens* of [4-^{14}C]sitosterol (**14b**) to 5α-stigmastonal (**14a**) and 5α-stigmast-7-en-3β-ol (**14c**), which was demonstrated by Smith and Goad (1975b) and confirmed by Voogt and Van Rheenen (1976a). Voogt and Van Rheenen (1976b) also reported the possibility that starfish can introduce a Δ^{22} bond into sterols.

An alternative route to 5α-cholest-7-en-3β-ol (**1c**) may operate through the production of 7-dehydrocholesterol (**1d**). Goodfellow (1974) found that [3α-^{3}H, 4-^{14}C]cholesterol (^{3}H : ^{14}C = 2.78) injected into *Asterias rubens* gave 5α-cholestanol (**1a**) with a ^{3}H to ^{14}C ratio of 0.50, in accord with the loss of the 3α-^{3}H at the 3-oxosteroid (**1h**) step. However, the 5α-cholest-7-en-3β-ol (**1c**) from this incubation had a ^{3}H to ^{14}C ratio of 1.00, indicating that only a portion could have been synthesized via the 5α-cholestanol (**1a**) and indicating an alternative route with retention of the 3α hydrogen

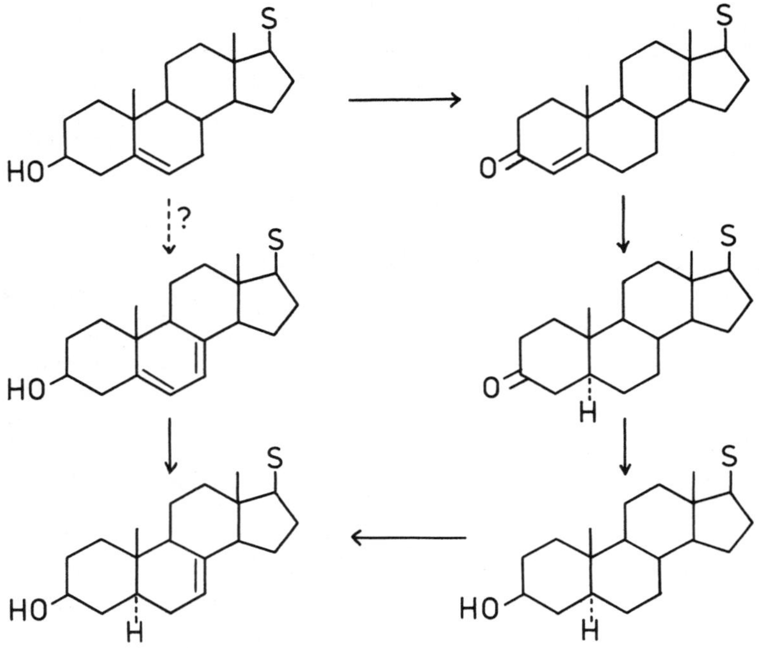

Fig. 9. The sequence(s) for conversion of Δ^5 sterols to Δ^7 sterols by starfish.

of cholesterol (**1b**). It was speculated that 7-dehydrocholesterol (**1d**) might be the intermediate in the alternative route, and injection of [3α-^{3}H,4-^{14}C]-cholesta-5,7-dien-3β-ol (**1d**) into *A. rubens* gave, after 96 hr, 5α-cholest-7-en-3β-ol (**1c**) with about 85% retention of tritium (Goodfellow and Goad, 1974). This result indicated direct reduction of the Δ^5 bond without prior production of a 3-oxosteroid (cf: Fig. 9) and is in contrast to the situation in mammals, in which the sequence $\Delta^5 \rightarrow \Delta^{5,7} \rightarrow \Delta^7$, a reversal of the cholesterol biosynthetic reactions, is thought not to occur (Frantz *et al.*, 1964). The utilization of $\Delta^{5,7}$ sterol may be of some significance in view of the presence of these sterol dienes in mollusks (Table 4), which constitute an important part of the food of many starfish. Also, there may be a requirement for $\Delta^{5,7}$ sterol as a precursor of vitamin D if this steroid is involved in calcium metabolism in relation to skeleton formation in these animals.

Goodfellow and Goad (1973) demonstrated the incorporation of injected [4-^{14}C]cholesterol (**1b**) into the sterol sulfates and fatty acyl esterified sterols of *Asterias rubens*. A more efficient incorporation into the sulfate fraction was noted when the [4-^{14}C]cholesterol was fed to the starfish in a gelatin capsule embedded in a scallop, which perhaps suggests that the pyloric caecae may be a major site of sterol sulfation (Goad, unpublished 1976). The radioactivity in the sulfated sterols was present mainly in unchanged cholesterol (**1b**) and its metabolite 5α-cholestanol (**1a**), but 5α-cholest-7-en-3β-ol (**1c**) was also labeled. The sterol sulfates recovered from *A. rubens* after [2-^{14}C]mevalonate injection contained no radioactivity in cholesterol (**1b**) or 5α-cholestanol (**1a**), but 5α-cholest-7-en-3β-ol (**1c**) was labeled (Goodfellow, 1974; Teshima and Goad, 1975). The incorporation of ^{35}S-labeled inorganic sulfate into *A. rubens* sterol sulfates was also shown by Goodfellow (1974). From these results it can be concluded that the starfish sterol sulfates are probably produced from two sterol sources. A considerable part of the dietary cholesterol (**1b**) may be sulfated unchanged by the animal, whereas the remainder may be converted to 5α-cholestanol (**1a**) and 5α-cholest-7-en-3β-ol (**1c**), portions of which can also be sulfated. *De novo* sterol synthesis provides a second, and possibly the main, pool, of 5α-cholest-7-en-3β-ol (**1c**) for sulfation.

Voogt and colleagues reported that in their studies on cholesterol (**1b**) and sitosterol (**14b**) metabolism in *A. rubens* extensive catabolism of these sterols appeared to occur, and radioactivity was incorporated into the neutral lipids containing triglycerides, into the polar lipids containing phospolipids, and into the expired carbon dioxide (Voogt and Schoenmakers, 1973; Voogt and Van Rheenen, 1976a,b). These authors did not examine the sterol sulfates, which could possibly have been present in the phospholipid fraction that they isolated. A reexamination of this point showed that the sterol sulfates are indeed the major labeled component of a crude phospholipid fraction isolated from *A. rubens* after [^{14}C]-cholesterol incubation (Sattar Khan and Goad, unpublished results). Traces of radioactivity were present in other compounds, but it is not yet known whether these are steroidal derivatives or the products of complete catabolism of sterol and reincorporation of the ^{14}C into nonsteroidal compounds.

H. Urochordata

Voogt and Van Rheenen (1975) reported the incorporation of [^{14}C]-acetate into tlc-purified sterols of the ascidians *Microcosmus sulcatus*, *Ciona intestinalis*, *Halocynthia papillosa*, and *Styela plicata*.

V. STEROID HORMONE METABOLISM

Steroid hormones have been reported in several animals from the phyla Porifera, Crustacea, Mollusca, Echinodermata, and Urochordata (Table

11). In some studies the identifications were based on paper chromatographic, histochemical, or radioimmunoassay results and should be regarded as tentative.

Steroid metabolism and steroid biochemistry of invertebrates were discussed in detail in a previous review (Goad, 1976) and are considered only briefly in this chapter. A highly simplified scheme of steroid metabolism is shown in Fig. 10, and Table 12 summarizes the steroid interconversions that have been recorded in marine animals of the invertebrate phyla. The most salient points are the demonstration of cholesterol (**1b**) conversion to C_{21} and C_{19} steroids and the conversion of progesterone (**43**) or pregnenolone (**41**) to corticosteroids and androgens, particularly testosterone (**53**), by gonad tissue preparations. These results establish in several phyla of invertebrates the presence of the enzyme systems required for steroid hormone elaboration (Lehoux and Sandor, 1970). However, it is notable that, although estrogens have been reported in several invertebrates (Table 11), there has been no demonstration of the production of either estrone (**51**) or estradiol (**52**) from C_{27}, C_{21}, or C_{19} steroid substrates.

The reports on steroid production in the mollusk *Aplysia depilans* by Lupo di Prisco and co-workers merit special attention. Gonad and hepatopancreas tissue of this gastropod were incubated with either [1-^{14}C]acetate or [4-^{14}C]cholesterol (Lupo di Prisco *et al.*, 1973). The metabolites isolated from the gonad after acetate incubation were pregnenolone (**41**), 17α-hydroxypregnenolone (**42**), progesterone (**43**), 17α-hydroxyprogesterone (**44**), dehydroepiandrosterone (**45**), androstenedione (**46**), cortisol (**50**), testosterone (**53**), and cortisone (**54**). These steroids, with the addition of 11-deoxycorticosterone (**48**), were also isolated from the acetate incubation with hepatic tissue. With cholesterol (**1b**) as the substrate the above steroids were isolated from each tissue with the exception of 20α-hydroxycholesterol and cortisone (**54**). The biosynthetic relationships of these steroids can be seen in Fig. 10. From the distribution of radioactivity in the products the authors suggested that the preferred route in the gonads may proceed via progesterone (**43**), whereas a route via dehydroepiandrosterone (**45**) might be more important in the hepatopancreas.

In subsequent studies with *A. depilans,* Lupo di Prisco and Dessi'Fulgheri (1974, 1975) presented evidence for a pathway to steroid hormones that does not involve cholesterol (**1b**) as an obligatory intermediate. [7α-^{3}H]Pregnenolone and [4-^{14}C]progesterone incubated together with either gonad or heptopancreas tissue gave labeled progesterone (**43**), 17α-hydroxyprogesterone (**44**), dehydroepiandrosterone (**45**), androstenedione (**46**), cortisol (**50**), and testosterone (**53**). With [2-^{14}C]mevalonic acid as the substrate both tissues gave low incorporation into progester-

TABLE 11

Steroid Hormones in Marine Invertebrates

Species	Tissue	Steroids	Reference
Phylum Porifera			
Haliclona rubens	Whole animal	3β-Hydroxy-pregn-5-en-20-one (**32b**), pregn-5-ene-3β,20α-diol (**33b**), pregn-5-ene-3β,20β-diol (**34b**)	Ballantine *et al.* (1977)
Phylum Mollusca			
Aplysia depilans	Gonads	Pregnenolone (**41**), progesterone (**43**), 17α-hydroxyprogesterone (**44**), testosterone (**53**), estrone (**51**), estradiol-17β (**52**)	Lupo di Prisco *et al.* (1973)
	Hepatic tissue	Pregnenolone (**41**), 17α-hydroxy-pregnenolone (**42**)	Lupo di Prisco *et al.* (1973)
Mytilus edulis	Gonads	Estrone (**51**), estradiol (**52**), testosterone (**53**)	Longcamp *et al.* (1974)
Pecten hericius	Ovaries	Estradiol-17β (**52**)	Boticelli *et al.* (1961)
Pecten maximus	Ovaries	Progesterone (**43**)	Saliot and Barbier (1971)
Sepia officinalis	Gonads	Progesterone (**43**)	Careau and Drosdowsky (1974)
Phylum Crustacea			
Crustacea species	Ovaries and eggs	Esterone (**51**), estradiol-17β (**52**)	Donahue (1948, 1952, 1957), Lisk (1961), Brodzicki (1963)
Phylum Echinodermata			
Asterias amurensis	Ovaries	Progesterone (**43**)	Ikegami *et al.* (1971)
Pisaster ochraceus	Ovaries	Progesterone (**43**), estradiol-17β (**52**)	Boticelli *et al.* (1960)
Strongylocentrotus franciscanus	Ovaries	Progesterone (**43**), estradiol-17β (**52**)	Boticelli *et al.* (1961)
Phylum Urochordata			
Ciona intestinalis	Gonads	Dehydroepiandrosterone (**45**), cortisone (**54**), cortisol (**50**)	Delrio *et al.* (1971)

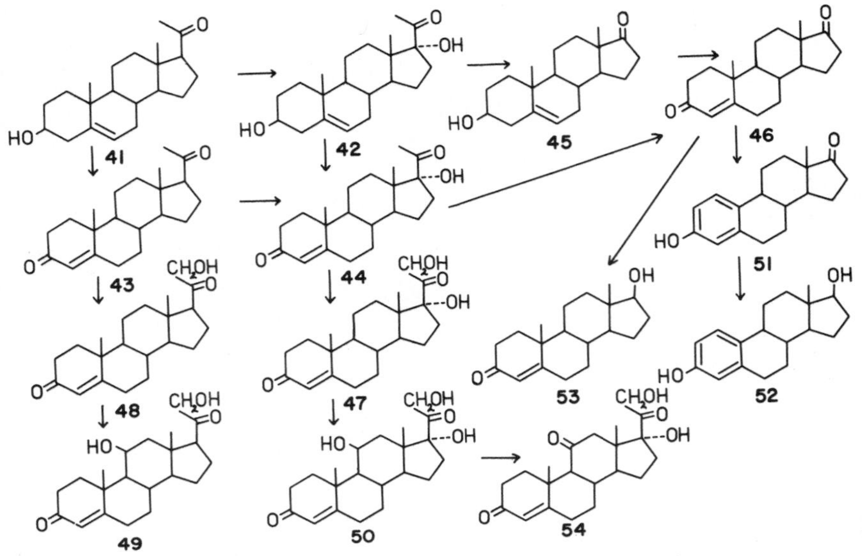

Fig. 10. A simplified scheme of steroid interconversions.

one (**43**), testosterone (**53**), 11-deoxycorticosterone (**48**), cortisol (**50**) and cortisone (**54**), but cholesterol was unlabeled. In contrast incubation with [4-^{14}C]demosterol (**2b**) gave radioactivity in cholesterol (**1b**), 20α-hydroxycholesterol, pregnenolone (**41**), progesterone (**43**), 17α-hydroxyprogesterone (**44**), dehydroepiandrosterone (**45**), androstenedione (**46**), cortisol (**50**), and testosterone (**53**) but not in 11-deoxycorticosterone (**48**) or cortisone (**54**). To explain these observations the authors suggested that two sequences operate to produce steroid hormones. One proceeds by the conventional route of desmosterol (**2b**) and cholesterol (**1b**) to pregnenolone (**41**), whereas the other route, leading particularly to the corticosteroids, proceeds from mevalonic acid by a pathway that does not require production of a C_{27} sterol precursor. Tait (1972, 1973) previously suggested that in mammalian tissues a C_{22} sterol may be produced as a steroid precursor. In this connection it is noteworthy that Kanazawa and Teshima (1971b) tentatively identified a C_{22} sterol (**21b**) in the clam *Tapes philippinarum*, and Djerassi *et al.* (1977) referred to unidentified C_{21} and C_{22} sterols from marine sources.

VI. ECDYSONE PRODUCTION IN CRUSTACEAE

In arthropods the molting process (ecdysis) is under the control of a group of steroid hormones referred to as the ecdysones (Highnam, 1970;

Rees, 1971; Horn, 1973; Karlson, 1973; Willig, 1974; Morgan and Poole, 1977). Although much of the interest in ecdysone biochemistry has centered on the Insecta, the role of these steroids in crustacean molting has received considerable attention in recent years. The literature up to the end of 1974 was covered in a previous review (Goad, 1976), and since then there have been more than 35 publications on crustacean ecdysone biochemistry. Much of this published work deals with the physiological or biochemical responses of crustaceans to administered ecdysones, and it is therefore beyond the scope of this chapter, which is restricted to the occurrence and biosynthesis of these steroids.

The structures of ecdysones reported in crustaceans are shown in Fig. 11, and their occurrence in crustacean species is listed in Table 13. Evidence for ecdysone (**55**) and ecdysterone (**56**) biosynthesis from [4-^{14}C]cholesterol was obtained with the lobster *Homarus americanus* (Gagosian *et al.*, 1974) and the crayfish *Orconectes limosus* (Willig and Keller, 1976). The transformation of ecdysone (**55**) into ecdysterone (**56**) by crustaceans was demonstrated first by King and Siddall (1969) and more recently by Lachaise *et al.* (1976) with the crab *Carcinus maenas*, in which the rate of conversion was dependent on the molting stage. The site of ecdysone (**55**) synthesis in crustaceans was established as the Y organ (Carlisle, 1965; Spaziani and Kater, 1973; Willig and Keller, 1976; Chang and O'Connor 1977). The major circulating hormone in a crab (*Pachygraspus crassipes*) was established as ecdysterone (**56**) by Chang *et al.* (1976). The conversion of ecdysone (**55**) to ecdysterone (**56**) does not apparently occur in the Y organ (Chang and O'Connor, 1977) but is mediated by several other tissues, particularly the testis (Chang *et al.*, 1976; Lachaise and Feyereisen, 1976).

VII. SAPONINS OF ECHINODERMS

A. Holothurins

The toxicity of sea cucumbers has been recognized for many years (Frey, 1951; Hyman, 1955; Bakus, 1968), and the active compounds, the greatest amount of which occurs in the Cuvierian glands of the animal, have been given the general name holothurins (Nigrelli, 1952; Yamanouchi, 1955). Earlier work on the structural elucidation of the holothurins and their pharmacological properties has been reviewed (Scheuer, 1969, 1973; Premuzic, 1971; Grossert, 1972; Goad, 1976).

Holothurin A was isolated from *Actinopyga agassizi,* and upon acid hydrolysis it gave sulfate, the sugars D-glucose, D-xylose, quinovose, and 3-*O*-methylglucose, and steroidal aglycones with a lanostane skeleton,

TABLE 12

Steroid Hormone Conversions in Marine Invertebrates

Species	Tissue	Precursor	Products	Reference
Phylum Crustacea				
Astacus astacus (freshwater species)	Ovaries, hepatopancreas	Androstenedione (**46**)	Testosterone (**53**)	Björkhem and Danielsson (1971)
Callinectes sapidus	Androgenic gland	Progesterone (**43**)	11-Deoxycorticosterone (**48**), 20α-hydroxy-pregn-4-en-3-one, androstenedione (**46**), testosterone (**53**)	Tcholakian and Eik-Nes (1969, 1971)
Callinectes sapidus	Whole animal	Pregnenolone (**41**)	Progesterone (**43**), androstenedione (**46**), 11-deoxycorticosterone (**48**)	Tcholakian and Eik-Nes (1971)
Callinectes sapidus	Vas deferens	Androstenedione (**46**)	Epitestosterone (17α-OH, **53**)	Blanchet *et al.* (1972)
			Estradiol-17α (17α-OH, **52**)	
		Estrone (**51**)		
		Progesterone (**43**)	11-Deoxycorticosterone (**48**)	
		17α-Hydroxyprogesterone (**44**)	11-Deoxycortisol (**47**)	
Carcinus maenus	Testis, vas deferens plus androgenic gland	Androstenedione (**46**)	Testosterone (**53**)	Blanchet *et al.* (1972)
		Dehydroepiandrosterone (**45**)	Androst-5-ene-3β,17β-diol	
		Estrone (**51**)	Estradiol-17β (**52**)	
Homarus americanus	Hepatopancreas, muscle, testis, vas deferens plus androgenic gland	Androstenedione (**46**)	Testosterone (**53**)	Gilgan and Idler (1967)
Panularis japonica	Hepatopancreas, ovaries, blood	Cholesterol (**1b**)	Progesterone (**43**), 17α-hydroxyprogesterone (**44**), androstenedione (**46**), testosterone (**53**), 11-deoxycorticosterone (**48**), corticosterone (**49**)	Kanazawa and Teshima (1971a)

trituberculatus			(**44**), testosterone (**53**), 11-deoxycorticosterone (**48**), 11-ketotestosterone	Kanazawa (1971f,g)
Phylum Mollusca				
Aplysia depilans	Gonads, hepatopancreas	Acetate, mevalonate, cholesterol (**1b**), desmosterol (**2b**), pregnenolene (**41**), or progesterone (**43**)	20α-Hydroxycholesterol, pregnenolone (**41**), 17α-hydroxypregnenolone (**42**), dehydroepiandrosterone (**45**), progesterone (**43**), 17α-hydroxyprogesterone (**44**), androstenedione (**46**), testosterone (**53**), cortisol (**50**), cortisone (**54**), 11-deoxycorticosterone (**48**)	Lupo di Prisco *et al.* (1973), Lupo di Prisco and Dessi' Fulgheri (1974, 1975)
Buccinum undatum	Hepatopancreas	Progesterone (**43**)	17α-Hydroxyprogesterone (**44**), androstenedione (**46**), testosterone (**53**), 11-deoxycorticosterone (**48**)	Teshima and Kanazawa (1971h)
Crepidula fornicata	Gonads, hepatopancreas	17α-Hydroxyprogesterone (**44**)	Androstenedione (**46**)	Bardon *et al.* (1971)
		Dehydroepiandrosterone (**45**)	Androstenedione (**46**) and testosterone (**53**)	
		Androstenedione (**46**)	Testosterone (**53**)	
		Testosterone (**53**)	Androstenedione (**46**)	
Crossostrea virginica	Sperm	Estradiol-17β (**52**)	Estrone (**51**)	Hathaway (1965)
Mytilus edulis	Gonads	Androstenedione (**46**)	Testosterone (**53**)	Longcamp *et al.* (1970, 1974)
		Pregnenolene (**41**)	Progesterone (**43**)	
		17α-Hydroxyprenenolene (**42**)	17α-Hydroxyprogesterone (**44**) and dehydroepiandrosterone (**45**)	
		Dehydroepiandrosterone (**45**)	Androstenedione (**46**) and androstenediol	
		Estradiol-17β (**52**)	Estrone (**51**)	

(*Continued*)

TABLE 12 (*Continued*)

Species	Tissue	Precursor	Products	Reference
Patella vulgata	Gonads	Pregnenolone (**41**)	Testosterone (**53**)	Lehoux *et al.* (1967)
		Progesterone (**43**)	Testosterone (**53**)	
Placopecten magellanicus	Gonads	17α-Hydroxyprogesterone (**44**)	Androstenedione (**46**)	Idler *et al.* (1969)
Sepia officinalis	Gonads	Pregnenolone (**41**)	17α-Hydroxypregnenolene (**42**), progesterone (**43**), androstenedione (**46**), dehydroepiandrosterone (**45**)	Careau and Drosdowsky (1974)
		Progesterone (**43**)	17α-Hydroxyprogesterone (**44**), 20α-hydroxypregn-4-en-3-one	
		Androstenedione (**46**)	Testosterone (**53**)	
Phylum Echinodermata				
Asterias rubens	Whole animal	Progesterone (**43**)	3β-Hydroxy-5α-pregnan-20-one (**32a**), 3β,6α-dihydroxy-5α-pregnan-20-one (**32t**), 5α-pregnane-3β,20α-diol (**33a**), 5α-pregnane-3β,20β-diol (**34a**), sulfates of **33t, 33a,** and **34b**	Gaffney and Goad (1973), Teshima *et al.* (1977)
		Cholesterol (**1b**)	5α-Cholestane-3β,6α-diol (**1t**), 3β-hydroxy-5α-pregnan-20-one (**32a**), 5α-pregnane-3β,20-diol (**33a, 34a**), sulfates of **1t, 32a,** and **33a** or **34a**	Teshima *et al.* (1977)
Asterias rubens	Gonads and pyloric ceca, tissue and homogenate	Progesterone (**43**)	5α-Pregnane-3,20-dione (**32h**) 3β-Hydroxy-5α-pregnan-20-one (**32a**)	Gaffney (1974), Fleming (1975), Teshima *et al.* (1977)

Dendraster excentricus	Eggs	Estrone (**51**)	Estradiol-17β (**52**)	Hathaway and Black (1969)
Marthasterias glacialis	Whole animal	Progesterone (**43**)	3β-Hydroxy-5α-pregnan-20-one (**32a**), 3β,6α-dihydroxy-5α-pregnan-20-one (**32t**)	Gaffney and Goad (1973)
Strongylocentrotus franciscanus	Gut	Estradiol-17β (**52**)	Estradiol 3-sulfate	Creange and Szego (1967)
Phylum Urochordata				
Ciona intestinalis	Ovary	Progesterone (**43**) and pregnenolone (**41**)	17α-Hydroxypregnenolone (**42**), dehydroepiandrosterone (**45**), androstenedione (**46**), testosterone (**53**), deoxycorticosterone (**48**), cortisone (**54**)	Delrio *et al.* (1971)
		Progesterone (**43**) and pregnenolone (**41**)	Dehydroepiandrosterone (**45**), testosterone (**53**), deoxycorticosterone (**48**)	
Styela plicata	Gonads	Progesterone (**43**)	5α-Pregnane-3,20-dione (**32h**), 3β-hydroxy-5α-pregnan-20-one (**32a**), 20α-hydroxypregn-4-en-3-one (**33h**), 20β-hydroxypregn-4-en-3-one (**34h**)	Colombo *et al.* (1977)

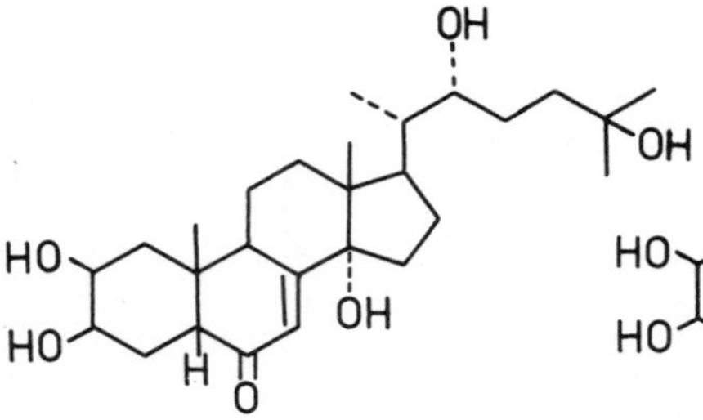

55

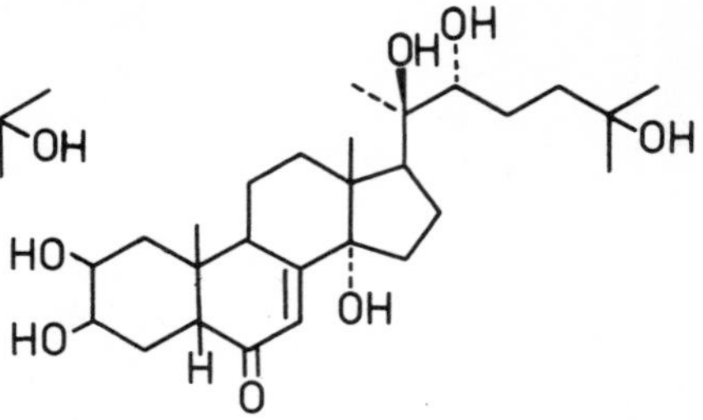

56

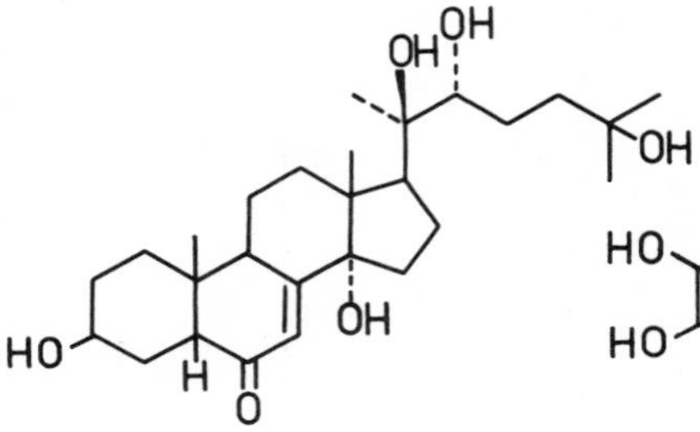

57

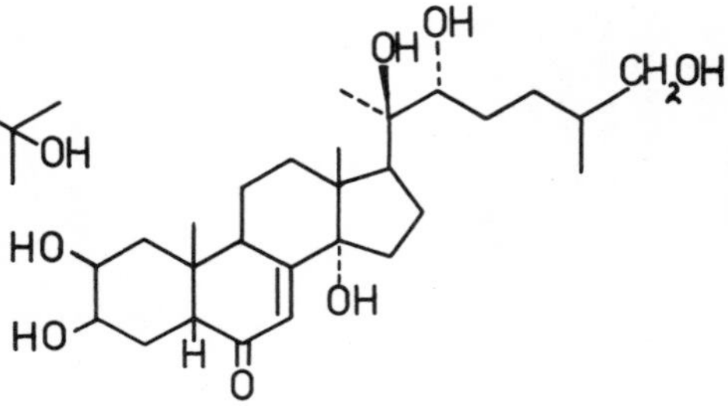

58

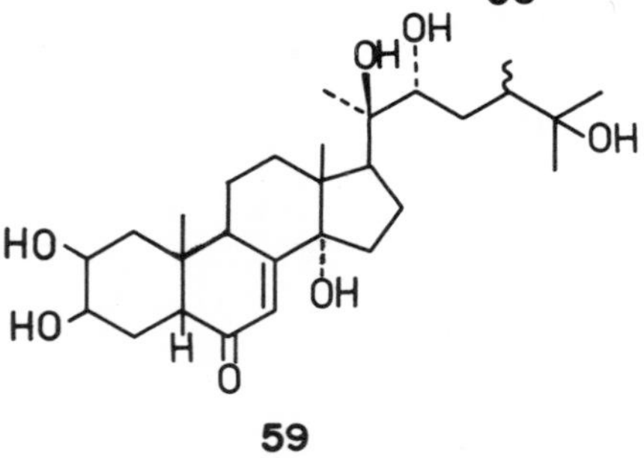

59

Fig. 11. Structures of ecdysones found in crustaceans.

22,25-oxidoholothurinogenin (**60**) and the 17-deoxy derivative (**61**) (Chanley *et al.*, 1959, 1966). Holothurin B isolated from *Holothuria vagabunda* and *H. lubria* gave only D-quinovose and D-xylose as the sugar components after hydrolysis (Yasumoto *et al.*, 1967). Subsequent studies on several species of sea cucumber resulted in the elucidation of a number of other steroidal aglycones (Table 14), all of which are related to lanostane (Roller *et al.*, 1969, 1970). The stereochemistry at C-20, which was unresolved in the earlier work, has now been established by X-ray analysis (Tan *et al.*, 1975; Ilin *et al.*, 1976), and it is shown in all structures of the

TABLE 14

Steroid Aglycones from Sea Cucumber Holothurins

Species	Trivial name of aglycone	Proposed nomenclature[a]	Reference
Actinopyga agassizi	22,25-Oxidoholothurinogenin (**60**)	22,25-Epoxyholosta-7,9(11)-diene-3β,17α-diol	Chanley *et al.* (1966)
	17-Deoxy-22,25-oxidoholothurinogenin (**61**)	22,25-Epoxyholosta-7,9(11)-dien-3β-ol	
Halodeima grisea	Griseogenin (**62**)	Holosta-7,9(11)-diene-3β,17α,22ξ-triol	Tursch *et al.* (1967)
Holothuria polii	Holothurinogenin (**63**)	Holosta-7,9(11)-diene-3β,17α-diol	Habermehl and Volkwein (1970)
Bohadschia koellikeri	Praslinogenin (**64**)	25-Methoxyholosta-7,9(11)-diene-3β,17α-diol	Roller *et al.* (1969), Tursch *et al.* (1970)
	Ternaygenin (**65**)	25-Methoxyholosta-7,9(11)-dien-3β-ol	
	Seychellogen (**66**)	Holosta-7,9(11)-dien-3β-ol	
	Koellikerigenen (**67**)	Holosta-7,9(11)-diene-3β,25-diol	
Stichopus chloronotus	23-Acetoxy-17-deoxy-7,8-dihydroholothurinogenin (**68**)	23-Acetoxyholost-9(11)-en-3β-ol	Rothberg *et al.* (1973)
Stichopus japonicus	Stichopogenin A_4 or holotoxinogen (**74**)	16-Oxoholost-9(11)-ene-3β,25-diol	Elyakov *et al.* (1969), Tan *et al.* (1975), Kitigawa *et al.* (1975b, 1976b)
	Holotoxinogen methyl ether (**75**)	16-Oxo-25-methoxyholost-9(11)-en-3β-ol	
Thelonota ananas	— (**77**)	23ξ-Acetoxyholost-8-en-3β-ol	Kelecom *et al.* (1976a)
	— (**78**)	23ξ-Acetoxyholosta-8,25-dien-3β-ol	

[a] See Habermehl and Volkwein (1971).

TABLE 13

Ecdysones of Crustaceans

Species	Compound	Reference
Balanus balanoides	*Ecdysone* (**55**), ecdysterone (**56**)	Bebbington and Morgan (1977)
Calinectes sapidus	*Ecdysterone* (**56**), callinecdysone A (**58**), callinecdysone B (**59**)	Faux *et al.* (1969)
Carcinus maenas	*Ecdysterone* (**56**)[a]	Andrieux *et al.* (1976)
Homarus americanus	*Ecdysterone* (**56**)	Gagosian *et al.* (1974)
Jasus lalandei	*Ecdysterone* (**56**), 2-deoxycrustecdysone (**57**)	Horn *et al.* (1966, 1968), Galbraith *et al.* (1968)
Orchestia gammarella pallas	*Ecdysterone* (**56**)	Blanchet *et al.* (1976)
Pachygrapsus crassipes	*Ecdysone* (**55**)	Chang *et al.* (1976)
	Ecdysterone (**56**)	Chang and O'Connor (1977)
Saphaeroma serratum	*Ecdysterone* (**56**)	Charmantier *et al.* (1976)

[a] Measured by radioimmunoassay.

aglycones presented in Fig. 12. Investigations of the distribution of holothurins in sea cucumbers have been reported by Elyakov *et al.* (1973, 1975a) and by Habermehl and Volkwein (1968). Habermehl and Volkwein (1971) also suggested a nomenclature system based on the name holostanol.

Recent work has revealed that several of the structures shown in Fig. 12 may not represent the true structures of the naturally occurring aglycones since the acid hydrolysis conditions to generate the aglycones may cause double-bond rearrangement, dehydration, or methylation of hydoxyl groups. For example, although the first structural studies by Chanley *et al.* (1959, 1966) produced aglycones with a $\Delta^{7,9(11)}$ diene system (**60**, **61**), subsequent milder hydrolysis conditions gave $\Delta^{9(11)}$-neoholothurinogenins with a 12α- or 12β-methoxyl group (e.g., **76**) (Chanley and Rossi, 1969a,b); even these compounds may also be artifacts since the parent saponin apparently does not have a methyl ether group (Chanley and Rossi, 1969a,b; Kelecom *et al.*, 1976a). Roller *et al.* (1969, 1970) presented evidence that methoxyl groups in holothurinogenins may be artifacts of methanolic acid hydrolysis.

The novel lanosta-5,8-diene system described by Elyakov *et al.* (1969) for stichopogen A_4 (**72**) and stichopogenin A_2 (**73**) isolated from *Stichopus japonicus* has been shown by two groups to be incorrect. Reexamination of the aglycone obtained from holotoxin, which is the saponin of *S. japonicus* (Shimada, 1969), permitted Tan *et al.* (1975) and Kitagawa *et al.* (1975b, 1976b) to revise the structure of stichopogenin A_4 to **74**. It should be noted that Tan *et al.* (1975) gave the name holotoxinogin to compound **74** in their paper. The need to reconsider whether a $\Delta^{9(11)}$ bond is indeed present in genuine holothurinogenins has been highlighted by a study on the production of artifacts during hydrolysis of the thelothurins obtained from *Thelonota ananas* (Kelecom *et al.*, 1976a,b). Aqueous hydrochloric acid hydrolysis of the saponins produced several aglycones with a $\Delta^{9(11)}$ bond (e.g., **68–71**), but mild aqueous acetic acid hydrolysis gave two aglycones with a Δ^8 bond (**77**, **78**).

The carbohydrate composition of the holothurins has been investigated in detail, with the result that the sugar sequences are known for some sea cucumber saponins. An early study by Chanley *et al.* (1961) suggested that the tetrasaccharide sequence in holothurin A is quinovose, 3-*O*-methylglucose, glucose, and xylose, with the last-named sugar linked to the 3β-hydroxyl of the steroid aglycone. Elyakov *et al.* (1975a) confirmed the presence of glucose, xylose, quinovose, and 3-*O*-methylglucose in holothurin A isolated from several species of sea cucumber and in stichoposide B from one species; however, holothurin B contained only

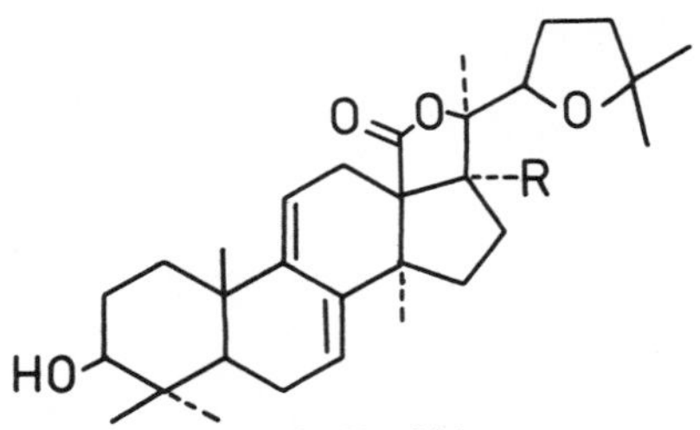

60 R=OH
61 R=H

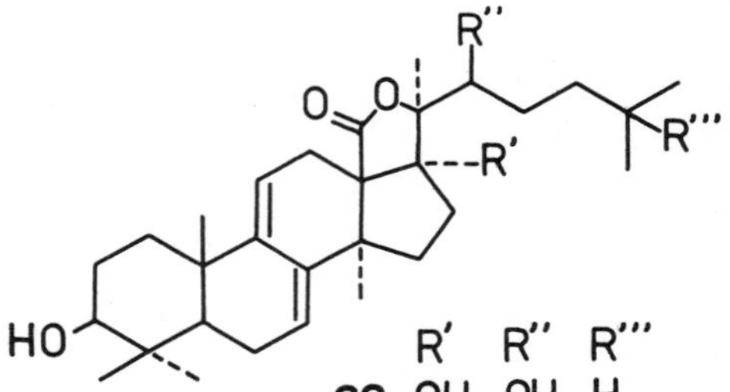

	R′	R″	R‴
62	OH	OH	H
63	OH	H	H
64	OH	H	OMe
65	H	H	OMe
66	H	H	H
67	H	H	OH

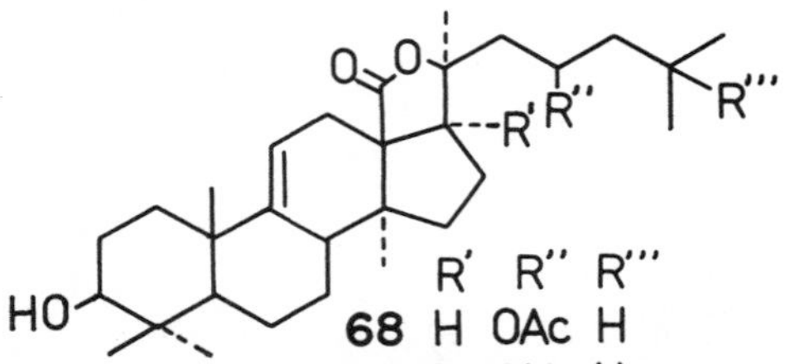

	R′	R″	R‴
68	H	OAc	H
69	H	OH	H
70	H	OH	OH
71	H	OAc	OH

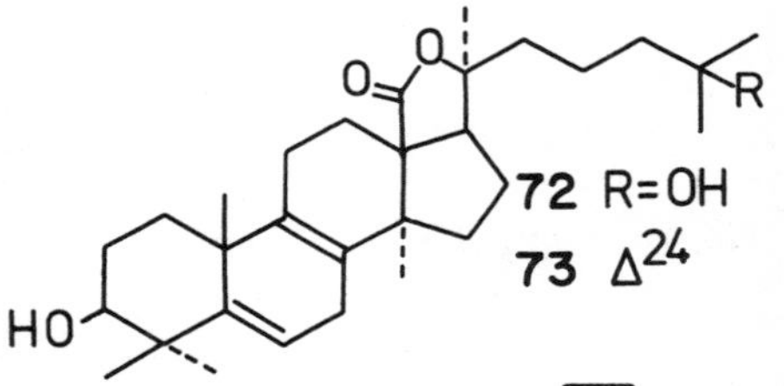

72 R=OH
73 Δ^{24}

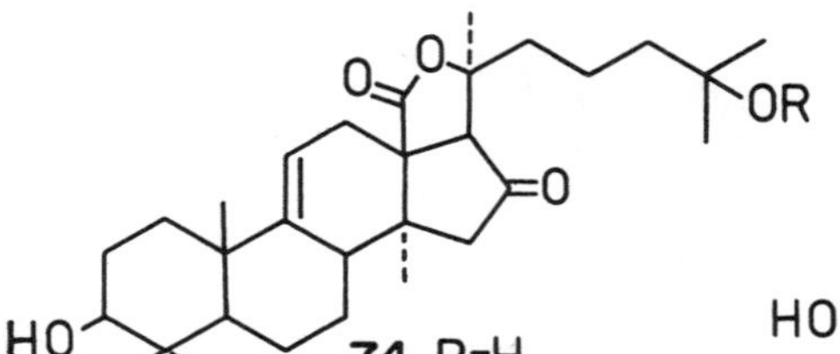

74 R=H
75 R=Me

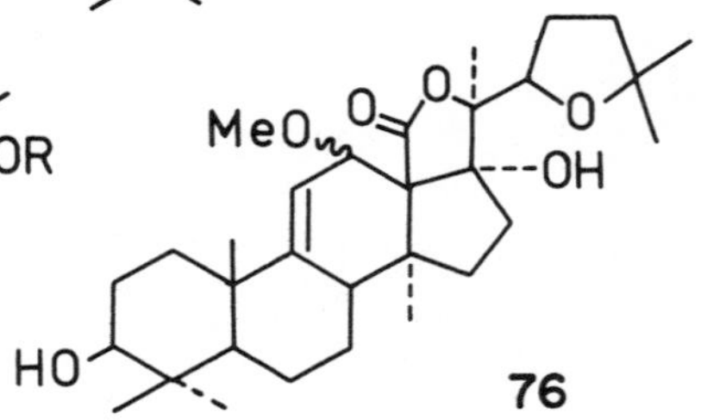

76

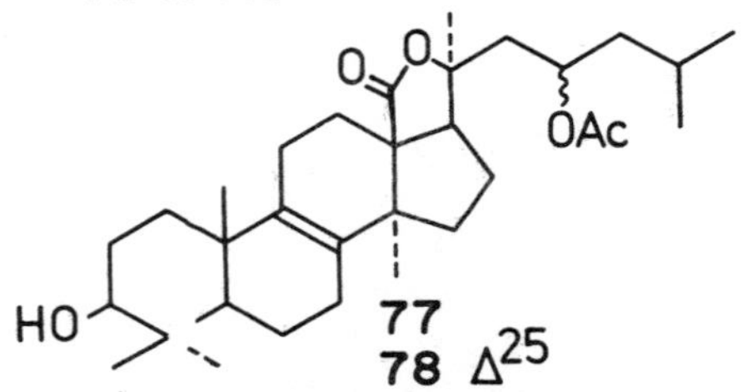

77
78 Δ^{25}

Fig. 12. Structures of the holothurin steroid aglycones isolated from sea cucumbers.

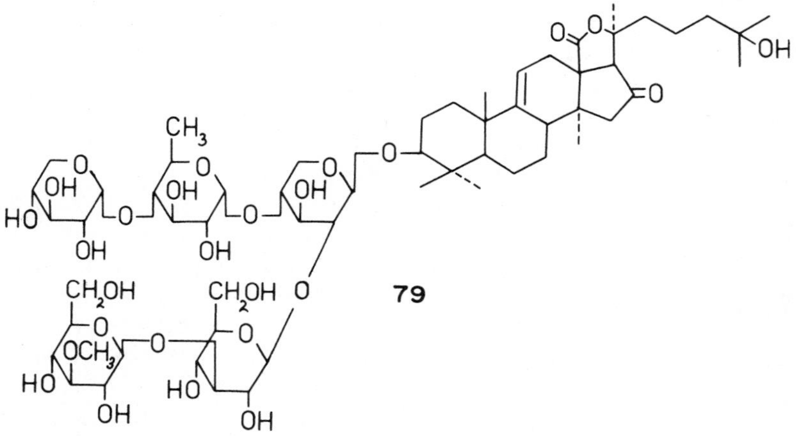

79

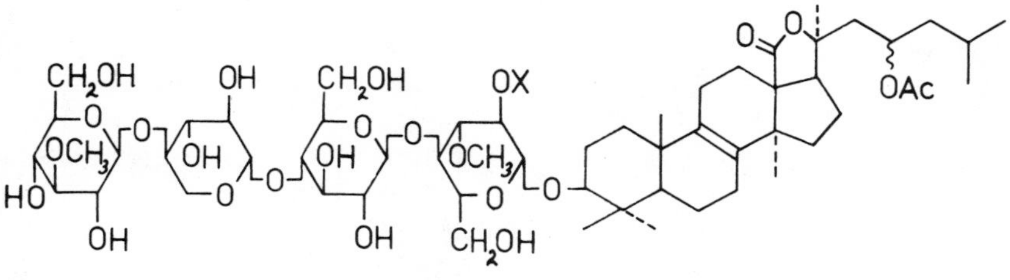

80

Fig. 13. Structures of holothurins.

xylose and quinovose, in agreement with the earlier report of Yasumoto *et al.* (1967).

The saponin isolated from *Stichopus japonicus* was separated into three components designated holotoxin A (main component), holotoxin B, and holotoxin C (Kitagawa *et al.*, 1974, 1976a). Holotoxin A has an oligosaccharide unit comprised of xylose, quinovose, 3-*O*-methylglucose, and glucose linked in the manner shown in structure **79**. Holotoxin B has an additional glucose unit attached at either the 3 or 4 position of the terminal xylose. The tetrasaccharide of thelothurins A and B from *Thelonota ananas* consists of two molecules of 3-*O*-methylglucose, xylose, and glucose linked together with an unknown constituent, as shown in structure **80** (Kelecom *et al.*, 1976c). It is notable that none of the recent examinations of holothurins have reported sulfate as a constituent.

There have been to date three investigations concerned with the biosynthesis of holothurins. Elyakov *et al.* (1975b) observed some incorporation of [^{14}C]acetate into the saponins of *Stichopus japonicus,* and Kelecom *et al.* (1976b) found very low incorporation of injected [1-^{14}C]acetate into thelothurins A and B by *Thelonota ananas.* In both studies most of the radioactivity resided in the steroid aglycone rather than in the sugar moieties, which led to the suggestion that, although the animals can synthesize the steroidal aglycone *de novo,* the carbohydrates are derived from a dietary source (Kelecom *et al.*, 1976b). Since the holothurinogenins are based on a lanostane skeleton, the most probable precursor is lanosterol (**2n**), which has been reported in holothuroids by Nomura *et al.* (1969a) and Habermehl and Volkwein (1968). Evidence for this point was provided by Sheikh and Djerassi (1976), who demonstrated the conversion of labeled lanosterol (**2n**) to holotoxinogenin (**74**) by *Stichopus californicus.*

Future studies should take into account possible seasonal fluctuations in the rate of holothurin biosynthesis in view of the evidence that suggests that the body content of these saponins varies during the year. Matsuno and Ishida (1969) found holothurins in all the tissues of *Holothuria leucospilota.* The highest concentrations occurred in the ovaries and Cuvier glands in July and August, but they were low at other times of the year; testes of this animal apparently contained very little holothurin. Elyakov *et al.* (1975b) and Levin and Stonik (1976) also noted seasonal changes in the concentrations of sea cucumber saponins.

B. Asterosaponins

Hashimoto and Yasumoto (1960) and Rio *et al.* (1963) were the first investigators to isolate toxic steroidal saponins, the asterosaponins, from starfish. These compounds are thought to be responsible, at least in part, for the escape responses elicited in mollusks and some other invertebrates when in close proximity to starfish (Feder and Christensen, 1966; Montgomery, 1967; Thomas and Gryffydd, 1971; Feder, 1972; Fange, 1963; Feder and Lasker, 1964; Feder and Arvidsson, 1967). The asterosaponins possess diverse physiological activity (Alexander and Russell, 1966; Owellen *et al.*, 1973; Cecil *et al.*, 1976; Goldsmith and Carlson, 1976; Mackie *et al.*, 1975) including antiviral activity (Shimizu, 1971).

The asterosaponins obtained from *Asterias amurensis* were separated into the major components asterosaponin A and asterosaponin B by Yasumoto and Hashimoto (1965). Acid hydrolysis of asterosaponin A gave sulfuric acid, a steroid aglycone, and 2 moles each of D-quinovose and D-fucose. Asterosaponin B also gave a steroid and sulfuric acid, but the sugars obtained were D-fucose, D-xylose, D-galactose, and 2 moles of D-quinovose.

Mackie *et al.* (1968) developed a bioassay for the asterosaponins based on the withdrawal response of the foot muscle of *Buccinum undatum*. This assay was used to follow the purification of the saponins of the starfish *Asterias rubens* and *Marthasterias glacialis,* which were found to be similar to the saponins isolated previously from other starfish by Yasumoto *et al.* (1964, 1966) and Yasumoto and Hashimoto, 1965, 1967). Acid hydrolysis of the saponin from *Marthasterias glacialis* gave sulfuric acid, fucose, quinovose, glucose, and a C_{27} steroid in the molar proportions 1 : 1 : 2 : 1 : 1 (Mackie and Turner, 1970). The major steroid aglycone, which was given the trivial name marthasterone, was shown to have the structure 3β,6α-dihydroxy-5α-cholesta-9(11),24-dien-3-one (**81**). 24,25-Dihydromarthasterone (**82**) and a minor component, 3β,6α-dihydroxy-5α-chol-9(11)-en-23-one (**83**), were also identified (Turner *et al.*, 1971; Smith *et al.*, 1973) (see Fig. 14).

The principal steroid aglycone recovered after acid hydrolysis of the saponins from *Acanthaster planci, Asterias amurensis,* and *Asterias forbesi* was found to be a pregnane derivative with the structure 3β,6α-dihydroxy-5α-pregn-9(11)-en-20-one (**84**) by Ikegami *et al.* (1972a,b), Sheikh *et al.* (1972a), Shimizu (1972) and ApSimon *et al.* (1973). This steroid, which was assigned the trivial name asterone by Gilgan *et al.* (1976), has been recovered from the asterosaponins of other starfish (Sheikh *et al.*, 1973; Habermehl and Christ, 1973; Elyakov *et al.*, 1976), and its chemical synthesis has been reported (Smith and Turner, 1972, 1973; Gurst *et al.*, 1973; ApSimon and Eenkhoorn, 1974).

The complexity of the steroid aglycone mixtures that are recovered from acid hydrolyzates of starfish saponins is apparent from a number of studies (Owellen *et al.*, 1973; Croft and Howden, 1974; Fleming *et al.*, 1976; Elyakov *et al.*, 1976; Howden *et al.*, 1977). Several C_{27} aglycones that have been characterized are listed in Table 15. A most significant recent contribution to asterosaponin chemistry has been the demonstration by Kitagawa *et al.* (1975a) that the C_{21} asterone (**84**) and possibly some of the C_{27} aglycones may be artifacts produced during the isolation procedure. Kitagawa *et al.* (1975a) obtained the aglycones of the saponins from *Acanthaster planci* by enzymatic hydrolysis and separated two compounds, thornasterol A (**90**) and thornasterol B (**91**), which were shown to be C_{27} compounds with a 20-hydroxyl group. Acid treatment of thornasterol A (**90**) produced asterone (**84**) and steroid **85**, thus suggesting that in previous studies these and possibly other aglycones may have been

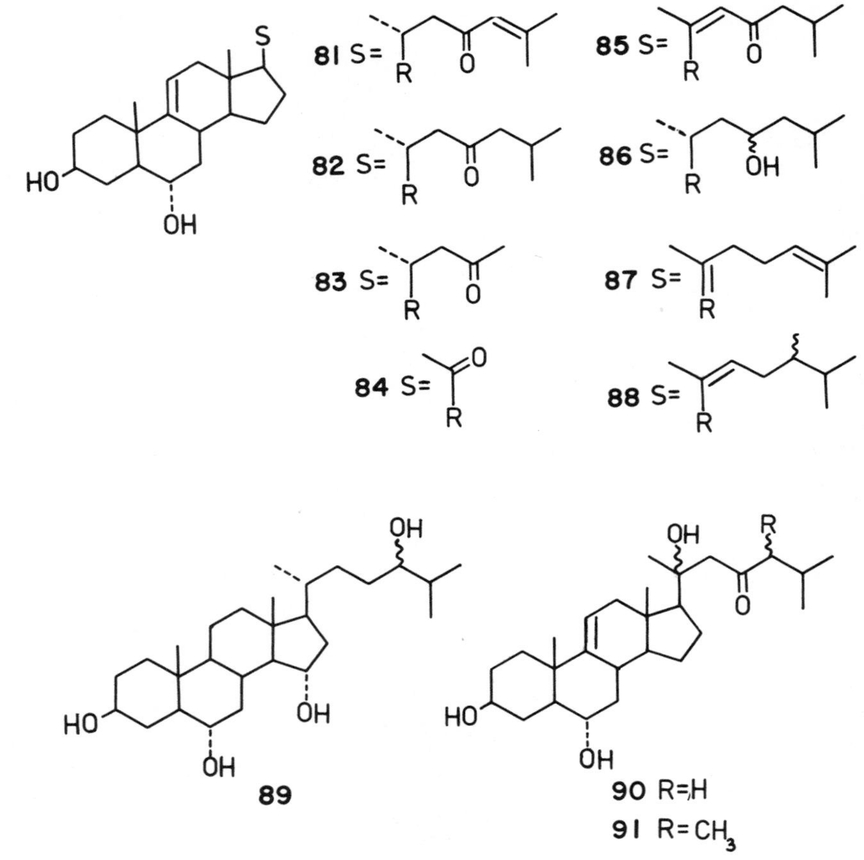

Fig. 14. Structures of the asterosaponin steroid aglycones isolated from starfish.

produced from the genuine algycones by the acid hydrolysis conditions employed. However, the facts that (1) progesterone (**32h**) has been reported in a starfish (Ikegami *et al*., 1971), (2) C-20 hydroxylation is a first step in the enzymatic cleavage of the cholesterol (**1b**) side chain to produce C_{21} steroids in other animals (Makin, 1975), and (3) cholesterol metabolism to pregnane derivatives has been observed in *Asterias rubens* (Teshima *et al*., 1977) suggest that pregnane aglycones may really occur in the asterosaponins and that further studies are warranted to establish their authenticity.

Sulfate in the asterosaponins is esterified to the steroid aglycone 3β-hydroxyl group (Ikegami *et al*., 1973b,c), whereas the sugars are linked by a glycosidic bond to the 6α-hydroxyl group of the steroid (Sheikh and

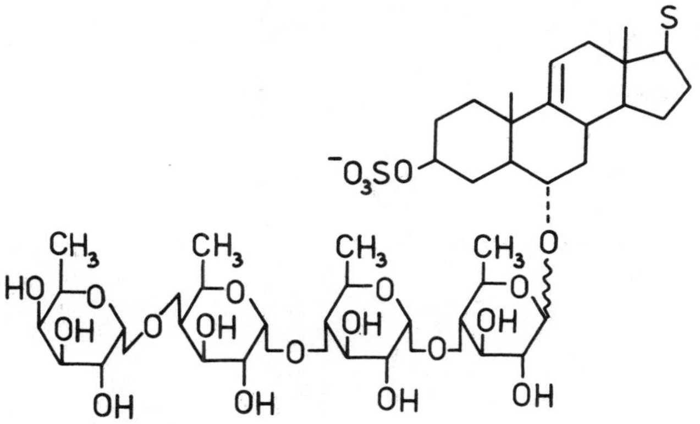

92

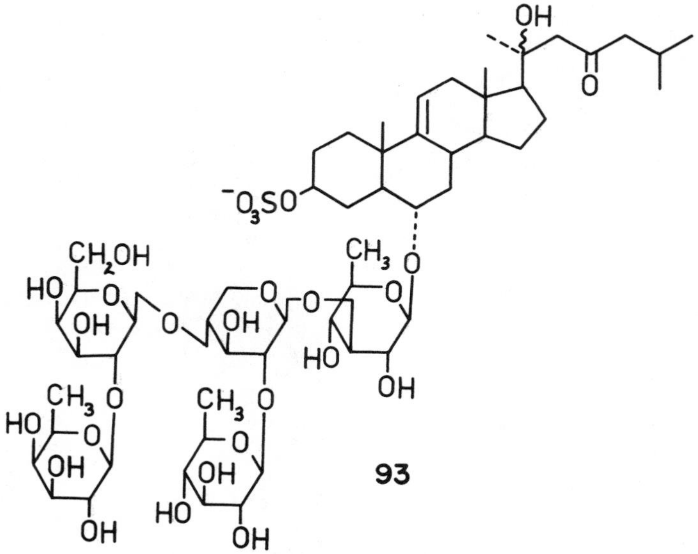

93

Fig. 15. Structures of asterosaponins.

Djerassi, 1973; Nicholson and Turner, 1976). The sugar sequences have been determined for asterosaponin A (**92**) from *A. amurensis* by Ikegami *et al.* (1972e,f,g) and for the *Acanthaster planci* saponin (**93**) by Kitagawa and Kobayashi (1977). The presence of the deoxy sugar fucose, together with various combinations of quinovose, xylose, glucose, and galactose, appears to be characteristic of all the saponins from different species of starfish so far examined (Elyakov *et al.*, 1976; Levina and Kapustina, 1975).

TABLE 15

Steroid Aglycones from Starfish Asterosaponins

Species	Compound	Reference
Marthasterias glacialis	Marthasterone (**81**), dihydromarthasterone (**82**), 3β,6α-dihydroxy-5α-chol-9(11)-en-23-one (**83**)	Turner *et al.* (1971), Smith *et al.* (1973)
Acanthaster planci, Asterias amurensis, Asterias forbesi	3β,6α-Dihydroxy-5α-pregn-9(11)-en-20-one (**84**)	Sheikh *et al.* (1972a), Ikegami *et al.* (1972b), Shimizu (1972), ApSimon *et al.* (1973)
Acanthaster planci, Asterias rubens	3β-Hydroxy-5α-cholesta-9(11),20(22)-dien-23-one (**85**)	Sheikh *et al.* (1972a), Habermehl and Christ (1973)
Asterias amurensis	5α-Cholest-9(11)-ene-3β,6α,23-triol (**86**)	Ikegami *et al.* (1972c, 1973a)
Acanthaster planci	5α-Cholesta-9(11),17(20),24-triene-3β,6α-diol (**87**), 24ξ-methyl-5α-cholesta-9(11),20-(22)-diene-3β,6α-diol (**88**)	Sheikh *et al.* (1972b)
Asterias amurensis	5α-Cholestane-3β,5α,15α,24-tetraol (**89**)	Kamiya *et al.* (1974)
Acanthaster planci	Thornasterol A (**90**), thornasterol B (**91**)	Kitigawa *et al.* (1975a)
Asterias rubens	3β,6α-Dihydroxy-5α-pregnan-20-one (**32t**)	Habermehl and Christ (1973)

Biosynthetic studies on the asterosaponins have so far been of a rather limited scope. Mackie *et al.* (1977) reported some evidence for the incorporation of radioactivity from [2-^{14}C]mevalonate and [4-^{14}C]cholesterol into the saponin aglycones of *Marthasterias glacialis,* but the incorporation was very low and the possibility exists that the radioactivity was largely in the sterol sulfates, which are known to be labeled by these substrates (see Section IV,G and below). Levina and Kapustina (1975) reported incorporation of [1,2-^{14}C]acetate and [4-^{14}C]cholesterol into asterone (**84**) by *Asterias amurensis.*

In studies with *Asterias rubens* and *Marthasterias glacialis* Gaffney and Goad (1973) observed the conversion of [4-^{14}C]progesterone (**32h**) into polar conjugates of 3β-hydroxy-5α-pregnan-20-one (**32a**) and 3β,6α-dihydroxy-5α-pregnan-20-one (**32t**), which have subsequently been identified as the 3β-sulfates (Teshima *et al.*, 1977) and which also include the sulfate of 5α-pregnane-3β,20ξ-diol (**33a** or **34a**). A very low level of incorporation of [4-^{14}C]progesterone into material with chromatographic properties similar to the asterosaponins was also noted. The observed conversion by *Asterias rubens* of [4-^{14}C]cholesterol (**1b**) to 5α-cholestane-3β,6α-diol (**1t**), which occurred both free and sulfated, may represent the first steps in asterosaponin production (Teshima *et al.*, 1977). The fact that two saponin aglycones (**88**, **91**) with a 24-methyl group have been obtained strongly suggests that dietary sterols contribute to the steroid precursors used for asterosaponin elaboration.

The distribution of the various asterosaponin aglycones in starfish species has been examined by Elyakov *et al.* (1976) and also by Mackie *et al.* (1977), who found that marthasterone (**81**) was most abundant in the order Forcipulata, family Asteriidae, subfamily Coscinasteriinae. Asterosaponins are found in all tissues of the animal. Yasumoto *et al.* (1966) found by a hemolytic assay that in *Asterias amurensis* the highest concentrations were in the stomach and gonads, whereas Mackie *et al.* (1977) with the same assay found in *Marthasterias glacialis* that the highest saponin concentrations were in the body wall, tube feet, and stomach. Both Yasumoto *et al.* (1966) and Mackie *et al.* (1977) have presented evidence for a seasonal fluctation in the content of starfish asterosaponin. In *Asterias amurensis* the concentration was low in winter but increased rapidly in June and July; similarly, in *Marthasterias glacialis* the saponin content increased to a maximum in the summer months. An improved method of asterosaponin extraction and a glc assay of the steroid aglycone released by acid hydrolysis were employed to show that the saponin content of *Asterias vulgaris* also fluctuates and is apparently highest in the April–June period, but geographical variations were also noted (Gilgan *et al.*, 1976; Burns *et al.*, 1977). The seasonal change and high gonad content of asterosaponin may be related to the suggested function of these compounds as a spawning inhibitor in starfish (Ikegami *et al.*, 1967, 1972d, 1976; Kanatani 1973; Ikegami, 1976). However, Mackie *et al.* (1977) suggested that the asterosaponins may have another primary role such as promoting sterol absorption in the pyloric cecae or as a chemical defense system.

APPENDIX: Sterols Reported in the Marine Environment

Structure	Trivial name	Systematic name
1a	Cholestanol	5α-Cholestan-3β-ol
1b	Cholesterol	Cholest-5-en-3β-ol
1c	Lathosterol	5α-Cholest-7-en-3β-ol
1d	7-Dehydrocholesterol	Cholesta-5,7-dien-3β-ol
1e	–	19-Nor-5α,10β-cholestan-3β-ol
1f	–	3β-Hydroxymethyl-A-nor-5α-cholestane
1g	–	5,8-Epidioxycholest-6-en-3β-ol
1h	Cholestenone	Cholest-4-en-3-one
1k	–	19-Nor-10β-cholest-5-en-3β-ol
1n	Dihydrolanosterol	5α-Lanost-8-en-3β-ol
1p	–	4,4-Dimethyl-5α-cholest-7-en-3β-ol
1q	Lophenol	4α-Methyl-5α-cholest-7-en-3β-ol
1r	Cycloartanol	4,4,14α-Trimethyl-19β,19-cyclo-5α-cholestan-3β-ol
1s	31-Norcycloartanol	4α,14α-Dimethyl-9β,19-cyclo-5α-cholestan-3β-ol
2b	Desmosterol	Cholesta-5,24-dien-3β-ol
2c	–	5α-Cholesta-7,24-dien-3β-ol
2n	Lanosterol	5α-Lanosta-8,24-dien-3β-ol
2r	Cycloartenol	4,4,14α-Trimethyl-9β,19-cyclo-5α-cholest-24-en-3β-ol
2q	–	4α-Methyl-5α-cholesta-7,24-dien-3β-ol
3a	–	(22*E*)-5α-Cholest-22-en-3β-ol
3b	22-Dehydrocholesterol	(22*E*)-Cholesta-5,22-dien-3β-ol
3c	–	(22*E*)-5α-Cholesta-7,22-dien-3β-ol
3d	–	(22*E*)-Cholesta-5,7,22-trien-3β-ol
3e	–	(22*E*)-19-Nor-5α,10β-cholest-22-en-3β-ol
3f	–	(22*E*)-3β-Hydroxymethyl-A-nor-5α-cholest-22-ene
4b	–	(22*Z*)-Cholesta-5,22-dien-3β-ol
5a	Campestanol	(24*R*)-24-Methyl-5α-cholestan-3β-ol
5b	Campesterol	(24*R*)-24-Methylcholest-5-en-3β-ol
5c	–	(24*R*)-24-Methyl-5α-cholest-7-en-3β-ol
5d	7-Dehydrocampsterol	(24*R*)-24-Methylcholesta-5,7-dien-3β-ol
5f/6f	–	3β-Hydroxymethyl-A-nor-5α,24ξ-ergostane
6a	Ergostanol	5α-Ergostan-3β-ol
6b	Dihydrobrassicasterol	Ergost-5-en-3β-ol or (24*S*)-24-methylcholest-5-en-3β-ol

Structure	Trivial name	Systematic name
6c	Fungisterol	Ergost-7-en-3β-ol
6d	22-Dihydroergosterol	Ergosta-5,7-dien-3β-ol
6e	–	19-Nor-5α,10β-ergostan-3β-ol
7a	–	(22*E*,24*S*)-24-Methyl-5α-cholest-22-en-3β-ol
7b	Pincsterol, crinosterol	(22*E*,24*S*)-24-Methylcholesta-5,22-dien-3β-ol
7c	Stellasterol	(22*E*,24*S*)-24-Methylcholesta-7,22-dien-3β-ol
7d	–	(22*E*,24*S*)-24-Methylcholesta-5,7,-22-trien-3β-ol
7f/8f	–	3β-Hydroxymethyl-A-nor-5α24ξ-ergost-22-ene
8a	–	(22*E*)-5α-Ergost-22-en-3β-ol or (22*E*,24*R*)-24-methyl-5α-cholest-22-en-3β-ol
8b	Brassicasterol	(22*E*)-Ergosta-5,22-dien-3β-ol
8c	5-Dihydroergosterol	(22*E*)-5α-Ergosta-7,22-dien-3β-ol
8d	Ergosterol	(22*E*)-Ergosta-5,7,22-trien-3β-ol
8e	–	(22*E*)-19-Nor-5α,10β-ergost-22-en-3β-ol
9b	Codisterol	(24*S*)-24-Methylcholesta-5,25-dien-3β-ol or ergosta-5,25-dien-3β-ol
–	–	24ξ-Methyl-5α-cholesta-7,22,25-trien-3β-ol
10a	24-Methylenecholestanol	5α-Ergost-24(28)-en-3β-ol
10b	24-Methylenecholesterol, chalinasterol	Ergosta-5,24(28)-dien-3β-ol
10c	Episterol	5α-Ergosta-7,24(28)-dien-3β-ol
10e	–	19-Nor-5α,10β-ergost-24(28)-en-3β-ol
10h	–	Ergosta-4,24(28)-dien-3-one
10r	24-Methylenecycloartanol	4,4,14α-Trimethyl-9β,19-cyclo-5α-ergost-24(28)-en-3β-ol
10q	24-Methylenelophenol	4α-Methyl-5α-ergosta-7,24(28)-dien-3β-ol
11a	Gorgostanol	(22*R*,23*R*,24*R*)-22,23-Methylene-23,-24-dimethylcholestan-3β-ol
11b	Gorgosterol	(22*R*,23*R*,24*R*)-22,23-Methylene-23,-24-dimethylcholest-5-en-3β-ol
11c	Acanthasterol	(22*R*,23*R*,24*R*)-22,23-Methylene-23,-24-dimethylcholest-7-en-3β-ol
11i	–	22,23-Methylene-4α,23,24ξ-trimethylcholestan-3β-ol
11j	–	22,23-Methylene-4α,23,24ξ-trimethylcholest-5-en-3β-ol
12b	Demethylgorgosterol	(22*R*,23*R*,24*R*)-22,23-Methylene-24-methylcholest-5-en-3β-ol
13b	Clionasterol	(24*S*)-24-Ethylcholest-5-en-3β-ol
13c	Dihydrochondrillastenol	(24*S*)-24-Ethylcholest-7-en-3β-ol

(*Continued*)

Structure	Trivial name	Systematic name
13d	7-Dehydroclionasterol	(24*S*)-24-Ethylcholesta-5,7-dien-3β-ol
13e/14e	–	19-Nor-(24ξ)-24-ethyl-5α,10β-cholestan-3β-ol
13f/14f		3β-Hydroxymethyl-A-nor-5α,24ξ-stigmastane
14a	Stigmastanol	5α-Stigmastan-3β-ol or (24*R*)-24-ethylcholestan-3β-ol
14b	Sitosterol	Stigmast-5-en-3β-ol
14c	Stigmast-7-enol, schottenol	Stigmast-7-en-3β-ol
14d	7-Dehydrositosterol	Stigmasta-5,7-dien-3β-ol
15b	Poriferasterol	(22*E*,24*R*)-24-Ethylcholesta-5,22-dien-3β-ol
15c	Chondrillasterol	(22*E*,24*R*)-24-Ethyl-5α-cholesta-7,-22-dien-3β-ol
15e/16e	–	(22*E*)-19-Nor-(24ξ)-24-ethyl-5α,-10β-cholest-22-en-3β-ol
15f/16f	–	3β-Hydroxymethyl-A-nor-5α,24ξ-stigmast-22-ene
16a	–	(22*E*)-Stigmast-22-en-3β-ol
16b	Stigmasterol	(22*E*)-Stigmasta-5,22-dien-3β-ol
16c	Spinasterol	(22*E*)-5α-Stigmasta-7,22-dien-3β-ol
16d	Corbisterol, 7-dehydrostigmasterol	(22*E*)-Stigmasta-5,7,22-trien-3β-ol
17b	Clerosterol	(24*S*)-24-Ethylcholesta-5,25-dien-3β-ol
18a	28-Isofucostanol	(24*Z*)-5α-Stigmast-24(28)-en-3β-ol
18b	28-Isofucosterol, avenasterol	(24*Z*)-Stigmasta-5,24(28)-dien-3β-ol
18c	Δ^7-Avenasterol	(24*Z*)-5α-Stigmasta-5,24(28)-dien-3β-ol
18q	24-Ethylidenelophenol	(24*Z*)-4α-Methyl-5α-stigmasta-7,24(28)-dien-3β-ol
19a	Fucostanol	(24*E*)-5α-Stigmast-24(28)-en-3β-ol
19b	Fucosterol	(24*E*)-Stigmasta-5,24(28)-dien-3β-ol
20b	–	(24*Z*)-24-Propylcholesta-5,24(28)-dien-3β-ol
21b	–	23,24-Dinorchola-5,20-dien-3β-ol
22a	–	24-Nor-5α-cholest-22-en-3β-ol
22b	–	24-Norcholesta-5,22-dien-3β-ol
22c	Asterosterol	24-Nor-5α-cholesta-7,22-dien-3β-ol
22e	–	19,24-Dinor-5α,10β-cholest-22-en-3β-ol
23b	–	24-Norcholest-5-en-3β-ol
24a	Patinosterol	(22*E*,24*S*)-27-Nor-24-methylcholest-22-en-3β-ol
24b	Occelasterol	(22*E*,24*S*)-27-Nor-24-methylcholesta-5,22-dien-3β-ol
24c	Amurasterol	(22*E*,24*S*)-27-Nor-24-methyl-5α-cholesta-7,22-dien-3β-ol

Structure	Trivial name	Systematic name
25b	Aplysterol	(24*R*,25*S*)-24,26-Dimethylcholest-5-en-3β-ol
26b	24,28-Didehydroaplysterol	(25*S*)-26-Methylergosta-5,24(28)-dien-3β-ol
27b	Calysterol	23,28-Cyclostigmasta-5,23-dien-3β-ol
28b	–	23-Ethylcholesta-5,23-dien-3β-ol
29b	–	Cholest-5-en-23-yn-3β-ol
35b	–	(22*E*,24*R*)-23,24-Dimethylcholesta-5,-22-dien-3β-ol
35i	Dinosterol	(22*E*,24*R*)-4α,23,24-Trimethylcholest-22-en-3β-ol
35j	Dehydrodinosterol	(22*E*,24*R*)-4α,23,24-Trimethylcholesta-5,22-dien-3β-ol
35l	Dinosterone	(22*E*,24*R*)-4α,23,24-Trimethylcholest-22-en-3-one
36b	–	23,24-Dimethylcholesta-5,23-dien-3β-ol
38b	–	(24*E*)-24-Propylcholesta-5,24(28)-dien-3β-ol
39b	Liagasterol	(23*Z*)-Cholesta-5,23-diene-3β,25-diol
40b	–	Cholesta-5,25-diene-3β,24ξ-diol

REFERENCES

Alcaide, A., Barbier, M., Potier, P., Manguer, A. M., and Teste, J. (1969). *Phytochemistry* **8**, 2301.

Alcaide, A., Viala, J., Pinte, F., Itoh, M., Nomura, T., and Barbier, M. (1971). *C. R. Acad. Sci., Ser. C* **273**, 1386.

Alexander, C. B., and Russell, F. E. (1966). *In* "Physiology of Echinodermata" (R. A. Boolootian, ed.), p. 529. Wiley (Interscience), New York.

Allais, J. P., Alcaide, A., and Barbier, M. (1973). *Experientia* **29**, 944.

Amico, V., Oriente, G., Piattelli, M., Tringali, C., Fattorusso, E., Magno, S., Mayol. L., Santacroce, C., and Sica, D. (1976). *Biochem. Syst. Ecol.* **4**, 143.

Anderson, R., Livermore, B. P., Kates, N., and Volcani, B. E. (1978). *Biochim. Biophys. Acta* **528**, 77.

Ando, T., and Barbier, M. (1975). *Biochem. Syst. Ecol.* **3**, 245.

Ando, T., Ferezou, J. P., and Barbier, M. (1975). *Biochem. Syst. Ecol.* **3**, 247.

Ando, T., Moueza, M., and Ceccaldi, H. J. (1976). *C. R. Seances Soc. Biol. Paris* **170**, 149.

Andrieux, N., Porcheron, P., Berreur-Bonnenfant, J., and Dray, F. (1976). *C. R. Acad. Sci., Ser. D* **283**, 1429.

ApSimon, J. W., and Eenkhoorn, J. A. (1974). *Can. J. Chem.* **52,** 4113.

ApSimon, J. W., Buccini, J. A., and Badripersaud, S. (1973). *Can. J. Chem.* **51**, 850.

Ashikaga, C. (1957). *Nippon Nogei Kagaku Kaishi* **31**, 115.

Austin, J. (1970). *Adv. Steroid Biochem. Pharmacol.* **1**, 73.

Bakus, G. J. (1968). *Mar. Biol.* **2,** 23.

Ballantine, J. A., Roberts, J. C., and Morris, R. J. (1975a). *Tetrahedron Lett.* p. 105.

Ballantine, J. A., Roberts, J. C., and Morris, R. J. (1975b). *J. Chromatogr.* **103**, 289.
Ballantine, J. A., Roberts, J. C., and Morris, R. J. (1976). *Biomed. Mass Spectrom.* **3**, 14.
Ballantine, J. A., Williams, K., and Burke, B. A. (1977). *Tetrahedron Lett.* p. 1547.
Bardon, O., Labet, P., and Drosdowsky, M. A. (1971). *Steroidologia* **2**, 366.
Baron, C., and Boutry, J. L. (1963). *C. R. Acad. Sci.* **256**, 4305.
Bean, G. A. (1973). *Adv. Lipid Res.* **11**, 193.
Beastall, G. H., Rees, H. H., and Goodwin, T. W. (1971). *Tetrahedron Lett.* p. 4935.
Beastall, G. H., Tyndall, A. M., Rees, H. H., and Goodwin, T. W. (1974). *Eur. J. Biochem.* **41**, 301.
Bebbington, P. M., and Morgan, E. D. (1977). *Comp. Biochem. Physiol. B* **56**, 77.
Bergmann, W. (1949). *J. Mar. Res.* **8**, 137.
Bergmann, W. (1958). *In* "Cholesterol" (R. P. Cook, ed.), p. 435. Academic Press, New York.
Bergmann, W. (1962). *In* "Comparative Biochemistry" (M. Florkin and H. S. Mason, eds.), Vol. 3, p. 103. Academic Press, New York.
Bergman, W., and Ottke, R. C. (1949). *J. Org. Chem.* **14**, 1085.
Björkhem, I., and Danielsson, H. (1971). *Comp. Biochem. Physiol., B* **38**, 315.
Björkman, L. R., Karlson, K. A., and Nilsson, K. (1972a). *Comp. Biochem. Physiol. B* **43**, 409.
Björkman, L. R., Karlson, K. A., Pascher, I., and Samuelsson, B. E. (1972b). *Biochim. Biophys. Acta* **270**, 260.
Blanchet, M.-F., Ozon, R., and Meusy, J.-J. (1972). *Comp. Biochem. Physiol. B* **41**, 251.
Blanchet, M.-F., Porcheron, P., and Dray, F. (1976). *C. R. Acad. Sci. Ser. D* **283**, 651.
Bligh, E. G., and Scott, M. A. (1966). *J. Fish Res. Board Can.* **23**, 1629.
Block, J. H. (1974). *Steroids* **23**, 421.
Bock, F., and Wetter, F. (1938). *Hoppe-Seyler's Z. Physiol. Chem.* **256**, 33.
Bodea, C., and Ciurdaru, V. (1968). *Rev. Roum. Biochim.* **5**, 87.
Bolker, H. I. (1967a). *Nature (London)* **213**, 904.
Bolker, H. I. (1967b). *Nature (London)* **213**, 905.
Bortolotto, M., Braekman, J. C., Daloze, D., and Tursch. B. (1976a). *Bull. Soc. Chim. Belg.* **85**, 27.
Bortolotto, M., Braekman, J. C., Daloze, D., Losman, D., and Tursch, B. (1976b). *Steroids* **28**, 461.
Boticelli, C. R., Hisaw, F. C., and Wotiz, H. H. (1960). *Proc. Soc. Exp. Biol. Med.* **103**, 875.
Boticelli, C. R., Hisaw, F. C., and Wotiz, H. H. (1961). *Proc. Soc. Exp. Biol. Med.* **106**, 887.
Boutry, J. L., and Barbier, M. (1974). *Mar. Chem.* **2**, 217.
Boutry, J. L., and Baron, C. (1967). *Bull. Soc. Chim. Biol.* **49**, 157.
Boutry, J. L., and Jacques, G. (1970). *Bull. Soc. Chim. Biol.* **52**, 349.
Boutry, J. L., Alcaide, A., and Barbier, M. (1971). *C. R. Acad. Sci., Ser. D* **272**, 1022.
Boutry, J. L., Barbier, M., and Ricard, M. (1976a). *J. Exp. Mar. Biol. Ecol.* **21**, 69.
Boutry, J. L., Bordes, M., and Barbier, M. (1976b). *Biochem. Syst. Ecol.* **4**, 201.
Braekman, J. C. (1977). *In* "Marine Natural Products Chemistry" (D. J. Faulkner and W. H. Fenical, eds.), p. 5. Plenum, New York.
Brodzicki, S. (1963). *Folia Histochem. Cytochem.* **1**, 259.
Burkhardt, G. N., Heilbron, I. M., Jackson, H., Pary, E. G., and Loven, J. A. (1934). *Biochem. J.* **28**, 1698.
Burns, B. G., Gilgan, M. W., Logan, V. H., Burnell, J., and ApSimon, J. W. (1977). *Anal. Biochem.* **81**, 196.

Cafieri, F., and de Napoli, L. (1976). *Gazz. Chim. Ital.* **106**, 761.
Cafieri, F., Fattorusso, E., Frigerio, A., Santacrole, C., and Sica, D. (1975). *Gazz. Chim. Ital.* **105**, 595.
Careau, S., and Drosdowsky, M. A. (1974). *Gen. Comp. Endocrinol.* **22**, 341.
Cargile, N., Edwards, H. N., and McChesney, J. D. (1975). *J. Phycol.* **11**, 457.
Carlisle, D. B. (1965). *Gen. Comp. Endocrinol.* **5**, 366.
Cecil, J. T., Ruggieri, G. D., and Nigrelli, R. F. (1976). *Invertebr. Tissue Cult.: Appl. Med., Biol. Agric. (Proc. Int. Conf., 4th, 1975), Mont. Gabriel, Quebec* p. 301.
Chang, E. S., and O'Conner, J. D. (1977). *Proc. Natl. Acad. Sci. U.S.A.* **74**, 615.
Chang, E. S., Sage, B. A., and O'Conner, J. D. (1976). *Gen. Comp. Endocrinol.* **30**, 21.
Chanley, J. D., and Rossi, C. (1969a). *Tetrahedron* **25**, 1897.
Chanley, J. D., and Rossi, C. (1969b). *Tetrahedron* **25**, 1911.
Chanley, J. D., Leeden, R., Wax, J., Nigrelli, R. F., and Sobotka, H. (1959). *J. Am. Chem. Soc.* **81**, 5180.
Chanley, J. D., Perlstein, J., Nigrelli, R. F., and Sobotka, H. (1961). *Ann. N.Y. Acad. Sci.* **90**, 902.
Chanley, J. D., Mezzetti, T., and Sobotka, H. (1966). *Tetrahedron* **22**, 1857.
Chardon-Loriaux, I., Morisaki, M., and Ikekawa, N. (1976). *Phytochemistry* **15**, 723.
Charmantier, G., Olle, M., and Trilles, J. P. (1976). *C. R. Acad. Sci., Ser. D* **283**, 1329.
Ciereszko, L. S., Johnson, M. A., Schmidt, R. W., and Koons, C. B. (1968). *Comp. Biochem. Physiol.* **24**, 899.
Collignon-Thiennot, F., Allais, J. P., and Barbier, M. (1973). *Biochimie* **55**, 579.
Colombo, L., Belvedere, P. C., and Breda, L. (1977). *Comp. Biochem. Physiol. B* **56**, 49.
Creange, J. E., and Szego, C. M. (1967). *Biochem. J.* **102**, 898.
Croft, J. A., and Howden, M. E. H. (1974). *Comp. Biochem. Physiol. B* **48**, 535.
Dawson, R. M. C., and Barnes, H. (1966). *J. Mar. Biol. Assoc. U.K.* **46**, 249.
Deffner, M. (1943). *Hoppe-Seyler's Z. Physiol. Chem.* **278**, 165.
Delrio, G., D'Istria, M., Milone, M., and Chieffe, G. (1971). *Experientia* **27**, 1348.
de Luca, P., de Rosa, M., Minale, L., and Sodano, G. (1972). *J. Chem. Soc. Perkin Trans. 1* p. 2132.
de Luca, P., de Rosa, M., Minale, L., Puliti, R., Sodano, G., Giordano, F., and Mazzarella, L. (1973). *J. Chem. Soc., Chem. Commun.* p. 825.
Dempsey, M. E. (1974). *Annu. Rev. Biochem.* **43**, 967.
de Rosa, M., Minale, L., and Sodano, G. (1973a). *Comp. Biochem. Physiol. B* **46**, 823.
de Rosa, M., Minale, L., and Sodano, G. (1973b). *Comp. Biochem. Physiol. B.* **45**, 883.
de Rosa, M., Minale, L., and Sodano, G. (1975a). *Experientia* **31**, 408.
de Rosa, M., Minale, L., and Sodano, G. (1975b). *Experientia* **31**, 758.
de Rosa, M., Minale, L., and Sodano, G. (1976). *Experientia* **32**, 1112.
de Zwaan, A., and Wijsman, T. C. M. (1976). *Comp. Biochem. Physiol. B* **54**, 313.
Djerassi, C., Carlson, R. M. K., Popov, S., and Varkony, T. H. (1977). *In* "Marine Natural Products Chemistry" (D. J. Faulkner and W. H. Fenical, eds.), p. 111. Plenum, New York.
Donahue, J. K. (1948). *Proc. Soc. Exp. Biol. Med.* **69**, 179.
Donahue, J. K. (1952). *State Maine Dep. Sea Shore Fish. Res. Bull.* No. 8.
Donahue, J. K. (1957). *State Maine Dep. Sea Shore Fish. Res. Bull.* No. 28. Cited in Lehoux and Sandor (1970).
Dorée, C. (1909). *Biochim. J.* **4**, 72.
Doyle, P. J., and Patterson, G. W. (1972). *Comp. Biochem. Physiol. B* **41**, 355.
Edmonds, C. G., Smith, A. G., and Brooks, C. J. W. (1977). *J. Chromatogr.* **133**, 372.

Eichenberger, W. (1976). *Chimia* **30**, 75.
Elyakov, G. B., Kuznetsova, T. A., Dzizenko, A. K., and Elkin, Y. N. (1969). *Tetrahedron Lett.* p. 1151.
Elyakov, G. B., Stonik, V. A., Levina, E. V., Splanke, V. P., Kuznetsova, T. A., and Levin, V. S. (1973). *Comp. Biochem. Physiol. B* **44**, 325.
Elyakov, G. B., Kuznetsova, T. A., Stonik, V. A., Levin, V. S., and Albores, R. (1975a). *Comp. Biochem. Physiol. B* **52**, 413.
Elyakov, G. B., Stonik, V. A., Levina, E. U., and Levin, V. S. (1975b). *Comp. Biochem. Physiol. B* **52,** 321.
Elyakov, G. B., Levina, E. V., Levin, V. S., and Kapustina, I. I. (1976). *Comp. Biochem. Physiol. B* **55**, 57.
Endre, E., and Canal, H. (1926). *C. R. Acad. Sci.* **183**, 152.
Englebrecht, J. P., Tursch, B., and Djerassi, C. (1972). *Steroids* **20**, 121.
Erdman, T. R., and Scheuer, P. J. (1975). *Lloydia* **38**, 359.
Erdman, T. R., and Thomson, R. H. (1972). *Tetrahedron* **28**, 5163.
Fagerlund, U. H. M. (1962). *Can. J. Biochem. Physiol.* **40**, 1839.
Fagerlund, U. H. M., and Idler, D. R. (1956). *J. Org. Chem.* **21**, 372.
Fagerlund, U. H. M., and Idler, D. R. (1957). *J. Am. Chem. Soc.* **79**, 6437.
Fagerlund, U. H. M., and Idler, D. R. (1959). *J. Am. Chem. Soc.* **81**, 401.
Fagerlund, U. H. M., and Idler, D. R. (1960). *Can. J. Biochem. Physiol.* **38**, 997.
Fagerlund, U. H. M., and Idler, D. R. (1961a). *Can. J. Biochem. Physiol.* **39**, 505.
Fagerlund, U. H. M., and Idler, D. R. (1961b). *Can J. Biochem. Physiol.* **39**, 1347.
Fange, R. (1963). *Sarsia* **10**, 19.
Fattorusso, E., Magno, S., Santacrose, C., and Sica, D. (1974). *Gazz. Chim. Ital.* **104**, 409.
Fattorusso, E., Magno, S., Mayol, L., Santacrose, C., and Sica, D. (1975a). *Tetrahedron* **31**, 1715.
Fattorusso, E., Magno, S., Mayol, L., Santacrose, C., and Sica, D. (1975b). *Gazz. Chim. Ital.* **105**, 635.
Fattorusso, E., Magno, S., Santacrose, C., Sica, D., Impellizzeri, G., Mangiafico, S., Oriente, G., Piattelli, M., and Scinto, S. (1975c). *Phytochemistry* **14**, 1579.
Fattorusso, E., Magno, S., Santacrose, C., Sica, D., Impellizzeri, G., Mangiafico, S., Piattelli, M., and Scinto, S. (1976). *Biochem. Syst. Ecol.* **4**, 135.
Faulkner, D. J., and Anderson, R. J. (1974). *In* "The Sea" (E. D. Goldberg, ed.), Vol. 5, p. 679. Wiley (Interscience), New York.
Faux, A., Horn, D. H. S., Middleton, E. J., Fales, H. M., and Lowe, M. E. (1969). *Chem. Commun.* p. 175.
Feder, H. M. (1972). *Sci. Am.* **227**, 93.
Feder, H. M., and Arvidsson, J. (1967). *Ark. Zool.* **19**, 369.
Feder, H. M., and Christensen, A. M. (1966). *In* "Physiology of Echinodermata" (R. A. Boolootian, ed.), p. 87, Wiley (Interscience), New York.
Feder, H. M., and Lasker, R. (1964). *Life Sci.* **3**, 1047.
Ferezou, J. P., Devys, M., and Barbier, M. (1972). *Experientia* **28**, 407.
Finer, J., Hirotsu, K., and Clardy, J. (1977). *In* "Marine Natural Products Chemistry" (D. J. Faulkner and W. H. Fenical, eds.), p. 147. Plenum, New York.
Fleming, R. (1975). Ph.D. Thesis, Univ. of Liverpool, Liverpool.
Fleming, W. J., Salathe, R., Wyllie, S. G., and Howden, M. E. H. (1976). *Comp. Biochem. Physiol. B* **53**, 267.
Flores, S. E., and Rosas, M. Y. (1966). *Bol. Inst. Oceanogr. Univ. Oriente* **5**, 116.
Frantz, I. D., and Schroepfer, G. J. (1967). *Annu. Rev. Biochem.* **36**, 691.
Frantz, I. D., Sanghvi, A. T., and Schroepfer, G. J. (1964). *J. Biol. Chem.* **239**, 1007.

Frey, D. G. (1951). *Copeia* p. 175.
Gaffney, J. (1974). Ph.D. Thesis, Univ. of Liverpool, Liverpool.
Gaffney, J., and Goad, L. J. (1973). *Biochem. J.* **138**, 309.
Gagosian, R. B. (1975). *Experientia* **31**, 878.
Gagosian, R. B. (1976). *Limnol. Oceanogr.* **21**, 702.
Gagosian, R. B., and Bourbonnierre, R. A. (1976). *Comp. Biochem. Physiol. B* **53**, 155.
Gagosian, R. B., Bourbonnierre, R. A., Smith, W. B., Couch, E. F., Blanton, C., and Novak, W. (1974). *Experientia* **30**, 723.
Galbraith, M. N., Horn, D. H. S., Middleton, E. J., and Hackney, R. J. (1968). *Chem. Commun.* p. 83.
Gaskell, S. J., and Eglinton, G. (1976). *Geochim. Cosmochim. Acta* **40**, 1221.
Gibbons, G. F., Goad, L. J., and Goodwin, T. W. (1967). *Phytochemistry* **6**, 677.
Gibbons, G. F., Goad, L. J., and Goodwin, T. W. (1968). *Phytochemistry* **7**, 983.
Gilgan, M. W., and Idler, D. R. (1967). *Gen. Comp. Endocrinol.* **9**, 319.
Gilgan, M. W., Pike, R. K., and ApSimon, J. W. (1976). *Comp. Biochem. Physiol. B* **54**, 561.
Goad, L. J. (1970). *In* "Natural Substances Formed Biologically from Mevalonic Acid" (T. W. Goodwin, ed.), p. 45. Academic Press, New York.
Goad, L. J. (1976). Unpublished observations.
Goad, L. J. (1975). *In* "The Biochemistry of Steroid Hormones" (H. L. J. Makin, ed.), p. 17. Blackwell, Oxford.
Goad, L. J. (1976). *In* "Biochemical and Biophysical Perspectives in Marine Biology" (D. C. Malins and J. R. Sargent, eds.), Vol. 3, p. 213. Academic Press, New York.
Goad, L. J. (1977). *In* "Lipids and Lipid Polymers in Higher Plants" (M. Tevini and H. K. Lichtenthaler, eds.), p. 146. Springer-Verlag, Berlin and New York.
Goad, L. J., and Goodwin, T. W. (1972). *Prog. Phytochem.* **3**, 113.
Goad, L. J., Rubinstein, I., and Smith, A. G. (1972a). *Proc. R. Soc., Ser. B* **180**, 223.
Goad, L. J., Knapp, F. F., Lenton, J. R., and Goodwin, T. W. (1972b). *Biochem. J.* **129**, 219.
Goad, L. J., Lenton, J. R., Knapp, F. F., and Goodwin, T. W. (1974). *Lipids* **9**, 582.
Goldsmith, L. A., and Carlson, G. P. (1976). *Food Drugs Sea, Proc. Conf., 4th, 1974* p. 354.
Goodfellow, R. M. (1974). Ph.D. Thesis, Univ. of Liverpool, Liverpool.
Goodfellow, R. M., and Goad, L. J. (1973). *Biochem. Soc. Trans.* **1**, 759.
Goodfellow, R. M., and Goad, L. J. (1974). Unpublished observations.
Goodwin. T. W. (1974). *Bot. Monogr.* **10**, 266.
Gosselin, L. (1965). *Arch. Int. Physiol. Biochim.* **73,** 543.
Gouseeva, N. N., Kandyak, R. P., and Nikolenko, I. A. (1973). *Biol. Morya* **30**, 134.
Grossert, J. S. (1972). *Chem. Soc. Rev.* **1**, 1.
Grossert, J. S., and Swee, T. (1977). *In* "Marine Natural Products Chemistry" (D. J. Faulkner and W. H. Fenical, eds.), Plenum, New York.
Grossert, J. S., Mathiaparanam, P., Hebb, G. D., Price, R., and Campbell, I. M. (1973). *Experientia* **29**, 258.
Grunwald, C. (1975). *Annu. Rev. Plant. Physiol.* **26**, 209.
Gupta, K. C., and Scheuer, P. J. (1968). *Tetrahedron* **24**, 5831.
Gupta, K. C., and Scheuer, P. J. (1969). *Steroids* **13**, 343.
Gurst, J. E., Sheikh, Y. M., and Djerassi, C. (1973). *J. Am. Chem. Soc.* **95**, 628.
Habermehl, G., and Christ, B. (1973). *Z. Naturforsch., Teil C* **28**, 225.
Habermehl, G., and Volkwein, G. (1968). *Naturwissenschaften* **55**, 83.
Habermehl, G., and Volkwein, G. (1970). *Justus Liebigs Ann. Chem.* **731**, 53.
Habermehl, G., and Volkwein, G. (1971). *Toxicon* **9**, 319.

Habermehl, G., Christ, B., and Krebs, H. C. (1976). *Naturwissenschaften* **63**, 42.
Hale, R. L., Leclerq, J., Tursch, B., Djerassi, C., Gross, R. A., Weinheimer, A. J., Gupta, K., and Scheuer, P. J. (1970). *J. Am. Chem. Soc.* **92**, 2179.
Hamilton, R. J., van den Heuvel, W. J. A., and Horning, E. C. (1964). *Experientia* **20**, 568.
Hashimoto, Y., and Yasumoto, T. (1960). *Nippon Suisan Gakkaishi* **26**, 1132.
Hatano, M. (1961). *Hokaido Daigaku Suisan Gakubu Kenkyu Iho Univ.* **11**, 218.
Hathaway, R. R. (1965). *Gen. Comp. Endocrinol.* **5**, 504.
Hathaway, R. R., and Black, R. E. (1969). *Gen. Comp. Endocrinol.* **12**, 1.
Hayashi, K., Yamada, M. (1973). *Nippon Suisan Gakkaishi* **39**, 809.
Hayashi, K., and Yamada, M. (1974). *Hokkaido Daigaku Suisan Gakubu Kenkyu Iho* **25**, 247.
Henze, M. (1904). *Hoppe-Seyler's Z. Physiol. Chem.* **41**, 109.
Henze, M. (1908). *Hoppe-Seyler's Z. Physiol. Chem.* **55**, 433.
Highnam, K. C. (1970). *Adv. Steroid Biochem. Pharmacol.* **1**, 1.
Horn, D. H. S. (1973). *In* "Naturally Occurring Insecticides" (M. Jacobson, ed.), p. 333 Dekker, New York.
Horn, D. H. S., Middleton, E. J., and Wunderlich, J. A. (1966). *Chem. Commun.* p. 339.
Horn, D. H. S., Fabbri, S. F., Hampshire, F., and Lowe, M. E. (1968). *Biochem. J.* **109**, 399.
Howden, M. E. H., Lucus, S. S., McDuffy, M., and Salathe, R. (1977). *Chem. Abst.* **86**, 2550d.
Hyman, L. H. (1955). "The Invertebrates: Echinodermata." McGraw-Hill, New York.
Idler, D. R., and Atkinson, B. (1976). *Comp. Biochem. Physiol. B.* **53**, 517.
Idler, D. R., and Fagerlund, U. H. M. (1955). *J. Am. Chem. Soc.* **77**, 4142.
Idler, D. R., and Fagerlund, U. H. M. (1957a). *Chem. Ind. (London)* p. 432.
Idler, D. R., and Fagerlund, U. H. M. (1957b). *J. Am. Chem. Soc.* **79**, 1988.
Idler, D. R., and Wiseman, P. (1968). *Comp. Biochem. Physiol.* **26**, 1113.
Idler, D. R., and Wiseman, P. (1970). *Comp. Biochem. Physiol.* **35**, 679.
Idler, D. R., and Wiseman, P. (1971a). *Int. J. Biochem.* **2**, 91.
Idler, D. R., and Wiseman, P. (1971b). *Comp. Biochem. Physiol. A* **38**, 581.
Idler, D. R., and Wiseman, P. (1971c). *Int. J. Biochem.* **2**, 516.
Idler, D. R., and Wiseman, P. (1972). *J. Fish. Res. Board Can.* **29**, 385.
Idler, D. R., Tamura, T., and Wanai, T. (1964). *J. Fish. Res. Board Can.* **21**, 1035.
Idler, D. R., Saito, A., and Wiseman, P. (1968). *Steroids* **11**, 465.
Idler, D. R., Sangalang, G. B., and Kanazawa, A. (1969). *Gen. Comp. Endocrinol.* **12**, 222.
Idler, D. R., Wiseman, P. M., and Safe, L. M. (1970). *Steroids* **16**, 451.
Idler, D. R., Safe, L. M., and MacDonald, E. F. (1971). *Steroids* **18**, 545.
Idler, D. R., Khalil, M. W., Gilbert, J. D., and Brooks, C. J. W. (1976). *Steroids* **27**, 155.
Ikegami, S. (1976). *J. Exp. Zool.* **198**, 359.
Ikegami, S., Tamura, S., and Kanatani, H. (1967). *Science* **158**, 1052.
Ikegami, S., Shirai, H., and Kamatani, H. (1971). *Dobutsugaku Zasshi* **80**, 26.
Ikegami, S., Kamiya, Y., and Tamura, S. (1972a). *Tetrahedron Lett.* p. 1601.
Ikegami, S., Kamiya, Y., and Tamura, S. (1972b). *Agric. Biol. Chem.* **36**, 1777.
Ikegami, S., Kamiya, Y., and Tamura, S. (1972c). *Tetrahedron Lett.* p. 3725.
Ikegami, S., Kamiya, Y., and Tamura, S. (1972d). *Agric. Biol. Chem.* **36**, 1087.
Ikegami, S., Horose, Y., Kamiya, Y., and Tamura, S. (1972e). *Agric. Biol. Chem.* **36**, 1843.
Ikegami, S., Hirose, Y., Kamiya, Y., and Tamura, S. (1972f). *Agric. Biol. Chem.* **36**, 2449.
Ikegami, S., Hirose, Y., Kamiya, Y., and Tamura, S. (1972g). *Agric. Biol. Chem.* **36**, 2453.
Ikegami, S., Kamiya, Y., and Tamura, S. (1973a). *Agric. Biol. Chem.* **37**, 367.

Ikegami, S., Kamiya, Y., and Tamura, S. (1973b). *Tetrahedron Lett.* p. 731.
Ikegami, S., Kamiya, Y., and Tamura, S. (1973c). *Tetrahedron* **29**, 1807.
Ikegami, S., Kamiya, Y., and Shirai, H. (1976). *Exp. Cell Res.* **103**, 233.
Ilin, S. G., Tarnopolskii, B. L., Safina, Z. I., Sobolev, A. N., Dzizenko, A. K., and Elyakov, G. B. (1976). *Dokl. Akad. Nauk SSSR* **230**, 860.
Johnson, T. W., and Sparrow, F. K. (1961). "Fungi in Oceans and Estuaries." Cramer, Weinheim.
Kamiya, Y., Ikegami, S., and Tamura, S. (1974). *Tetrahedron Lett.* p. 655.
Kanatani, H. (1973). *Int. Rev. Cytol.* **35**, 253.
Kanazawa, A., and Teshima, S. (1971a). *Nippon Suisan Gakkaishi* **37**, 891.
Kanazawa, A., and Teshima, S. (1971b). *Nippon Suisan Gakkaishi* **37**, 675.
Kanazawa, A., and Teshima, S. (1971c). *J. Ocean Soc. Jpn* **27**, 207.
Kanazawa, A., and Yoshioka, M. (1971a). *Proc. Int. Seaweed Symp., 7th* p. 502.
Kanazawa, A., and Yoshioka, M. (1971b). *Nippon Suisan Gakkaishi* **37**, 397.
Kanazawa, A., Shimaya, M., Kawasaki, M., and Kashiwada, K. I. (1970). *Nippon Suisan Gakkaishi* **36**, 949.
Kanazawa, A., Tanaka, N., Teshima, S., and Kashiwada, K. I. (1971a). *Nippon Suisan Gakkaishi* **37**, 211.
Kanazawa, A., Tanaka, N., Teshima, S., and Kashiwada, K. I. (1971b). *Nippon Suisan Gakkaishi* **37**, 1015.
Kanazawa, A., Yoshioka, M., and Teshima, S. (1971c). *Nippon Suisan Gakkaishi* **37**, 899.
Kanazawa, A., Yoshioka, M., and Teshima, S. (1972). *Kagoshima Daigaku Suisan Gakubu Kiyo* **21**, 103.
Kanazawa, A., Teshima, S., and Ando, T. (1973). *Kagoshima Daigaku Suisan Gakubu Kiyo* **22**, 21.
Kanazawa, A., Teshima, S., Tomita, S., and Ando, T. (1974a). *Nippon Suisan Gakkaishi* **40**, 1077.
Kanazawa, A., Teshima, S., and Tomita, S. (1974b). *Nippon Suisan Gakkaishi* **40**, 1257.
Kanazawa, A., Teshima, S., Ando, T., and Tomita, S. (1974c). *Nippon Suisan Gakkaishi* **40**, 729.
Kanazawa, A., Guary, J. C. B., and Ceccaldi, H. J. (1976a). *Comp. Biochem. Physiol. B* **54**, 205.
Kanazawa, A., Teshima, S., Ando, T., and Tomita, S. (1976b). *Mar. Biol.* **34**, 53.
Kanazawa, A., Ando, T., and Teshima, S. (1977a). *Nippon Suisan Gakkaishi* **43**, 83.
Kanazawa, A., Teshima, S., and Ando, T. (1977b). *Comp. Biochem. Physiol. B* **57**, 317.
Karlson, P. (1973). *In* "Invertebrate Endocrinology and Hormone Heterophylly" (W. J. Burdette, ed.), p. 43. Springer-Verlag, Berlin and New York.
Kelecom, A., Daloze, D., and Tursch, B. (1976a). *Tetrahedron* **32**, 2313.
Kelecom, A., Daloze, D., and Tursch, B. (1976b). *Tetrahedron* **32**, 2353.
Kelecom, A., Tursch, B., and Vanhaelen, M. (1976c). *Bull. Soc. Chim. Belg.* **85**, 277.
Khalil, M. W., and Idler, D. R. (1976). *Comp. Biochem. Physiol. B.* **55**, 239.
Kind, C. A., and Fasolino, E. M. (1945). *J. Org. Chem.* **10**, 286.
Kind, C. A., and Goldberg, M. H. (1953). *J. Org. Chem.* **18**, 203.
Kind, C. A., and Herman, S. C. (1948). *J. Org. Chem.* **13**, 867.
Kind, C. A., and Meigs, R. A. (1955). *J. Org. Chem.* **20**, 1116.
Kind, C. A., Salter, S. G., and Vinci, A. (1948). *J. Org. Chem.* **13**, 538.
King, D. S., and Siddall, J. B. (1969). *Nature(London)* 955.
Kirk, R. W., and Catalfomo, P. (1970). *Phytochemistry* **9**, 595.
Kita, M., and Toyama, Y. (1959). *Mem. Fac. Eng., Nagoya Univ.* **11**, 216.

Kita, M., and Toyama, Y. (1960). *Nippon Kagaku Zasshi* **81**, 485.
Kitagawa, I., and Kobayashi, M. (1977). *Tetrahedron Lett.* p. 859.
Kitagawa, I., Sugawara, T., and Yosioka, I. (1974). *Tetrahedron Lett.* p. 4111.
Kitagawa, I., Kobayashi, M., Sugawara, T., and Yasioka, I. (1975a). *Tetrahedron Lett.* p. 967.
Kitagawa, I., Sugawara, T., Yosioka, I., and Kuriyama, K. (1975b). *Tetrahedron Lett.* p. 963.
Kitagawa, I., Sugawara, T., and Yosioka, I. (1976a). *Chem. Pharm. Bull.* **24**, 275.
Kitagawa, I., Sugawara, T., Yosioka, I., and Kuriyama, K. (1976b). *Chem. Pharm. Bull.* **24**, 266.
Knights, B. A. (1970). *Phytochemistry* **9**, 903.
Kobayashi, M., and Mitsuhashi, H. (1974a). *Steroids* **24**, 399.
Kobayashi, M., and Mitsuhashi, H. (1974b). *Tetrahedron* **30**, 2147.
Kobayashi, M., and Mitsuhashi, H. (1975). *Steroids* **26**, 605.
Kobayashi, M., Tsuru, R., Todo, K., and Mitsuhashi, H. (1972). *Tetrahedron Lett.* p. 2935.
Kobayashi, M., Nishizawa, M., Todo, T., and Mitsuhashi, H. (1973a). *Chem. Pharm. Bull.* **21**, 323.
Kobayashi, M., Tsuru, R., Todo, K., and Mitsuhashi, H. (1973b). *Tetrahedron* **29**, 1193.
Kobayashi, M., Todo, K., and Mitsuhashi, H. (1974). *Chem. Pharm. Bull.* **22**, 236.
Kohlmeyer, J., and Kohlmeyer, E. (1971). *Mycologia* **63**, 831.
Kritchevsky, D., Tepper, S. A., Di Tullio, N. W., and Holmes, W. L. (1967). *J. Food Sci.* **32**, 64.
Lachaise, F., and Feyereisen, R. (1976). *C. R. Acad. Sci., Ser. D* **283**, 1445.
Lachaise, F., Lagueax, M., Feyereisen, R., and Hoffmann, J. A. (1976). *C. R. Acad. Sci., Ser. D* **283**, 943.
Lee, R. F., Nevenzel, J. C., and Lewis, A. G. (1974). *Lipids* **9**, 891.
Lehoux, J. G., and Sandor, T. (1970). *Steroids* **16**, 141.
Lehoux, J. G., Lusis, O., and Sandor, T. (1967). *Can. Med. Assoc. J.* **96**, 352.
Leulier, A., and Policard, A. (1930). *Seances Soc. Biol. Paris* **103**, 82.
Lévi, C., and Lévi, P. (1965). *J. Microsc. (Paris)* **4**, 60.
Levin, V. S., and Stonik, V. A. (1976). *Biol. Morya* p. 73. [*Chem. Abstr.* **85**, 75292K (1976).]
Levina, E. V., and Kapustina, I. I. (1975). *Tezisy Dokl.-Uses. Simp. Bioorg. Khim.* p. 23. [*Chem. Abstr.* **85**, 90481q (1976).]
Ling, N. C., Hale, R. C., and Djerassi, C. (1970). *J. Am. Chem. Soc.* **92**, 5281.
Lisk, R. D. (1961). *Can. J. Biochem. Physiol.* **39**, 659.
Longcamp, D., Drosdowsky, M., and Lubet, P. (1970). *C. R. Acad. Sci., Ser. D* **271**, 1564.
Longcamp, D., Lubet, P., and Drosdowsky, M. (1974). *Gen. Comp. Endocrinol.* **22**, 116.
Low, E. M. (1955). *J. Mar. Res.* **14**, 199.
Lupo di Prisco, C., and Dessi'Fulgheri, F. (1974). *Gen. Comp. Endocrinol.* **22**, 340.
Lupo di Prisco, C., and Dessi'Fulgheri, F. (1975). *Comp. Biochem. Physiol.* **50**, 191.
Lupo di Prisco, C., Dessi'Fulgheri, F., and Tomasucci, M. (1973). *Comp. Biochem. Physiol. B* **45**, 303.
MacDonald, E. S., and Wilber, C. G. (1955). *Biol. Bull. (Woods Hole, Mass.)* **109**, 363.
Mackie, A. M., and Turner, A. B. (1970). *Biochem. J.* **117**, 543.
Mackie, A. M., Lasker, R., and Grant, R. T. (1968). *Comp. Biochem. Physiol.* **26**, 415.
Mackie, A. M., Singh, H. T., and Fletcher, T. C. (1975). *Mar. Biol.* **29**, 307.
Mackie, A. M., Singh, H. T., and Owen, J. M. (1977). *Comp. Biochem. Physiol. B* **56**, 9.
Madri, P. P., Claus, C., Kunen, S. M., and Moss, E. E. (1967). *Life Sci.* **6**, 889.
Makin, H. L. J. (1975). "The Biochemistry of Steroid Hormones." Blackwell, Oxford.

Marsden, J. R. (1976). *Comp. Biochem. Physiol. B* **53**, 225.
Marumo, H., Nakajima, H., and Tomiyama, S. (1955). *Jpn. Patent 979*. [*Chem. Abstr.* **50**, 17490 (1956).]
Masada, Y., Hashimoto, K., Inoue, T., Fujioka, M., and Nagata, S. (1973). *Shitsuryo Bunseki* **21**, 109.
Mason, W. T. (1972). *Biochim. Biophys. Acta* **280**, 538.
Matsumoto, T., and Tamura, T. (1955a). *Nippon Kagaku Zasshi* **76**, 951.
Matsumoto, T., and Tamura, T. (1955b). *Nippon Kagaku Zasshi* **76**, 1413.
Matsumoto, T., and Toyama, Y. (1944). *Nippon Kagaku Zasshi* **65**, 258.
Matsuno, T., and Ishida, T. (1969). *Experientia* **25**, 1261.
Matsuno, T., Nagata, S., and Mizutani, K. (1972a). *Nippon Suisan Gakkaishi* **38**, 144.
Matsuno, T., Nagata, S., and Hashimoto, K. (1972b). *Nippon Suisan Gakkaishi* **38**, 1261.
Middlebrook, R. E., and Lane, C. E. (1968). *Comp. Biochem. Physiol.* **24**, 507.
Minale, L. (1976). *Pure Appl. Chem.* **48**, 7.
Minale, L. (1977). *In* "Marine Natural Products" (P. J. Scheuer, ed.), Vol. 1, Academic Press, New York.
Minale, L., and Sodano, G. (1974a). *J. Chem. Soc., Perkin Trans. 1* p. 1888.
Minale, L., and Sodano, G. (1974b). *J. Chem. Soc., Perkin Trans. 1* p. 2380.
Minale, L., and Sodano, G. (1977). *In* "Marine Natural Products Chemistry" (D. J. Faulkner and W. H. Fenical, eds.), p. 87. Plenum, New York.
Minale, L., Cimino, G., de Stefano, S., and Sodano, G. (1976). *Fortschr. Chem. Org. Naturst.* **33**, 1.
Moldowan, J. M., Tursch, B. M., and Djerassi, C. (1974). *Steroids* **24**, 387.
Moldowan, J. M., Tan, W. L., and Djerassi, C. (1975). *Steroids* **26**, 107.
Montgomery, D. H. (1967). *Veliger* **9**, 359.
Morgan, E. D., and Poole, C. F. (1977). *Comp. Biochem. Physiol. B* **57**, 99.
Morisaki, M., Ohtaka, H., Okubayashi, M., Ikekawa, N., Horie, Y., and Nakasone, S. (1972). *J. Chem. Soc., Chem. Commun.* p. 1275.
Morisaki, M., Kidooka, S., and Ikekawa, N. (1976). *Chem. Pharm. Bull.* **24**, 3214.
Mulheirn, L. J. (1973). *Tetrahedron Lett.* p. 3175.
Mulheirn, L. J., and Ramm, P. J. (1972). *Chem. Soc. Rev.* **1**, 259.
Muscatine, L. (1974). *In* "Coelenterate Biology" (L. Muscatine and H. M. Lenhoff, eds.), p. 359. Academic Press, New York.
Nes, W. R., Krevitz, K., and Behzadan, S. (1976). *Lipids* **11**, 118.
Nicholson, S. H., and Turner, A. B. (1976). *J. Chem. Soc., Perkin Trans. 1* p. 1357.
Nigrelli, E. F. (1952). *Zoologica* **37**, 89.
Nomura, T., Tsuchiya, Y., Andre, D., and Barbier, M. (1969a). *Nippon Suisan Gakkaishi* **35**, 293.
Nomura, T., Tsuchiya, Y., Andre, D., and Barbier, M. (1969b). *Nippon Suisan Gakkaishi* **35**, 299.
Nomura, T., Itoh, M., Viala, J., Alcaide, A., and Barbier, M. (1972). *Nippon Suisan Gakkaishi* **38**, 1365.
Oppenheimer, C. H. (1963). "Symposium on Marine Microbiology." Thomas, Springfield, Illinois.
Orcutt, D. M., and Patterson, G. W. (1975). *Comp. Biochem. Physiol. B* **50**, 579.
Orlando Munoz, M., Atkinson, B., and Idler, D. R. (1976). *Comp. Biochem. Physiol. B* **54**, 231.
Owellen, R. J., Owellen, R. G., Gorog, M. A., and Klein, D. (1973). *Toxicon* **11**, 319.
Patterson, G. W. (1971). *Lipids* **6**, 120.

Patterson, G. W. (1974). *Comp. Biochem. Physiol. B* **47**, 453.
Pennock, J. F. (1977). *In* "Biochemistry of Lipids II," International Review of Biochemistry Vol. 14. (T. W. Goodwin, ed.), p. 153. Univ. Park Press, Baltimore, Maryland.
Petering, H. G., and Waddell, J. (1951). *J. Biol. Chem.* **191**, 765.
Pettler, P. J., Lockley, W. J. S., Rees, H. H., and Goodwin, T. W. (1974). *J. Chem. Soc., Chem. Commun.* p. 844.
Piretti, M. V., and Viviani, R. (1976). *Comp. Biochem. Physiol. B* **55**, 229.
Popov, S., Carlson, R. M. K., Wegmann, A. M., and Djerassi, C. (1976a). *Steroids* **28**, 699.
Popov, S., Carlson, R. M. K., Wegmann, A. M., and Djerassi, C. (1976b). *Tetrahedron Lett.* p. 3491.
Premuzic, E. (1971). *Fortschr. Chem. Org. Naturst.* **29**, 417.
Rees, H. H. (1971). In "Aspects of Terpenoid Chemistry and Biochemistry" (T. W. Goodwin, ed.), p. 181. Academic Press, New York.
Reiswig, H. M. (1971). *Biol. Bull. (Woods Hole, Mass.)* **141**, 568.
Rio, G. J., Ruggieri, G. D., Stempien, M. F., and Nigrelli, R. F. (1963). *Am. Zool.* **3**, 554.
Rodwell, V. W., McNamara, D. J., and Shapiro, D. J. (1973). *Adv. Enzymol. Relat. Areas Mol. Biol.* **38**, 373.
Roller, P., Djerassi, C., Cloetens, R., and Tursch, B. (1969). *J. Am. Chem. Soc.* **91**, 4918.
Roller, P., Tursch, B., and Djerassi, C. (1970). *J. Org. Chem.* **35**, 2585.
Roth, F. J., Ahearn, D. G., Fell, J. W., Meyers, S. P., and Meyer, S. A. (1962). *Limnol. Oceanogr.* **7**, 178.
Rothberg, I., Tursch, B. M., and Djerassi, C. (1973). *J. Org. Chem.* **38**, 209.
Rubinstein, I. (1973). Ph.D. Thesis, Univ. of Liverpool, Liverpool.
Rubinstein, I., and Goad, L. J. (1974a). *Phytochemistry* **13**, 481.
Rubinstein, I., and Goad, L. J. (1974b). *Phytochemistry* **13**, 485.
Rubinstein, I., and Goad, L. J. (1973). Unpublished observations.
Rubinstein, I., Goad, L. J., Clague, A. D. H., and Mulheirn, L. J. (1976). *Phytochemistry* **15**, 195.
Salaque, A., Barbier, M., and Lederer, E. (1966). *Comp. Biochem. Physiol.* **19**, 45.
Saliot, A., and Barbier, M. (1971). *Biochimie* **53**, 265.
Saliot, A., and Barbier, M. (1973). *J. Exp. Mar. Biol. Ecol.* **13**, 207.
Scheuer, P. J. (1969). *Fortschr. Chem. Org. Naturst.* **27**, 322.
Scheuer, P. J. (1973). "Chemistry of Marine Natural Products." Academic Press, New York.
Schmitz, F. J. (1977). *In* "Marine Natural Products" (P. J. Scheuer, ed.), Vol. 1, Academic Press, New York.
Schmitz, F. J., and Pattabhiraman, T. (1970). *J. Am. Chem. Soc.* **92**, 6073.
Schmitz, F. J., Campbell, D. C., and Kubo, I. (1976). *Steroids* **28**, 211.
Sheikh, Y. M., and Djerassi, C. (1973). *Tetrahedron Lett.* p. 2927.
Sheikh, Y. M., and Djerassi, C. (1974). *Tetrahedron* **20**, 4095.
Sheikh, Y. M., and Djerassi, C. (1976). *J. Chem. Soc., Chem. Commun.* p. 1057.
Sheikh, Y. M., Djerassi, C., and Tursch, B. M. (1971). *Chem. Commun.* p. 217.
Sheikh, Y. M., Tursch, B. M., and Djerassi, C. (1972a). *J. Am. Chem. Soc.* **94**, 3278.
Sheikh, Y. M., Tursch, B., and Djerassi, C. (1972b). *Tetrahedron Lett.* p. 3721.
Sheikh, Y. M., Kaisin, M., and Djerassi, C. (1973). *Steroids* **22**, 835.
Shimada, S. (1969). *Science* **163**, 1462.
Shimadate, T., Rosenstein, F. U., and Kircher, H. W. (1977). *Lipids* **12**, 241.
Shimizu, Y. (1971). *Experientia* **27**, 1188.
Shimizu, Y. (1972). *J. Am. Chem. Soc.* **94**, 4051.
Shimizu, Y., Alain, M., and Kobayashi, A. (1976). *J. Am. Chem. Soc.* **98**, 1059.

Shimizu, Y., Alain, M., Oshima, Y., Buckley, L. J., Fallon, W. E. Kasai, H., Miura, I., Gullo, V. P., and Nakanishi, K. (1977). *In* "Marine Natural Products Chemistry" (D. J. Faulkner and W. H. Fenical, eds.), p. 261. Plenum, New York.
Smith, A. G., and Goad, L. J. (1971a). *Biochem. J.* **123**, 671.
Smith, A. G., and Goad, L. J. (1971b). *FEBS Lett.* **12**, 233.
Smith, A. G., and Goad, L. J. (1974). *Biochem. J.* **142**, 421.
Smith, A. G., and Goad, L. J. (1975a). *Biochem. J.* **146**, 25.
Smith, A. G., and Goad, L. J. (1975b). *Biochem. J.* **146**, 35.
Smith, A. G., Goodfellow, R., and Goad, L. J. (1972). *Biochem. J.* **128**, 1371.
Smith, A. G., Rubinstein, I., and Goad, L. J. (1973). *Biochem. J.* **135**, 443.
Smith, D. S. H., and Turner, A. B. (1972). *Tetrahedron Lett.* p. 5263.
Smith, D. S. H., and Turner, A. B. (1975). *J. Chem. Soc., Perkin Trans. 1* p. 1751.
Smith, D. S. H., Turner, A. B., Mackie, A. M. (1973). *J. Chem. Soc., Perkin Trans. 1* p. 1745.
Spaziani, E., and Kater, S. B. (1973). *Gen. Comp. Endocrinol.* **20**, 534.
Steudler, P. A., Schmitz, F. J., and Ciereszko, L. J. (1977). *Comp. Biochem. Physiol. B* **56**, 385.
Sucrow, W., Slopianka, M., and Kircher, W. (1976). *Phytochemistry* **15**, 1533.
Svoboda, J. A., Hutchins, R. F. N., Thompson, M. J., and Robbins, W. E. (1969). *Steroids* **14**, 469.
Svoboda, J. A., Thompson, M. J. and Robbins, W. E. (1971). *Nature (London)* **230**, 57.
Tait, A. D. (1972). *Biochem. J.* **128**, 467.
Tait, A. D. (1973). *Steroids* **22**, 239.
Takagi, T., and Toyama, Y. (1956a). *Mem. Fac. Eng., Nagoya Univ.* **8**, 164.
Takagi, T., and Toyama, Y. (1956b). *Mem. Fac. Eng., Nagoya Univ.* **8**, 177.
Takagi, T., and Toyama, Y. (1958). *Mem. Fac. Eng., Nagoya Univ.* **10**, 84.
Takagi, T., Maeda, T., and Toyama, Y. (1956). *Mem. Fac. Eng., Nagoya Univ.* **8**, 169.
Tamura, T., Kokuma, K., and Matsumoto, T. (1956). *Nippon Kagaku Zasshi* **77**, 987.
Tamura, T., Truscott, B., and Idler, D. R. (1964a). *J. Fish. Res. Board Can.* **21**, 1519.
Tamura, T., Wainai, T., Truscott, B., and Idler, D. R. (1964b). *Can. J. Biochem.* **42**, 1331.
Tan, W. J., Djerassi, C., Fayos, J., and Clardy, J. (1975). *J. Org. Chem.* **40**, 466.
Tanaka, T., and Toyama, Y. (1957a). *Mem. Fac. Eng., Nagoya Univ.* **9**, 116.
Tanaka, T., and Toyama, Y. (1957b). *Mem. Fac. Eng., Nagoya Univ.* **9**, 122.
Tanaka, T., and Toyama, Y. (1959). *Mem. Fac. Eng., Nagoya Univ.* **11**, 204.
Taylor, D. L. (1969). *J. Cell Sci.* **4**, 751.
Tcholakian, R. K., and Eik-Nes, K. B. (1969). *Gen. Comp. Endocrinol.* **12**, 171.
Tcholakian, R. K., and Eik-Nes, K. B. (1971). *Gen. Comp. Endocrinol.* **17**, 115.
Teshima, S. (1971a). *Comp. Biochem. Physiol. B* **39**, 815.
Teshima, S. (1971b). *Nippon Suisan Gakkaishi* **37**, 671.
Teshima, S. (1972). *Kagoshima Daigaku Suisan Gakubu Kiyo* **21**, 69.
Teshima, S., and Goad, L. J. (1975). Unpublished observations.
Teshima, S., and Kanazawa, A. (1971a). *Nippon Suisan Gakkaishi* **37**, 68.
Teshima, S., and Kanazawa, A. (1971b). *Nippon Suisan Gakkaishi* **37**, 63.
Teshima, S., and Kanazawa, A. (1971c). *Nippon Suisan Gakkaishi* **37**, 720.
Teshima, S., and Kanazawa, A. (1971d). *Comp. Biochem. Physiol. B* **38**, 597.
Teshima, S., and Kanazawa, A. (1971e). *Comp. Biochem. Physiol. B* **38**, 603.
Teshima, S., and Kanazawa, A. (1971f). *Nippon Suisan Gakkaishi* **37**, 524.
Teshima, S., and Kanazawa, A. (1971g). *Gen. Comp. Endocrinol.* **17**, 152.
Teshima, S., and Kanazawa, A. (1971h). *Nippon Suisan Gakkaishi* **37**, 529.
Teshima, S., and Kanazawa, A. (1972a). *Nippon Suisan Gakkaishi* **38**, 1305.

Teshima, S., and Kanazawa, A. (1972b). *Nippon Suisan Gakkaishi* **38**, 1299.
Teshima, S., and Kanazawa, A. (1973a). *Nippon Suisan Gakkaishi* **39**, 1309.
Teshima, S., and Kanazawa, A. (1973b). *Kagoshima Daigaku Suisan Gakubu Kiyo* **22**, 15.
Teshima, S., and Kanazawa, A. (1973c). *Comp. Biochem. Physiol. B* **44**, 881.
Teshima, S., and Kanazawa, A. (1974). *Comp. Biochem. Physiol. B* **47**, 555.
Teshima, S., and Kanazawa, A. (1975). *Comp. Biochem. Physiol. B* **52**, 437.
Teshima, S., and Kanazawa, A. (1976). *Nippon Suisan Gakkaishi* **39**, 813.
Teshima, S., Kanazawa, A., and Ando, T. (1971). *Kagoshima Daigaku Suisan Gakubu Kiyo* **20**, 131.
Teshima, S., Kanazawa, A., and Ando, T. (1972). *Comp. Biochem. Physiol. B* **41**, 121.
Teshima, S., Kanazawa, A., and Ando, T. (1974a). *Kagoshima Daigaku Suisan Gakubu Kiyo* **23**, 105.
Teshima, S., Kanazawa, A., and Ando, T. (1974b). *Comp. Biochem. Physiol. B* **47**, 507.
Teshima, S., Kanazawa, A., and Ando, T. (1974c). *Nippon Suisan Gakkaishi* **40**, 63.
Teshima, S., Kanazawa, A., and Haruhito, O. (1974d). *Nippon Suisan Gakkaishi* **40**, 1015.
Teshima, S., Ceccaldi, H., Patrois, J., and Kanazawa, A. (1975). *Comp. Biochem. Physiol. B* **50**, 485.
Teshima, S., Kanazawa, A., and Ando, T. (1976a). *Nippon Suisan Gakkaishi* **42**, 997.
Teshima, S., Kanazawa, A., and Okamoto, H. (1976b). *Nippon Suisan Gakkaishi* **42**, 1273.
Teshima, S., Fleming, R., Gaffney, J., and Goad, L. J. (1977). *In* "Marine Natural Products Chemistry" (D. J. Faulkner and W. H. Fenical, eds.), p. 133. Plenum, New York.
Thomas, E. G., and Gryffydd, Ll. D. (1971). *Mar. Biol.* **10**, 87.
Thompson, M. H. (1964). *Fish. Ind. Res.* **2**, 11.
Thompson, M. J., Dutky, S. R., Patterson, G. W., and Gouden, E. L. (1972). *Phytochemistry* **11**, 1781.
Toyama, Y. (1958). *Fette, Seifen, Anstrichm.* **60**, 909.
Toyama, Y., and Shibano, F. (1943). *Nippon Kagaku Zasshi* **64**, 322.
Toyama, Y., and Tanaka, T. (1953). *Bull. Chem. Soc. Jpn.* **26**, 497.
Toyama, Y., and Tanaka, T. (1954). *Bull. Chem. Soc. Jpn.* **27**, 264.
Toyama, Y., and Tanaka, T. (1956a). *Mem. Fac. Eng., Nagoya Univ.* **8**, 29.
Toyama, T., and Tanaka, T. (1956b). *Mem. Fac. Eng., Nagoya Univ.* **8**, 40.
Toyama, Y., Takagi, T., and Tanaka, T. (1955a). *Mem. Fac. Eng., Nagoya Univ.* **7**, 1.
Toyama, Y., Tanaka, T., and Maeda, T. (1955b). *Mem. Fac. Eng., Nagoya Univ.* **7**, 145.
Trench, R. K. (1969). *Nature (London)* **222**, 1071.
Trench, R. K. (1971a). *Proc. R. Soc., Ser. B* **177**, 225.
Trench, R. K. (1971b). *Proc. R. Soc., Ser. B* **177**, 237.
Trench, R. K. (1971c). *Proc. R. Soc., Ser. B* **177**, 251.
Tsuda, K., and Sakai, K. (1960). *Chem. Pharm. Bull.* **8**, 554.
Tsuda, K., Sakai, K., Tanabe, K., and Kishida, Y. (1960). *J. Am. Chem. Soc.* **82**, 1442.
Tsujimoto, M., and Koyanagi, H. (1934a). *J. Soc. Chem. Ind. Jpn. Suppl.* **37**, 81B.
Tsujimoto, M., and Koyanagi, H. (1934b). *J. Soc. Chem. Ind. Jpn. Suppl.* **37**, 85B.
Tsujimoto, M., and Koyanagi, H. (1934c). *J. Soc. Chem. Ind. Jpn. Suppl.* **37**, 436B.
Turner, A. B., Smith, D. S. H., and Mackie, A. M. (1971). *Nature (London)* **233**, 209.
Tursch, B., de Souza Guinaraes, I. S., Gilbert, B., Aplin, R. T., Duffield, A. M., and Djerassi, C. (1967). *Tetrahedron* **23**, 761.
Tursch, B., Cloetens, R., and Djerassi, C. (1970). *Tetrahedron Lett.* p. 467.
Tursch, B., Hootele, C., Kaisin, M., Losman, D., and Karlsson, R. (1976). *Steroids* **27**, 137.
Vacelet, J. (1967). *J. Microsc. (Paris)* **6**, 237.
van Aarem, H. E., Vonk, H. J., and Zandee, D. I. (1964). *Arch. Int. Physiol. Biochim.* **72**, 606.

Van den Oord, A. H. A. (1964). *Comp. Biochem. Physiol.* **13**, 461.
Vanderah, D. J., and Djerassi, C. (1977). *Tetrahedron Lett.* p. 683.
van der Vliet, J. (1948). *Recl. Trav. Chim. Pays-Bas* **67**, 265.
Viala, J., Devys, M., and Barbier, M. (1972). *Bull. Soc. Chim. Fr.* p. 3626.
Voogt, P. A. (1967a). *Arch. Int. Physiol. Biochim.* **75**, 492.
Voogt, P. A. (1967b). *Arch. Int. Physiol. Biochim.* **75**, 809.
Voogt, P. A. (1968). *Arch. Int. Physiol. Biochim.* **76**, 721.
Voogt, P. A. (1969). *Comp. Biochem. Physiol.* **31**, 37.
Voogt, P. A. (1971a). *Arch. Int. Physiol. Biochim.* **79**, 391.
Voogt, P. A. (1971b). *Comp. Biochem. Physiol. B* **39**, 139.
Voogt, P. A. (1972a). *In* "Chemical Zoology" (M. Florkin and B. T. Scheer, eds.), Vol. 4, p. 245. Academic Press, New York.
Voogt, P. A. (1972b). *Neth. J. Zool.* **22**, 489.
Voogt, P. A. (1972c). *Comp. Biochem. Physiol. B* **41**, 831.
Voogt, P. A. (1972d). *Neth. J. Zool.* **22**, 59.
Voogt, P. A. (1972e). *Arch. Int. Physiol. Biochim.* **80**, 883.
Voogt, P. A. (1972f). *Comp. Biochem. Physiol. B* **43**, 457.
Voogt, P. A. (1973a). *Arch. Int. Physiol. Biochim.* **81**, 871.
Voogt, P. A. (1973b). *Arch. Int. Physiol. Biochim.* **81**, 401.
Voogt, P. A. (1973c). *Int. J. Biochem.* **4**, 479.
Voogt, P. A. (1973d). *Int. J. Biochem.* **4**, 42.
Voogt, P. A. (1973e). *Comp. Biochem. Physiol. B* **45**, 593.
Voogt, P. A. (1974). *Neth. J. Zool.* **24**, 22.
Voogt, P. A. (1975a). *Comp. Biochem. Physiol. B* **50**, 499.
Voogt, P. A. (1975b). *Comp. Biochem. Physiol. B* **50**, 505.
Voogt, P. A. (1976). *Neth. J. Zool.* **26**, 84.
Voogt, P. A., and Over, J. (1973). *Comp. Biochem. Physiol. B* **45**, 71.
Voogt, P. A., and Schoenmakers, H. J. (1973). *Comp. Biochem. Physiol. B* **45**, 509.
Voogt, P.A., and Van der Horst, D. J. (1972). *Arch. Int. Physiol. Biochim.* **80**, 293.
Voogt, P. A., and Van Rheenen, J. W. A. (1973a). *Experientia* **29**, 1070.
Voogt, P. A., and Van Rheenen, J. W. A. (1973b). *Neth. J. Sea Res.* **6**, 409.
Voogt, P. A., and Van Rheenen, J. W. A. (1974). *Comp. Biochem. Physiol. B* **47**, 131.
Voogt, P. A., and Van Rheenen, J. W. A. (1975). *Arch. Int. Physiol. Biochim.* **83**, 563.
Voogt, P. A., and Van Rheenen, J. W. A. (1976a). *Comp. Biochem. Physiol. B* **54**, 473.
Voogt, P. A., and Van Rheenen, J. W. A. (1976b). *Comp. Biochem. Physiol. B* **54**, 479.
Voogt, P. A., Van de Ruit, J. M., and Van Rheenen, J. W. A. (1974). *Comp. Biochem. Physiol. B* **48**, 47.
Wainai, T., Tamura, T., Truscott, B., and Idler, D. R. (1964). *J. Fish Res. Board Can.* **21**, 1543.
Walton, M. J., and Pennock, J. F. (1972). *Biochem. J.* **127**, 471.
Whitney, J. O. (1967). Cited in Idler and Wiseman (1971a).
Whitney, J. O. (1969). *Mar. Biol.* **3**, 134.
Wilbur, K. M. (1972). *In* "Chemical Zoology" (M. Florkin and B. T. Scheer, eds.), Vol. 7, p. 103. Academic Press, New York.
Willig, A. (1974). *Fortschr. Zool.* **22**, 55.
Willig, A., and Keller, R. (1976). *Experientia* **32**, 936.
Withers, N. W., Tuttle, R. C., Holz, G. G., Beach, D. H., Goad, L. J., and Goodwin, T. W. (1978). *Phytochemistry.* In press.
Wootton, J. M., and Wright, L. D. (1962). *Comp. Biochem. Physiol.* **5**, 253.
Yamanouchi, T. (1955). *Publ. Seto Mar. Biol. Lab.* **4**, 184.

Yasuda, S. (1966). *Yukagaku* **15**, 50.
Yasuda, S. (1970). *Yukagaku* **19**, 1014.
Yasuda, S. (1973). *Comp. Biochem. Physiol. B* **44**, 41.
Yasuda, S. (1974a). *Comp. Biochem. Physiol. B* **48**, 225.
Yasuda, S. (1974b). *Comp. Biochem. Physiol. B* **49**, 361.
Yasuda, S. (1975). *Comp. Biochem. Physiol. B* **50**, 399.
Yasuda, S. (1976). *Nippon Suisan Gakkaishi* **42**, 1307.
Yasumoto, T., and Hashimoto, Y. (1965). *Agric. Biol. Chem.* **29**, 804.
Yasumoto, T., and Hashimoto, Y. (1967). *Agric. Biol. Chem.* **31**, 368.
Yasumoto, T., Watanabe, T., and Hashimoto, Y. (1964). *Nippon Suisan Gakkaishi* **30**, 357.
Yasumoto, T., Tanaka, M., and Hashimoto, Y. (1966). *Nippon Suisan Gakkaishi* **32**, 673.
Yasumoto, T., Nakamura, K., and Hashimoto, Y. (1967). *Agric. Biol. Chem.* **31**, 7.
Yonge, C. M. (1953). *Proc. Zool. Soc. London* **123**, 551.
Yoshizawa, T., and Nagai, Y. (1974). *Jpn. J. Exp. Med.* **44**, 465.
Zagalsky, P. F., Cheesman, D. F., and Ceccaldi, H. J. (1967). *Comp. Biochem. Physiol.* **22**, 851.
Zandee, D. I. (1962). *Nature (London)* **195**, 814.
Zandee, D. I. (1964). *Nature (London)* **202**, 1335.
Zandee, D. I. (1966). *Arch. Int. Physiol. Biochim.* **74**, 435.
Zandee, D. I. (1967a). *Comp. Biochem. Physiol.* **20**, 811.
Zandee, D. I. (1967b). *Arch. Int. Physiol. Biochim.* **75**, 487.

Chapter 3

Diterpenoids

WILLIAM FENICAL

MARINE NATURAL PRODUCTS

ISBN 0-12-624002-7

I. INTRODUCTION

The diterpenoids are C_{20} compounds which, if carefully dissected, can be recognized to consist of four C_5 isoprene units (2-methyl-1,3-butadiene) linked in a symmetrical head-to-tail fashion. This configuration, whose consistency throughout nature forms the basis of the "biogenetic isoprene rule" (Ruzicka, 1959), led to the original proposal that the acyclic diterpenoid geranylgeraniol (**1**) is, as its pyrophosphate,

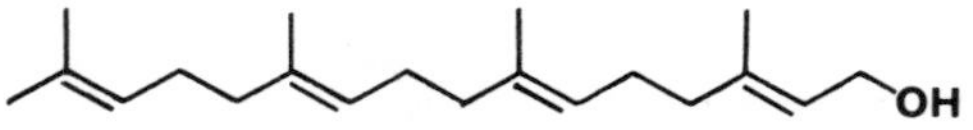

1 Geranylgeraniol

the precursor for a host of more complex diterpenoids (Ruzicka *et al.*, 1953). Only recently, however, has geranylgeraniol been found in nature (Kinzer *et al.*, 1966) and its biological conversion to complex diterpenoids elucidated by ^{14}C- and ^{13}C-labeling studies. Complete references to biogenetic studies and structure citations for many of the diterpenoids can be found in the excellent reviews of Nakanishi *et al.* (1974), Devon and Scott (1972), Hanson (1971), and McCrindle and Overton (1969).

When one compares the approximately 90 marine diterpenoids with the hundreds of diterpenoid structures reported from terrestrial sources, some interesting conclusions can be drawn. With some notable exceptions, a large number of the terrestrial diterpenoids (probably a majority) are derived by a proton-induced cyclization of all-*trans*-geranylgeraniol pyrophosphate (**2**), which is in a folded, all-chair conformation. This cyclization proceeds in a concerted trans–anti-trans fashion, generating the intermediate cation **3** with trans ring juncture and cis side chain stereochemistry (Scott *et al.*, 1964). Subsequent modification of this cation leads to the production of at least 30 isolated diterpenoid skeletons (Devon and Scott, 1972).

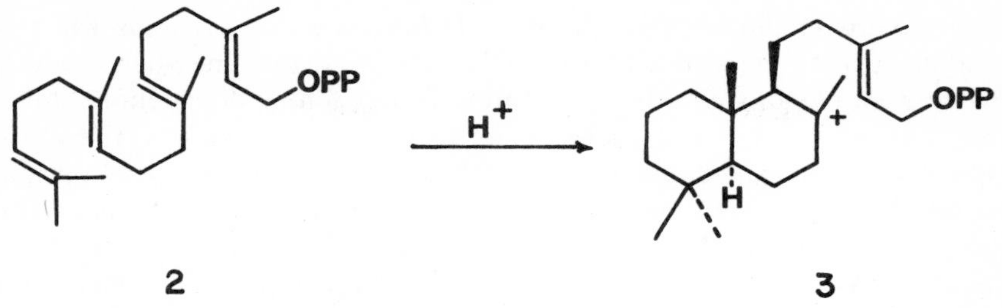

In contrast, only two groups of marine diterpenoids, the bromolabdanes (Section VI) and the sponge-derived isoagathic acid derivatives (Section V), can be rationalized as products of the same general biogenetic scheme. More often, marine diterpenoids are acyclic modifications or derivatives, such as the catabolic products reported in Section VII and the biogenetically mixed examples in Section VIII, or they are more complex cyclic systems produced by mechanisms of cyclization not previously encountered in nature. The high probability of finding new diterpene skeletons, coupled with the potential for the development of new biologically active compounds, renders the investigation of marine diterpenoids a lucrative field.

In reviewing the marine-derived diterpenoids, I have adopted the opinion that a significant number of readers will utilize this chapter as a reference tool to complement active research. Therefore, I have attempted to compile as complete a list of diterpenoids as possible, including numerous examples that have been communicated to me before publication. For each compound I have summarized, when available, data on pertinent physical and spectral properties, which should allow for rapid comparisons of these diterpenoids with newly isolated compounds.

My basic approach is to present the marine diterpenoids in order of their increasing complexity, beginning with the acyclic examples and progressing through monocyclic and bicyclic compounds to more complex polycyclic systems. The bromoditerpenoids have been separated and discussed independently because of their unique biogenesis. Sections on the catabolic products of diterpenoids and those of mixed biogenetic origin have also been included since they provide insight into the fate of diterpenoids in secondary metabolism. Rather than commenting extensively on the biological aspects of this research, I have organized this chapter to emphasize the derivation of marine diterpenoids from acyclic precursors and have pointed out some logical but untested biogenetic schemes.

The diterpenoids were, until recently, considered rare components of the marine environment. In "Chemistry of Marine Natural Products," published in 1973, only four examples of this prominent class of compounds were included. The literature today, however, reveals the tremendous growth in marine diterpenoid research, with somewhere over 90 new compounds described, representing 20 or so new diterpenoid skeletons. The origins of these substances in the sea are largely the more tropical seaweeds and coelenterates of the order Alcyonacea (sea fans and soft corals), with a few examples being isolated from the sponges. The more than casual discoveries of unprecedented and unexpected diterpenoid ring systems in marine organisms are likely to provide the basis for continued growth in this field.

II. ACYCLIC DITERPENOIDS

A. Phytol and Phytadienes

Although geranylgeraniol is rarely a significant component of plants and animals and has not been observed in marine sources, its partially saturated analog, *trans*-phytol (**4**), is a ubiquitous constituent of both marine

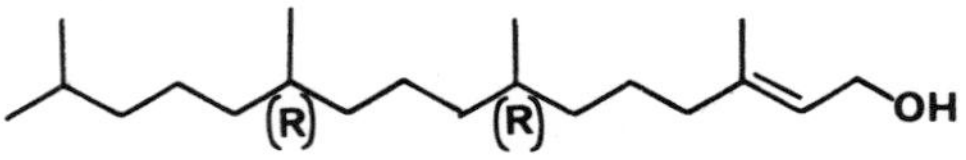

(R) (R) OH

4 *trans*-Phytol

and terrestrial plants. Phytol universally occurs as an ester of the propionic acid side chain in chlorophylls a and b and is usually isolated only after saponification of the pigment extract. The total synthesis of **4** from optically active precursors (Burrell *et al.*, 1966) has shown the 7*R*, 11*R* configurations for the natural product. In an investigation of the nonsaponified extract of the benthic red seaweed *Gracilaria andersoniana* (Rhodophyta), Sims and Pettus (1976) observed unusual amounts of free phytol. Using high-pressure liquid chromatography, **4** was isolated, but more interestingly *cis*-phytol (**5**) was also observed. The geometric isomers of phytol, having been previously synthesized by Burrell *et al.* (1966), could be readily assigned by interpretation of their respective nuclear magnetic resonance (nmr) spectra. Although a majority of protons in **4** and **5** are superimposable, the C-1 methylene groups are influenced by

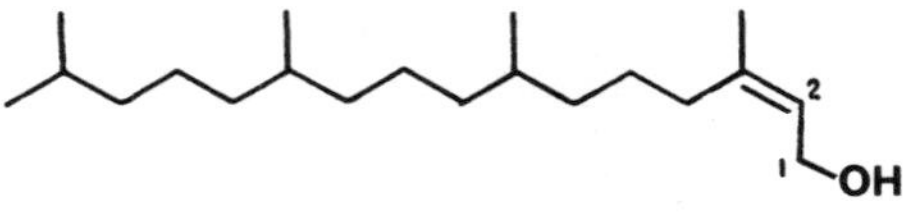

2 1 OH

C-1 δ value	
trans	4.05
cis	4.48

5 *cis*-Phytol

steric interactions, and their chemical shifts vary significantly. This represents the first reported isolation of *cis*-phytol from a natural source.

In nature plant-produced phytol is clearly transmitted along the food web to herbivorous animals. In the ocean this process must begin with the unicellular plants, the phytoplankton, and their predators, the microscopic invertebrates or zooplankton. Although phytol has not been reported to be a component of zooplankton, Blumer and Thomas (1965) isolated from a Gulf of Maine mixed collection of zooplankton a series of closely related hydrocarbons, the phytadienes (**6–9**). These olefins were

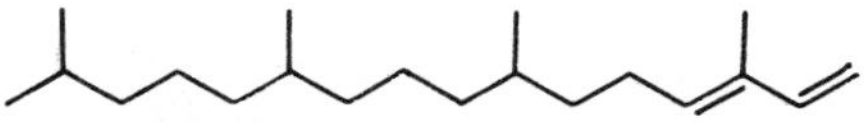

6 1,3-*trans*-Phytadiene

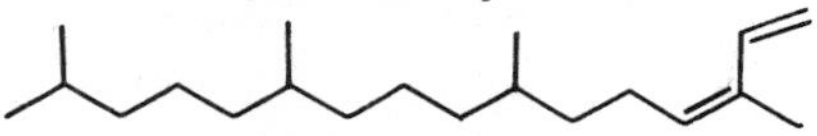

7 1,3-*cis*-Phytadiene

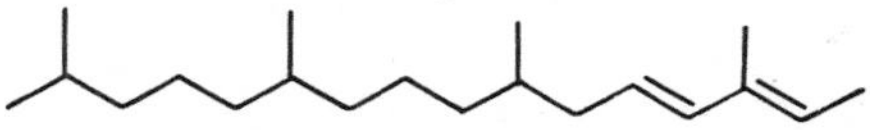

8 2,4-*trans*-Phytadiene

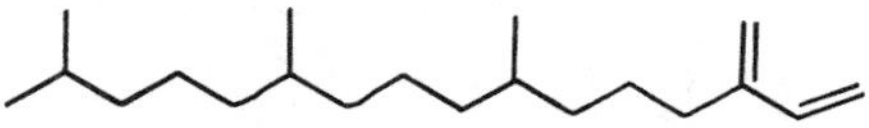

9 Neophytadiene

isolated by preparative gas chromatography and characterized by a combination of spectral analysis and chemical modification. In support of the origin of these dienes in plant-produced phytol, treatment of the latter with oxalic acid under mild dehydrating conditions yields a mixture of the phytadienes similar in composition to that observed in nature.

B. Crinitol

Although geranylgeraniol has yet to be isolated from a marine source, 9-hydroxygeranylgeraniol (crinitol, **10**) has been isolated from the

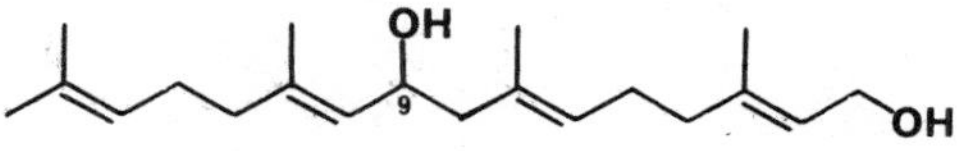

OH 9 OH

10 Crinitol

10: $[\alpha]_D$ $-3°$; ^1H nmr:δ4.95–5.55(4H,m), 4.24(1H,q,J = 7), 3.98(2H,d,J = 7), 2.65(2H,m), 1.70–1.56(15H,s) (incomplete)

Mediterranean brown seaweed *Cystoseira crinita* (Phaeophyta) collected in Sicily (Fattorusso *et al.*, 1976a). The basic structure of crinitol was determined by a comparison of the major sodium in liquid ammonia reduction product, 2,6,10,14-tetramethylhexadeca-2,6,10,14-tetraene, with an authentic sample. The secondary hydroxyl group was positioned at C-9 on the basis of the multiplicity and coupling of the methine proton in the nmr spectrum of **10**. Furthermore, the mass spectral fragmentation of crinitol illustrated prominent fragments which support positioning the secondary hydroxyl at C-9. In the same study a degraded C_{14} compound, oxocrinol, of obvious terpene origin, was isolated (see Section VII).

C. *Halichondria* Isonitrile

It has recently been recognized that a characteristic of certain marine sponges (Porifera) is the synthesis of metabolites that possess the isonitrile functionality. Isonitriles are rare components of terrestrial organisms and have been isolated only from fungi of the genera *Penicillium* and *Trichoderma* (Rothe, 1950; Achenbach, 1976). In marine sponges at least five genera are known to produce isonitriles (for a recent review, see Minale *et al.*, 1976), and curiously all of these metabolites are of a sesquiterpenoid or diterpenoid nature. In the diterpenoids two compounds have been described, an acyclic compound and a tetracyclic molecule discussed in Section V,C.

Along with a sesquiterpenoid isonitrile of the amorphane ring system, Burreson and Scheuer (1974; Burreson *et al.*, 1975) isolated and characterized the geranyllinaloyl isonitrile (**11**), as well as the corresponding formamide (**12**) and isothiocycanate (**13**) derivatives from a Hawaiian *Halichondria* species. The structure elucidation of these interesting

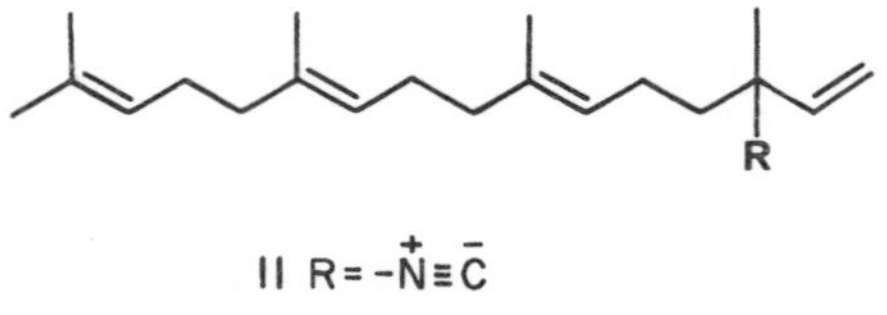

R

11 R= $-\overset{+}{N}\equiv\overset{-}{C}$

12 R= -NHCHO

13 R= -N=C=S

11: $[\alpha]_D$ +15°; ir (CCl_4): 2140, 990, 930 cm^{-1}; ^{13}C nmr: δ5.5–5.1(5H,m), 1.99(3H,bs), 1.67(3H,s), 1.62(3H,s), 1.59(3H,s), 1.47(3H,t,*J* = 4.5), 1.27(m); ^{13}C nmr: 155.6(t,*J* = 1.5), 138.1(d), 136.2(s), 134.8(s), 131.0(s), 123.9(d), 123.9(d), 122.3(d), 114.0(t), 62.5(t), 41.5(t), 39.5(t), 29.7(t), 28.4(q), 26.8(t), 26.8(t), 25.6(q), 22.8(t), 17.7(q), 16.0(q), 16.0(q)

metabolites was afforded by their interconversion, which included hydrolysis of **11** to yield **12** and desulfurization of **13** with refluxing aniline to give **11**. Treatment of the isonitrile **11** with lithium in ethylamine afforded a reductive elimination of the isonitrile with concomitant isomerization of the olefin system to yield the hydrocarbon **14**. Further, ozonolysis of **11**,

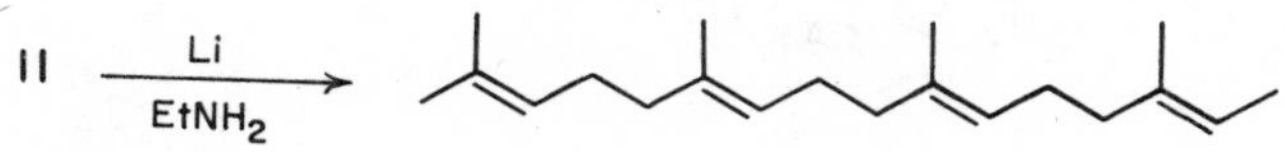

11 $\xrightarrow[EtNH_2]{Li}$

14

followed by reductive work-up and treatment with 2,4-DNPH reagent, led to the isolation of the 2,4-DNPH derivatives of formaldehyde, acetone and 4-ketopentanal, thus substantiating the geranyl–linalool skeleton.

D. Trifarin

The cosmopolitan green alga *Caulerpa* (Chlorophyta), which generally occurs in tropical and subtropical waters, is avoided by the majority of its proximate herbivorous fishes and invertebrates. This avoidance may be induced by a series of *Caulerpa* metabolites including nitrogen compounds (Doty and Santos, 1966; Santos, 1970; Maiti and Thomson, 1977) and triterpenes (Santos and Doty, 1971). It has also been recently recognized that various *Caulerpa* species produce simple diterpenoids that are clearly related to geranylgeraniol. For example, a simple acyclic derivative, trifarin (**15**), has been discovered in hexane extracts of *C. trifaria* col-

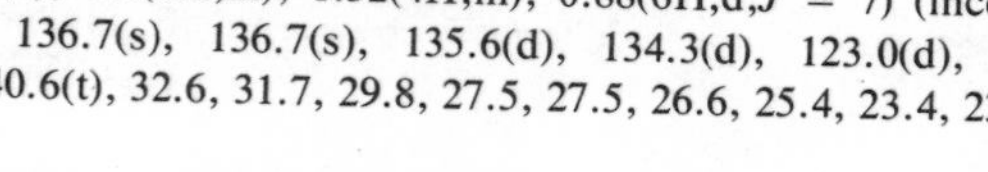

AcO 15 14 OAc

15 Trifarin

15: ^{1}H nmr: δ7.33(1H,d,J = 12), 7.19(1H,s), 5.83(1H,d,J = 12), 5.17(2H,m), 2.12(6H,s), 2.1(1H,m), 1.62(6H,s), 1.6(1H,m), 1.32(4H,m), 0.88(6H,d,J = 7) (incomplete). ^{13}C nmr: 167.7(s), 167.3(s), 136.7(s), 136.7(s), 135.6(d), 134.3(d), 123.0(d), 121.2(s), 119.8(d), 113.3(dd), 39.0(d), 40.6(t), 32.6, 31.7, 29.8, 27.5, 27.5, 26.6, 25.4, 23.4, 23.0, 20.6, 20.6, 16.1

lected in Tasmania (Blackman and Wells, 1978). Trifarin is an interesting modification of geranylgeraniol in that selective hydrogenation of the C-14–C-15 olefin has occurred and the 1,4-diacetoxy-1,3-butadiene moiety is unprecedented among natural compounds. The assignment of *E,E* stereochemistry for the diacetoxybutadiene group was supported by a comparison of the spectral characteristics of trifarin with the model compounds (*E*, *E*)-1,4-diacetoxy-2-methylbutadiene (**16**) (*Z*, *Z*)-1,4-diacetoxy-2-methylbutadiene (**17**). The Australian investigators have also described a monocyclic diterpenoid from *Caulerpa* (see Section III,A).

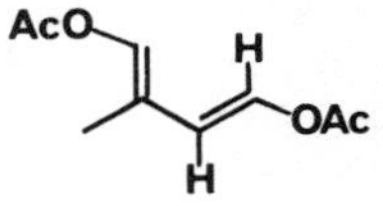

AcO H OAc H

16

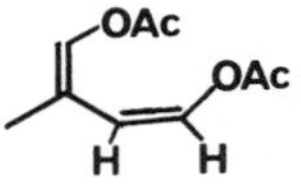

OAc OAc H H

17

III. MONOCYCLIC DITERPENOIDS

The monocarbocyclic diterpenoids are an interesting group of compounds that are mainly produced by marine macroalgae (the seaweeds) and coelenterates of the order Alcyonacea (sea fans and soft corals). The seaweeds elaborate monocyclic systems possessing 6-, 9-, and 10-membered rings, whereas the alcyonacean-derived monocycles are generally of the known but otherwise rare thunbergane or cembrane (14-membered ring) skeleton. In each of these monocyclic metabolites, a logical but speculative biogenesis from geranylgeraniol can be envisioned via a proton-induced olefin cyclization.

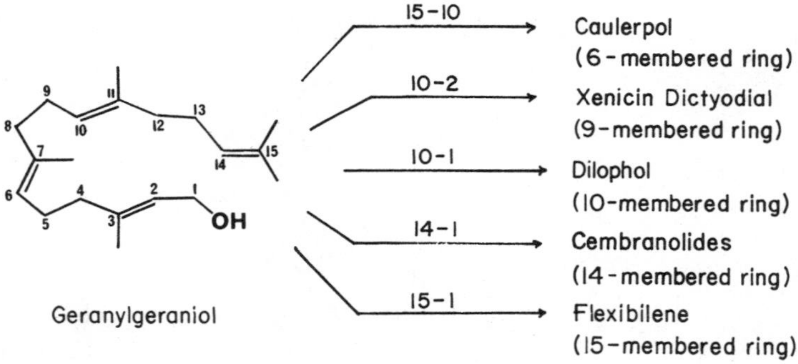

Geranylgeraniol

15–10 → Caulerpol (6-membered ring)
10–2 → Xenicin Dictyodial (9-membered ring)
10–1 → Dilophol (10-membered ring)
14–1 → Cembranolides (14-membered ring)
15–1 → Flexibilene (15-membered ring)

A. Caulerpol

Perhaps the simplest examples of monocyclic diterpenoids are the algal metabolites caulerpol (**18**) and the corresponding acetate **19**, which were reported by Blackman and Wells (1976) from the Tasmanian green seaweed *Caulerpa brownii*.

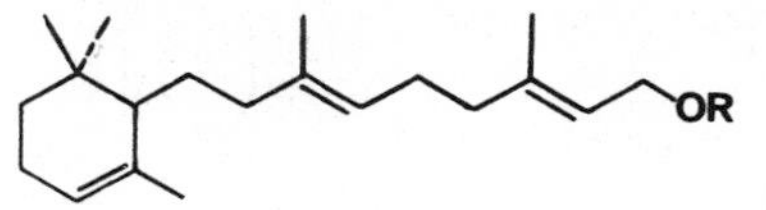

18 R=H,Caulerpol
19 R=Ac

18: $[\alpha]_D^{MeOH}$ −84.8°; ir: 3170, 1000 cm^{-1}; ^1H nmr: (CCl_4) δ5.35(1H,m), 5.26(1H,m), 5.08(1H,m), 4.01(2H,d,J = 7.5), 1.97(9H,m), 1.65(6H,s), 1.58(3H,s), 1.40(4H,m), 0.91(3H,s), 0.85(3H,s); cmr: 139.3(s), 136.9(s), 123.8(d), 123.8(d), 120.0(d), 59.3(t), 49.1, 40.6, 39.7, 32.7, 31.8, 30.0, 27.5, 26.5, 23.2, 16.2

The structure of caulerpol, as well as that of its acetate, were proposed on the basis of oxidative ozonolysis yielding levulinic acid and by a careful comparison of the spectral characteristics of caulerpol with those of farnesol (**20**), and the C_{10} terpene model compound **21**. Caulerpol can readily be recognized as being produced by cyclization chemistry charac-

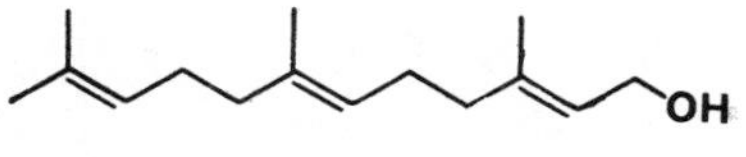

20

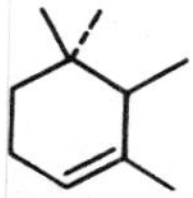

21

teristically found in the carotenoid pigments consisting of proton-induced C-15–C-10 closure. In that regard, caulerpol is closely related to retinol or vitamin A (**22**), an essential precursor for the *in vivo* production of the

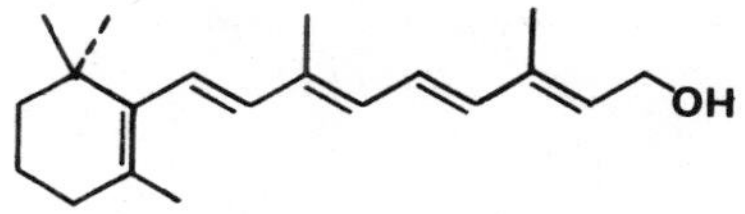

22 Retinol

corresponding aldehyde, retinene. Retinene is an essential component of rhodopsin, the protein complex that is the basis for visual chemistry in mammals.

B. Xenianes: Xenicin and Dictyodial

It is known but infrequently observed that metabolities of the same terpenoid skeleton can be isolated from plants and animals. A recent example of this situation is the soft coral metabolite xenicin and a related compound, dictyodial, which was isolated from the brown seaweed *Dictyota*,

During a careful investigation of the lipid constituents of the Australian soft coral *Xenia elongata* (order Alcyonacea), The University of Oklahoma group (Vanderah *et al.*, 1977) isolated an unusual tetraacetate, xenicin (**23**), which on inspection of its spectral characteristics appeared

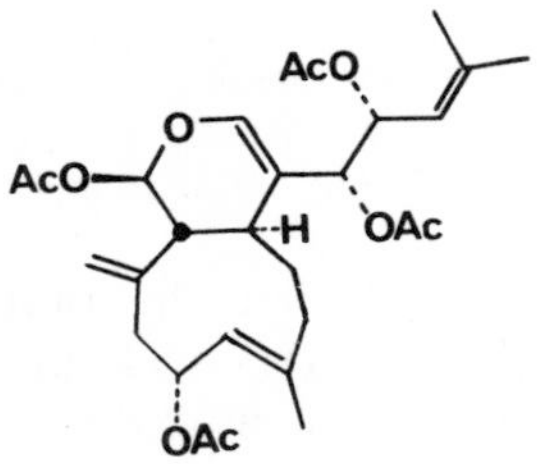

23 Xenicin

23: mp 141.5°–142.3°; $[\alpha]_D$ −36.7° ($CHCl_3$); ir (KBr): 2980, 2940, 2860, 1735, 1700(sh), 1665, 1635, 1440, 1375, 1235, 1205, 1180, 1155, 1005, 930, 870 cm^{-1}: ^1H nmr ($CDCl_3$): δ6.58(1H,d,*J* = 2), 5.87(1H,d,*J* = 2), 5.82(1H,t,*J* = 10), 5.70(1H,bt,*J* = 8–9), 5.38(1H,d,*J* = 10), 5.27(1H,bd,*J* = 8–9), 5.08(1H,bd,*J* = 10), 4.82(1H,m), 2.19(1H,m), 2.08(3H,s), 2.06(6H,s), 2.04(3H,s), 1.88(1H,m) (incomplete)

to possess a new diterpenoid skeleton. The final structure for this compound was determined, including absolute stereochemistry, by a single-crystal X-ray diffraction study. Xenicin contains a nine-membered ring, which can formally be produced via cyclization between C-2 and C-10 of an acyclic precursor. This cyclization, however, represents an unfavorable (anti-Markownikoff) and unprecedented biochemical cyclization. Curiously, extracts of *X. elongata* from other locations were devoid of xenicin, perhaps suggesting that the diterpenoid is not produced by the coral itself but by symbiotic microalgae associated with coelenterates, which are known as zooxanthellae. This continuing question is valid for most of the compounds produced by alcyonaceans, which, particularly in tropical waters, are the hosts of a significant mass of zooxanthellae. The zooxanthellae are difficult to classify, and various "species" may exist and be geographically dispersed.

In 1973, investigation of the Gulf of California brown seaweed *Dictyota flabellata* resulted in the isolation of several new oxygenated diterpenoids (Robertson and Fenical, 1977). Along with some bicyclic compounds (see Section IV), a unique monocyclic dialdehyde was discovered which exhibited substantial antibacterial activity. This compound, subsequently named dictyodial (Fenical *et al.*, 1978), could not be crystallized, and its total structure elucidation by spectrochemical methods was not possible due to several zero couplings between critical protons in its nmr spectrum. Lithium aluminum hydride reduction of dictyodial gave, without rearrangement, modest yields of the corresponding diol, which, although isolated as a viscous oil, crystallized after lengthy storage in the cold. Initially, X-ray diffraction studies of the diol were frustrating, and a solution was not obtained due to symmetry problems in the crystal. However, more recently, with an improved computer program, Finer and Clardy (1978) were able to solve the diol structure, and hence the structure of dictyodial, with relative stereochemistry only, could be assigned as **24**. A thorough nmr analysis of **24** at 360 MHz and of **25** at 220 MHz, with

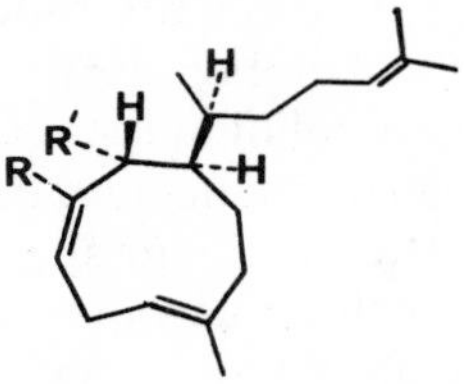

24 R = R' = CHO, Dictyodial
25 R = R' = CH_2OH

24: Oil, ir (film): 2740, 1685, 1725, 770, 790 cm^{-1}; ^1H nmr ($CDCl_3$): δ 10.20(1H,d), 9.33(1H,d), 6.95(1H,dd), 5.35(1H,m), 5.08(1H,m), 3.28(1H,m), 3.05(1H,m), 2.96(1H,m), 1.77(3H,s), 1.66(3H,s), 1.57(3H,s), 1.20(2H,m), 0.89(3H,d), 1.6–3.5 (various bands)

and without the addition of lanthanide shift reagent, allowed the assignment of all protons except the three illustrated. These protons, the relationship of which is critical for the assignment of ring size and side chain position, are unfortunately disposed at dihedral angles near 90° and do not couple. In addition, the endocyclic *trans*-olefin was not immediately recognized and could not be assigned from spectral data. However, inspection of the thin-film infrared spectral features of **24** illustrated the classic doublet bands at 770 and 790 cm^{-1}, which are characteristic of trans-trisubstituted olefins. The common origin of xenicin and dictyodial is perhaps further substantiated by the comparable stereochemistry of the *trans*-olefin and the substituents.

C. Dilophol

Metabolites of the biologically related brown algal genera *Dictyota, Pachydictyon,* and *Dilophus* are discussed in various sections of this

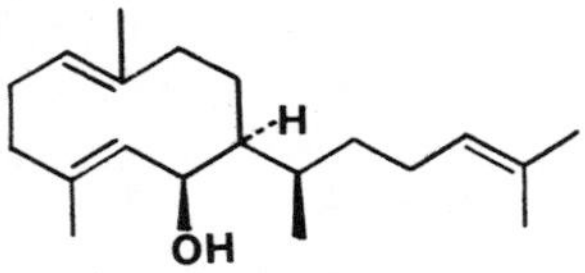

26 Dilophol

26: Oil, $[\alpha]_D$ −4.3°, ir: 3100–3300 cm^{-1}, MS: $M^+ - H_2O$, *m/e* 272; ^1H nmr (CCl_4): δ1.70(3H,s), 1.60(6H,s), 1.46(3H,s), 0.98(3H,d,*J* = 6); ^1H nmr (C_6H_6): 5.25(1H,bt,*J* = 7), 5.02(1H,d,*J* = 9), 4.87(1H,bs), 4.55(1H,m) (incomplete)

chapter, and they have been separated according to their respective ring systems. These metabolites are, however, closely related, and in many cases their interconversion can be readily rationalized. A potential precursor for many of the bicyclic metabolites discussed in Section IV is the monocyclic 10-membered ring alcohol, dilophol (**26**), which was described by Italian investigators from the Sicilian alga *Dilophus ligulatus* (Amico *et al.*, 1976; Fattorusso, 1977). Mass and nmr spectrometry fixed dilophol as a monocyclic compound with three carbon–carbon double bonds. Furthermore, oxidation of **26** with $KMnO_4$–$NaIO_4$ (Lemieux–von Rudloff reagent) yielded levulinic acid, proving the cyclic system to be at least of medium size. The final assignment of this unique diterpenoid as a C_{20} analog of the 10-membered ring sesquiterpene germacrene was supported by its selenium-induced aromatization to yield 1,4-dimethyl-7-(1,5-dimethylhex-5-enyl)azulene (**27**). Analogous aromatizations of germa-

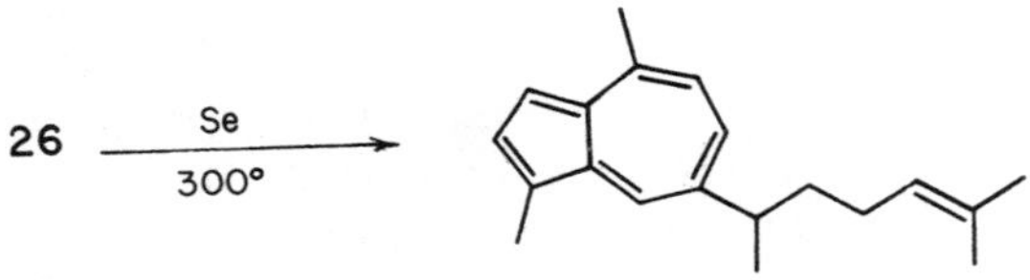

crenes involving ring closure to azulene have been reported (Treibs, 1953). The stereochemistry of the hydroxyl group and side chain was assigned a *cis* relationship because of the magnitude of coupling (2 Hz) between the corresponding methine ring protons, as well as the magnitude of shift observed for the side chain secondary methyl on complexation with $Eu(fod)_3$.

D. Dictyoxepin

Another formally monocarbocyclic metabolite from *Dictyota*, which appears logically to be derived from dilophol, is the recently reported

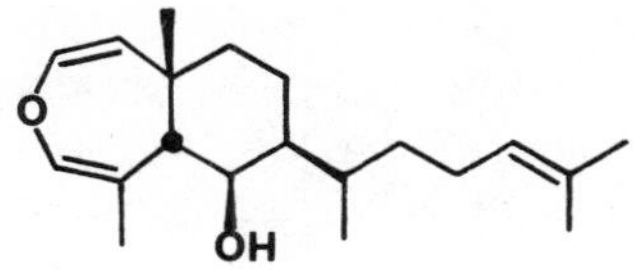

28 Dictyoxepin

28: ir(CCl_4): 3600, 1670, 1650, 1290, 1240, 1195, 1110, 1040 cm^{-1}; ^1H nmr: δ6.37(1H,m), 6.15(1H,d,J = 8), 5.26(1H,m), 4.27(1H,dd,J = 8,2), 4.22(1H,dd,J = 11,4), 1.80(3H,d,J = 1), 1.73(3H,s), 1.67(3H,s), 0.96(3H,d,J = 6) (incomplete): ^{13}C nmr: 140.7, 139.4, 131.0, 125.4, 119.5, 2.85(1H,d,J = 9.5), 1.52(3H,s), 1.15(3H,s), 0.85(3H,d,J = 6.5) (incomplete)

compound dictyoxepin (**28**) (Sun *et al.*, 1977), which was isolated from the Hawaiian alga *D. acutiloba*. The structure of dictyoxepin, including absolute stereochemistry, was determined by X-ray diffraction experiments on the corresponding crystalline *p*-bromophenylurethane derivative. Isolated along with **28** was a noncrystalline bicyclic diterpenoid dictyolene (**79**) (see Section IV for complete details). The relation of these two previously

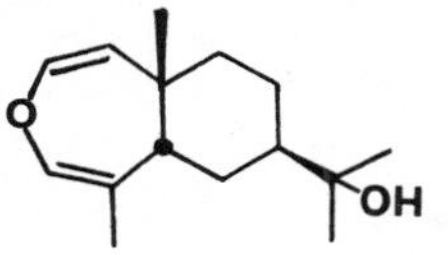

29

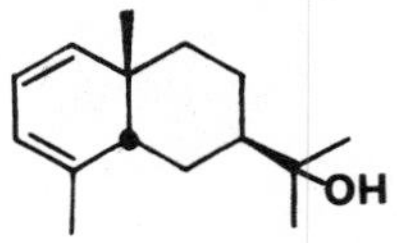

30

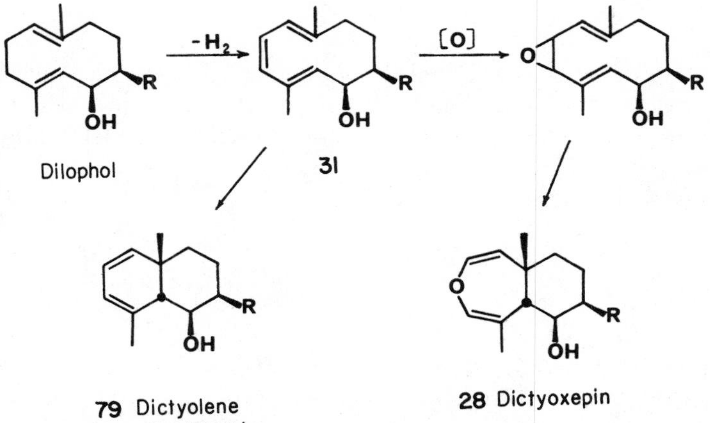

Dilophol

31

79 Dictyolene
(See Sec. IV, A)

28 Dictyoxepin

unknown diterpenoid skeletons to the analogous sesquiterpenes occidentol (**29**) and 7-epioccidentalol (**30**) strongly suggests a similar biogenesis, potentially involving the intermediacy of a dilophol type of 10-membered ring compound which proceeds via dehydrogenation through a 1,3,5-cyclodecatriene system (**31**). An electrocyclic ($6\pi \rightarrow 4\pi, 2\sigma$) ring closure of **31** then generates dictyolene, and epoxidation of the central olefin gives a *cis*-divinyl epoxide, which, via a Cope rearrangement ([3,3]-sigmatropic shift), yields dictyoxepin.

E. Elemene-Type Diterpenoids from Soft Coral

Still another group of monocyclic diterpenoids (**32–35**), which could biogenetically be derived from a dilophol-like 10-membered ring intermediate, has been isolated by the Australian group at the Roche Institute from an indigenous soft coral *Lobophytum* species (Wells, 1978). These

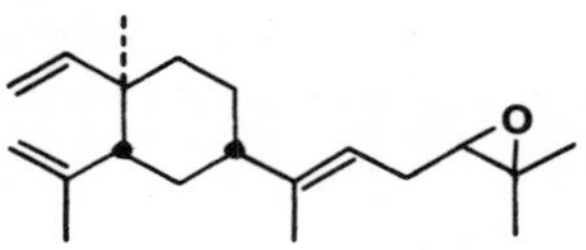

32

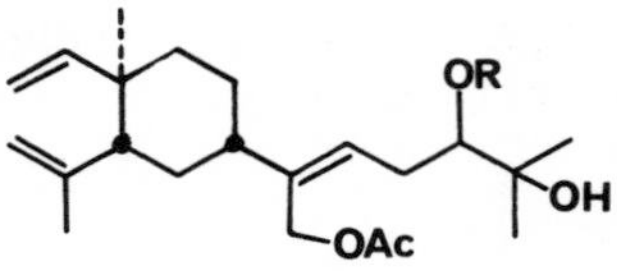

OR OH OAc

33 R = H
34 R = Ac

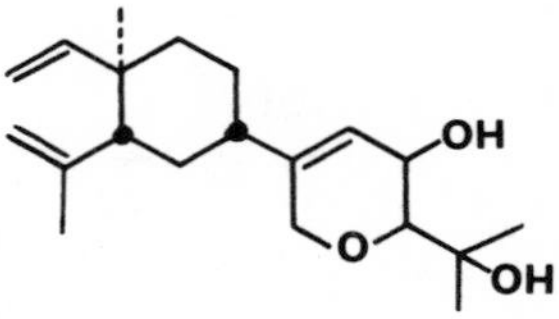

OH O OH

35

compounds appear to be related to dilophol (**26**) and may be generated by a Cope rearrangement in much the same way that the sesquiterpene β-elemene (**36**) is related to germacrene (**37**). All substituents of the

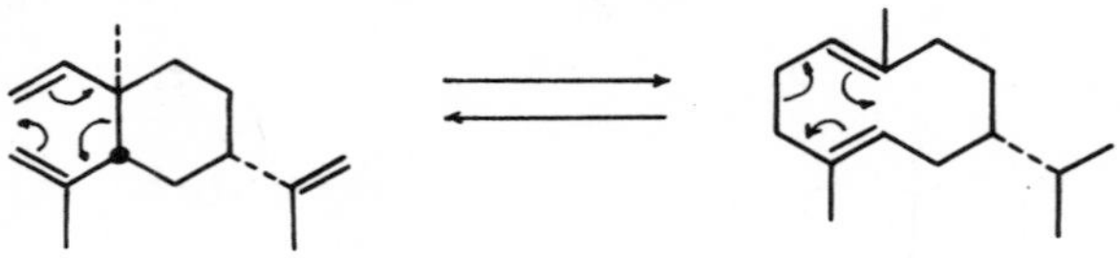

36 **37**

cyclohexane rings in **32–35** were assigned equatorial conformations, presumably by analysis of nmr data. However, the details of this work have not yet been published.

F. Marine Cembrane Derivatives

Cyclization of a geranylgeraniol-derived precursor between carbons 1 and 14 to yield the 14-membered diterpenoid ring named cembrane or thunbergane is a known but infrequent phenomenon in terrestrial organisms. The simplest example of this process results in the production of the unsaturated parent hydrocarbon cembrene (**38**), which has been iso-

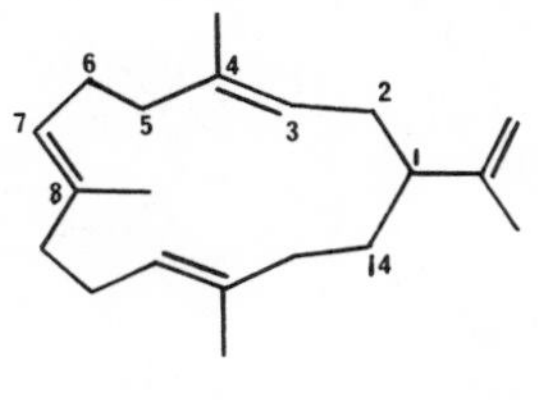

38 Cembrene

lated mainly from the essential oil of pine trees (*Pinus*) (Kobayashi and Akiyoshi, 1962). More recently cembrene has been described as the active trail pheromone of the termite *Nasutitermes exitiosus* (Birch *et al.*, 1972; Patil *et al.*, 1973), and a biomimetic-type synthesis of the racemic hydrocarbon has been reported (Kodama *et al.*, 1975).

In reporting the structures of various cembrene-based compounds I have adopted a convention suggested by Weinheimer (1978). Specifically, all cembrene rings are drawn as shown in cembrene (**38**). As an unsymmetrical marker the unsaturation (or a vestige of unsaturation such as an epoxide) usually extant at C-8 is assigned Δ^7 rather than Δ^8. In most cases this convention adequately describes and unifies most cembrene derivatives. Slight modifications are necessary, however, to accommodate cis double bonds.

In the marine environment the production of cembrane derivatives is at this time known in only two groups of coelenterates, the sea fans (order Gorgonacea) and the related soft corals (order Alcyonacea), which comprise the subclass Octocorallia. Since these organisms abound in tropical waters such as the Caribbean and Indo-Pacific regions and are easily collected in shallow waters, they represent next to the sponges perhaps the most thoroughly investigated group of marine invertebrates. Their extensive investigation has also been prompted by the high levels of biological activity found in the cembrane lactone derivatives (cembranolides), which are common metabolites of this group.

Investigation of the cembrene-based diterpenoids from the sea fans or gorgonians was initiated in the early and mid-1960's by the University of Oklahoma group. Resulting from their efforts were the first examples of marine cembranolides from various species of sea fans from the Caribbean. *Plexaura crassa* was among the first goronians to be investigated and, after extensive chemical studies (Weinheimer *et al.*, 1968), the structure of crassin acetate (**39**) was elucidated by X-ray analysis of the corresponding *p*-iodobenzoate ester (**40**) (Houssain and van der Helm,

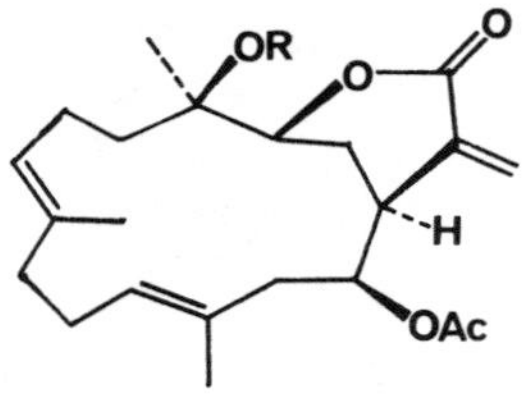

39 R = H, Crassin acetate
40 R = $-COC_6H_4-I(p)$

1969). Subsequently, crassin acetate was also isolated from four gorgonians of the genus *Pseudoplexaura,* namely, *P. porosa, P. flagellosa, P. wagenaari,* and *P. crucis*. The lactone possesses substantial antineoplastic activity (Weinheimer and Matson, 1975).

Eunicea mammosa was also among the first gorgonians to be investigated, and collections made in the Bahamas yielded the oxygen-bridged lactone eunicin (**41**). The structure of eunicin was proposed initially on chemical and spectral grounds (Weinheimer *et al.*, 1968) and confirmed by X-ray crystallographic analysis (Houssain *et al.*, 1968) of the corresponding iodoacetate (**42**). Interestingly, when collections of *E. mammosa* from

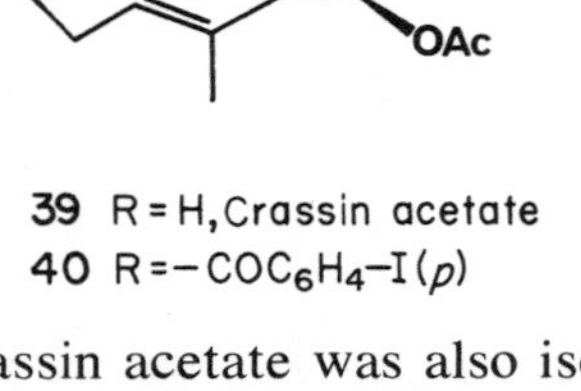

41 R = H, eunicin
42 R = $-COCH_2I$

41: mp 154°–155.5°; $[\alpha]_D$ −89.4°; ir: 3623, 1765, 1664 cm^{-1}; ^1H nmr: δ6.40(1H,d,J = 3.5), 5.70(1H,d,J = 3.5), 5.07(1H,t,J = 7.5), 4.43(1H,dd,J = 9.5, 7.8), 3.25(1H,dd,J = 11,2.5), 2.85(1H,d,J = 9.5), 1.52(3H,s), 1.15(3H,s), 0.85(3H,d,J = 6.5) (incomplete)

other Caribbean locations were investigated, related but nonidentical cembranolides were isolated. From Jamaican *E. mammosa* the same group isolated the related cembranolide jeunicin (**43**), which was unam-

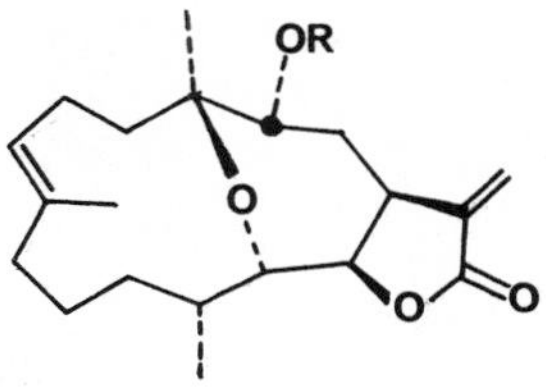

43 R = H, Jeunicin
44 R = $COC_6H_4Br(p)$

biguously assigned structure **43** by crystallographic analysis of the *p*-iodobenzoate derivative (**44**) (van der Helm *et al.*, 1976; Ciereszko and Karns, 1973). Several other related cembranolides, which have been labeled ceunicin and more recently peunicin (Ciereszko, 1977), have also been isolated from *Eunicea*. The structures for these compounds have not yet appeared in the literature.

From the related Caribbean gorgonian *Eunicea palmeri,* the Oklahoma researchers later described the structure of another cembranolide, eupalmerin acetate (**45**) (Weinheimer *et al.*, 1972). The acetate (**45**) is

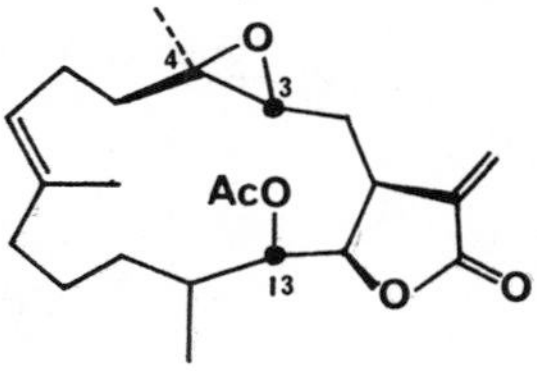

45 Eupalmerin acetate

clearly a potential precursor for both eunicin (**41**) and jeunicin (**43**) via a nucleophilic transannular epoxide displacement at either C-3 or C-4 with the proximate hydroxyl at C-13.

The most recent work by the Oklahoma group has been the report (Weinheimer *et al.*, 1977) of the structure of a new nonlactonic cembrane, asperdiol (**46**), which was isolated from the gorgonians *Eunicea asperula* and *E. tourneforti.* Asperdiol is a unique compound in that it is the first cembrane lacking the lactone function which displays significant *in vitro* antitumor activity in the National Cancer Institute's KB, PS, and LE test systems.

Soft corals have yielded a greater number of cembrene derivatives than have gorgonians. The parent hydrocarbon cembrene (**38**) was reported by the Tursch group (Tursch, 1976; Herin and Tursch, 1976) to be a component of the Moluccan soft coral *Sinularia flexibilis.* The Roche group in

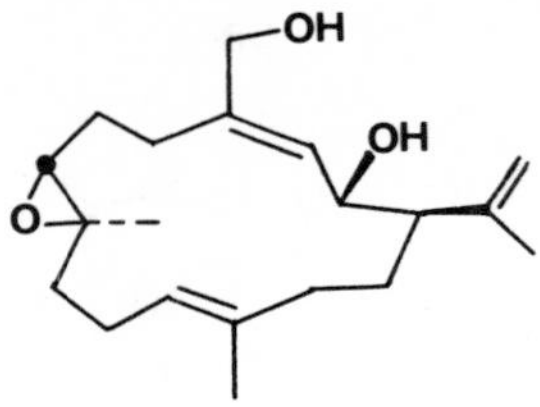

46 Asperdiol

46: mp 109°–110°; $[\alpha]_D$ −87°; ir (KBr): 3450, 1645 cm^{-1}; ^1H nmr ($CDCl_3$): δ5.45(1H,bd,J = 8), 5.14(1H,bt,J = 7), 4.94(1H,bs), 4.75(1H,bs), 4.50(1H,dd,J = 5,8), 4.05(2H,bs), 2.70(1H,dd,J = 4,6), 1.77(3H,s), 1.62(3H,bs), 1.20(3H,s)

Australia (Wells, 1978) isolated from a *Lobophytum* species the isocembrene derivative **47**, but no details have been published.

Other simple cembrene derivatives have been isolated from several soft

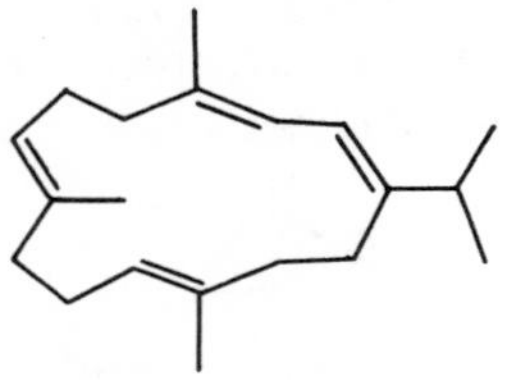

47

corals. Schmitz and co-workers (1974) described nephthenol (**48**) and epoxynephthenol acetate (**49**) as metabolites of the soft coral *Nephthea* from Enewetrak. The structures of **48** and **49** were assigned on the basis of spectral features of the two compounds and their oxidative degradation to the lactone fragments **51** and **52**. Nephthenol gave levulinic acid and the known lactone homoterpenyl methyl ketone (**51**). Epoxynephthenol (**50**) yielded (**51**) and the lactone **52**, which was identified by its spectral features. In the oxidation of **50** sufficient amounts of **51** were isolated for rotation measurements. Since **51** showed $[\alpha]_D$ −41°, which compared favorably with the literature value of −59° for (−)-homoterpenyl methyl ketone, and since the asymmetric center in the levorotatory compound is known to be *R*, the absolute stereochemistry in both **49** and **50** must also be *R*.

In a later study of the soft coral *Litophyton viridis*, the Tursch group (Tursch *et al.*, 1975a) also isolated nephthenol and showed that the levorotatory enantiomer has the *R* configuration at C-1. In the same study a simple dihydroxy compound, 2-hydroxynephthenol (**53**), was described; the structure is based largely on spectral and chemical evidence. The

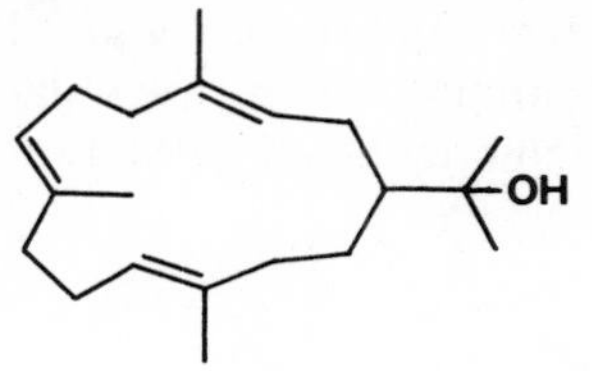

48 Nephtenol

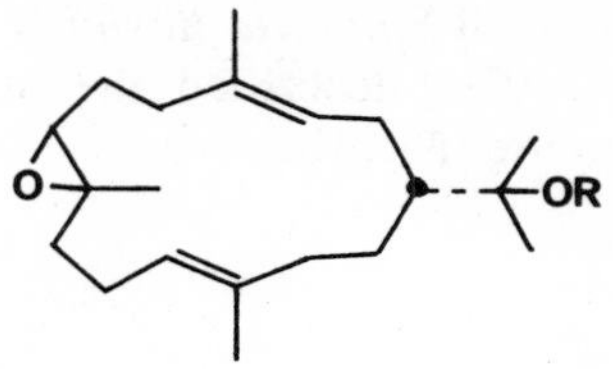

49 R = Ac, epoxynephtenol acetate
50 R = H (synthetic)

48: bp ~96°/0.03 mm, ir: 3450, 1655 cm^{-1}; ^{1}H nmr: δ5.9(3H,bm), 2.3–1.9(12H,m), 1.20(6H,s), 1.60(9H,s)

49: bp ~96°/0.03 mm, $[\alpha]_D$ −20.7°; ir: 3100, 1658, 1730, 1255 cm^{-1}; ^{1}H nmr ($CDCl_3$): δ5.18(2H,m), 2.84(1H,t,J = 5), 1.97(3H,s), 1.67(3H,bs), 1.55(3H,bs), 1.48(6H,s), 1.30(3H,s) (incomplete)

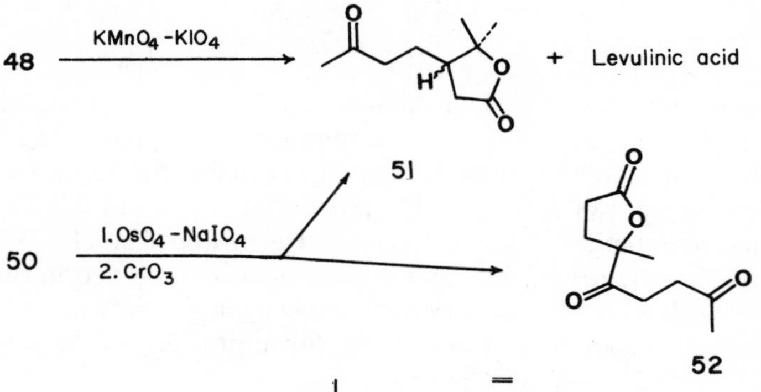

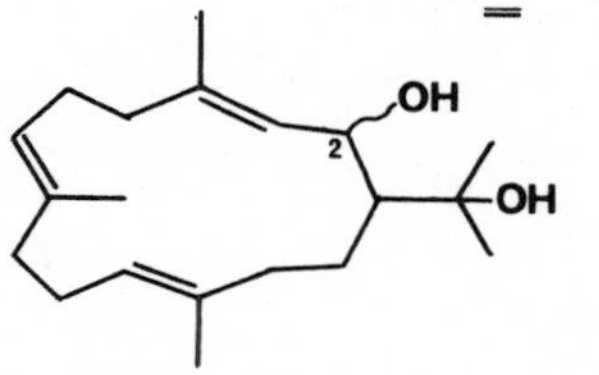

48 —$KMnO_4$-KIO_4→ 51 + Levulinic acid

50 —1. OsO_4-$NaIO_4$ 2. CrO_3→ 51 + 52

53 2-Hydroxynephthenol

53: mp 98°–99°; $[\alpha]_D$ −104°; MS: M^+−H_2O, *m/e* 288; ir (KBr): 3320, 1665 cm^{-1}; ^{1}H nmr: δ5.26(1H,bd,J = 9), 4.94(2H,m), 4.50(1H,t,J = 9,11), 3.50(2H,bm), 1.7(3H,s), 1.56(6H,bs), 1.28(3H,s), 1.26(3H,s)

relative proximity of hydroxyls in **53** was since the corresponding 2-ketone underwent a facile retro-aldol reaction.

The soft coral genus *Sinularia* has been the subject of numerous studies in Hawaii, Indonesia, and, more recently, Australia. From Indonesian collections of *Sinularia flexibilis,* the Tursch group (Tursch *et al.*, 1975b; Tursch, 1976) described the structure of an interesting new lactone, sinulariolide (**54**), the structure of which (including absolute stereochem-

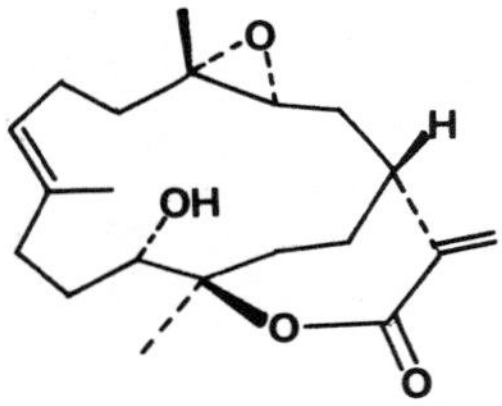

54 Sinulariolide

54: mp 170°–173°; $[\alpha]_D$ +76°; ir (film): 3460, 1720, 1630 cm^{-1}; uv: 212 nm (ϵ = 6000); ^{1}H nmr: δ6.30(1H,s), 5.50(1H,s), 5.20(1H,m), 4.15(1H,m), 3.00(1H,dd), 1.63(3H,bs), 1.35(3H,s), 1.22(3H,s) (incomplete): ^{13}C nmr ($CDCl_3$): 168.8(s), 144.3(s), 134.8(s), 126.4(d), 123.9(t) (incomplete)

istry) was unambiguously established by X-ray crystallography. Sinulariolide is unique among the cembranolides in possessing a seven-membered ring lactone (ϵ-lactone) system, probably formed via a carboxylate displacement of a C-11–C-12 epoxide precursor. In addition to **54**, the major metabolite of the soft coral, Herin and Tursch (1976) later reported the structure of four minor components. Included in this group were two derivatives of sinulariolide: the 6-hydroxyl-substituted lactone **55** and the corresponding 11-ketone **56**. Structures **55** and **56** were rigor-

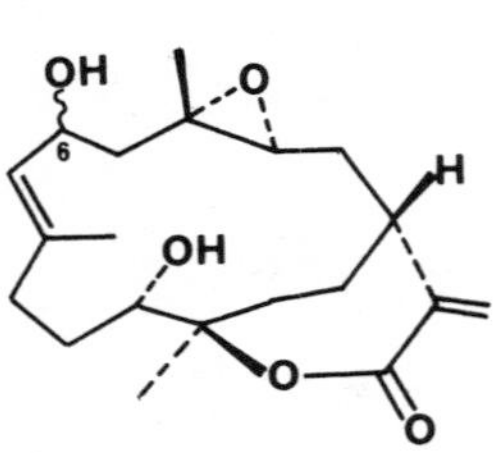

55 6-Hydroxysinulariolide

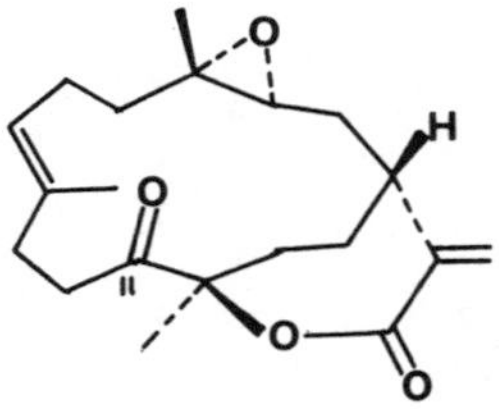

56 11-Dehydrosinulariolide

55: mp 192°–194°; $[\alpha]_D$ +54.5°; MS: M$^+$, *m/e* 350; ir (film): 3460, 1715, 1635 cm^{-1}; uv (MeOH): 213 nm (ϵ = 5000): ^{1}H nmr: δ6.27(1H,s), 5.47(1H,s), 5.30(1H,d,*J* = 9), 4.54(1H,m), 4.05(1H,m), 2.95(1H,dd,*J* = 3,10), 1.67(3H,bs), 1.34(3H,s), 1.23(3H,s) (incomplete)

56: mp 120°; $[\alpha]_D$ +87°; MS: M$^+$, *m/e* 332; ir (KBr): 1710, 1625 cm^{-1}; ^{1}H nmr: δ6.30(1H,s), 5.50(1H,s), 5.07(1H,m), 1.65(3H,bs), 1.50(3H,s), 1.20(3H,s) (incomplete)

ously determined by interconversion with sinulariolide and its degradation products. Two other compounds identified in this work were the parent hydrocarbon, cembrene A (**38**) (*R* configuration), and a 3,4,11,12-diepoxide (**57**). The absolute stereochemistry of **57** was determined by an

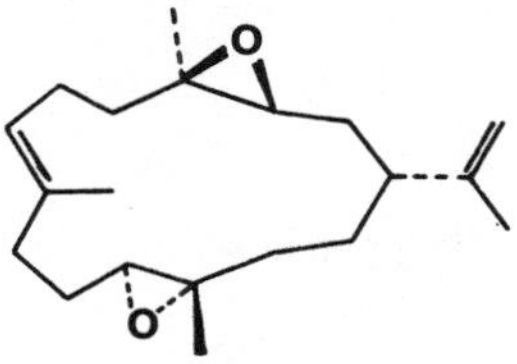

57

57: mp 66°–68°; $[\alpha]_D$ +63°; MS: M^+ *m/e* 289; ir (KBr): 2940, 1645, 1380, 885 cm^{-1}; ^1H nmr: δ5.25(1H,m), 4.82(1H,bs), 4.70(1H,bs), 2.65–3.0(2H,m), 1.67(6H,bs), 1.28(6H,s) (incomplete)

apparent stereospecific reduction to all-*trans*-cembrene with a zinc–copper couple and by careful optical and chemical analysis of the diols produced by lithium aluminum hydride reduction.

The same species of soft coral, but collected along the Great Barrier Reef, was investigated by the Roche group (Kazlauskas *et al.*, 1978a). Australian *Sinularia flexibilis* was also shown to contain, but in minor quantities, the metabolites **38**, **54**, and **56**. Somewhat surprisingly, the major compounds were the related α-methylene-δ-lactones flexibilide (**58**) and dihydroflexibilide (**59**). The structure of flexibilide was generated,

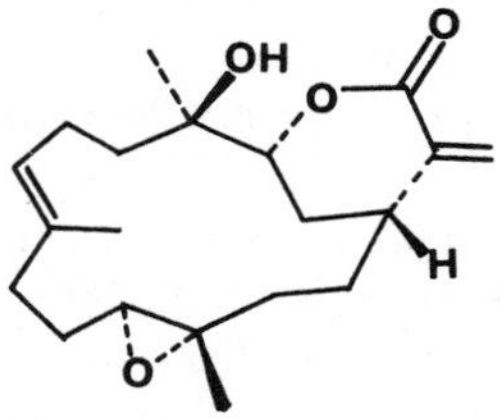

58 Flexibilide

58: mp 147°–148°; $[\alpha]_D$ −115°; ir (KBr): 3460, 1730, 1620 cm^{-1}; ^1H nmr: δ6.50(1H,d,*J* = 2.2), 5.72(1H,d,*J* = 2), 5.28(1H,bt,*J* = 7), 3.98(1H,d,*J* = 11), 2.82(1H,dd,*J* = 8,2), 1.68(3H,s), 1.46(3H,s), 1.32(3H,s) (incomplete); ^{13}C nmr: 166.9(s), 140.2(s), 134.2(s), 127.4(t), 125.8(d), 82.6(d), 73.9(s), 62.8(d), 58.9(s), 38.9(t), 35.9(t), 34.7, 33.8, 32.7(t), 27.8(t), 25.3, 24.7(q), 22.7(t), 15.4(q), 15.4(q)

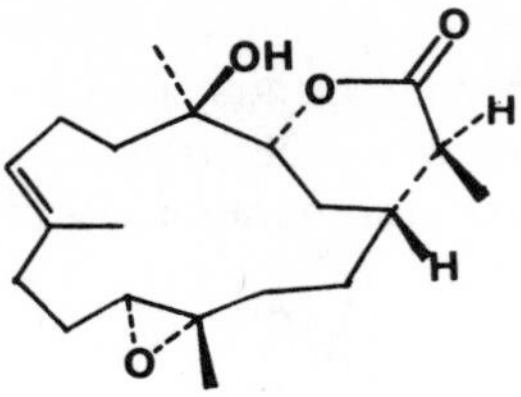

59 Dihydroflexibilide

59: mp 108°–109°; $[\alpha]_D$ −44°; ir (KBr): 3500, 1740 cm^{-1}; ^1H nmr: δ5.30(1H,6t), 4.02(1H,d,J = 11), 2.90(1H,dd,J = 8,6), 1.66(3H,s), 1.44(3H,s), 1.23(3H,s), 1.35(3H,d,J = 7); ^{13}C nmr: 174.5(s), 133.5(s), 126.0(d), 83.5(d), 74.2(s), 62.8(d), 58.8(s), 42.4(d), 38.4(t), 36.8, 35.7, 34.7(t), 31.0(t), 27.6, 25.3, 24.3, 23.0(t), 17.1(q), 16.4(q), 15.4(q)

without absolute stereochemistry, by single-crystal X-ray diffraction, whereas the dihydro derivative **59** was initially related to **58** by hydride reduction of both lactones to the identical epoxytriol **60**. The

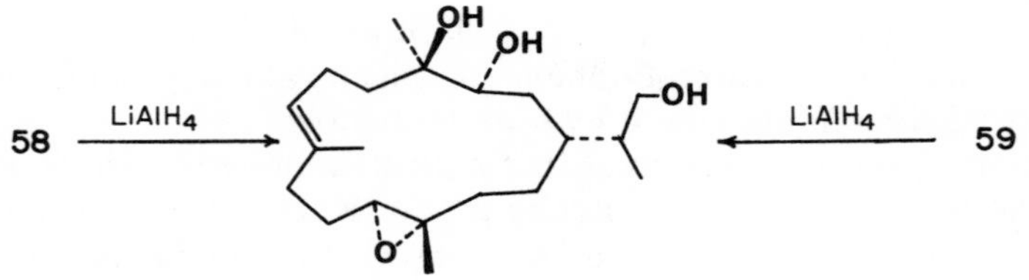

58 —$LiAlH_4$→ 60 ←$LiAlH_4$— 59

60

stereochemistry of the ^1H proton in **59** was predicted as pseudoequatorial on the basis of ^1H nmr data but was later corrected to pseudoaxial on the basis of a subsequent X-ray analysis.

Included in this discussion of the genus *Sinularia* must be the work of the Scheuer group (Missakian *et al.*, 1975) on the Hawaiian coral *S. abrupta*. Contrary to the consistency one might expect in studies of soft corals of the same genus, the Scheuer group reported as the major metabolite an interesting furanocembranolide, pukalide (**61**). The structure of pukalide, devoid of stereochemistry, was proposed on the basis of detailed interpretation of proton and carbon nmr data. The β-carbomethoxyfuran moiety in **61** is a unique feature not observed in other coelenterate cembrene metabolites. The uniqueness of pukalide and the fact that it is crystalline would suggest an X-ray crystallography study for confirmation of structure and elucidation of stereochemistry at a minimum of four centers.

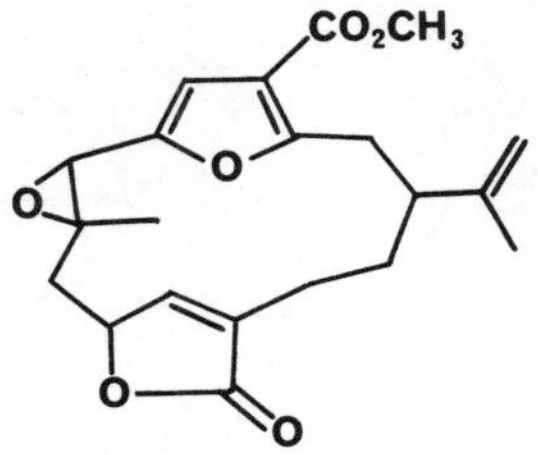

61 Pukalide

61: mp 204°–206°; $[\alpha]_D$ +44°; ir: 3140, 3020, 1760, 1715, 1580, 1270, 1230, 1080, 905, 890, 870, 830 cm^{-1}; ^1H nmr: δ7.06(1H,bs), 6.33(1H,s), 5.20(1H,s), 4.91(2H,m), 4.06(1H,m), 3.78(3H,s), 3.56(1H,ddd,*J* = 11,11,3.5), 2.95(1H,dd,*J* = 18,11), 2.8(1H,dd,*J* = 18,3.5), 2.5(1H,dd,*J* = 15,3.5), 2.35(2H,m), 2.15(1H,dd,*J* = 15,3.5), 1.80(3H,bs), 1.05(3H,s); ^{13}C nmr: 173.7(s), 163.8(s), 160.0(s), 148.2(s), 148.2(d), 145.8(s), 137.3(s), 113.9(s), 112.9(t), 106.4(d), 77.8(d), 57.0(s), 55.0(d), 51.2(q), 40.7(d), 40.0(t), 32.5(t), 32.5(t), 22.8(t), 19.8(q), 18.7(q)

Species of the soft coral genus *Sarcophytum* have now been investigated for cembranolide chemistry from three locations: the Red Sea, the Phoenix Islands (central Pacific), and the Australian Great Barrier Reef. The first structure reported was that of the epoxy cembranolide sarcophine (**62**), which crystallized from extracts of *Sarcophytum glaucum*

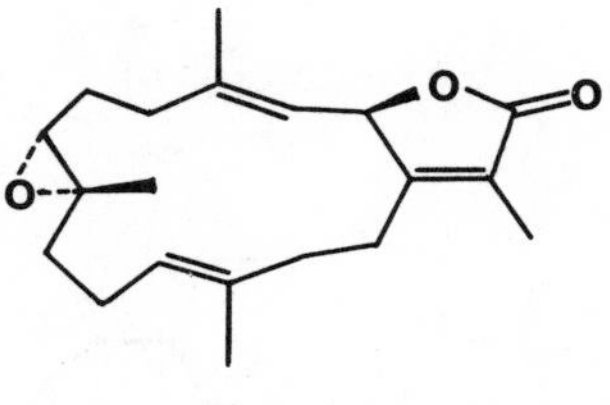

62 Sarcophine

62: mp 133°–134°; $[\alpha]_D$ +92°; ir (KBr): 3000, 2950, 2940, 2920, 2850, 1750, 1675, 1450, 1390, 1270, 1180, 1100, 1060, 985, 935, 900, 860, 840 cm^{-1}; ^1H nmr: δ5.55(1H,dq,*J* = 10,1.5,1), 5.15(1H,bt,*J* = 5), 5.05(1H,dq,*J* = 10,1,1), 2.68(1H,t,*J* = 4), 1.89(3H,d,*J* = 1), 1.85(3H,d,*J* = 1.5), 1.63(3H,s), 2.37(2H,m), 1.2(3H,s) (incomplete); ^{13}C nmr: 173.0, 162.3, 143.9, 135.4, 125.0, 122.9, 120.8, 78.8, 61.4, 59.9, 38.9, 37.4, 36.3, 27.4, 25.3, 23.3, 17.2, 16.1, 15.5, 8.9

collected in the Red Sea (Bernstein *et al.*, 1974; Kashman, 1977). The structure of sarcophine was obtained by single-crystal X-ray diffraction analysis. In a later publication (Kashman *et al.*, 1974), several related structures (**63–67**) were reported as minor components of the same ex-

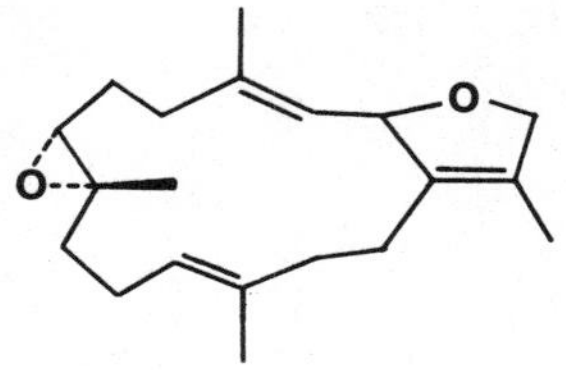

63

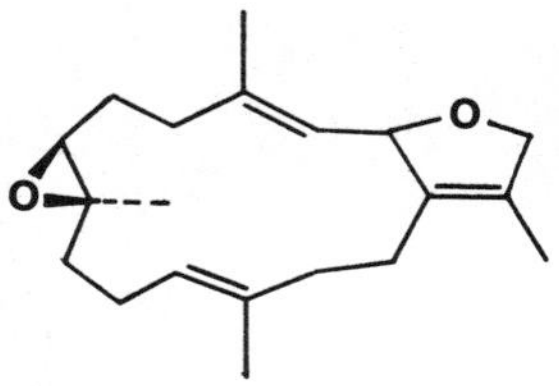

64

tract. Among the most abundant were the dihydrofurans **63** and **64**, which may be formally considered simple lactone reduction products of sarcophine, disregarding stereochemical implications. Two other compounds (**65** and **66**) were shown to be two diastereomers of sarcophine, largely on spectral grounds. Finally, trace amounts of a conjugated diene were isolated, and its structure was proposed to be the alcohol **67** on spectral grounds.

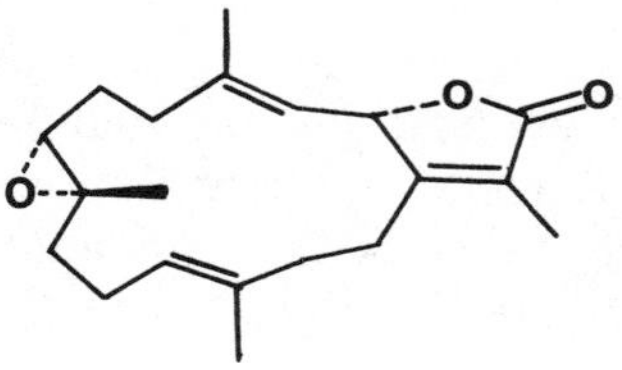

65

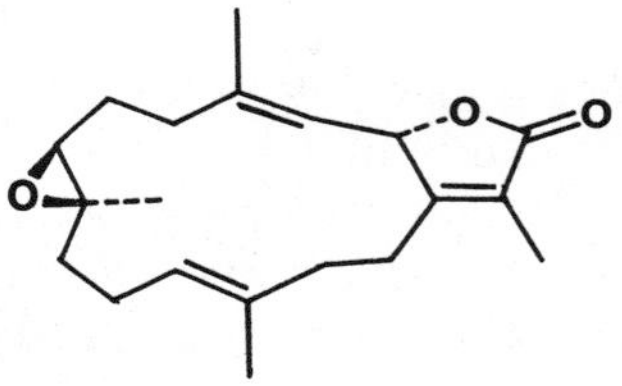

66

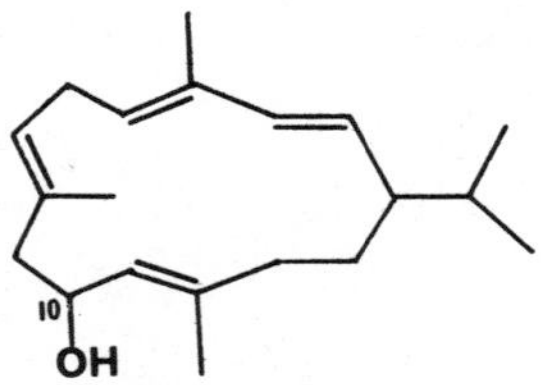

67

Extracts of a Great Barrier Reef collection of an unidentified *Sarcophytum* species contained the cembrene epoxide **68**, the structure of which was ascertained by X-ray crystallography (Coll *et al.*, 1978). This compound is closely related to the metabolite epoxynephtenol acetate (**49**) from the close relative *Nephtea*. Epoxide **68** can, in principle, be converted to **49** by hydrolysis of the acetate ester and dehydrogenation between C-1 and C-2.

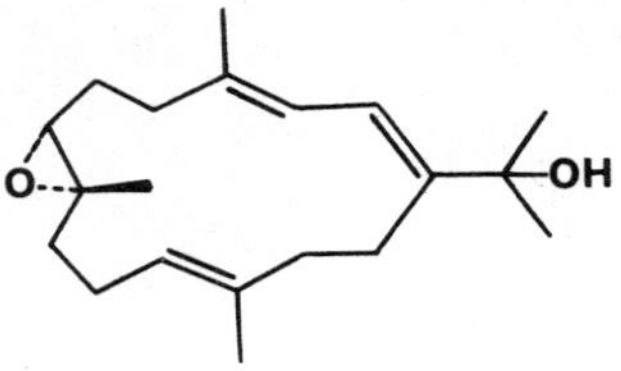

68

68: mp 105°–106°; $[\alpha]_D$ +61.8°; ir (KBr): 3510 cm^{-1}; uv: 252 nm (ϵ = 22,000); ^1H nmr ($CDCl_3$): δ6.38(1H,d,J = 11), 5.96(1H,d,J = 11), 5.16(1H,bt,J = 6), 2.88(1H,dd,J = 4.4,6), 1.78(3H,s), 1.64(3H,s), 1.36(6H,s), 1.26(3H,s); ^{13}C nmr: 148.5(s), 136.6(s), 136.0(s), 124.8(d), 120.6(d), 118.3(d), 73.6(s), 62.2(d), 59.9(s), 41.6(t), 38.6(t), 35.5(t), 29.7(q), 26.3(t), 26.0(t), 23.2(t), 17.7(q), 17.7(q), 17.0(q), 16.0(q)

Some additions to the list of cembrene derivatives from *Sarcophytum* came recently from my colleagues at Scripps (Ravi and Faulkner, 1978). Several ketones (**69–72**) and two new epoxides (**73** and **74**) were isolated and their structures elucidated by extensive chemical and spectral analysis. These compounds were isolated in large amounts from an unidentified *Sarcophytum* species collected in the South Pacific at Canton Island (Phoenix Island group).

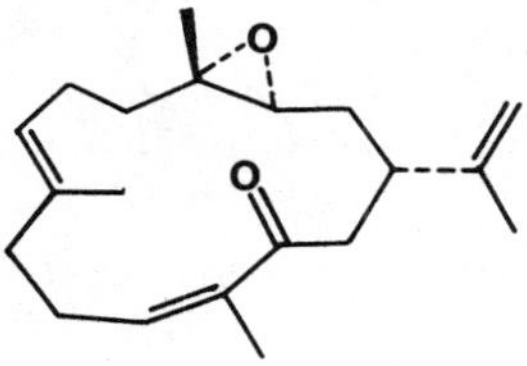

69 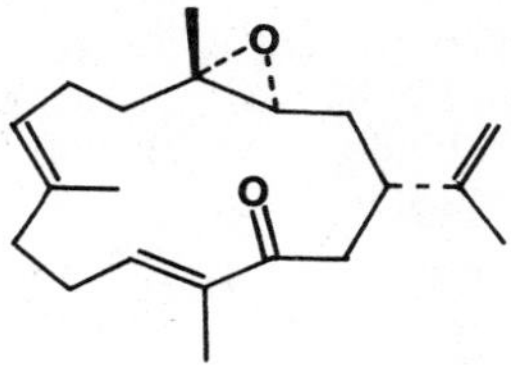 **70**

69: ir: 1690, 1642, 918 cm^{-1}; uv: 236 nm (ϵ = 3000); ^1H nmr: δ5.68(1H,t,J = 6.5), 5.08(1H,t,J = 7), 4.85(1H,s), 4.74(1H,s), 2.84(3H,m), 2.64(1H,t,J = 6.5), 1.84(3H,s), 1.80(3H,s), 1.61(3H,s), 1.21(3H,s); ^{13}C nmr: 186.7(s), 147.2(s), 137.3(s), 135.4(s), 134.5(d), 124.1(d), 110.5(t), 60.4(d), 59.5(s), 44.9(t), 39.0(t), 38.4(t), 38.2(d), 31.1(t), 29.2(t). 23.54(t), 21.7(q), 20.5(q), 16.7(q), 16.7(q)

70: ir: 1690, 1650, 910 cm^{-1}; uv: 234 nm (ϵ = 9500); ^1H nmr: δ6.64(1H,t,J = 7), 5.18(1H,t,J = 7), 4.74(1H,s), 4.64(1H,s), 3.09(1H,dd,J = 7,14), 2.76(1H,dd,J = 5,7), 2.64(1H,bq), 1.75(3H,s), 1.73(3H,s), 1.64(3H,s), 1.22(3H,s); ^{13}C nmr: 189.0(s), 147.4(s), 143.4(d), 137.3(s), 134.2(s), 125.5(d), 110.8(t), 61.7(d), 59.8(s), 43.7(d), 40.9(t), 38.6(t), 38.2(t), 32.5(t), 24.1(t), 19.4(q), 16.6(q), 15.7(q), 11.4(q)

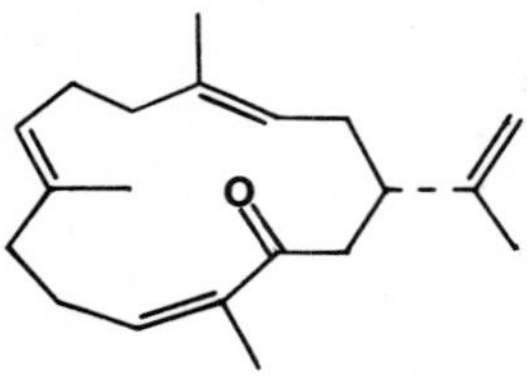

71

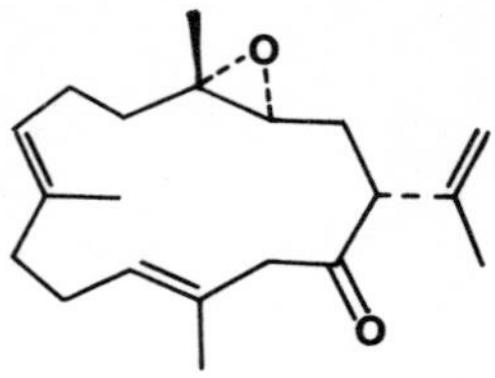

72

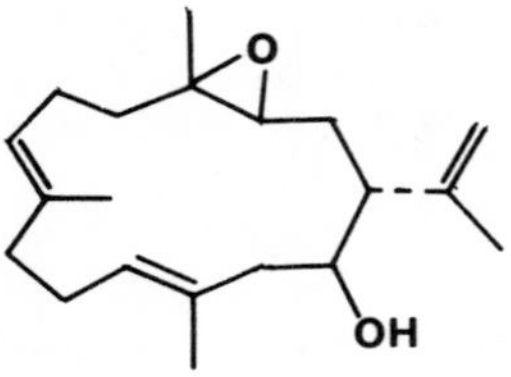

73

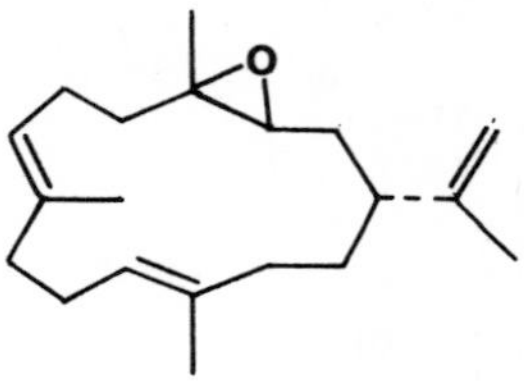

74

71: ir: 2865, 1692, 1650, 921 cm^{-1}; uv: 234 nm (ϵ = 3200); ^{1}H nmr: δ5.61(1H,t,J = 7.5), 5.02(2H,m), 4.77(1H,s), 4.69(1H,s), 2.71(3H,m), 1.83(3H,s), 1.78(3H,s), 1.62(3H,s), 1.55(3H,s); ^{13}C nmr: 181.5(s), 148.2(s), 136.4(s), 135.4(s), 134.5(d), 134.4(s), 124.3(d), 123.2(d), 109.1(t), 45.3(t), 40.5(d), 39.0(t), 38.7(t), 32.2(t), 29.2(t), 24.0(t), 21.0(q), 20.2(q), 16.7(q), 15.1(q)

72: mp 65°–66°; ir: 2865, 1650, 1706, 915 cm^{-1}; uv: 217 nm (ϵ = 1070); ^{1}H nmr: δ5.02(1H,t), 4.88(1H,t), 4.73(1H,s), 4.66(1H,s), 2.86(1H,t), 2.61(3H,m), 1.71(3H,s), 1.60(3H,s), 1.40(3H,s), 1.35(3H,s); ^{13}C nmr: 209.6(s), 148.4(s), 136.2(s), 132.1(s), 127.7(d), 122.3(d), 110.2(t), 63.6(s), 58.5(d), 39.4(t), 39.0(d), 37.0(t), 36.6(t), 30.7(t), 25.4(t), 24.7(t), 21.7(q), 15.5(q), 15.0(q), 12.4(q)

73: ir: 3340, 2865, 918 cm^{-1}; ^{1}H nmr: δ5.12(1H,t), 5.09(1H,t), 5.00(1H,dd,J = 5,7), 4.79(1H,s), 4.70(1H,s), 2.92(1H,t,J = 7), 2.04(3H,s), 1.69(3H,s), 1.64(3H,s), 1.55(3H,s), 1.20(3H,s); ^{13}C nmr: 169.8(s), 147.4(s), 135.7(s), 132.1(s), 126.6(d), 122.4(d), 110.6(t), 73.6(d), 59.6(s), 58.03(d), 44.7(d), 38.4(t), 35.9(t), 32.1(t), 29.0(t), 24.5(t), 23.8(t), 20.4(q), 18.7(q), 15.6(q), 14.7(q)

74: ir: 3030, 2958, 1648, 1443, 1383, 932 cm^{-1}; ^{1}H nmr: δ5.07(2H,m), 4.77(1H,s), 4.64(1H,s), 2.64(1H,dd,J = 5,7), 1.71(3H,s), 1.64(3H,s), 1.53(3H,s), 1.21(3H,s); ^{13}C nmr: 147.5(s), 135.0(s), 132.6(s), 126.2(d), 124.4(d), 110.0(t), 61.0(s), 61.0(d), 45.0(d), 38.7(t), 36.3(t), 32.6(t), 29.8(t), 24.7(t), 24.4(t), 24.2(t), 21.0(q), 17.2(q), 14.9(q), 14.9(q)

Finally, two independent studies of soft corals of the genus *Lobophytum* have appeared. From extracts of the Indonesian coral *L. cristagalli* Tursch and co-workers (1974) isolated and solved, by spectrochemical and X-ray diffraction methods, the structure of lobophytolide (**75**) with relative stereochemistry as shown.

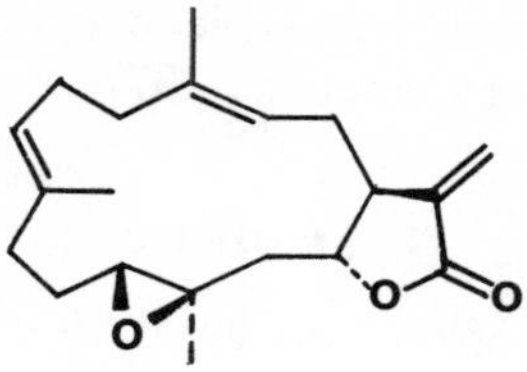

75 Lobophytolide

75: mp 137°–138°; $[\alpha]_D$ +7°; ir: 1755, 1670 cm^{-1}; uv: 208 nm (ϵ = 9080); ^1H nmr ($CDCl_3$): δ5.02(2H,m), 3.94(1H,q), 2.82(1H,m), 2.80(1H,t,J = 6), 2.66(1H,m), 2.22(8H,m), 1.66(3H,s), 1.56(3H,s), 1.29(3H,s) (incomplete)

From collections of an undescribed *Lobophytum* species from the Gulf of Eilat (Red Sea), Kashman and Groweiss (1977) isolated a new epoxycembranolide and proposed the name lobolide and the structure **76**.

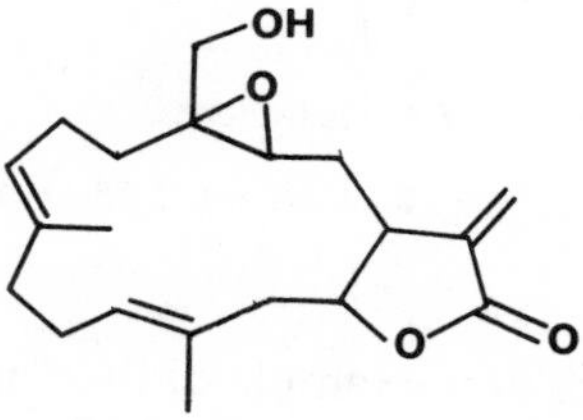

76 Lobolide

76: mp 114°–115°; $[\alpha]_D$ −58°; ir (CCl_4): 1770, 1740, 1660 cm^{-1}; uv: (MeOH): 209 nm (ϵ = 7600); ^1H nmr ($CDCl_3$): δ6.28(1H,d,J = 3.1), 5.93(1H,d,J = 2.8), 5.18(1H,bq), 5.04(1H,bt), 4.34(1H,d,J = 12), 4.17(1H,ddd,J = 9.7,6.7,2.9), 3.93(1H,d,J = 12), 2.94(1H,dd,J = 6.9,3.8), 2.85(1H,m), 2.12(3H,s), 1.88(1H,dt,J = 14.9,3.8), 1.72(3H,bs), 1.61(3H,bs), 1.52(1H,dd,J = 14.9, 6.9); ^{13}C nmr ($CDCl_3$): 170.4(s), 169.4(s), 139.7(s), 135.0(s), 130.2(d), 129.5(s), 124.5(d), 122.7(t), 79.6(d), 64.3(t), 61.9(d), 60.4(s), 45.3(t), 45.0(d), 38.7(t), 32.4(t), 31.8(t), 24.8(t), 23.2(t), 20.5(q), 17.0(q), 15.6(q)

The structure of lobolide was based on extensive nmr analysis including interpretation of lanthanide-induced shift (LIS) data. Microozonolysis to yield levulinaldehyde precluded positioning the epoxide in the C-7–C-8 position. The oxidized methyl group is an unfamiliar feature of cembranolides, but it is also found in the previously mentioned gorgonian metabolite asperdiol (**46**). Secondary metabolites, and particularly the cembrene derivatives, are found in large amounts in the sensitive tissues of gorgonians and soft corals, and it has been proposed that they may function as chemical defense agents. A number of the cembranolides have indeed been shown to be toxic to fish (Neeman *et al.*, 1974; Ciereszko and Karns, 1973; Kashman and Groweiss, 1977).

G. Flexibilene

During an investigation of the soft coral *Sinularia flexibilis* the Tursch group (Herin *et al.*, 1976) isolated, in addition to the aforementioned cembrene derivatives (**54–57**), a hydrocarbon fraction rich in two major diterpenoids. Preparative gas chromatography separated the major hydrocarbon, identified as (−)-cembrene A, and gave small amounts of an isomeric diterpene, flexibilene (**77**). Flexibilene possesses the 15-

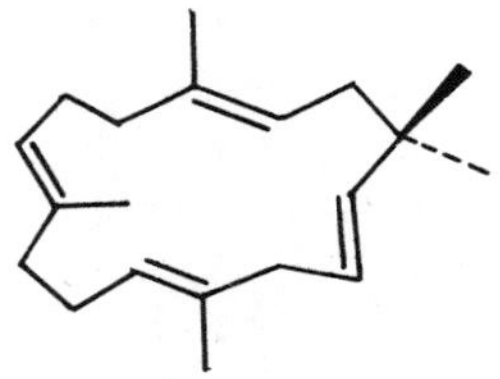

77 Flexibilene

77: ir: 1380, 1360, 970 cm^{-1}; ^1H nmr: δ5.1(5H,m), 2.55(2H,m), 2.05(10H,m), 1.57(9H,bs), 0.98(6H,s)

membered ring and is the first example of this unique diterpene skeleton. Structure **77** was securely assigned to this hydrocarbon on the basis of the interpretation of a series of unambiguous spectral features. Flexibilene can occur by a direct cyclization of geranylgeraniol between C-1 and C-15. However, it is also possible that a bicyclic intermediate, such as the known cembrene derivative casbene (**78**), could isomerize to yield flexibilene.

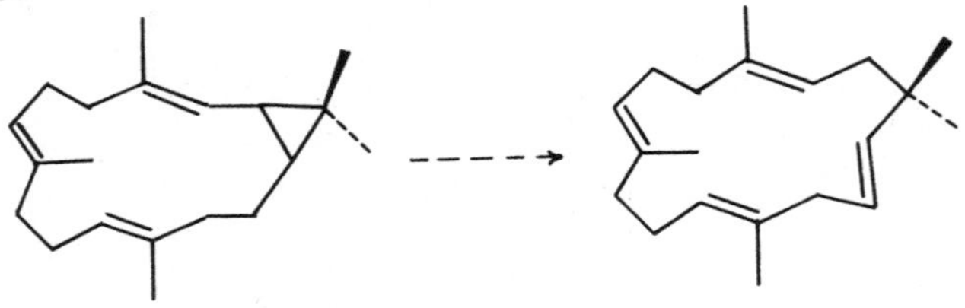

78

77

IV. BICYCLIC DITERPENOIDS

It is convenient to consider the genesis of the bicyclic diterpenoids by two consecutive ring closure reactions. Furthermore, gross structural and

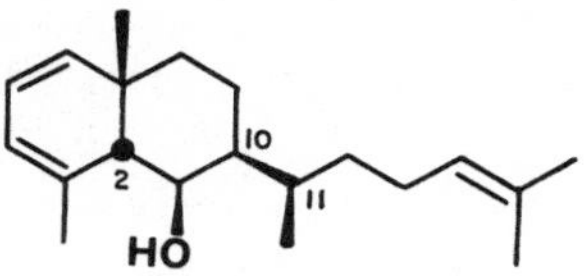

79 Dictyolene

79: ir (CCl_4): 3600, 3030, 1650, 1630, 1600 cm^{-1}; uv (EtOH): 209, 269 nm (ϵ = 4200, 4000); ^1H nmr: δ5.73(1H,dd,J = 9.5,5), 5.66(1H,bm), 5.15(1H,bt,J = 6), 5.29(1H,d,J = 9.5), 4.09(1H,bs), 1.90(3H,s), 1.73(3H,s), 1.64(3H,s), 1.07(3H,s), 0.95(3H,d,J = 6) (incomplete); ^{13}C nmr: 135.6, 119.1, 121.9, 72.0 (incomplete)

stereochemical considerations indicate in many cases that initial cyclization produces a medium or large ring and that subsequent cyclization occurs in a transannular fashion from the more rigidly held cyclic system. With this viewpoint in mind, I have organized the bicyclic compounds in this section according to their relationships to the larger ring precursors discussed in Section III.

A. Dictyolene

In their study of the diterpenoids from the Hawaiian brown alga *Dictyota acutoloba* Erickson and co-workers (Sun *et al.*, 1977) isolated the formally monocyclic compound dictyoxepin (**28**) (Section III,D). At the same time a bicyclic conjugated diene, called dictyolene, was isolated, which, on the basis of its spectral similarity to and biogenetic relationship with **28**, was proposed to have structure **79**. Bridgehead stereochemistry of this molecule was assigned cis on the basis of an 11% NOE enhancement of the proton (C-2) and bridgehead methyl resonances, and the relative stereochemistries at C-10 and C-11 were derived by careful analysis of ^1H nmr spectra. Dictyolene can logically be related to the 10-membered ring precursor dilophol (**26**) by further C-2–C-7 closure, as discussed in Section III.

B. *Dictyota* Bicyclo[8.1.0] Compounds

Recent and as yet unpublished work at the Roche Institute in Australia (Kazlauskas *et al.*, 1978d) has yielded a series of very interesting bicyclic diterpenoids that possess unprecedented bicyclo[8.1.0]undecane skeletons (**80–84**). These unusual compounds were isolated from an undescribed species of the widely distributed brown alga *Dictyota* and were

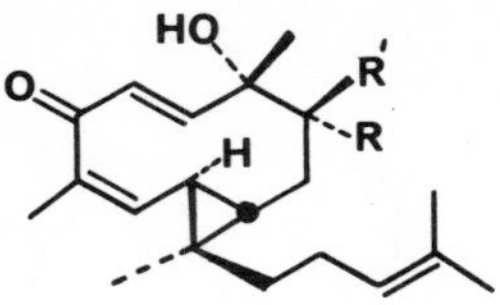

80 R′=H, R=OAc

81 R′=R=H

82 R′=OAc, R=H

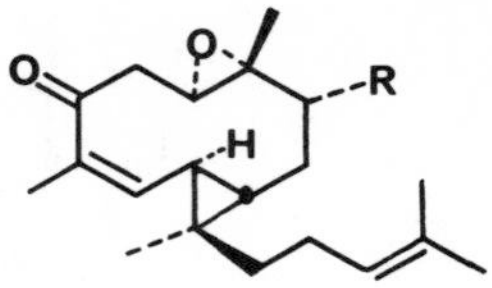

83 R=H

84 R=OAc

elucidated by an X-ray analysis of the cross-conjugated dienone **80**. The biogenesis of these *Dictyota* metabolites logically follows that of the related compound containing a 10-membered ring (dilophol, **26**), complemented by subsequent cyclopropane formation between C-1 and C-11. The absolute stereochemistry of **80** was not obtained, and spectral details for these compounds are not yet available.

C. Pachydictyol A and Dictyols

Brown seaweeds of the family Dictyotaceae were recognized from early studies in Japan (Takaoka and Ando, 1951) to contain interesting secondary metabolites. My early work in this field begain in 1971 by the initiation of what has become a continuing study of the biologically related seaweeds *Pachydictyon, Dictyota, Dilophus,* and others. These seaweeds, which have a similar appearance and are differentiated only by microscopic anatomical features, produce a variety of unprecedented monocyclic and bicyclic diterpenoids, examples of which are discussed in Sections III and IV. The first compound that we encountered from this group of seaweeds (Hirschfeld *et al.*, 1973) was the substituted hydroazulene diterpene alcohol pachydictyol A (**85**). This new structural type was isolated from extracts of the Pacific seaweed *Pachydictyon coriaceum,* collected at numerous southern California locations. Chemical and spectral studies of **85** led to the realization that pachydictyol A possessed a new terpenoid skeleton. The alcohol itself did not crystallize, but the corresponding *p*-bromophenylurethane derivative (**86**) formed excellent crystals, and its total structure was solved by X-ray crystallographic methods. Pachydictyol A has more recently been isolated from a number of related seaweeds, and it is probably the most frequently encountered metabolite in this group.

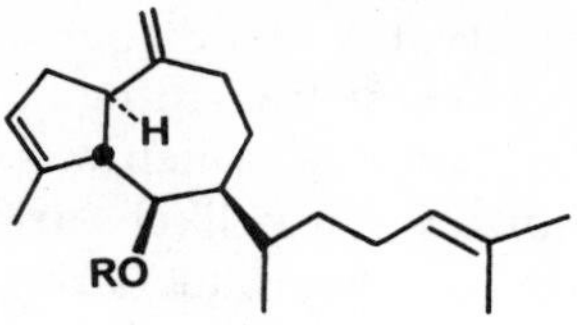

85 R=H Pachydictyol A

86 R=–CO–NH–C_6H_4Br (*p*)

85: Oil; $[\alpha]_D$ +106°; ir (film): 3484, 1634, 890 cm^{-1}; ^1H nmr ($CDCl_3$): δ5.30(1H,m), 5.10(1H,t,*J* = 7), 4.72(2H,s), 3.85(1H,bs), 1.75(3H,bs), 1.67(3H,s), 1.60(3H,s), 0.97(3H,d,*J* = 7) (incomplete); ^{13}C nmr ($CDCl_3$): 152.5(s), 141.5(s), 131.2(s), 124.9(d), 124.0(d), 107.2(t), 75.4(d), 60.8(d), 47.8(d), 46.5(d), 40.5(t), 35.2(d), 34.3(t), 25.9(t), 25.6(q), 23.8(t), 17.6(q), 17.6(q), 15.7(q)

In a more recent study (Robertson and Fenical, 1977) the epoxide derivative of pachydictyol A (**87**) was isolated from extracts of the Gulf of California seaweed *Dictyota flabellata*. Lithium aluminum hydride reduction of **87** gave a single crystalline diol (**88**), which was further converted,

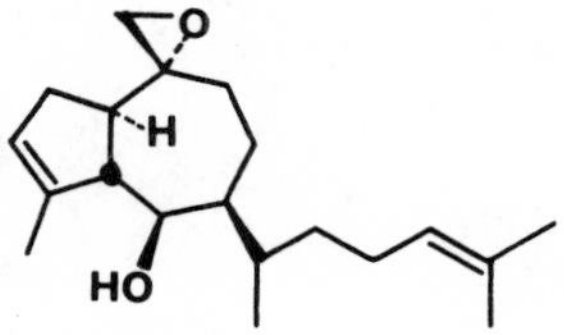

87 Pachydictyol A epoxide

87: $[\alpha]_D$ +28.6°; ir ($CHCl_3$): 3450, 1600, 884 cm^{-1}; ^1H nmr (CCl_4): δ5.18(1H,bs), 5.07(1H,dd,*J* = 7,7), 3.84(1H,dd,*J* = 8,3), 2.62(1H,d,*J* = 5), 2.48(2H,m), 2.34(1H,d,*J* = 5), 1.74(3H,s), 1.66(3H,s), 1.57(3H,s), 1.20(2H,m), 1.91–1.36(9H,m), 0.98(3H,d,*J* = 7); ^{13}C nmr ($CDCl_3$): 141.2(s), 131.3(s), 125.0(d), 124.2(d), 74.6(d), 62.2(s), 58.0(t), 50.6, 48.8, 43.9, 39.7, 35.4, 34.9, 31.2, 25.9, 25.9, 20.6, 17.7, 17.5, 15.7

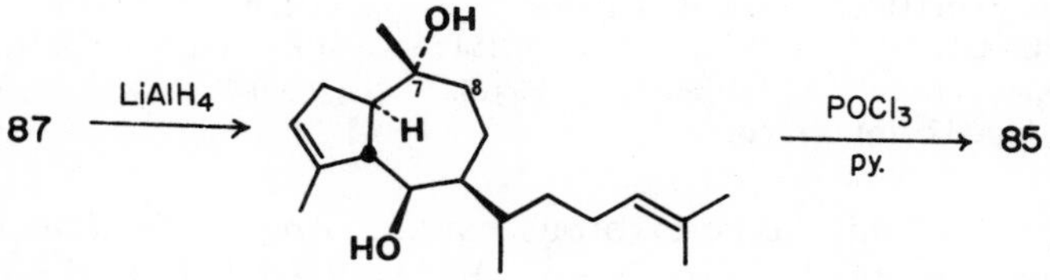

87 —$LiAlH_4$→ 88 —$POCl_3$, py.→ 85

88

but in low yield, to pachydictyol A (**85**) by dehydration with phosphorus oxychloride in pyridine (py). These transformations rigorously defined all asymmetric centers of **87** except C-7, which was eliminated. The alcohol stereochemistry at C-7, however, was likely to be cis to the bridgehead proton at C-6, which is based on the products obtained from a dehydration reaction. Since phosphorus oxychloride dehydrations are known to proceed via a trans-diaxial elimination (Barton *et al.*, 1956) and the Δ^7 olefin was the exclusive endocyclic product, the C-7 hydroxyl in **88** must be trans-diaxial only to the axial proton at C-8. These predictions were later confirmed since the C-7 epimeric alcohol (dictyol C) was found as a natural product in various collections of the Atlantic alga *Dictyota dichotoma* (Danise *et al.*, 1977; Faulkner *et al.*, 1977).

The rather widely distributed Mediterranean alga *Dictyota dichotoma* (in this case var. *implexa*) has been shown to be the source of five diterpene alcohols closely related to pachydictyol A. Working independently, Minale and Riccio (1976) and the Fattorusso group (Fattorusso *et al.*, 1976b) discovered and described two new diterpenoids: dictyol A (**89**) and dictyol B (**90**). Minale and Riccio had isolated these compounds not

89 Dictyol A

90 Dictyol B

89: mp 85°–86°; $[\alpha]_D$ +86.2°; ir: 3400 cm^{-1}; ^1H nmr ($CDCl_3$): δ5.46(2H,bs), 5.09(1H,t,J = 7), 4.62(1H,d,J = 7), 4.36(2H,s), 4.06(1H,m), 3.13(1H,m), 2.85(1H,d), 1.86(3H,s), 1.68(3H,s), 1.61(3H,s), 0.99(3H,d,J = 6) (incomplete); ^{13}C nmr ($CDCl_3$): 151.2(s), 142.8(s), 131.9(s), 124.6(d), 123.9(d), 121.4(d), 85.8(d), 75.0(t), 74.6(d), 61.3(d), 48.4(d), 45.0(d), 35.3(t), 34.2(d), 26.4(t), 25.6(t), 25.6(q), 17.7(q), 17.4(q), 15.8(q)

90: mp 110°–114°; $[\alpha]_D$ +73.5°; ir: 3340, 3040, 1640, 896 cm^{-1}; ^1H nmr (CCl_4): δ5.21(1H,m), 5.08(1H,bs), 5.05(1H,t,J = 7), 4.81(1H,bs), 4.03–3.65(2H,m), 1.75(3H,s), 1.67(3H,s), 1.59(3H,s), 1.00(3H,d,J = 6); ^{13}C nmr ($CDCl_3$): 154.5(s), 140.9(s), 131.5(s), 124.6(d), 123.8(d), 104.0(t), 76.4(d), 74.9(d), 61.3(d), 43.8(d), 43.0(d), 35.1(t), 35.0(d), 33.8(t), 33.4(t), 25.7(t), 25.7(q), 17.7(q), 17.5(q), 15.6(q)

from algae but from the herbivorous sea hare *Aplysia depilans* collected near Naples, and the Fattorusso group had isolated dictyols A and B from *D. dichotoma* var. *implexa* in Sicily. Inspection of the gut contents of *A. depilans* clearly indicated, however, that the sea hare had been feeding mainly on *Dictyota dichotoma*. Sea hares are well known to develop distinct species preferences in their algal diets. *Aplysia californica*, for example, prefers red algae of the genera *Laurencia* and *Plocamium* (Stallard and Faulkner, 1974), whereas the Hawaiian sea hare *Stylocheilus longicauda* prefers the blue-green alga *Lyngbya majuscula* and concentrates several *Lyngbya* metabolites in its digestive organ (Dalietos and Moore, 1977). On the basis of Minale's discovery of *Dictyota* metabolites in *A. depilans* and Vanderah and Faulkner's (1976) observation of pachydictyol A in digestive glands of the Pacific sea hare *A. vaccaria*, it appears that sea hares have developed feeding preferences for macroalgae of at least three major divisions (Rhodophyta, Phaeophyta, Cyanophyta).

When Minale and Fattorusso discovered that both laboratories were working on identical diterpenoids, the groups later joined forces for publication of the minor metabolites from *D. dichotoma* var. *implexa* (Danise *et al.*, 1977). Two additional relatives of pachydictyol A, dictyol C (**91**) and dictyol D (**92**), were described and their structures were based on

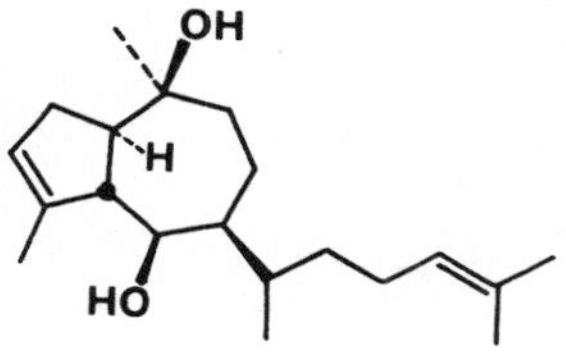

91 Dictyol C

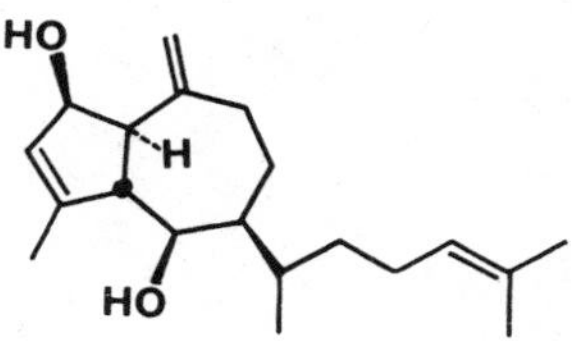

92 Dictyol D

91: mp 68°; $[\alpha]_D$ −16.6°; ^{1}H nmr ($CDCl_3$): δ5.26(1H,bs), 5.14(1H,t,J = 6), 3.87(1H,dd,J = 9,3), 2.74(1H,bs), 2.21(1H,m), 1.85(3H,s), 1.70(3H,s), 1.62(3H,s), 1.22(3H,s), 1.00(3H,d,J = 6) (incomplete) ^{13}C nmr ($CDCl_3$): 142.7(s), 131.3(s), 124.7(d), 123.3(d), 74.5(d), 72.5(s), 52.8(d), 50.0(d), 49.2(d), 46.6(t), 34.8(t), 34.5(d), 32.9(t), 30.0(q), 25.7(q), 25.6(t), 19.7(t), 17.7(q), 17.5(q), 16.3(q)

92: $[\alpha]_D$ −80°; ^{1}H nmr ($CDCl_3$): δ5.56(1H,bs), 5.11(1H,bs), 5.11(1H,bs), 4.86(1H,bs), 4.50(1H,bs), 3.86(1H,bd,J = 9), 2.94(1H,bt,J = 10), 2.62(1H,bs), 1.92(3H,s), 1.70(3H,s), 1.61(3H,s), 0.98(1H,d,J = 6) (incomplete); ^{13}C nmr ($CDCl_3$): 150.3(s), 147.8(s), 131.4(s), 125.6(d), 124.6(d), 111.8(t), 75.1(d), 74.5(d), 52.6(d), 52.4(d), 44.4(d), 35.6(t), 35.1(t), 34.3(d), 25.7(q), 25.5(t), 22.7(t), 17.7(q), 17.3(q), 15.8(q)

careful spectral analysis and interconversion with known compounds. Dictyol C (**91**) was dehydrated ($POCl_3$/py) to yield pachydictyol A (**85**). Dictyol D (**92**) gave on reduction (lithium in ethylamine) the derivative **93**, which was also produced under the same conditions from pachydictyol A (**85**). In the same publication a related compound, dictyol E (**94**), was described, which was isolated not from *Dictyota* but from the closely related alga *Dilophus ligulatus*. Accompanying dictyol E in *D. ligulatus* was the monocyclic diterpenoid dilophol (**26**) (Section III,C).

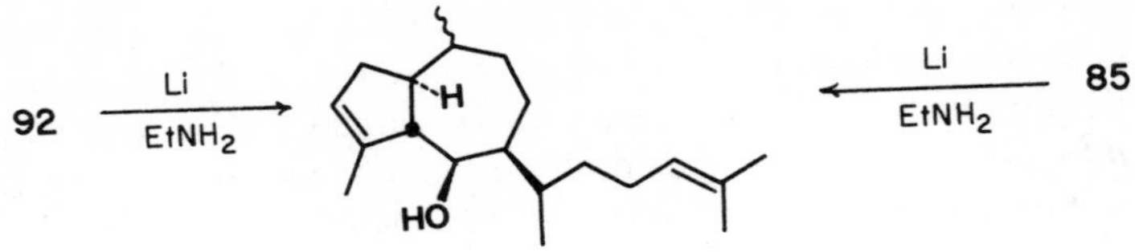

92 $\xrightarrow[EtNH_2]{Li}$ 93 $\xleftarrow[EtNH_2]{Li}$ 85

The nonvarietal form of *Dictyota dichotoma*, collected in southern Wales, was shown to contain quite similar diterpenoids. Two known compounds, pachydictyol A (**85**) and dictyol C (**91**), were isolated along with the corresponding acetate of dictyol B (**90**) and a new diol labeled dictyotadiol (**95**). The structure of this modification of the dictyols (with relative stereochemistry only) was determined by X-ray diffraction methods (Faulkner *et al.*, 1977).

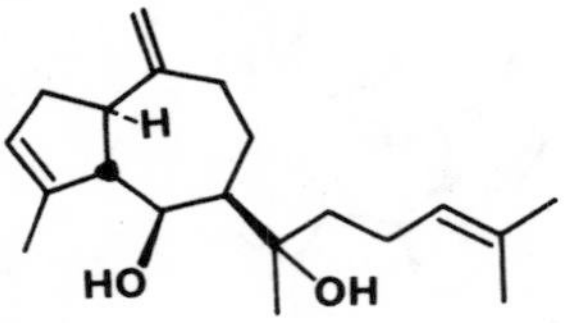

94 Dictyol E

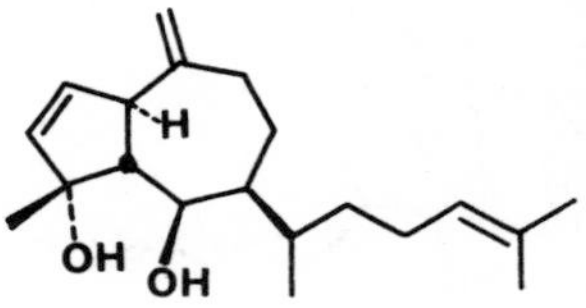

95 Dictyotadiol

94: $[\alpha]_D$ +26.8°: ^{1}H nmr ($CDCl_3$): 5.32(1H,bs), 5.14(1H,t,J = 6), 4.75(2H,bs), 4.18(1H,bd,J = 9), 1.82(3H,s), 1.70(3H,s), 1.62(3H,s), 1.23(3H,s); ^{13}C nmr ($CDCl_3$): 151.9(s), 140.9(s), 131.7(s), 124.4(d), 124.2(d), 107.5(t), 76.2(s), 74.4(d), 60.3(d), 48.7(d), 46.2(d), 41.0(t), 40.5(t), 33.8(t), 25.7(q), 25.4(q), 23.3(t), 21.7(t), 17.6(q), 15.8(q)

95: mp 150°–151°; $[\alpha]_D$ +1.3°; ir (Nujol): 3500 cm^{-1}; ^{1}H nmr (Me_2CO-d_6): δ5.74(2H,s), 5.15(1H,t,J = 7), 4.54(2H,bs), 4.48(1H,m), 3.20(1H,d,J = 8), 1.68(3H,s), 1.60(3H,s), 1.44(3H,s), 1.00(3H,d,J = 7) (incomplete)

The perhydroazulene ring system as discussed here for pachydictyol A and the dictyols is well known in the sesquiterpenoids and has been predicted to arise in the C_{15} system by a transannular cyclization of germacrene (**37**). In an analogous fashion the genesis of pachydictyol A derivatives can be envisioned from dilophol (**26**) via a C-2–C-6 closure.

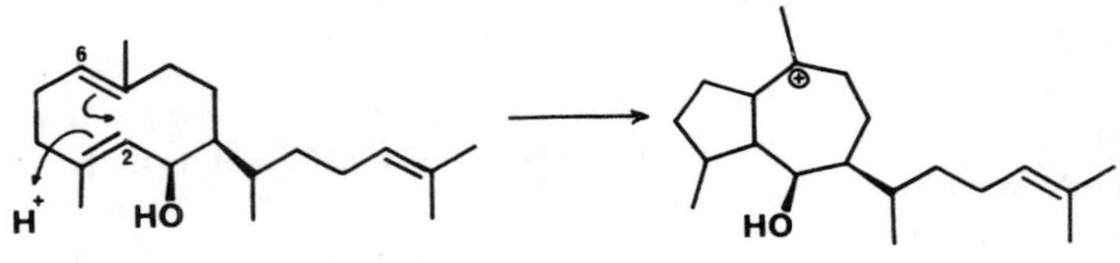

26

D. Acetoxycrenulatin

Our recent work with various *Dictyota* species has resulted in the isolation and partial characterization of a bicyclic cyclopropane-containing compound, acetoxycrenulatin (**96**), which was first isolated

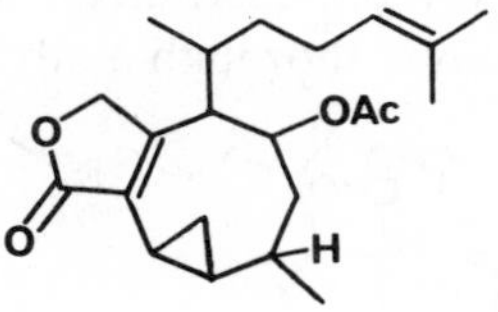

96 Acetoxycrenulatin

96: $[\alpha]_D$ +13°; uv (MeOH): 227 nm (ϵ = 11,500); ir (CCl_4): 1760, 1735 cm^{-1}; ^1H nmr ($CDCl_3$): δ5.48(1H,bt,*J* = 4), 5.07(1H,dd,*J* = 7.5,7.5), 4.85(1H,dd,*J* = 16,2.5), 4.74(1H,dd,*J* = 16,2.5), 3.23(1H,d,*J* = 7.5), 2.03(3H,s), 1.70(3H,s), 1.59(3H,s), 1.02(3H,d,*J* = 7.5), 0.99(3H,d,*J* = 6), 0.89(1H,m), 0.39(1H,m) (incomplete); ^{13}C nmr ($CDCl_3$): 174.2(s), 169.7(s), 166.5(s), 132.4(s), 128.8(s), 123.5(d), 72.2(d), 71.5(t), 47.4(d), 43.9(t), 35.7, 32.8, 29.2, 25.9, 25.7, 25.5, 23.4, 21.3, 17.7, 17.1, 10.3, 8.4(t)

from the Gulf of California alga *D. crenulata* (McEnroe *et al.*, 1977). The proposed structure of acetoxycrenulatin is based on detailed analysis of spectral data and a number of chemical conversions. Unfortunately neither **96** nor any of its derivatives were crystalline, which so far has precluded an unambiguous structure elucidation. An obvious relationship exists between acetoxycrenulatin and the monocyclic compound dictyodial (**24**), also from *Dictyota* (Section III,B). The cyclopropane moiety in **96** may be thought of as the product of a proton-induced transformation of the homoconjugated diene in **24**.

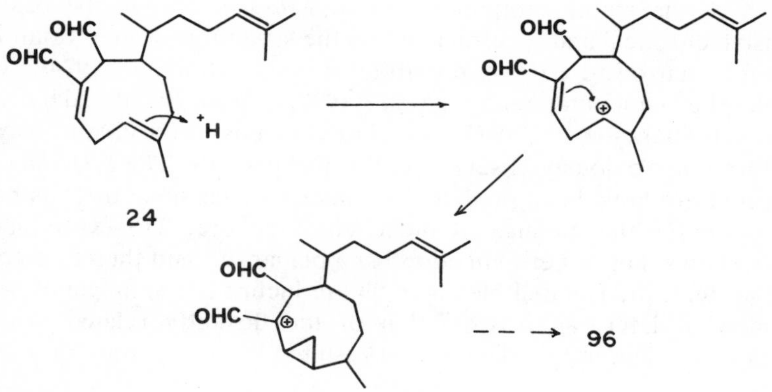

E. *Dictyota* Bicyclo[4.3.0] Compounds

Current research at the Roche Research Institute of Marine Pharmacology in Australia (Wells, 1978) has resulted in the determination of the structure of an unusual dialdehyde alcohol (**97**) and its corresponding acetate (**98**) from an undescribed Australian *Dictyota* species. The structure of **97** was determined by a thorough analysis of ^{1}H nmr and ^{13}C nmr

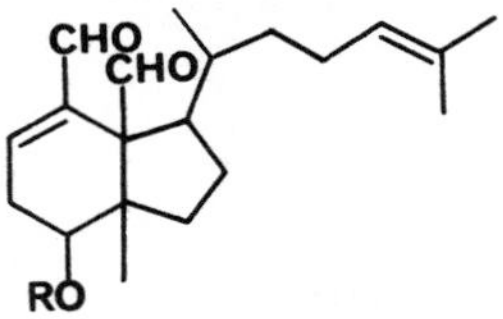

97 R=H
98 R=Ac

data, including lanthanide shift and double resonance experiments. Details of this work and spectral data are not yet available. As in acetoxycrenulatin (**96**), the dialdehydes **97** and **98** can be related to the macrocycle dictyodial (**24**) by a transannular ring closure, in this case between C-2 and C-7.

F. Briarein A, Stylatulide, and Ptilosarcone

Some years ago the Oklahoma group (Hyde, 1966) announced the isolation of a series of new highly oxygenated diterpenoids which were shown by mass spectrometry to contain one chlorine atom. These compounds were components of the Caribbean gorgonian *Briareum asbestinum*. During the same period the Tursch group in Brussels (Hale *et al.*, 1970) encountered the same compounds, the "briareins," during an investigation of the sterol components of *B. asbestinum*. On the basis of extensive chemical and spectral analysis the structures of the "briareins" were first narrowed to two possible types (Bartholome, 1974). More recently the major compound briarein A (**99**) has been elucidated by X-ray diffraction (Burks *et al.*, 1977) and shown to possess a unique bicyclic structure not previously observed in the diterpenoids. The 1,3-diene system could not have been predicted in this molecule since there is no uv absorption for the conjugated diene chromophore. The X-ray model clearly shows that rather than adopting a planar (0°) and therefore conjugated system, the isolated olefins in **99** are inclined at an angle of 68.9°.

Somewhat later, an investigation of the distantly related sea pen *Stylatula* (Coelenterata, Alcyonacea) resulted in the isolation of a series

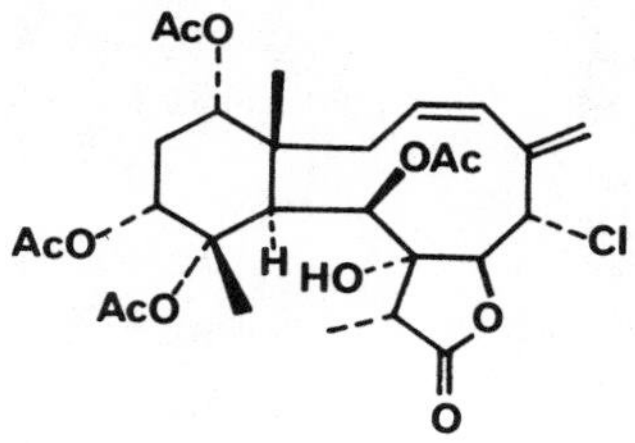

99 Briarein A

99: ^{1}H nmr ($CDCl_3$): δ6.12(1H,d,*J* = 10), 6.00(1H,d,*J* = 12), 5.84(1H,dd,*J* = 12,10), 5.71(1H,dd,*J* = 3.4,2.8), 5.60(1H,dd,*J* = 1,1), 5.58(1H,s), 5.48(1H,dd,*J* = 1,1), 5.12(1H,ddd,*J* = 3.7,1,1), 4.87(1H,d,*J* = 3.7), 4.87(1H,dd,*J* = 3,2.5), 3.14(1H,ddd,*J* = 7,7,7), 3.10(1H,s), 2.35(1H,ddd,*J* = 17,2.8,2.5), 1.99(1H,ddd,*J* = 17,3.4,3), 1.54(3H,s), 1.40(3H,s), 1.34(3H,d,*J* = 7)

of closely related metabolites, the major component being the briarein A-related epoxide stylatulide (**100**). The structure of stylatulide was determined by X-ray methods, without absolute stereochemistry (Wratten *et al.*, 1977a). Stylatulide showed significant toxicity (LD_{100} = 0.5 ppm) against the marine copepod *Tisbe furcata johnstonii*.

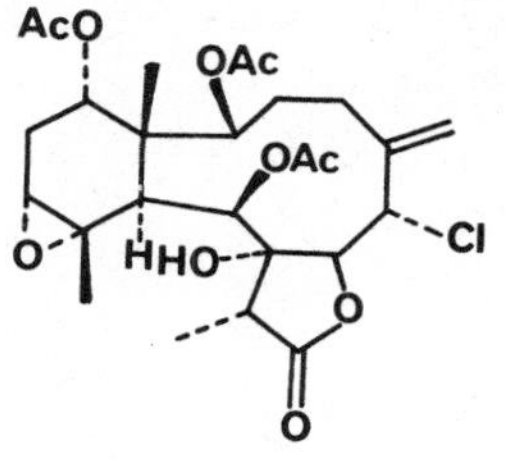

100 Stylatulide

100: mp 179°–181°; $[\alpha]_D$ +65°; ir: 3500, 1780 cm^{-1}; ^{1}H nmr ($CDCl_3$): δ6.00(1H,bs), 5.93(1H,d,*J* = 9), 5.79(1H,bs), 5.50(1H,bs), 4.90(1H,d,*J* = 6.5), 4.71(1H,d,*J* = 4), 4.63(1H,m), 3.36(1H,s), 3.18(1H,q,*J* = 7), 3.04(1H,s), 2.97(1H,d,*J* = 4), 2.59(1H,m), 2.4(2H,m), 2.27(1H,m), 1.31(3H,d,*J* = 18), 1.29(3H,s), 1.10(3H,s)

Before the isolation of stylatulide Wekell (1974) had reported the presence of a toxin, ptilosarcone, from the Pacific sea pen *Ptilosarcus gurneyi*, the molecular composition of which was confirmed to be $C_{28}H_{37}O_{10}Cl$ by mass spectrometry. After firm structure data and nmr spectra became available for **99** and **100**, the Scripps group (Wratten *et al.*, 1977b) carefully compared this toxin with the *Stylatula* and *Briareum* metabolites.

This comparison showed that the toxin was amazingly similar, and by double resonance experiments structure **101** was assigned to ptilosarcone.

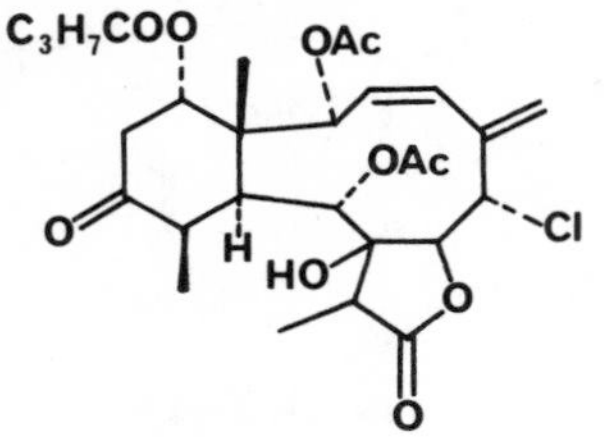
C3H7COO
OAc
OAc
Cl
H
HO
O
O
O

101 Ptilosarcone

101: ir (CH_2Cl_2): 3559, 2959, 2890, 1783, 1732, 1361, 1229, 1217 cm^{-1}; ^1H nmr ($CDCl_3$): δ6.26(1H,d,J = 9), 6.13(1H,dd,J = 1,1), 5.96(1H,d,J = 12), 5.93(1H,dd,J = 2,1), 5.70(1H,dd,J = 12,9), 5.50(1H,d,J = 5.5), 5.22(1H,ddd,J = 3.7,2,1), 5.22(1H,dd,J = 4.2,2.5), 4.94(1H,d,J = 3.7), 3.10(1H,dd,J = 6.2,5.5), 2.91(1H,dd,J = 15,2.5), 2.78(1H,m,J = 6.2,7,7,7,1), 2.35(1H,ddd,J = 15,4.2,1), 2.58(1H,ddd,J = 7,7,7), 1.38(3H,d,J = 7), 1.23(3H,s), 1.22(3H,d,J = 7)

The structures of briarein A (**99**), stylatulide (**100**), and ptilosarcone (**101**) are composed of unprecedented but regular terpenoid bicyclo[8.4.0]-tetradecane skeletons. This skeletal arrangement requires two biogenetic cyclizations of a geranylgeraniol precursor: a C-1–C-14 cyclization and a C-2–C-7 cyclization. It is tempting to suggest that the first step in the

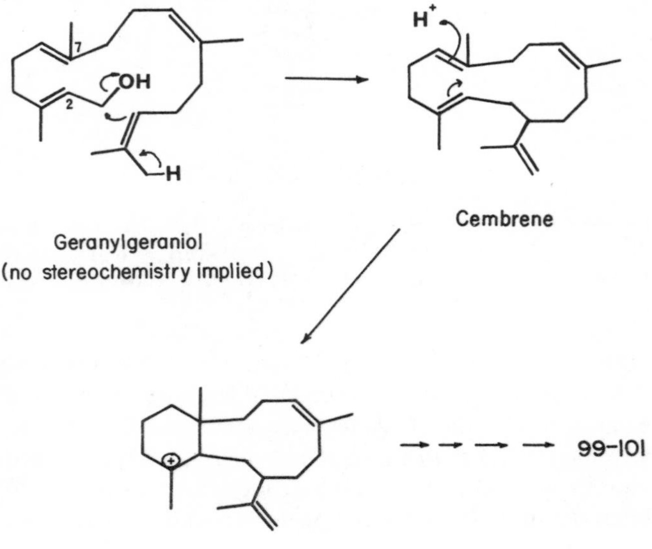
7
OH
2
H
H+
Cembrene
Geranylgeraniol
(no stereochemistry implied)
99–101

biogenesis of these compounds is a C-1–C-14 cyclization to produce cembrene, which then can further cyclize between C-2 and C-7. Intermediacy of the large ring compound is perhaps supported by the fact that many cembrene derivatives have been isolated from other alcyonaceans. Furthermore, inspection of a molecular model of cembrene indicates that a favorable folded-chair conformation can be adopted which allows close proximity of C-2 and C-7.

G. Eunicellin and Derivatives

A bicyclic diterpenoid molecule, which may also be a potential cyclization product of cembrene, is the tetraacetate eunicellin (**102**). Eunicellin was isolated in low yield from the Mediterranean gorgonian *Eunicella stricta,* and was one of the first sea fan metabolites to be described (Kennard *et al.*, 1968). The bicyclo[8.4.0]tetradecane skeleton, which varies by alkyl substitution from the same skeletal framework found in compounds **99–101**, elucidated by X-ray diffraction analysis of the corresponding dibromide (**103**). Since the dibromide could be reconverted to eunicellin by zinc reduction, the structure of **102** is certain.

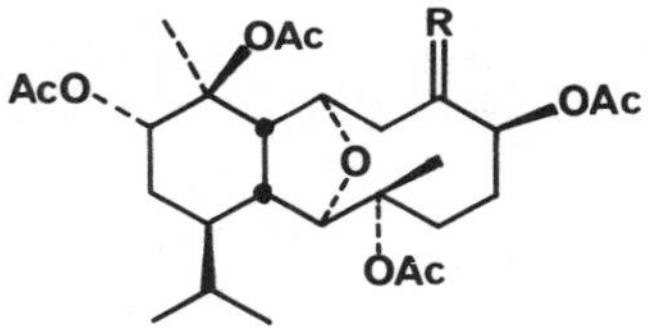

102 R = CH_2, Eunicellin
103 R = (Br)CH_2Br

102: mp 189°–188°; $[\alpha]_D$ −36°; ^{1}H nmr: δ5.43(1H,m), 5.27(1H,m), 1.59(3H,s), 1.56(3H,s), 0.92(3H,d), 0.79(3H,d) (incomplete)

Since 1968 eunicellin has remained the only example of this unique diterpene system. Recently, however, in extensive studies of the soft corals from the Australian Barrier Reef, the Roche group (Kazlauskas *et al.*, 1978b) encountered several derivatives of eunicellin. Their studies of an unidentified *Cladiella* species resulted in the description of two new metabolites, cladiellin (**104**) and acetoxycladiellin (**105**), which are formally related to **104** by addition of acetic acid to the exomethylene. All structural elements (relative stereochemistry only) of the crystalline compound acetoxycladiellin (**105**) were determined by X-ray crystallography. Cladiellin (**104**) was then shown to be the six-membered ring exomethylene derivative of **105** by a two-step conversion to the tetrol **106**. The tetrol produced was identical to the tetrol obtained, under the same reaction conditions, from acetoxycladiellin, thus establishing six identical centers of relative stereochemistry. In both conversions, treatment with *m*-chloroperbenzoic acid gave an initial *trans*-epoxide, which was quickly solvolyzed by the inherent by-product *m*-chlorobenzoate. The secondary ester was then reduced with $LiAlH_4$ to yield the vicinal diol.

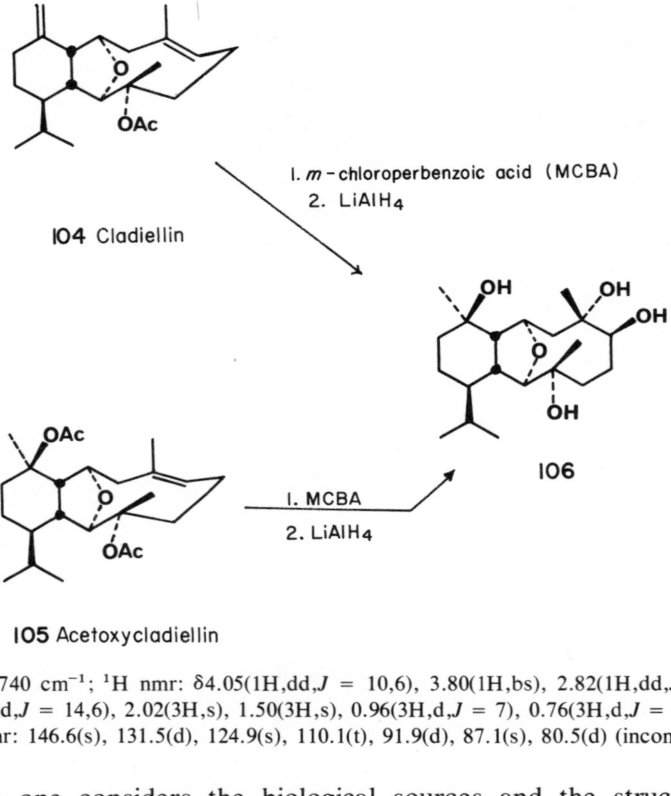

1. *m*-chloroperbenzoic acid (MCBA)
2. $LiAlH_4$

104 Cladiellin

106

1. MCBA
2. $LiAlH_4$

105 Acetoxycladiellin

104: ir: 1740 cm^{-1}; ^1H nmr: δ4.05(1H,dd,*J* = 10,6), 3.80(1H,bs), 2.82(1H,dd,*J* = 10,8), 2.46(1H,dd,*J* = 14,6), 2.02(3H,s), 1.50(3H,s), 0.96(3H,d,*J* = 7), 0.76(3H,d,*J* = 7) (incomplete); cmr: 146.6(s), 131.5(d), 124.9(s), 110.1(t), 91.9(d), 87.1(s), 80.5(d) (incomplete)

When one considers the biological sources and the structures of eunicellin, cladiellin, and acetoxycladiellin, it is tempting to postulate that the genesis of these compounds involves a 14-membered cembrene-like intermediate. Formally this transformation involves initial C-1–C-14 cyclization of geranylgeraniol and subsequent rearrangement to give **47**. Isocembrene (**47**) has been isolated as a natural product from the soft coral genus *Lobophytum*, and it is discussed in Section III,F. From the intermediate **47** a proton-induced cyclization between C-1 and C-10 generates the eunicellin ring system. It is important to point out that the

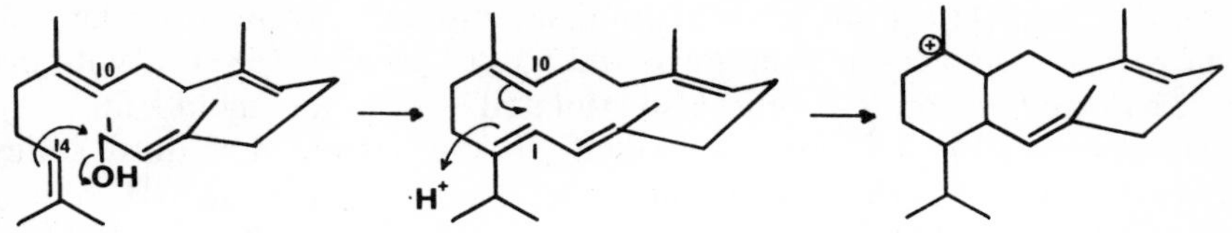

Geranylgeraniol 47 Isocembrene

reverse sequence, an initial C-1–C-10 cyclization followed by a C-1–C-14 ring closure, is equally feasible and involves a dilophol-like intermediate (see Section III,C).

H. Dolabelladienes

Although the secondary metabolites found in most sea hares (Opisthobranchia, Mollusca) have been clearly shown to originate in their algal diets, current work by my Scripps colleagues (Ireland *et al.*, 1976; Ireland and Faulkner, 1977) has resulted in the isolation and characterization of an interesting group of new bicyclic diterpenoids from a sea hare, which could not be traced to obvious algal sources. The digestive organs of the nocturnal Gulf of California opisthobranch *Dolabella californica* were found to be a rich source of a series of 14 related diol and triol acetates, the dolabelladienes (**107–119**). As a basis for the thorough structure elucidations of these metabolites, an X-ray study was completed on one of the major isolates, the alcohol **107**. Furthermore, compounds in the series **107–109** were related by mutual interconversion ($LiAlH_4$ reduction). A

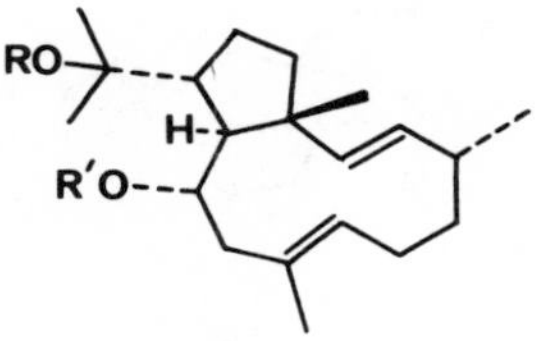

107 R = H, R′ = Ac (X-ray)
108 R = R′ = H
109 R = R′ = Ac

107: mp 78°; $[\alpha]_D$ −101°; ir ($CHCl_3$): 3500, 1740 cm^{-1}; ^1H nmr ($CDCl_3$): δ5.22(1H,dd,*J* = 16,9), 5.10(1H,t,*J* = 7), 5.07(1H,d,*J* = 16), 4.81(1H,dt,*J* = 9,1,1), 2.05(3H,s), 1.62(3H,s), 1.25(3H,s), 1.18(3H,s), 0.94(3H,d,*J* = 7), 0.82(3H,s); ^{13}C nmr (C_6D_6); 168.9, 135.8, 131.2, 128.3, 127.4, 73.0, 71.9, 56.1, 50.0, 47.7, 46.1, 40.1, 38.3, 36.4, 32.5, 27.9, 27.3, 23.9, 22.0, 20.0, 18.8

second group (**110–114**), all containing additional rearranged tertiary allylic alcohol and acetate functions, and a third group (**115–118**), with the unrearranged systems, were also isolated. A minor metabolite, the monoalcohol **119**, was also described. In the second and third series (**110–114** and **115–118**) all compounds were converted to ($LiAlH_4$ reduction) and compared with the triols **110** and **115**. In order to establish the

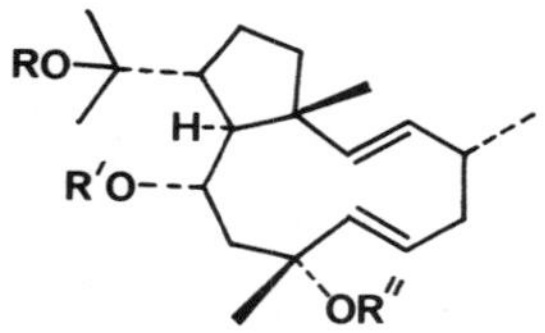

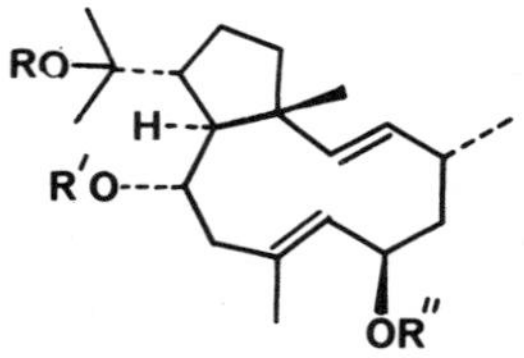

110 R=R'=R''=H
111 R=Ac, R'=R''=H
112 R=R'=Ac, R''=H
113 R=H, R'=R''=Ac
114 R=R'=R''=Ac

115 R=R'=R''=H
116 R=Ac,R'=R''=H
117 R=R'=H, R''=Ac
118 R=H, R'=R''=Ac

110: mp 168°–169°; $[\alpha]_D$ −86°; ir ($CHCl_3$): 3500 cm^{-1}; ^1H nmr ($CDCl_3$): δ5.75(1H,d,J = 16), 5.32(1H,dd,J = 16,10), 5.09(1H,m), 4.92(1H,d,J = 16), 3.89(1H,d,J = 9), 2.57(1H,bs), 2.34(2H,m), 2.16(2H,m), 1.95(2H,m), 1.25(3H,s), 1.23(3H,s), 1.16(3H,s), 1.05(3H,d,J = 7), 0.95(3H,s); ^{13}C nmr ($CDCl_3$): 143.1, 137.4, 132.4, 125.5, 75.0, 72.7, 67.0, 57.1, 56.8, 49.4, 47.1, 38.2, 37.9, 34.1, 31.5, 29.3, 26.0, 23.2, 16.6, 16.4

115: mp 157°–158°; $[\alpha]_D$ −29°; ir ($CHCl_3$): 3500 cm^{-1}; ^1H nmr ($CDCl_3$): δ5.33(1H,dd,J = 16,7), 5.08(1H,d,J = 9), 4.90(1H,d,J = 16), 4.50(1H,m), 2.53(1H,m), 2.41(1H,m), 2.14(1H,d,J = 11), 1.93(1H,m), 1.70(3H,s), 1.27(3H,s), 1.20(3H,s), 1.05(3H,d,J = 7), 0.98(3H,s) (incomplete)

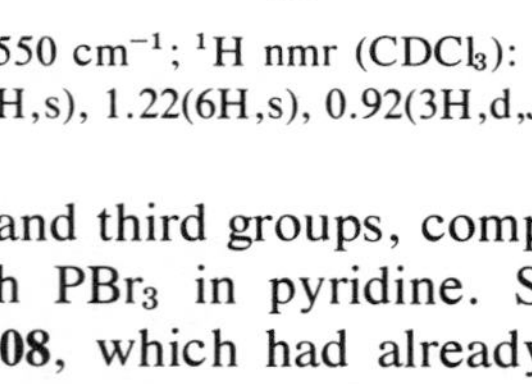

119

119: $[\alpha]_D$ −75.1°; ir ($CHCl_3$): 3550 cm^{-1}; ^1H nmr ($CDCl_3$): δ5.16(2H,m), 5.02(1H,t,J = 7), 2.09(4H,m), 1.82(3H,m), 1.45(3H,s), 1.22(6H,s), 0.92(3H,d,J = 7), 0.89(3H,s) (incomplete)

structures of the second and third groups, compound **112** was converted to the bromide **120** with PBr_3 in pyridine. Sodium reduction of the bromide gave the diol **108**, which had already been related with **107**. Treatment of the same bromide with silver acetate in aqueous tetrahydrofuran (THF) and subsequent $LiAlH_4$ reduction gave two isomeric triols

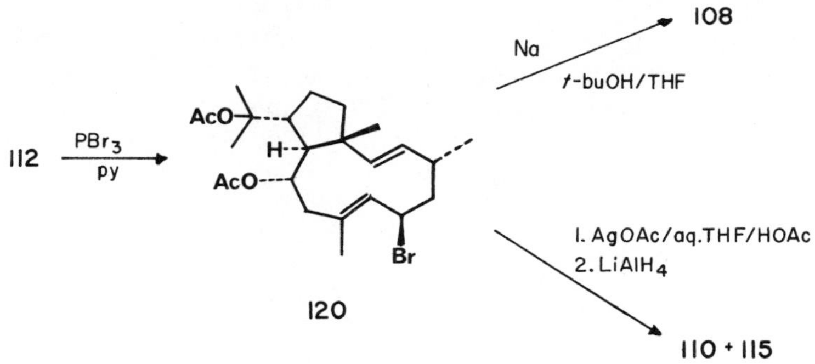

120

(**110** and **115**), thus relating the second and third groups with compound **107**. The remaining structure (**119**) was proposed on the basis of spectral grounds, and the stereochemical features of the allylic alcohol and ester functions in **110–118** were based on analysis of LIS data.

The dolabelladienes are closely related to another *Dolabella* metabolite, dolatriol (Section IV,B). These compounds are indeed unique diterpenoids which may be rationalized as products of regular terpene synthesis. Formally the dolabelladiene ring system results from C-1–C-11 and C-10–C-14 cyclizations of geranylgeraniol, which may proceed, as the authors point out, by a concerted process as shown below.

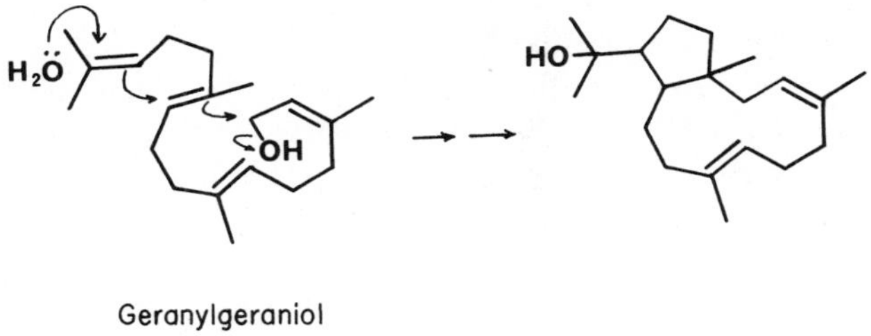

Geranylgeraniol

V. TRICYCLIC AND TETRACYCLIC DITERPENOIDS

The tri- and tetracyclic diterpenoids isolated from marine sources, mainly from sponges (Porifera), consist of only a few structural types. Only one example of a tetracyclic diterpenoid has been described, and this substance may be of debatable terpenoid origin.

A. Isoagatholactone and Derivatives

In their extensive studies of the natural products chemistry of Mediterranean sponges, the Minale group investigated several collections of the common bath sponge *Spongia officinalis* (Minale *et al.*, 1976). Although some collections yielded C_{21} and C_{25} linear furanoterpenes, another collection was curiously devoid of these compounds and contained instead the true diterpenoid isoagatholactone (**121**) (Cimino *et al.*, 1974). Spectral features for the lactone, including characteristic ^{1}H nmr signals and mass spectral fragmentations, indicated the basic structure **121**, which contains

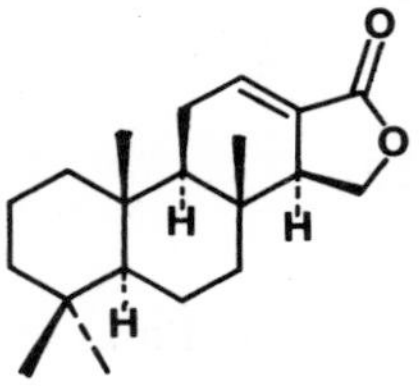

121 Isoagatholactone

121: mp 153°–155°; $[\alpha]_D$ +6.3°; ir ($CHCl_3$): 1760, 1690 cm^{-1}; uv (C_6H_{12}): 222 nm (ϵ = 6000); ^{1}H nmr (CCl_4): δ6.69(1H,q,J = 3), 4.24(1H,t,J = 9), 3.93(1H,t,J = 9), 2.75(1H,m), 2.23(2H,bm), 0.94(3H,s), 0.88(3H,s), 0.84(3H,s), 0.79(3H,s)

the carbon skeleton of isoagathic acid, an acid-catalyzed cyclization product of the naturally occurring bicyclic diterpene agathic acid (Ruzicka and Hosking, 1930). This structure, including absolute stereochemistry, was confirmed by conversion of isoagatholactone to the synthetic alcohol **122**. Alcohol **122**, with identical absolute stereochemistry, was also produced from the well-known bicyclic diterpenoid grindelic acid (**123**) in a process involving seven intermediate steps; this transformation rigorously relates isoagatholactone with a structure of established absolute configuration.

Collections of a common but unidentified *Spongia* species made along the Australian Barrier Reef have also recently been reported to yield compounds related to isoagatholactone (Kazlauskas *et al.*, 1978c). From several complex extracts the Australian group isolated a series of related diols and triols, the spongiadiols and spongiatriols, as well as their respective acetate esters. Contrary to the functionality found in isoagatholactone, the Australian *Spongia* metabolites were characterized as furanoditerpenoids, each possessing a carbonyl group in the A ring. The structures of two groups of compounds, the diols spongiadiol (**125**) and epispongiadiol (**127**) and the triols spongiatrol (**129**) and epispongiatriol (**131**), and the acetates **124**, **126**, **128**, and **130** have been elucidated as a result of extensive spectral analysis and an X-ray diffraction study of the major metabolite, spongiatriol triacetate (**128**). Interestingly, the A ring in **128** was found to exist in a boat conformation with carbons 1, 2, 4, and 5

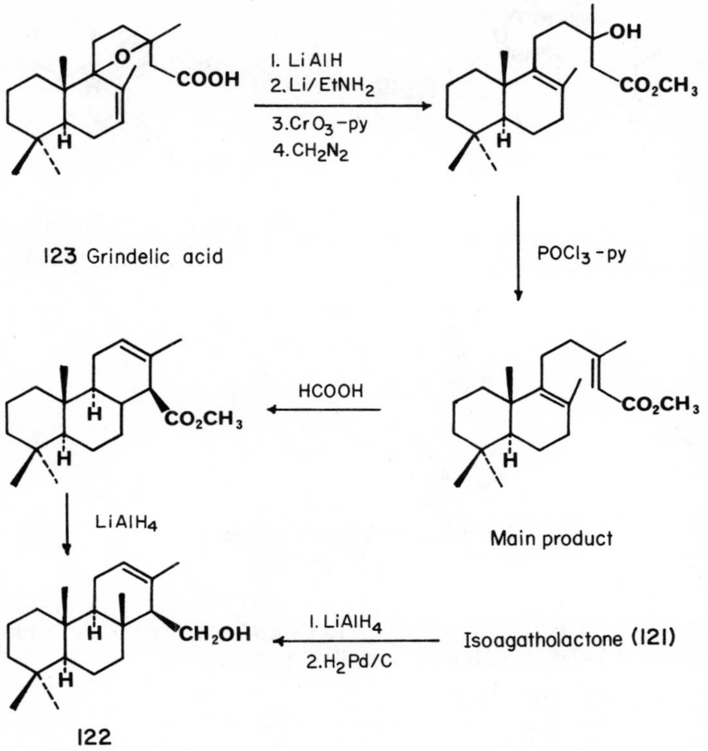
COOH
1. LiAlH
2. Li/EtNH2
3. CrO3-py
4. CH2N2
OH
CO2CH3
123 Grindelic acid
POCl3-py
HCOOH
CO2CH3
CO2CH3
LiAlH4
Main product
CH2OH
1. LiAlH4
2. H2 Pd/C
Isoagatholactone (121)
122

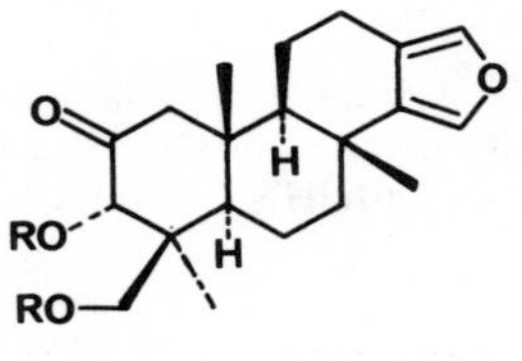

124 R = Ac, Spongiadiol diacetate

125 R = H, Spongiadiol

124: mp 131.5°–133°; $[\alpha]_D$ +14.5; tr: 1745, 1725 cm^{-1}; uv (MeOH): 219 nm (ϵ = 5113); ^1H nmr ($CDCl_3$): δ7.11(1H,d,J = 1), 7.05(1H,m,J = 1,1.5), 5.46(1H,s), 4.03(2H,s), 2.62(1H,d,J = 16.5), 2.21(1H,d,J = 16.5), 2.83(1H,m), 2.45(1H,m), 2.16(3H,s), 2.08(3H,s), 1.26(3H,s), 1.24(3H,s), 1.04(3H,s); ^{13}C nmr ($CDCl_3$): 205.5(s), 170.4(s), 169.7(s), 137.0(d), 136.2(s), 135.2(d), 119.1(s), 77.4(d), 65.9(t), 56.5(d), 54.0(t), 53.7(d), 43.8(s), 40.3(t), 39.9(s), 34.2(s), 25.5(q), 21.2, 20.8, 20.7, 20.5, 19.8, 19.3(q), 19.2(q)

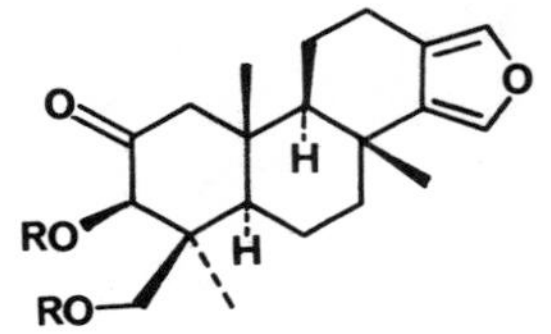

126 R=Ac, *epi*-Spongiadiol diacetate
127 R=H, *epi*-Spongiadiol

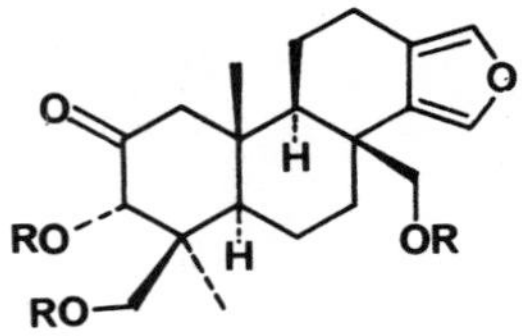

128 R=Ac, Spongiatriol triacetate
129 R=H, Spongiatriol

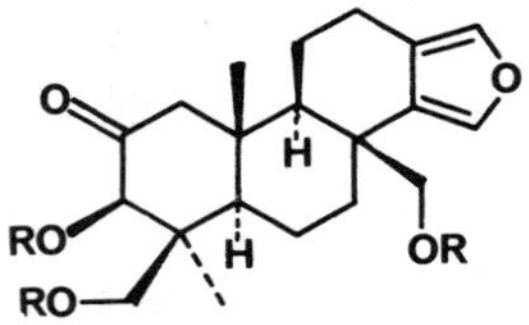

130 R = Ac, *epi*-Spongiatriol triacetate
131 R = H, *epi*-Spongiatriol

126: mp 184°–185.4°; $[\alpha]_D$ +40°; ir: 1745, 1725 cm^{-1}; ^1H nmr ($CDCl_3$): δ7.06(1H,s), 7.03(1H,d,*J* = 0.5), 4.95(1H,s), 4.05(2H,s), 2.66(1H,d,*J* = 12), 2.60(2H,m), 2.19(1H,d,*J* = 12), 2.15(3H,s), 2.04(3H,s), 1.21(3H,s), 0.95(3H,s) (incomplete)

128: mp 157.5°–158.5°; $[\alpha]_D$ +11.8°; ir: 1720–1750 cm^{-1}; uv (MeOH): 217 nm (ε = 5116); ^1H nmr ($CDCl_3$) δ7.13(1H,d,*J* = 7.10(1H,m), 5.46(1H,s), 4.25(2H,q,*J* = 10.5), 4.0(2H,s), 2.81(1H,m), 2.65(1H,d,*J* = 16.5), 2.4(1H,m), 2.21(1H,d,*J* = 16.5), 2.15(3H,s), 2.08(3H,s), 2.02(3H,s), 1.30(3H,s), 1.03(3H,s); ^{13}C nmr ($CDCl_3$): 205.0(s), 170.5(s), 170.4(s), 169.7(s), 137.9(d), 137.2(d), 129.1(s), 119.2(s), 77.1(d), 65.6(t), 63.9(t), 56.9(d), 53.9(t), 53.7(d), 43.7(s), 39.4(s), 38.1(s), 35.1(t), 21.2, 20.9, 20.5, 19.9, 19.3, 18.8

130: mp 181.7°–182.7°; ir: 1750–1720 cm^{-1}; ^1H nmr ($CDCl_3$): δ7.13(2H,bs), 4.99(1H,s), 4.42(1H,d,*J* = 10), 4.07(1H,d,*J* = 10), 4.06(2H,s), 2.70(1H,d,*J* = 13), 2.50(2H,m), 2.30(1H,m), 2.16(3H,s), 2.07(3H,s), 2.01(3H,s), 1.21(3H,s), 0.99(3H,s) (incomplete)

being coplanar. The free diols and triols (**125**, **127** and **129**, **131**) were easily related to the acetates by interconversion with acetic anhydride in pyridine. The absolute configuration of **128** was obtained not by crystallographic methods but by measurement of the circular dichroism (CD) and optical rotary dispersion (ORD) curves. A strong positive maximum at 290 nm in the CD spectrum and a strong positive Cotton effect centered at 291 nm in the ORD spectrum could be correlated with previous treatments of this A-ring conformation, which allowed the assignment of absolute stereochemistry as drawn in structure **128**. Although spongiadiol (**125**), epispongiadiol (**127**), and epispongiatriol (**131**) were not interconverted chemically with the rigorously elucidated **128**, the assignments of structure were strongly supported by extensive ^{1}H nmr and ^{13}C nmr analysis including interpretation of LIS data.

As discussed above, the carbon skeleton of isoagatholactone and the spongiadiols and spongiatriols is a regular terpenoid system already known as a synthetic product from the naturally occurring diterpenoid agathic acid. Among the marine- derived diterpenoids this skeletal system is perhaps the most closely related to the majority of the terrestrially derived diterpenoids. The cyclization chemistry involved in producing this sytem is stereospecific and involves well-known steps from a folded, all-*trans*-geranylgeraniol precursor.

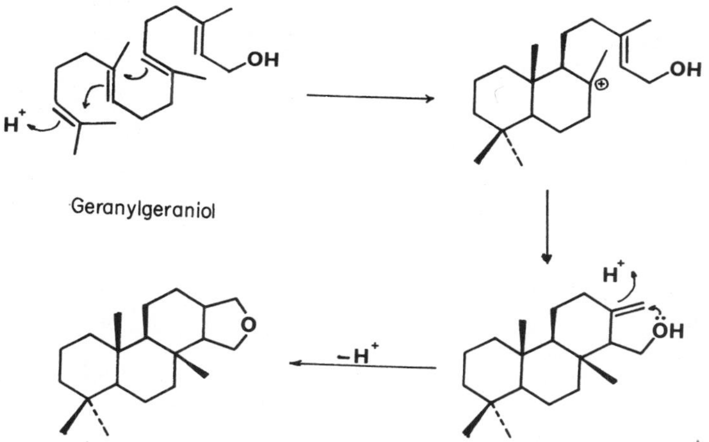

Geranylgeraniol

B. Dolatriol

Coincidentally with the work on isoagatholactone two independent studies were reported on the digestive gland components of sea hares of the genus *Dolabella* (Opisthobranchia, Mollusca). The Faulkner group at Scripps reported the structure of a new bicyclic diterpenoid (**107**) representing the dolabelladiene group; it was isolated from *D. californica,* an abundant opisthobranch from the southern Gulf of California (Ireland *et al.*, 1976). This compound and some related structures are discussed in Section IV,H. The Pettit group at Arizona State University, following the antineoplastic activity of several sea hare extracts, isolated and described two new tricyclic diterpenoids, dolatriol (**132**) and dolatriol 6-acetate (**133**), which appear to show activity in the National Cancer Institute's P-388 leukemia bioassay. These compounds were isolated in low yields

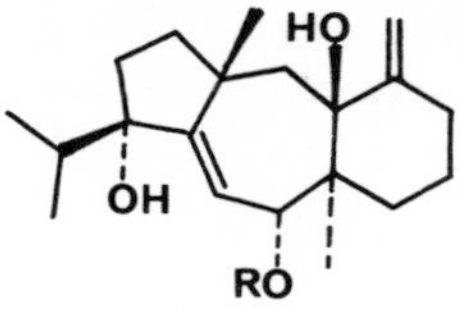

132 R = H, Dolatriol
133 R = Ac, Dolatriol 6-acetate

132: mp 235°–236°; ir (KBr): 3500, 3410, 1638, 909 cm^{-1}; ^1H nmr ($CDCl_3$): δ5.45(1H,d,J = 6), 4.97(1H,d,J = 6), 4.89(1H,s), 4.73(1H,s), 1.19(3H,s), 1.01(3H,d,J = 7), 0.84(3H,d,J = 7), 0.78(3H,s) (incomplete)

133: mp 210°–212°; ir (KBr): 3570, 3439, 1733, 1638, 899 cm^{-1}; ^1H nmr ($CDCl_3$): δ6.16(1H,d,J = 6), 5.28(1H,d,J = 6), 4.93(1H,s), 4.78(1H,s), 2.6(1H,m), 2.09(3H,s), 1.19(3H,s), 1.03(3H,d,J = 7), 0.89(3H,s), 0.85(3H,d,J = 7) (incomplete)

(0.0005 and 0.001%) from extracts of the digestive organs of the Indian Ocean sea hare *Dolabella auricularia,* and their structures were determined by an X-ray crystallographic study of **133** (Pettit *et al.*, 1976). In mild acid, **133** was readily converted to dolatriol.

The dolatriol carbon skeleton represents a new, but regular terpenoid skeleton that can logically be derived from geranylgeraniol via the intermediacy of the bicyclic dolabelladiene ring system. Protonation and transannular cyclization between C-2 and C-7 of a precursor such as the known dolabelladiene derivative **119** yields the dolatriol skeleton directly.

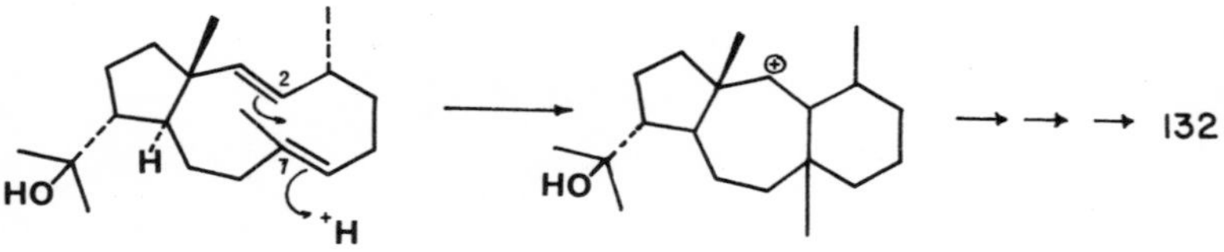

119 → → → → 132

C. Diisocyanoadociane

From a biogenetic viewpoint perhaps the most interesting polycyclic compound yet to be described from marine sources is the diisocyanide diisocyanoadociane (**134**), which was isolated by the Roche group from the Australian Barrier Reef sponge *Adocia* sp. (Baker *et al.*, 1976). Diisocyanoadociane was a massive component (2%) of *Adocia* sp. along with several minor isocyanides, which have not yet been described. The

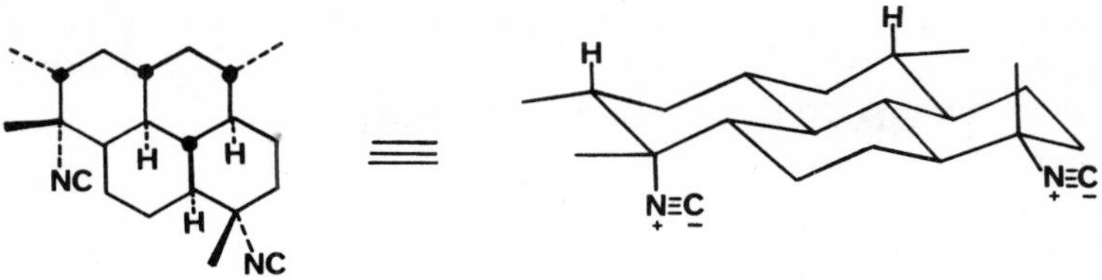

134 Diisocyanoadociane

134: $[\alpha]_D$ +47.4°; ir: 2140, 2130 cm^{-1}; ^{1}H nmr: δ1.37(3H,m), 1.29(3H,m), 1.06(3H,d,*J* = 6), 0.88(3H,d,*J* = 6) (incomplete)

spectral features of diisocyanoadociane fixed the molecular formula ($C_{22}H_{32}N_2$) and indicated the presence of isocyanide functionality but gave no meaningful structural information. The ^{1}H nmr spectrum, for example, was featureless below δ2.2. Fortunately, the compound crystallized from hexane, and a successful X-ray study provided the substituted perhydropyrene structure **134** (relative stereochemistry only) for diisocyanoadociane.

The origin of diisocyanoadociane from a diterpenoid precursor is not obvious since the molecule is nonterpenoid. The Roche group had, however, pointed out a biogenetic scheme for the production of **134** from geranylgeraniol which involves only a single methyl migration. It is hoped that the minor, as yet unreported metabolites from *Adocia* will provide insight into this question.

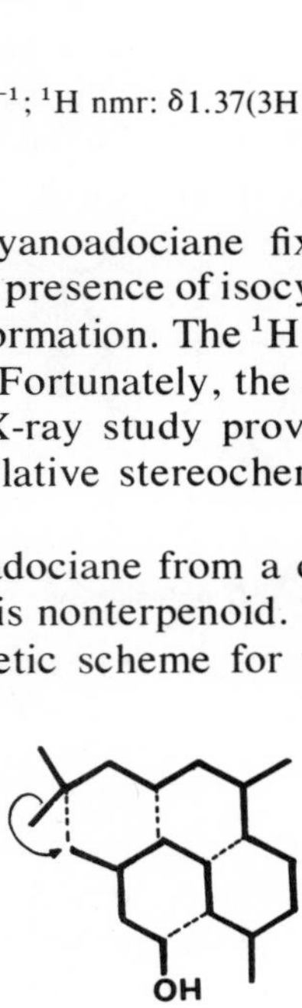

VI. BROMODITERPENOIDS AND DERIVATIVES

The marine-derived diterpenoids that contain one or more strategically placed bromine atoms are considered here in a separate section because of their unique biogenesis. To date brominated diterpenoids have been isolated only from certain species of red seaweeds (Rhodophyta) belonging to the unrelated genera *Laurencia* and *Sphaerococcus*. This section does not include the chlorine-containing diterpenoids from variou coelenterates (Section IV,F), since in those metabolites the chlorine aton appears to possess no biogenetic significance.

A. Aplysin-20 and Concinndiol

Yamamura and Hirata (1971) were the first to recognize the dietar origin of metabolites isolated from the digestive organs of herbivorous se hares (Opisthobranchia). Their studies of the northern Japanese mollus *Aplysia kurodai* resulted initially in the isolation of several sequiter penoids (Yamamura and Hirata, 1963) and later in the isolation of bromine-containing diterpenoid, aplysin-20 (**135**). Aplysin-20 was recog

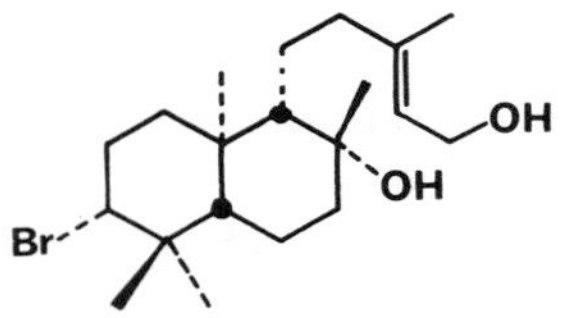

135 Aplysin-20

135: mp 146°–147°; MS: M^+ -18, *m/e* 352; ir: 3500, 3300, 1675 cm^{-1}; ^1H nmr: δ5.41(1H,bt, = 7), 4.16(2H,d,J = 7), 3.85(1H,m,J = 8.5,3.1), 2.2–1.3(16H,m), 1.70(3H,bs), 1.16(3H,s), 1.08(3H,s), 1.00(3H,s), 0.96(3H,s)

nized by spectral analysis to be related to several known diterpenoids, bu its total structure elucidation was rigorously accomplished by X-ray crys tallography (Matsuda *et al.*, 1967; Yamamura and Hirata, 1971). Sinc several of the sesquiterpenoids isolated with **135** had already been showr by Irie and co-workers (Irie, 1969) to be produced by the red seawee *Laurencia,* aplysin-20 was also predicted to be an algal metabolite.

Several years later insight into the origin of aplysin-20 was provided by the University of California at Riverside group, who investigated the halogen-containing metabolites of the Australian red alga *Laurencia concinna* (Sims *et al.*, 1973). The major compound obtained was a bromoditerpenoid, concinndiol (**136**), which was initially recognized to be closely related to aplysin-20. The final structure elucidation, obtained by X-ray crystallography, yielded the absolute stereochemistry of **136** and clearly showed that *Laurencia* was the probable source for aplysin-20. Although this is a reasonable assumption, the elusive aplysin-20 has not yet been isolated from *Laurencia* despite investigations of over 40 worldwide species. Concinndiol has also been observed as a major component in the California alga *L. snyderiae*.

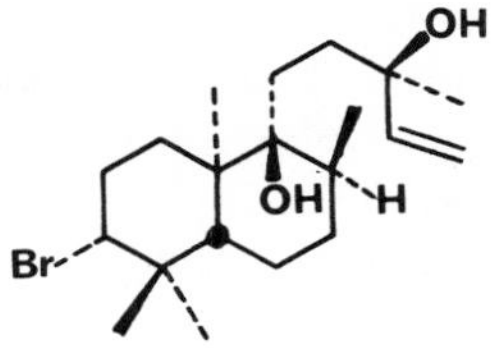

136 Concinndiol

136: mp 122°–123°; ^{1}H nmr ($CDCl_3$): δ5.88(1H,dd,J = 17,10), 5.15(1H,dd,J = 17,1), 5.07(1H,dd,J = 10,1), 4.04(1H,dd,J = 12,6), 2.2(2H,m), 1.98(1H,bs), 1.82(1H,bs), 1.8–1.2(12H,m), 1.25(3H,s), 1.05(3H,s), 0.93(3H,s), 0.81(3H,d,J = 6) (incomplete)

The key to the biogenesis of aplysin-20 and concinndiol must clearly lie in the bromonium ion-induced cyclization of geranylgeraniol and geranyllinalool, respectively. Although all facets of biosynthesis based on halonium ion (X^+) are not yet understood, structure and biomimetic synthesis studies of marine metabolites indicate that Br^+ (or its equivalent) induces ring closure reactions of suitable acyclic precursors. One can therefore surmise that bromonium ion-induced cyclizations are analogous to proton-induced biosynthesis. Hence, the biogenesis of the labdane derivatives **135** and **136** can be predicted to occur via typical carbonium ion mechanisms.

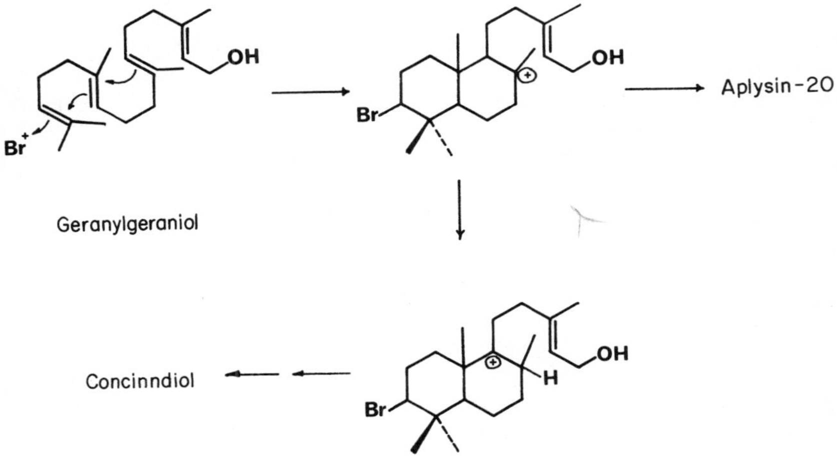

Geranylgeraniol → Aplysin-20

Concinndiol

B. Neoconcinndiol Hydroperoxide

Included in the group of marine metabolites derived from halogen metabolism is an interesting hydroperoxide, neoconcinndiol hydroperoxide (**137**). The hydroperoxide was isolated as a minor component from extracts of the red alga *Laurencia snyderiae* along with major amounts of concinndiol (**136**) (Howard *et al.*, 1977). The structure of neoconcinndiol hydroperoxide (**137**), although derived by X-ray crystallography, required spectral analysis for total structure elucidation. In the

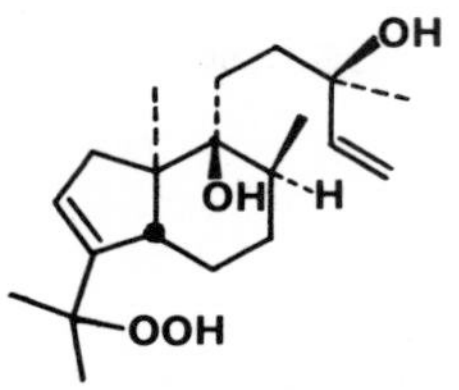

137 Neoconcinndiol hydroperoxide

137: mp 158°–159°; $[\alpha]_D$ −35°; ir (KBr): 3500–3400 cm^{-1}; ^1H nmr (Me_2CO-d_6): δ9.80(1H,bs), 5.89(1H,dd,*J* = 15,10), 5.43(1H,m), 5.19(1H,dd,*J* = 15,2), 4.94(1H,dd,*J* = 10,2), 3.61(1H,bs), 2.90(1H,bs), 3.11(1H,dd,*J* = 14,4), 2.61(1H,dd,*J* = 14,4), 1.29(6H,s), 1.22(3H,s), 0.89(3H,d,*J* = 7) (incomplete)

crystal the terminal vinyl group was badly disordered, and its assignment rests on characteristic ^1H nmr spectral features. The hydroperoxide moiety could, however, be unambiguously assigned, with an overall crystallographic residual for this structure of 4.9%. Hydroperoxides are rare natural products, which probably reflects their instability rather than their absence in cellular metabolism. It appears, for example, that peroxides and hydroperoxides are important intermediates in prostaglandin synthesis (Hamberg *et al.*, 1974, and references cited therein). Hydroperoxide (**137**) is the first example of a naturally occurring diterpene hydroperoxide, but a steroidal hydroperoxide (3α,22α-dihydroxy-7α-hydroperoxy-Δ^5-stigmastene) has been reported (Fischer and Mägerlein, 1960). Also, **137** represents a new diterpenoid skeletal class.

The strong structural similarities (including relative stereochemistry at five asymmetric centers) between the hydroperoxide **137** and concinndiol (**136**), and the fact that they co-occur in the same alga, point to a biogenesis involving the elimination of bromine. A solvolytic A-ring contraction reaction, well precedented in the solvolysis of 3β-tosyltriterpenes and characteristic of 4,4-dimethyl substitution, accounts for the production of the cyclopentane A ring. The hydroperoxide moiety could then be introduced by many methods, one of which is the reaction of singlet-state oxygen with an intermediate isopropylidene-containing precursor.

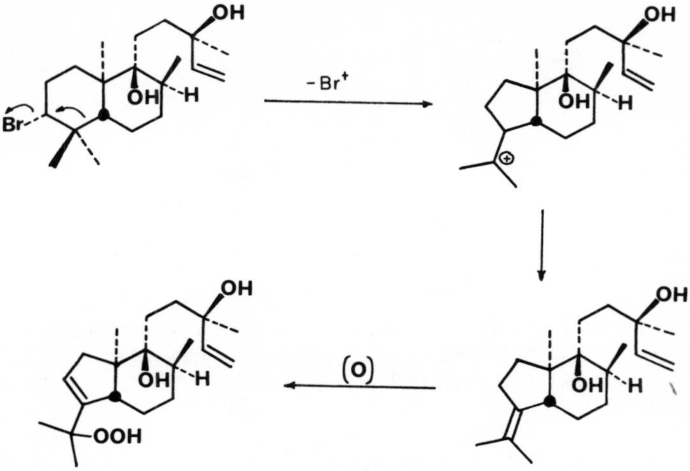

C. Irieols

In the course of our studies of the halogen-containing compounds from various species of *Laurencia,* we recognized that an undescribed species from the Gulf of California contained several related compounds of a new diterpenoid class. Both the binomial of this new species, *Laurencia irieii* (Norris, 1978), and the chemical nomenclature were coined as a tribute to the pioneering work in this field of Professor Irie in Hokkaido. Two crystalline members of this group of compounds, iriediol (**138**) and irieol A (**140**), were initially analyzed by X-ray crystallography (Fenical *et al.*,

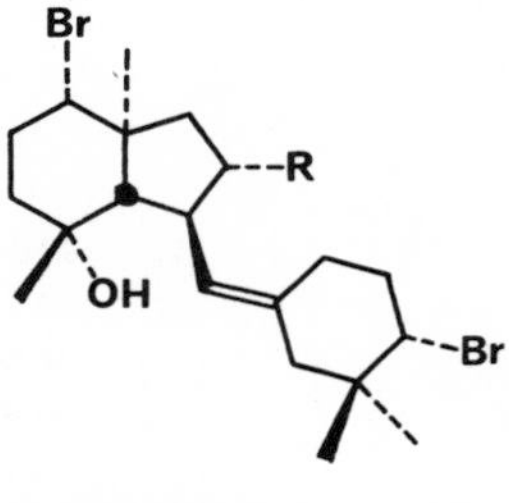

138 R = OH, Iriediol
139 R = H, Iriediol

138: mp 103°–105°; $[\alpha]_D$ −18.3°; ^{1}H nmr ($CDCl_3$): δ5.05(1H,d,*J* = 10), 4.14(1H,dd,*J* = 9,4), 3.91(1H,dd,*J* = 12,4), 3.84(1H,m), 3.06(1H,m), 2.68(1H,m), 2.34(1H,m), 2.2–1.5(13H,m), 1.41(3H,s), 1.20(3H,s), 1.07(3H,s), 0.93(3H,s); ^{13}C nmr ($CDCl_3$): 134.7(s), 129.4(d), 78.4(d), 71.5(s), 65.8(d), 64.5(d), 59.8(d), 50.6(t), 48.9(t), 47.2(d), 46.9(s), 42.9(t), 38.4(s), 33.8(t), 31.2(q), 30.8(t), 29.6(q), 28.4(t), 23.1(q), 17.9(q)

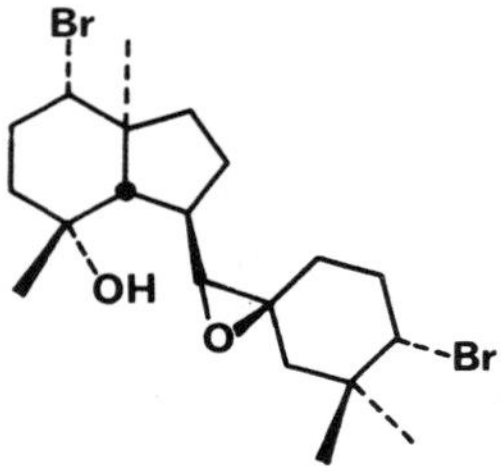

140 Irieol A

140: mp 142°–144°; $[\alpha]_D$ 0° (fortuitous); ^{1}H nmr ($CDCl_3$): δ4.36(1H,bs), 3.97(1H,dd,J = 12,5), 2.61(1H,d,J = 10), 2.5–1.3(17H,m), 1.27(1H,s), 1.24(3H,s), 1.20(3H,s), 1.09(3H,s); ^{13}C nmr ($CDCl_3$): 71.1, 67.9, 66.1, 64.3, 64.1, 59.7, 43.0, 41.1, 40.0, 37.1, 36.1, 34.9, 31.8, 30.9, 30.8, 29.4, 25.7, 24.8, 24.5, 16.2

1975), which gave the gross structures and relative stereochemistries. Since small differences were being measured in the crystallographic study, the assignments of absolute stereochemistry were tenuous. Subsequent to our initial work we described by chemical methods the absolute stereochemistry of these compounds and elucidated the structures of several related compounds, the irieols B through G (**141–146**), from the same alga.

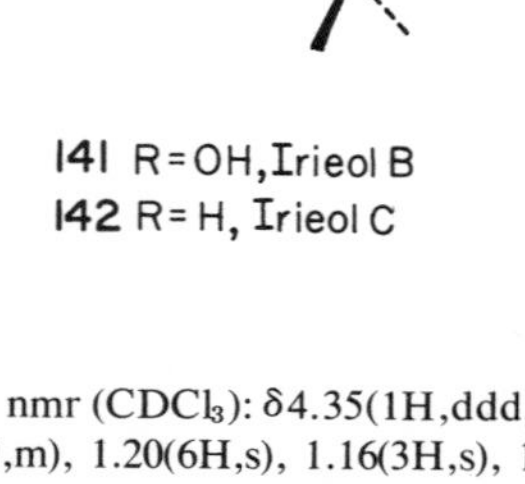

141 R=OH, Irieol B
142 R=H, Irieol C

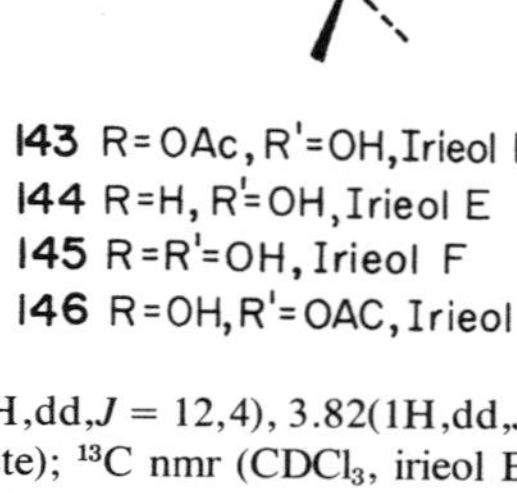

143 R=OAc, R'=OH, Irieol D
144 R=H, R'=OH, Irieol E
145 R=R'=OH, Irieol F
146 R=OH, R'=OAC, Irieol G

141: ^{1}H nmr ($CDCl_3$): δ4.35(1H,ddd,J = 12,8,4), 3.89(1H,dd,J = 12,4), 3.82(1H,dd,J = 12,4), 2.46(1H,m), 1.20(6H,s), 1.16(3H,s), 1.03(3H,s) (incomplete); ^{13}C nmr ($CDCl_3$, irieol B-acetate): 73.5(d), 72.6(s), 71.7(s), 65.3(d), 64.3(d), 57.7(d), 51.3(t), 49.1(t), 45.4(s), 43.3(t), 39.7(t), 38.4(t), 36.6(s), 36.4(d), 32.4(q), 32.1(q), 30.8(t), 30.3(t), 23.0(q), 17.7(q)

143: ^{1}H nmr ($CDCl_3$): δ5.36(1H,ddd,J = 12,8,4), 3.98(1H,dd,J = 12,4), 3.91(1H,dd,J = 12,4), 3.82(1H,d,J = 10), 2.77(1H,dd,J = 12,8), 2.07(3H,s), 1.44(3H,s), 1.26(3H,s), 1.20(3H,s), 1.05(3H,s) (incomplete); ^{13}C nmr ($CDCl_3$): 169.3(s), 78.6(d), 75.6(d), 75.3(s), 71.7(s), 65.2(d), 64.2(d), 54.1(d), 50.2(t), 47.4(s), 44.9(t), 44.0(t), 41.1(d), 36.6(s), 35.4(t), 32.7(q), 31.4(q), 30.6(t), 30.2(t), 22.9(q), 21.7(q), 18.1(q)

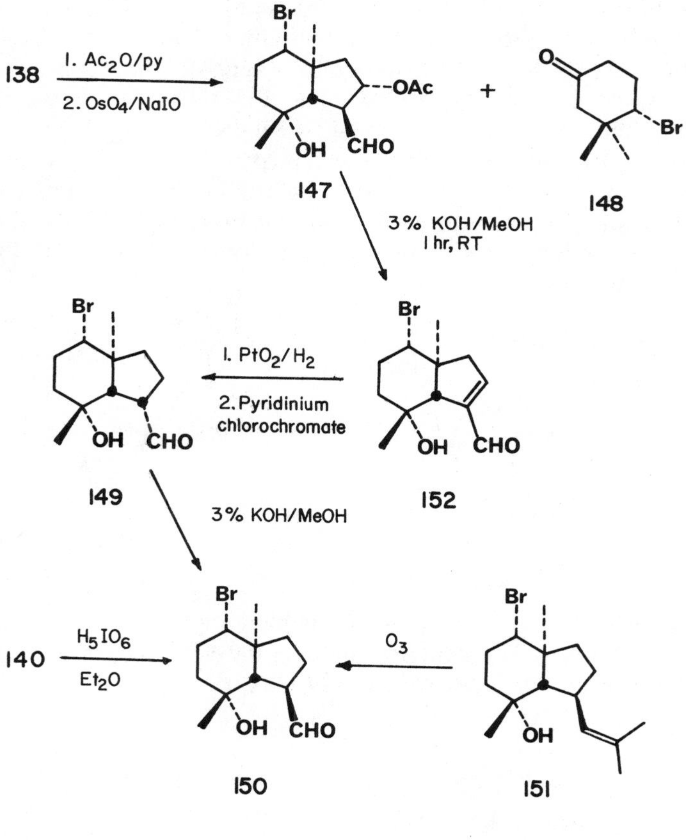

To determine the absolute stereochemistry of the X-ray structures of iriediol (**138**) and irieol A (**140**), we converted each to the aldehyde **150**. The aldehyde showed spectral characteristics including optical rotation ($[\alpha]_D$ +6.3°) that were comparable to those from the same aldehyde obtained from ozonolysis of the sesquiterpenoid oppositol (**151**). Oppositol had earlier been shown to possess the absolute configuration as drawn in **151** (Hall *et al.*, 1973). To confirm this assignment and as a method of assigning the absolute stereochemistry to irieol (**139**) and irieols D, E, and F (**143–145**), we oxidatively cleaved each compound to yield optically active ($[\alpha]_D$+31.9°) 4-bromo-3,3-dimethylcyclohexanone (**148**). Ketone **148**, as isolated from each compound, showed a positive Cotton effect. This can be interpreted, by the octant rule and the assumption that the bromine atom in **148** is more stable in an equatorial position, to indicate that the asymmetric center at C-4 must have *S* configuration. It is interesting in this case that the asymmetric center at C-4 is disposed in the plane of the chromophore and therefore does not contribute to its CD spectrum. Rather the positive Cotton effect is a product of the contribution from the rigidly held *gem*-dimethyl group at C-3.

The structures of the remaining metabolites were elucidated by comparing mutual oxidation products with those from iriediol and irieol A (Howard and Fenical, 1978). Treatment of irieol E (**144**) with periodate gave the aldehyde **150** directly, and analogous treatment of irieol D (**143**) and irieol F (**145**) gave, after base work-up, aldehydes that were identical to **152**. These transformations confirmed the structures and absolute stereochemistries of **143**, **144**, and **145**. Irieol G (**146**) could be converted to irieol F by mild saponification with 3% KOH in methanol. The structures of irieol B (**141**) and C (**142**) have been assigned solely on the basis of spectral features since these metabolites lack functionality at the bridging methylene and could not be cleaved to smaller fragments.

The irieols are composed of regular diterpenoid skeletons of an unprecedented nature. For the genesis of the irieols from geranylgeraniol a C-1–C-9 closure must occur as well as two bromonium ion-induced closures, one involving C-3–C-8 and one involving C-15 and the C-18 methyl

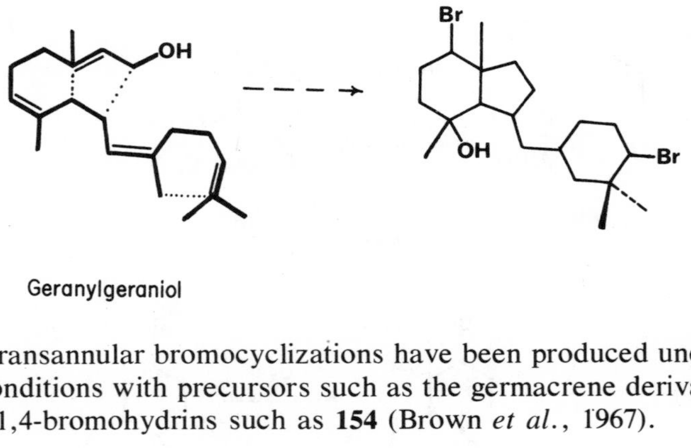

OH Br OH Br

Geranylgeraniol

group. Transannular bromocyclizations have been produced under laboratory conditions with precursors such as the germacrene derivative **153** to yield 1,4-bromohydrins such as **154** (Brown *et al.*, 1967).

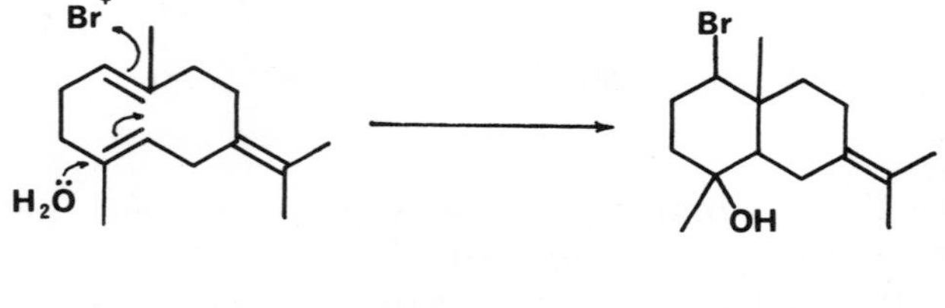

Br^+ $H_2\ddot{O}$ Br OH

153 154

D. Obtusadiol

Because of our continuing interest in the metabolites from red algae of the genus *Laurencia,* we have explored several collections of the Mediterranean alga *L. obtusa* (Howard and Fenical, 1978). The chemical components of this species were found to vary greatly from locale to locale. One extract from collections of this alga made near Athens was the only source of an interesting new dibromoditerpenoid, obtusadiol, to which we can now assign structure **155**. The structure of obtusadiol was obtained by

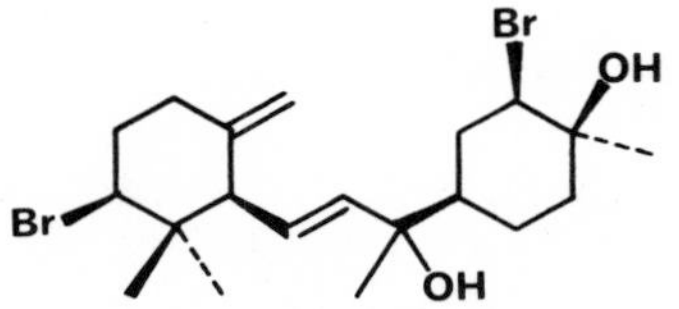

155 Obtusadiol

155: ir ($CHCl_3$): 3500–3400 cm^{-1}; ^1H nmr ($CDCl_3$): δ5.80(1H,dd,J = 15,10), 5.56(1H,d,J = 15), 4.86(1H,s), 4.64(1H,s), 4.18(2H,dd,J = 12,5), 2.54(1H,d,J = 10), 1.34(3H,s), 1.29(3H,s), 1.09(3H,s), 0.96(3H,s) (incomplete); ^{13}C nmr ($CDCl_3$): 147.2(s), 130.1(d), 126.8(d), 110.3(t), 74.3(s), 70.3(s), 66.4(d), 66.2(d), 56.5(d), 49.7(d), 41.1(s), 37.5(t), 36.6(t), 34.9(t), 34.9(t), 30.4(q), 29.5(q), 26.1(q), 22.0(t), 16.1(q)

combined spectral analysis and chemical transformation, which gave insight into the relative stereochemistry at several centers. In a partially selective osmilation–periodate cleavage reaction the exocyclic olefin in **155** was cleaved to the six-membered ring ketone **156** (ν c=o = 1715 cm^{-1}) and the substituted cyclohexyl methyl ketone fragment **157**. Also, treatment of obtusadiol with mild base caused elimination of bromine and ring contraction of the bromohydrin-containing ring to give the methyl ketone **158**, which was formed as a mixture of equilibrated epimers at the cyclopentane carbon. These transformations defined both rings in obtusadiol as six-membered and the bromohydrin moiety as axial–equatorial (cis). Furthermore, ^1H nmr coupling constants and attempted acetylation showed that the bromine atoms in **155** were secondary and equatorial.

Obtusadiol can be related to another *Laurencia* metabolite, β-snyderol (**159**) (Howard and Fenical, 1976), by virtue of the common feature of one ring. However, obtusadiol is unique among the bicyclic diterpenoids and represents a new structure class. The genus *Laurencia* can be credited at this point with providing six new terpenoid structural classes: for ses-

156 157

1. OsO_4(cat.)
2. $NaIO_4$(1 eq.)

155

3 % KOH/MeOH

158

159 β- Snyderol

155

quiterpenoids and two diterpenoids. The projected biogenesis of the obtusadiol ring system is straightforward and involves a C-1–C-6 cyclization of geranylgeraniol and a bromonium ion-induced cyclization at C-15–C-10.

E. Sphaerococcenol A and Bromosphaerols

Along with *Laurencia* one species of the red algal genus *Sphaerococcus* produces bromoditerpenoids. In two concurrent studies, my group and the Fattorusso group in Naples isolated and identified the major metabolites of the Mediterranean alga *Sphaerococcus coronopifolius* (Sphaerococceae). Our collections of *S. coronopifolius* were made in Spain, whereas those of Fattorusso were made in Sicily. The major diterpenoid from the Spanish collection was the α,β-unsaturated ketone sphaerococcenol A (**160**), the structure of which was elucidated by an X-ray crystallographic study (Fenical *et al.*, 1976).

The Sicilian collection also contained **160**, but in addition a new dibromoditerpenoid, bromosphaerol (**161**), was isolated as a viscous oil (Fattorusso *et al.*, 1976c). Spectral analysis of **161** gave substantive structural data which, however, were insufficient for a definite structure assignment.

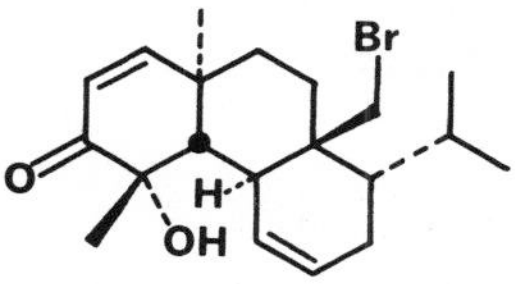

160 Sphaerococcenol A

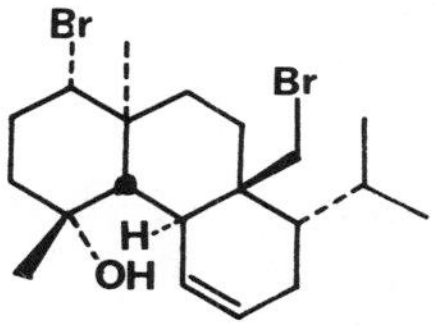

161 Bromosphaerol

160: mp 184°–185°; $[\alpha]_D$ −93°; ir: 1680 cm^{-1}; uv (MeOH): 228 nm (ε = 7510); ^1H nmr ($CDCl_3$): δ6.80(1H,d,*J* = 10), 6.06(1H,d,*J* = 10), 6.03(1H,d,*J* = 10), 5.73(1H,m), 3.98(1H,d,*J* = 11), 3.70(1H,d,*J* = 11), 3.13(1H,s), 2.96(1H,m), 2.3–1.3(9H,m), 1.34(3H,s), 1.09(3H,s), 0.98(3H,d,*J* = 7), 0.94(3H,d,*J* = 7); ^{13}C nmr ($CDCl_3$): 200.7(s), 159.7(d), 126.2(d), 125.3(d), 122.3(d), 73.1(s), 39.9(d), 38.0(t), 37.6(s), 34.6(d), 33.4(t), 30.9, 29.1, 23.7, 23.6, 22.4(q), 20.2(q), 19.2(q), 17.3(q)

161: $[\alpha]_D$ +0.5°; ir: 3600–3380 cm^{-1}; ^1H nmr: δ5.76(2H AB system,*J* = 10), 3.81(2H,AB,*J* = 9.5), 4.00(1H,dd,*J* = 9,4), 1.38(3H,s), 1.31(3H,s), 0.97(3H,d,*J* = 6), 0.90(3H,d,*J* = 6) (incomplete)

For suitable derivatization for spectral analysis bromosphaerol was dehydrated with $POCl_3$ in pyridine to give the olefin **162**. Treatment of the olefin with diazabicyclononene (DBN) at 100° for 3 days yielded the nonconjugated triene **163**. Poor but satisfactory crystals of **163** were

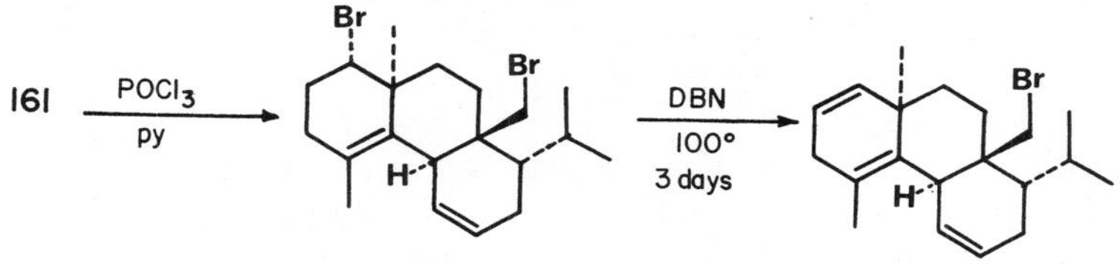

161 $\xrightarrow{POCl_3,\ py}$ 162 $\xrightarrow{DBN,\ 100°,\ 3\ days}$ 163

162

163 (X-ray)

obtained from methanol and subsequently analyzed by X-ray crystallography, yielding confirmatory evidence for the structure of bromosphaerol.

The Fattorusso group also reported the structure of a related diol, bromosphaerodiol (**164**), which is a minor component of extracts of *S. coronopifolius* (Cafieri *et al.*, 1977). The structure of bromosphaerodiol

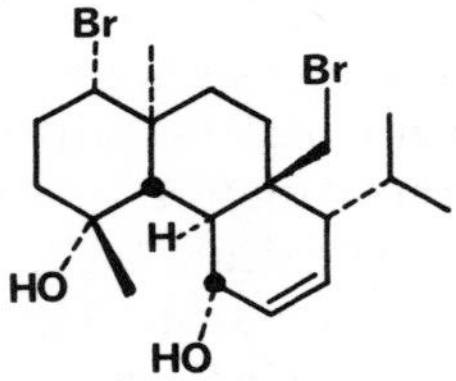

164 Bromosphaerodiol

164: mp 170°–172°; $[\alpha]_D$ −44.3°; ir (KBr): 3400 cm^{-1}; ^1H nmr ($CDCl_3$): δ5.80(2H,bs), 4.22(1H,m), 3.90(2H,AB,J = 9), 3.52(1H,bd,J = 11), 1.32(3H,s), 1.27(3H,s), 1.02(3H,d,J = 6.5), 0.97(3H,d,J = 6.5)

was rigorously established by synthesis of this metabolite from bromosphaerol (**161**). Treatment of bromosphaerol with a modified Cornforth reagent (CrO_3–pyridine–CH_2Cl_2) at 25° for 3 days gave low yields of the ketone **165** (~5% conversion) and starting material. Sodium borohydride reduction of **165** gave a single diol that was identical to bromosphaerodiol (**164**). The secondary hydroxyl group in bromosphaerodiol could be as-

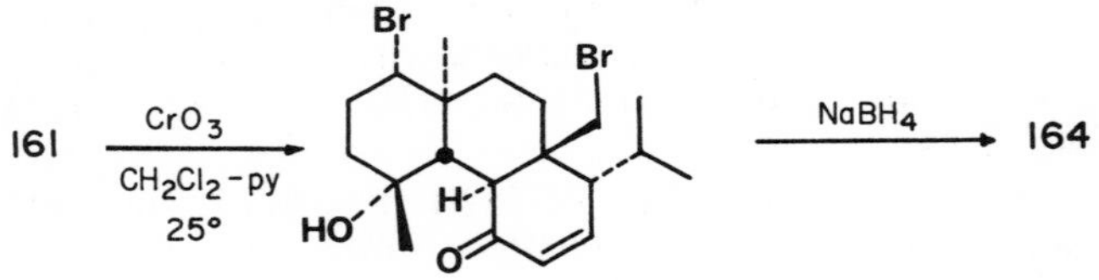

161 —CrO_3, CH_2Cl_2-py, 25°→ 165 —$NaBH_4$→ 164

165

signed as quasi-equatorial since the α proton showed a coupling to the adjacent bridgehead proton of 11 Hz. Coupling of this magnitude strongly suggests an axial–axial proton interaction with a dihedral angle of about 180°. The absolute configuration at the secondary alcohol carbon was suggested to be *S* on the basis of the Horeau esterification method (Horeau, 1961, 1962).

The diterpenoids from *Sphaerococcus* represent a new class of irregular diterpenoids which appear to be substantially rearranged. A clue to the biogenesis of these diterpenoids may lie in the existence of the bromomethyl group, which appears out of place when one dissects this ring system into several obvious isoprene units. It would therefore appear that at least one methyl migration is required to rationalize this skeletal system. One possible biogenetic scheme, which accounts for the production of the bromomethyl group and methyl migration via a cyclopropane-containing intermediate, is shown below. Even with this proposal the acyclic C_{20} precursor must contain one tail-to-tail isoprene linkage. Perhaps studies of the minor metabolites from *Sphaerococcus* will provide insight into the biogenesis of this novel ring system.

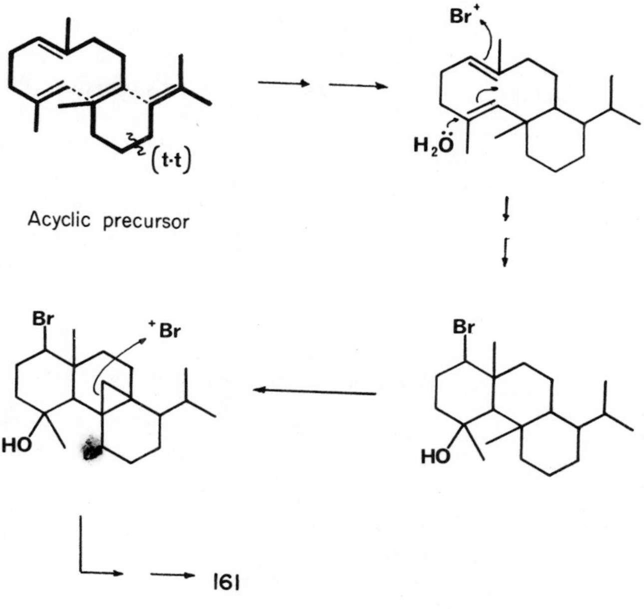

(t-t)

Acyclic precursor

Br^+

$H_2\ddot{O}$

Br

HO

Br

^+Br

HO

161

VII. COMPOUNDS FROM DITERPENOID CATABOLISM

Several compounds from marine sources can be grouped together and considered here as degradation products of the diterpenoids. These compounds are composed of C_{19}, C_{18}, and C_{14} acyclic carbon skeletons and are suggested on biogenetic and structural grounds to be diterpenoid products.

A. C_{19} Compounds: Pristane and Zamene

The related C_{19} hydrocarbons pristane (**166**) and zamene (**167**) have been known since the early 1900's to be constituents of shark and whale

166 Pristane

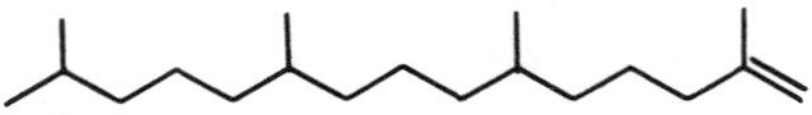

167 Zamene

liver oil (Tsujimoto, 1917), but it was not until some 30 years later that the structures of these simple hydrocarbons became fully certain (Tsujimoto, 1935; Sørensen and Mehlum, 1948; Sørensen and Sørensen, 1949; Christensen and Sørensen, 1951). Although the source of these C_{19} hydrocarbons has not been clearly demonstrated, Blumer *et al.* (1963) isolated large amounts of pristane (**166**) from extracts of a mixed collection of three species of zooplankton of the genus *Calanus*. One considers the isolation of the phytadienes from comparable sources, it would seem reasonable to predict that pristane and zamene are products reflecting the environmental fate of phytol.

B. C_{18} Compounds: Farnesylacetone and Derivatives

A recent report described the isolation of farnesylacetone (**168**) and its hexahydro derivative **169** from the androgenic gland and hemolymph of the male crab *Carcinus maenas* (Ferezou *et al.*, 1977). The androgenic gland in crustaceans is responsible for the induction and differentiation of male sexual characteristics. Guided by bioassay results, Ferezou *et al.* fractionated the extract of 5000 glands several times, ultimately obtaining an active fraction, which was analyzed by gas chromatographic–mass spectrometric techniques. The mass spectra indicated a 10:1 mixture of the hexahydro derivative **169** to farnesylacetone (**168**). Coinjection of

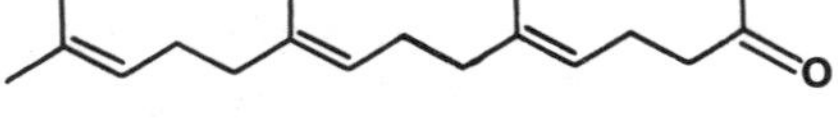

168

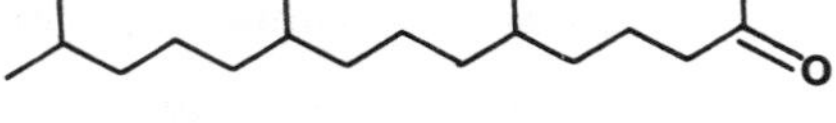

169

authentic samples of **168** and **169** and comparison of mass spectral data confirmed the structures of these two ketones. *In vitro* bioassays of **168** and **169** showed that these compounds possessed two of the three forms of hormonal activity recognized for the crude extract. Bioassay of the synthetic ketone **169,** which possessed the 6*R*,10*R* conformation as prepared from naturally occurring phytol, showed identical biological activity.

Two derivatives of farnesylacetone, the epoxide **170** and the dione **171**, were isolated as major lipid components from extracts of the Australian brown alga *Cystophora moniliformis* (Kazlauskas *et al.*, 1978a). The major metabolite, the epoxide **170**, was determined to be identical to a synthetic epoxide reported earlier that shows juvenile hormone activity (Ikan *et al.*, 1973) and was further shown to possess the C-13=*R* absolute configuration. The minor algal metabolite was readily assigned the dione structure **171** on the basis of characteristic mass spectrometric, and ^{1}H nmr, spectral features. Although this is the first observation of hormonally active lipids from marine algae, their presence in some species may be linked to chemical defense.

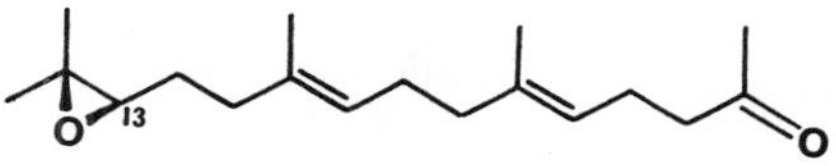

170

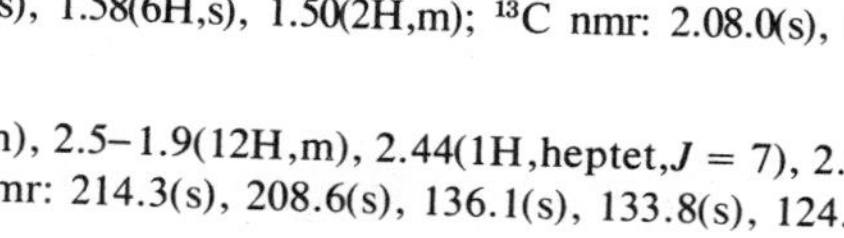

171

170: $[\alpha]_D$ −3.2°; ir: 1710 cm^{-1}; ^{1}H nmr: δ5.04(1H,bt), 4.96(1H,bt), 2.48(1H,dd,*J* = 5.5,5.5), 2.4–1.9(10H,m), 2.00(3H,s), 1.58(6H,s), 1.50(2H,m); ^{13}C nmr: 2.08.0(s), 63.9(d), 58.0(s), (incomplete)

171: ^{1}H nmr: δ5.02(2H,m), 2.5–1.9(12H,m), 2.44(1H,heptet,*J* = 7), 2.08(3H,s), 1.56(6H,bs), 1.02(6H,d,*J* = 7); ^{13}C nmr: 214.3(s), 208.6(s), 136.1(s), 133.8(s), 124.5(d), 122.7(d) (incomplete)

C. C_{14} Compounds: Oxocrinol

As a result of their extensive studies of the chemistry of numerous Italian seaweeds, the Fattorusso group described two acyclic terpenoids from extracts of the brown alga *Cystoseira crinita* (Fattorusso *et al.*,

1976a). One of these compounds, crinitol (**10**), which is formally the 9-hydroxyl derivative of geranylgeraniol, is discussed in detail in Section II,B. Detected along with crinitol, and therefore suspected to originate in diterpenoid biosynthesis, was the C_{14} fragment oxocrinol acetate (**172**).

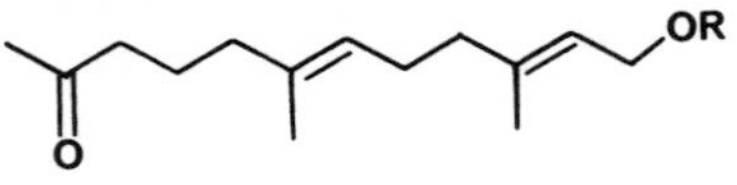

OR

O

172 R=Ac, Oxocrinol acetate
173 R=H, Oxocrinol

173: ir (CCl_4): 3300, 1720 cm^{-1}; ^{1}H nmr (CCl_4): δ5.30(1H,bt,J = 7), 5.06(1H,bt,J = 6), 3.98(2H,d,J = 7), 2.53(1H,bs), 2.30(2H,t,J = 7), 2.02(3H,s), 1.64(3H,bs), 1.58(3H,bs)

Hydrolysis of the extract allowed the isolation of free oxocrinol, which could be rigorously described by spectral methods. The structure was confirmed chemically by oxidative ozonation of the acetate **172** to yield, after CH_2N_2 esterification, levulinic acid methyl ester (**174**), 2,6-heptanedione (**175**), and acetylglycolic acid methyl ester (**176**). Oxocrinol

CO_2CH_3 + O O O

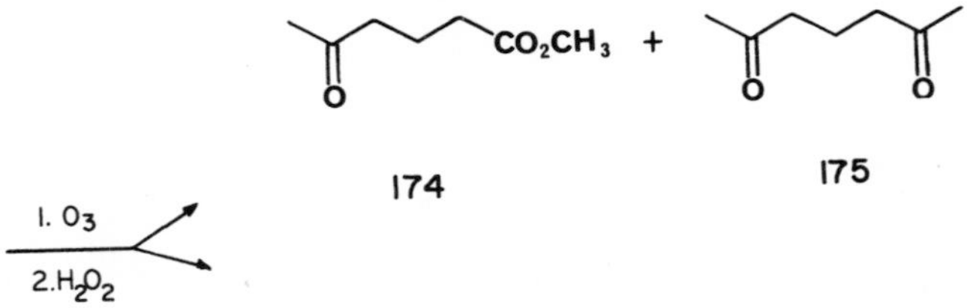

174 **175**

172 1. O_3 2. H_2O_2

$CH_2(OAc)COOCH_3$

176

could be produced by oxidative cleavage of the C-11–C-12 bond in geranylgeraniol, which is a conceivable process involving a C-11 hydroperoxide intermediate.

VIII. DITERPENOIDS OF MIXED BIOGENESIS

A number of interesting and more complex compounds that contain diterpenoid components have been reported from various marine organisms. In general, these compounds are the products of mixed biogenesis and are composed of a diterpenoid that is bonded (generally through C-1) to either an aromatic moiety or a carbohydrate residue. In the majority of the metabolites so far reported the diterpenoid component is a geranylgeranyl group or a closely related acyclic derivative.

A. Geranylgeranylglycerol

While investigating several brown seaweeds of the family Dictyotaceae, Italian investigators at the universities of Naples and Catania discovered a most unusual polar lipid in extracts of *Dilophus fasciola* (Amico *et al.*, 1977). Spectroscopic analysis showed the compound to be an alcohol and to possess terpenoid elements. A molecular formula of $C_{23}H_{40}O_3$ strongly suggested that a three-carbon unit had somehow been added to geranylgeraniol. By a series of chemical transformations and spectral analyses, the structure of this compound was shown to be the unprecedented ether (−)-(*R*)-1-*O*-geranylgeranylglycerol (**177**). Acetylation gave the more easily analyzed diacetate **178**, the ^{1}H nmr spectrum of which illustrated that the protons of an all-*trans*-geranylgeraniol skeleton were

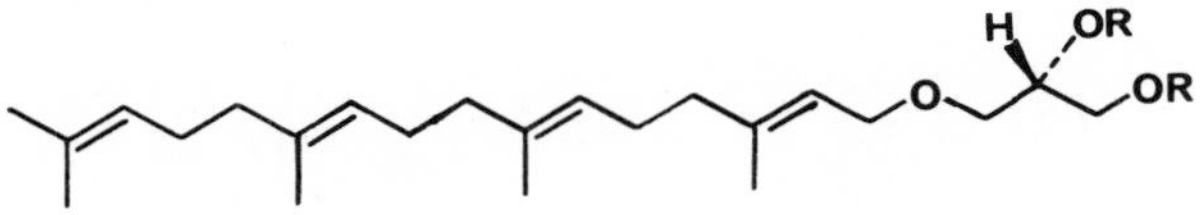

177 R=H
178 R=Ac

178: $[\alpha]_D$ −8.4°; ir (film): 1750 cm^{-1}; ^{1}H nmr (CCl_4): δ5.29(1H,bt), 5.10(3H,m), 4.98(1H,m), 4.20(2H,AB,J = 14), 3.94(2H,d,J = 7), 2.03(12H,m), 2.01(3H,s), 1.98(3H,s), 1.67(6H,s), 1.60(9H,s)

intact. Furthermore, since two acetates could be formed from three oxygen atoms and there were no remaining hydroxyl groups, the compound must be an ether. The ether linkage could be placed at C-1 of glycerol, since **177** gave a positive periodate reaction for a vicinal diol. The gross structure of **177** was also suggested by the results of a sodium in ammonia reduction, which gave all-*trans*-2,6,10,14-tetramethylhexadeca-2,6,10,14-tetraene (**14**), confirmed as the tetraene by comparison with an authentic sample. By Mislow's method (Green *et al.*, 1966) for the evalua-

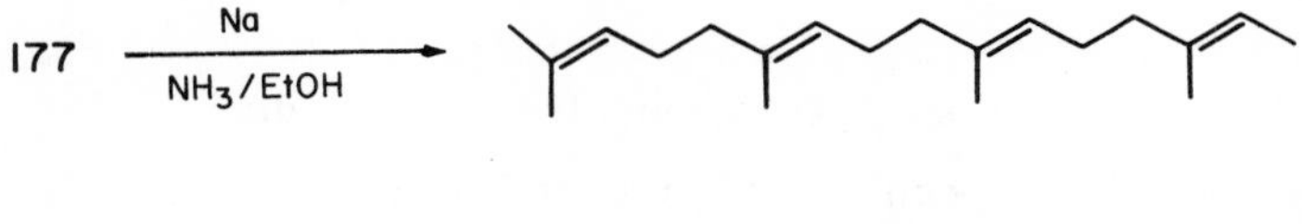

177 —Na / NH_3/EtOH→

14

tion of absolute stereochemistry the asymmetric center in **177** was determined as *R*. Interestingly, this unusual lipid is a product of carbohydrate and terpenoid biosynthesis. It is also interesting that the *R* (C-2) configuration is enantiomeric to that which is usually found in natural lipids.

B. Geranylgeranylquinone and Derivatives

Marine sponges contain large amounts of prenylated aromatic compounds of both a benzenoid and furanoid nature (Minale, 1976). Three compounds enter this discussion, since the tetraprenyl substituent is geranylgeranyl. Although many compounds possessing larger terpenoid groups are known, sponge compounds containing the geranylgeranyl substituent have been reported only from the Mediterranean sponge *Ircinia muscarum* (Cimino *et al.*, 1972). On the basis of well-established spectral features the Minale group described the three related structures **179–181**. Three hundred grams of dry sponge yielded 0.8 gm of the quinone **179**, 16.5 gm of the quinol **180**, and 5.2 gm of the acid **181**, which indicates their

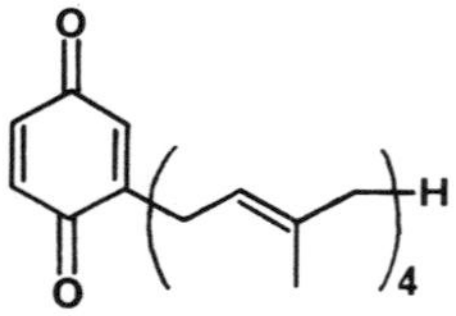

O O H 4

179

R HO H 4

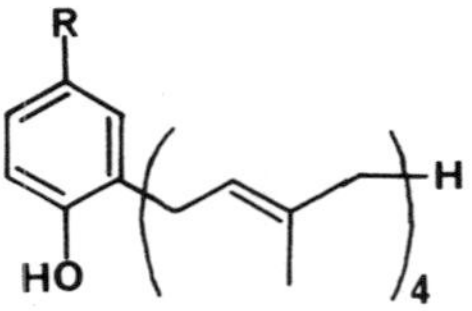

180 R=OH
181 R=COOH

179: uv: 293 nm (log ϵ = 3.64); ir: 1660, 1600 cm^{-1}; ^{1}H nmr: δ6.64(2H,bs), 6.43(1H,bs), 1.64(6H,s), 1.58(9H,s) (incomplete)

180: mp 47°–48°; ir: 3350, 1500, 1445, 910, 785, 730 cm^{-1}; uv: 293 nm (log ϵ = 3.64); ^{1}H nmr: δ6.46(3H,m) (incomplete)

181: mp 61°–63°; ir: 3420, 1670, 1600 cm^{-1}; uv (MeOH): 257 nm (log ϵ = 4.1); ^{1}H nmr: δ7.89(2H,m), 6.81(1H,d,J = 8) (incomplete)

large concentrations in *I. muscarum*. The cooccurrence of functionalities in **179–181** strongly suggests that the *p*-hydroxybenzoic acid derivative **181** is the precursor of **179** and **180**, analogous to ubiquinone biogenesis (Threlfall and Whistance, 1971).

C. Larval Attractants from *Sargassum*

Many seemingly specific symbiotic and parasitic relationships exist between marine organisms. In many cases the symbiont or parasite is expected to recognize the host organism by virtue of the chemoreception of a host-produced chemical. In few cases, however, have these predictions been confirmed by the isolation of the attractant compound and the demonstration of its biological activity. One successful example was reported in which the attractant for the settling larvae of the symbiotic hydrozoan *Coryne uchidai* was isolated from its algal host *Sargassum tortile* (Kato *et al.*, 1975a). A bioassay to guide the chromatographic separation of the attractant led to a complex mixture, from which the phenolic ethers δ-tocotrienol (**182**) and epoxy-δ-tocotrienol (**183**) were

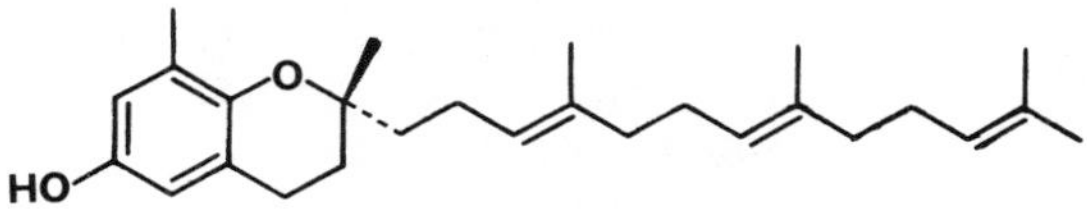

182 δ-Tocotrienol

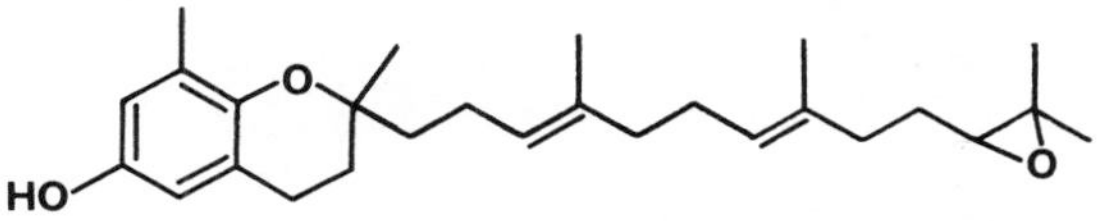

183 Epoxy δ-tocotrienol

182: ^{1}H nmr: δ6.32(1H,d,J = 3), 6.20(1H,d,J = 3), 5.02(3H,m), 2.63(2H,t,J = 7), 2.05(3H,s), 1.71(2H,t,J = 7), 1.64(3H,bs), 1.56(9H,bs), 1.22(3H,s) (incomplete)

183: uv (MeOH): 295 nm (ϵ = 2090); ^{1}H nmr: δ6.34(1H,m), 6.20(1H,m), 5.07(2H,m), 2.58(3H,m), 2.06(3H,s), 1.58(6H,bs), 1.25(3H,s), 1.22(6H,s) (incomplete)

isolated. Both compounds were effective in attracting and inducing larval settling. On the basis of comparable spectral features, **182** was confirmed as δ-tocotrienol, which had earlier been isolated from *Hevea latex* and was structurally described (Whittle *et al.*, 1966). The structure of the corresponding epoxide was suggested by spectral features and later confirmed in a total synthesis of racemic **182** and **183** (Kato *et al.*, 1975b). The stereochemistry at the two asymmetric centers in **183** was not determined.

D. Taondiol and Atomaric Acid

The diterpenoids of mixed biogenesis already described have the common feature of possessing uncyclized terpenoid skeletons. A single example of polycyclic, mixed-biogenesis diterpenoids exists in the algal

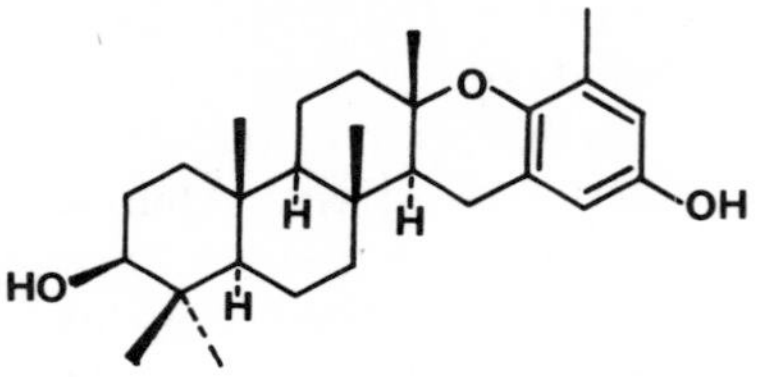

184 Taondiol

184: mp 283°–284°; $[\alpha]_D$ −76°; ir (KBr): 3540, 3340, 1620, 1500, 860, 800 cm^{-1}; uv (EtOH): 298 nm (ϵ = 3860); ^1H nmr ($CDCl_3$): δ6.44(1H,d,J = 2.3), 6.38(1H,d,J = 2.3), 4.24(1H,bs), 3.2(1H,m), 2.6(2H,bd,J = 8), 2.11(3H,s), 1.14(3H,s), 1.00(3H,s), 0.90(3H,s), 0.88(3H,s), 0.82(3H,s) (incomplete)

metabolites taondiol (**184**) and atomaric acid (**185**), which were isolated from separate collections of the Atlantic brown seaweed *Taonia atomaria* (Dictyotaceae) (Gonzalez *et al.*, 1971, 1974). The structure of taondiol was initially proposed largely on the basis of its characteristic spectral features and by comparison with similar structures such as α-tocopherol (vitamin E). Later, however, a total biomimetic-type synthesis of taondiol methyl ether was reported in confirmation of structure **185** (Gonzalez *et*

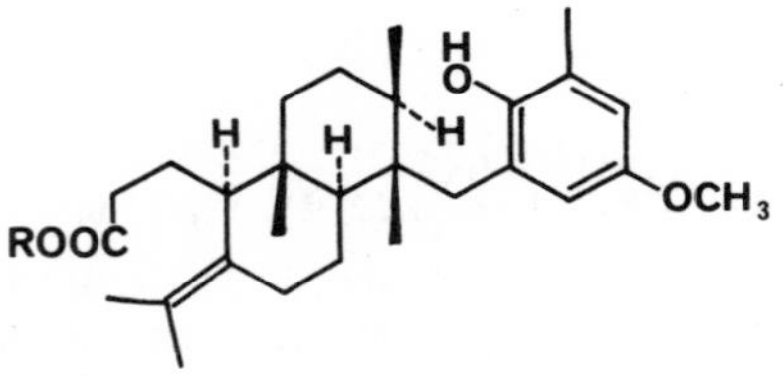

185 R=H, Atomaric acid
186 R=CH ,Atomaric acid methyl ester

186: mp 112°–115°; ir: 3620, 1740, 1620, 1490, 860; uv: 293 nm (ϵ = 3270); ^1H nmr ($CDCl_3$): δ6.68(1H,d,J = 3), 6.49(1H,d,J = 3), 4.34(1H,bs), 3.72(3H,s), 3.64(3H,s), 2.86(1H,d,J = 14), 2.25(1H,d,J = 14), 2.22(3H,s), 1.70(3H,s), 1.68(3H,s), 1.16(3H,d,J = 7), 1.04(3H,s), 0.96(3H,s) (incomplete)

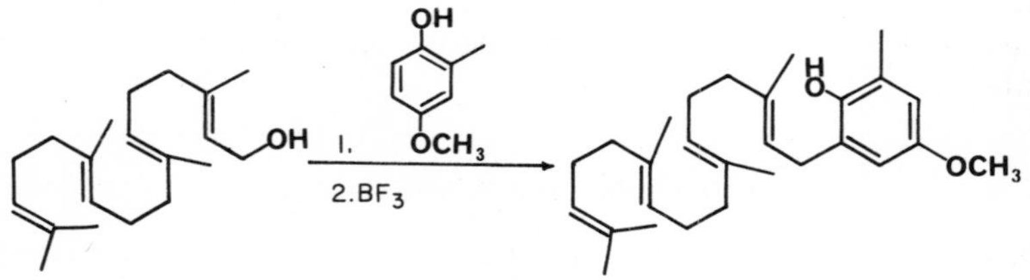

1. (2-methyl-4-methoxyphenol) 2. BF_3

Geranylgeraniol → 187

al., 1973). Although not rigorously shown, the stereochemistry of the polycyclic system of taondiol was assumed from the established mechanism of cyclization of all-*trans*-geranylgeraniol. The structure of atomaric acid was established as a result of its chemistry and by analysis of spectral data derived from several derivatives. Biogenetically atomaric acid may be related to taondiol via a proton-induced opening of the chroman ring

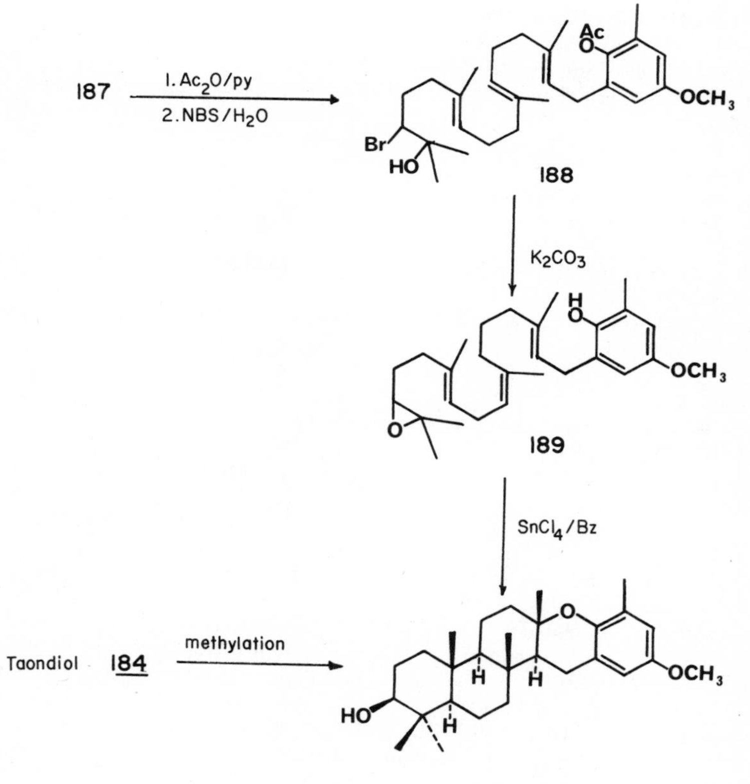

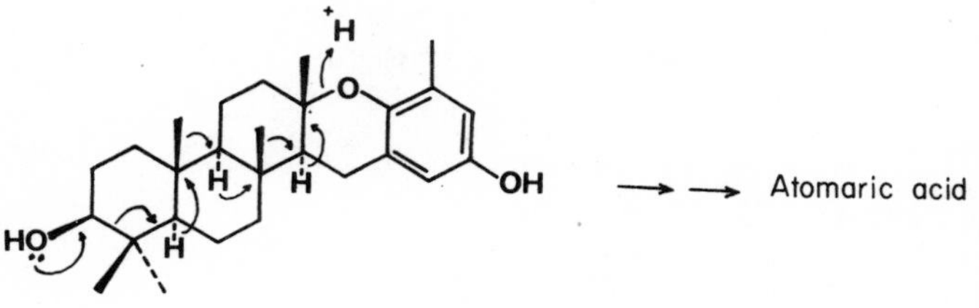

187 — 1. Ac_2O/py, 2. NBS/H_2O → 188

188 — K_2CO_3 → 189

189 — $SnCl_4$/Bz → 190

Taondiol 184 — methylation → 190

→ → Atomaric acid

followed by a series of presumably concerted methyl and hydrogen shifts. There is precedent for the oxidative cleavage of the C-3–C-4 bond in diterpenoids, but this transformation is apparently the first example of this cleavage accompanied by extensive skeletal rearrangement.

REFERENCES

Achenbach, H. (1976). *Chem. Pharm. Bull.* **24**, 832.

Amico, V., Oriente, G., Piattelli, M., Tringali, C., Fattorusso, E., Magno, S., and Mayol, L. (1976). *J. Chem. Soc., Chem. Commun.* p. 1024.

Amico, V., Oriente, G., Piattelli, M., Tringali, C., Fattorusso, E., Magno, S., and Mayol, L. (1977). *Experientia*.

Baker, J. T., Wells, R. J., Oberhänsli, W. E., and Hawes, G. B. (1976). *J. Am. Chem. Soc.* **98,** 4010.

Bartholome, C. (1974). Thesis, Univ. Libre de Bruxelles, Brussels.

Barton, D. H. R., Campus-Neves, A. S., and Cookson, R. C., Jr. (1956). *J. Chem. Soc.* p. 3500.

Bernstein, J., Schmeuli, U., Zadock, E., Kashman, Y., and Neeman, I. (1974). *Tetrahedron* **30,** 2817.

Birch, A. J., Brown, W. V., Corrie, J. E. T., and Moore, B. P. (1972). *J. Chem. Soc., Perkin Trans. 1,* p. 2653.

Blackman, A. J., and Wells, R. J. (1976). *Tetrahedron Lett.* p. 2729.

Blackman, A. J., and Wells, R. J. (1978). Work in progress.

Blumer, M., and Thomas, A. W. (1965). *Science* **147,** 1148.

Blumer, M., Mullin, M. M., and Thomas, D. W. (1963). *Science* **140,** 974.

Brown, E. D., Solomon, M. D., Sutherland, J. K., and Torre, A. (1967). *J. Chem. Soc., Chem. Commun.* p. 111.

Burks, J. E., van der Helm, D., Change, C. Y., and Ciereszko, L. S. (1977). *Acta Crystallogr., Sect. B* **33,** 704–709.

Burrell, J. W. K., Garwood, R. F., Jackman, L. M., Oskay, E., and Weedon, B. C. L. (1966). *J. Chem. Soc. C.* p. 2144.

Burreson, B. J., and Scheuer, P. J. (1974). *J. Chem. Soc., Chem. Commun.* p. 1035.

Burreson, B. J., Christophersen, C., and Scheuer, P. J. (1975). *Tetrahedron* **31,** 2015.

Cafieri, F., de Napoli, L., Fattorusso, E., Impellizzeri, G., Piattelli, M., and Sciuto, S. (1977). *Experientia* **32,** 1549–50.

Christensen, P. K., and Sørensen, N. A. (1951). *Acta Chem. Scand.* **5**, 751.

Ciereszko, L. S. (1977). *In* "Marine Natural Products Chemistry" (D. J. Faulkner and W. H. Fenical eds.), pp. 1–3. Plenum, New York.

Ciereszko, L. S., and Karns, T. K. B. (1973). *In* "Biology and Geology of Coral Reefs" (O. A. Jones and R. Endean, eds.), pp. 183–203. Academic Press, New York.

Cimino, G., de Stefano, S., and Minale, L. (1972). *Experientia* **28,** 1401.

Cimino, G., de Rosa, D., de Stefano, S., and Minale, L. (1974). *Tetrahedron* **30,** 645.

Coll, J. C., Hawes, G. B., Liyanage, N., Wells, R. B., and Oberhänsli, W. E. (1978). Work in progress.

Dalietos, D., and Moore, R. E. (1977). Unpublished observations.

Danise, B., Minale, L., Riccio, R., Amico, V., Oriente, G., Piattelli, M., Tringali, C., Fattorusso, E., Magno, S., and Mayol, L. (1977). *Experientia* **33,** 413.

Devon, T. K., and Scott, A. I. (1972). "Handbook of Naturally Occurring Compounds," Vol. 2, "Terpenes." Academic Press, New York.

Doty, M. S., and Santos, G. A. (1966). *Nature (London)* **211**, 990.
Fattorusso, E. (1977). *In* "Marine Natural Products Chemistry" (D. J. Faulkner and W. H. Fenical, eds.), pp. 165–178. Plenum, New York.
Fattorusso, E., Magno, S., Mayol, L., Santacroce, C., Sica, D., Amico, V., Oriente, G., Piattelli, M., and Tringali, C. (1976a). *Tetrahedron Lett.* p. 937.
Fattorusso, E., Magno, S., Mayol, L., Santacroce, C., Sica, D., Amico, V., Oriente, G., Piattelli, M., and Tringali, C. (1976b). *J. Chem. Soc., Chem. Commun.* p. 575.
Fattorusso, E., Magno, S., Santacroce, C., Sica, D., di Blasio, B., and Pedone, C. (1976c). *Gazz. Chim. Ital.* **106,** 779.
Faulkner, D. J., Ravi, B. N., Finer, J., and Clardy, J. (1977). *Phytochemistry* **16**, 991.
Fenical, W., Howard, B., Gifkins, K., and Clardy, J. (1975). *Tetrahedron Lett.* p. 3983.
Fenical, W., Finer, J., and Clardy, J. (1976). *Tetrahedron Lett.* p. 731.
Fenical, W., Moore, R. E., Finer, J., and Clardy, J. (1978). Work in progress.
Ferezou, J. P., Berreur-Bonnenfant, J., Meusy, J. J., Barbier, M., Suchy, M., and Wipf, H. K. (1977). *Experientia* **33,** 290.
Finer, J., and Clardy, J. (1978). Work in progress.
Fisher, F. G., and Mägerlein, H. (1960). *Justus Liebigs Ann. Chem.* **636,** 88.
Gonzalez, A. G., Darias, J., and Martín, J. D. (1971). *Tetrahedron Lett.* p. 2729.
Gonzalez, A. G., Martín, J. D., and Rodriguez, M. L. (1973). *Tetrahedron Lett.* p. 3657.
Gonzalez, A. G., Darias, J., Martín, J. D., and Norte, M. (1974). *Tetrahedron Lett.* p. 3951.
Green, M. M., Axelrod, M., and Mislow, K. (1966). *J. Am. Chem. Soc.* **88,** 861.
Hale, R. L., Leclercq, J., Tursch, B., Djerassi, C., Gross, R. A., Weinheimer, A. J., Gupta, K., and Scheuer, P. J. (1970). *J. Am. Chem. Soc.* **92,** 2179.
Hall, S. S., Faulkner, D. J., Fayos, J., and Clardy, J. (1973). *J. Am. Chem. Soc.* **95,** 7187.
Hamberg, M., Svenson, J., Wakabayashi, T., and Samuelson, B. (1974). *Proc. Natl. Acad. Sci. U.S.A.* **71,** 345.
Hanson, J. R. (1971). *Fortschr. Chem. Org. Naturst.* **29,** 395.
Herin, M., and Tursch, B. (1976). *Bull. Soc. Chim. Belg.* **85,** 707.
Herin, M., Colin, M., and Tursch, B. (1976). *Bull. Soc. Chim. Belg.* **85,** 801.
Hirschfeld, D. R., Fenical, W., Lin, G. H. Y., Wing, R. M., Radlick, P., and Sims, J. J. (1973). *J. Am. Chem. Soc.* **95,** 4049.
Horeau, A. (1961). *Tetrahedron Lett.* p. 506.
Horeau, A. (1962). *Tetrahedron Lett.* p. 965.
Houssain, M. B., and van der Helm, D. (1969). *Recl. Trav. Chim. Pays-Bas* **88,** 1413.
Houssain, M. B., Nicholas, A. F., and van der Helm, D. (1968). *J. Chem. Soc., Chem. Commun.* p. 385.
Howard, B. M., and Fenical, W. (1976). *Tetrahedron Lett.* p. 41–44.
Howard, B. M., and Fenical, W. (1978). *Tetrahedron Lett.* p. 2453–6.
Howard, B. M., Fenical, W., Finer, J., Hirotsu, K., and Clardy, J. (1977). *J. Am. Chem. Soc.* **99,** 6440.
Hyde, R. W. (1966). Ph.D. Thesis, Univ. of Oklahoma, Norman, Oklahoma.
Ikan, R., Baedeker, M. J., and Kaplan, I. R. (1973). *Nature (London)* **244**, 154.
Ireland, C., and Faulkner, D. J. (1977). *J. Org. Chem.* **42,** 3157.
Ireland, C., Faulkner, D. J., Finer, J., and Clardy, J. (1976). *J. Am. Chem. Soc.* **98,** 4664.
Irie, T. (1969). *Nippon Kagaku Zasshi* **90,** 1179.
Kashman, Y. (1977). *In* "Marine Natural Products Chemistry" (D. J. Faulkner and W. H. Fenical, eds.), pp. 17–22. Plenum, New York.
Kashman, Y., and Groweiss, A. (1977). *Tetrahedron Lett.* p. 1159.
Kashman, Y., Zadock, E., and Neeman, I. (1974). *Tetrahedron* **30,** 3615.
Kato, T., Kumanireng, A. S., Ichinose, I., Kitahara, Y., Kakinuma, Y., Nishihira, M., and Kato, M. (1975a). *Experientia* **31,** 433.

Kato, T., Kumanireng, A. S., Ichinose, I., Kitahara, Y., Kakinuma, Y., and Kato, Y. (1975b). *Chem. Lett.* p. 335.

Kazlauskas, R., Murphy, P. T., and Wells, R. J. (1978a). Work in progress.

Kazlauskas, R., Murphy, P. T., Wells, R. J., and Schönholzer, P. (1978b). *Tetrahedron Lett.* (submitted for publication).

Kazlauskas, R., Murphy, P. T., Wells, R. J., Oberhänsli, W. E., and Noack, K. (1978c). Work in progress.

Kazlauskas, R., Murphy, P. T., Wells, R. J., and Blount, J. F. (1978d). Work in progress.

Kennard, O., Watson, D. G., di Sanseverino, L. R., Tursch, B., Bosmans, R., and Djerassi, C. (1968). *Tetrahedron Lett.* p. 2879.

Kinzer, G. W., Page, T. F., and Johnson, R. R. (1966). *J. Org. Chem.* **31,** 1797.

Kobayashi, H., and Akiyoshi, S. (1962). *Bull. Chem. Soc. Jpn.* **35**, 1044.

Kodama, M., Matsuki, Y., and Ito, S. (1975). *Tetrahedron Lett.* p. 3065.

McCrindle, R., and Overton, K. H. (1969). *In* "The Diterpenoids, Triterpenoids and Sesterterpenoids. Chemistry of Carbon Compounds" (E. H. Rodd, ed.), Vol. 2b, Ch. 14. Elsevier, Amsterdam.

McEnroe, F. J., Robertson, K. J., and Fenical, W. (1977). *In* "Marine Natural Products Chemistry" (D. J. Faulkner and W. H. Fenical, eds.), pp. 179–190. Plenum, New York.

Maiti, B. C., and Thomson, R. H. (1977). *In* "Marine Natural Products Chemistry" (D. J. Faulkner and W. H. Fenical, eds.), pp. 159–163. Plenum, New York.

Matsuda, H., Tomiie, Y., Yamamura, S., and Hirata, Y. (1967). *J. Chem. Soc., Chem. Commun.* p. 898.

Minale, L., and Riccio, R. (1976). *Tetrahedron Lett.* p. 2711.

Minale, L., Cimino, G., de Stefano, S., and Sodano, G. (1976). *Fortschr. Chem. Org. Naturst.* **33,** 1.

Missakian, M. G., Burreson, B. J., and Scheuer, P. J. (1975). *Tetrahedron* **31,** 2513.

Nakanishi, K., Goto, T., Ito, S., Natori, S., and Nozoe, S. (1974). "Natural Products Chemistry," Vol. 1. Kodansha, Tokyo; Academic Press, New York.

Neeman, I., Fishelson, L., and Kashman, Y. (1974). *Toxicon* **12,** 593.

Norris, J. N. (1978). Work in progress.

Patil, V. D., Nayar, U. R., and Dev, S. (1973). *Tetrahedron* **29,** 341.

Pettit, G., Ode, R. H., Herald, C. L., von Dreele, R. B., and Michel, C. (1976). *J. Am. Chem. Soc.* **98,** 4677.

Ravi, B. N., and Faulkner, D. J. (1978). Work in progress.

Robertson, K. J., and Fenical, W. (1977). *Phytochemistry* **16,** 1071.

Rothe, W. (1950). *Pharmazie* **5,** 190.

Ruzicka, L. (1959). "History of the Isoprene Rule." Chem. Soc., London.

Ruzicka, L., and Hosking, J. R. (1930). *Helv. Chim. Acta* **13,** 1402.

Ruzicka, L., Eschenmoser, A., and Heusser, H. (1953). *Experientia* **9,** 357.

Santos, G. A. (1970). *J. Chem. Soc. C.* p. 842.

Santos, G. A., and Doty, M. S. (1971). *Lloydia* **34,** 88.

Schmitz, F. J., Vanderah, D. J., and Ciereszko, L. S. (1974). *J. Chem. Soc., Chem. Commun.* p. 407.

Scott, A. I., McCapra, F., Comer, F., Sutherland, S. A., Young, D. W., Si, G. M., and Ferguson, F. (1964). *Tetrahedron* **20,** 1339.

Sims, J. J., and Pettus, J. A., Jr. (1976). *Phytochemistry* **15,** 1076.

Sims, J. J., Lin, G. H. Y., Wing, R. M., and Fenical, W. (1973). *J. Chem. Soc., Chem. Commun.* p. 470.

Sørensen, J. S., and Sørensen, N. A. (1949). *Acta Chem. Scand.* **3,** 939.

Sørensen, N. A., and Mehlum, J. (1948). *Acta Chem. Scand.* **2,** 140.

Stallard, M. O., and Faulkner, D. J. (1974). *Comp. Biochem. Physiol. B.* **49,** 25.

Sun, H. H., Waraszkiewicz, S. M., Erickson, K. L., Finer, J., and Clardy, J. (1977). *J. Am. Chem. Soc.* **99,** 3516.

Takaoka, M., and Ando, Y. (1951). *Nippon Kagaku Zasshi* **72,** 999.

Threlfall, D. R., and Whistance, G. R. (1971). *In* "Aspects of Terpenoid Chemistry and Biochemistry" (T. W. Goodwin, ed.), p. 357. Academic Press, New York.

Treibs, W. (1953). *Justus Liebigs Ann. Chem.* **576,** 116.

Tsujimoto, M. (1917). *J. Ind. Eng. Chem.* **9,** 1098. [*Chem. Abstr.* **10,** 1602 (1916).]

Tsujimoto, M. (1935). *Bull. Chem. Soc. Jpn.* **10,** 149.

Tursch, B. (1976). *Pure Appl. Chem.* **48,** 1.

Tursch, B., Braekman, J. C., Daloze, D., Herin, M., and Karlsson, R. (1974). *Tetrahedron Lett.* p. 3769.

Tursch, B., Braekman, J. C., and Daloze, D. (1975a). *Bull. Soc. Chim. Belg.* **84,** 767–774.

Tursch, B., Braekman, J. C., Daloze, D., Herin, M., Karlsson, R., and Losman, D. (1957b). *Tetrahedron* **31,** 129.

Vanderah, D., and Faulkner, D. J. (1976). Unpublished observations.

Vanderah, D. J., Steudler, P. A., Ciereszko, L. S., Schmitz, F. J., Ekstrand, J. D., and van der Helm, D. (1977). *J. Am. Chem. Soc.* **99,** 5780.

van der Helm, D., Enwall, E. L., Weinheimer, A. J., Karns, T. K. B., and Ciereszko, L. S. (1976). *Acta Crystallogr., Sect. B* **32,** 1558.

Weinheimer, A. J. (1978). Work in progress.

Weinheimer, A. J., and Matson, J. A. (1975). *Lloydia* **38,** 378.

Weinheimer, A. J., Middlebrook, R. E., Bledsoe, J. O., Marsico, W. E., and Karns, T. K. B. (1968). *J. Chem. Soc., Chem. Commun.* p. 384.

Weinheimer, A. J., Rehm, S. J., and Ciereszko, L. S. (1972). *Nat. Prod. Symp., 4th, Jamaica.*

Weinheimer, A. J., Matson, J. A., van der Helm, D., and Poling, M. (1977). *Tetrahedron Lett.* p. 1295.

Wekell, C. (1974). *Proc. Food-Drugs Sea Symp., 4th, Mar. Technol. Soc., Washington, D.C.* pp. 324–330.

Wells, R. J. (1978). Work in progress.

Whittle, K. J., Dunphy, P. J., and Pennock, J. F. (1966). *Biochem. J.* **100,** 138.

Wratten, S. J., Faulkner, D. J., Hirotsu, K., and Clardy, J. (1977a). *J. Am. Chem. Soc.* **99,** 2824.

Wratten, S. J., Fenical, W., Faulkner, D. J., and Wekell, J. C. (1977b). *Tetrahedron Lett.* p. 1559.

Yamamura, S., and Hirata, Y. (1963). *Tetrahedron* **19,** 1485.

Yamamura, S., and Hirata, Y. (1971). *Bull. Chem. Soc. Jpn.* **44,** 2560.

Chapter 4

Terpenoids from Coelenterates

B. TURSCH, J. C. BRAEKMAN, D. DALOZE, and M. KAISIN

1. COELENTERATES

The name Coelenterata* in its present context combines the two animal phyla Cnidaria and Ctenophora. The latter consists of marine (mostly pelagic) animals, and they will not be treated here since they do not yield any of the chemicals under consideration in this chapter. The phylum Cnidaria forms a very large group of aquatic animals, most of them marine. They are acoelomate Metazoa, diploblastic, and often with radial symmetry. Cnidaria have no anus; their internal cavity opens into a mouth often surrounded by tentacles bearing specialized stinging organs (the

* For an excellent zoological introduction to the coelenterates, see Bouillon (*15*).

MARINE NATURAL PRODUCTS

ISBN 0-12-624002-7

nematocysts or cnidocysts). An amazing variety of fixed, free-swimming, solitary, or colonial forms, ranging from very large animals to microscopic organisms, are known. Cnidaria are greatly diversified, as can be seen from the following classification:

Phylum Cnidaria
- Class Hydrozoa
  - Subclass Hydroida and Hydromedusa (jellyfish, hydra, hydroids, fire coral)
  - Subclass Siphonophora (e.g., *Physalia*)
- Class Scyphozoa (jellyfish, e.g., *Chironex*)
- Class Anthozoa
  - Subclass Alcyonaria or Octocorallia (e.g., soft corals, gorgonians)
  - Subclass Zoantharia or Hexacorallia (e.g., sea anemones, hermatypic corals)

Terpenoids have been isolated thus far only from the subclass Alcyonaria. However, many of the other coelenterate groups have not yet been properly investigated and could thus constitute a future source of these compounds.

Progress in marine natural products chemistry is still largely conditioned by the availability of suitable material and, when considered as chemical producers, coelenterates fall into three distinct groups. Some groups contain mainly seawater (e.g., jellyfish), others mainly calcium carbonate (e.g., hermatypic corals). Some (e.g., octocorals) produce large amounts of extractible organic matter, which is naturally favored by organic chemists. None of the collecting and processing difficulties encountered with the first two groups occurs with octocorals, as can be illustrated by the fact that the dichloromethane extract of Melanesian alcyonaceans commonly amounts to 5–8% of the dry weight.

The subclass Alcyonaria includes the groups tabulated on next page.

Alcyonaria are widely distributed but have a marked preference for tropical waters, where they constitute a very conspicuous feature of the coral reefs.

As can be seen from Table 1, Gorgonacea are especially abundant in the tropical western Atlantic, whereas Alcyonacea are very dominant on the Indo-Pacific reefs, where in many instances they are probably the largest single contributor to the biomass. Furthermore, these two orders are often represented by quite large colonies within easy reach of the collector because of their occurrence in rather shallow waters. It is thus not surprising that gorgonians and soft corals have been the subject of most of the organic chemical research carried out on coelenterates.

Subclass Alcyonaria (or Octocorallia)	Occurrence of Terpenoids
Superorder Protoalcyonaria; small, solitary polyps	?
Superorder Synalcyonaria; colonial forms	
Order Stolonifera; polyps connected but not fused, e.g., *Clavularia, Tubipora*	+
Order Telestacea; dendriform colonies, e.g., *Telesto*	?
Order Alcyonacea (soft corals); small, distinct polyps fused into a common mass of coenchyme, e.g., *Alcyonium, Sarcophyton, Xenia*	++
Order Coenothecalia (blue coral); one living genus: *Heliopora*	−(?)
Order Gorgonacea (sea fans, sea whips); coenchyme covering a plantlike axial skeleton, e.g., *Eunicea, Gorgonia, Pseudoplexaura*	++
Order Pennatulacea (sea pens); very long axial polyp bearing smaller lateral polyps, e.g., *Ptilosarcus*	+

Most coelenterates are considered carnivorous (*85*). However, a great number of them, including most tropical Alcyonaria, contain large numbers of intracellular symbiotic algae, the zooxanthellae. These unicellular plants represent the vegetative stage of dinoflagellates and appear as minute spheres 8–12 μm in diameter. They are often extremely abundant in animal tissues, up to 30,000 algae per cubic millimeter of animal tissue having been observed (*159*). For comprehensive discussions of the symbiotic algae problem, see Taylor (*120*) and Yonge (*159*).

Zooxanthellae are not confined to coelenterates and are present in certain sponges, mollusks, and other marine animals. Despite their great ecological importance, they are poorly known. Although a number of studies have dealt with different aspects of their symbiosis with various hosts, the systematics of zooxanthellae still present many problems. One might not be able to discern closely related taxa by means of classic morphology and ultrastructural criteria, especially in the very similar encysted stages present in coelenterate tissues. Some authors have assumed that all symbiotic xanthellae found in marine cnidarians belong to the same species, namely, *Symbiodinium microadriaticum* (Freudenthal) (= *Gymnodinium microadriaticum* Freudenthal), whereas others suggest the existence of at least a variety of forms (*112*). In the second hypothesis it would be most interesting to know whether these symbiotic associations between animal polyps and algae are of a specific nature.

Important metabolic exchanges between the partners of these associations have been established (*120*). In addition, it has been suggested that coelenterates could feed directly on their symbiotic algae, as is the case

TABLE 1

Geographical Distribution of the Major Orders of Alcyonaria

Region	Species distribution (%)					Total reported species (Ref.)
	Alcyonacea	Gorgonacea	Pennatulacea	Stolonifera	Telestacea	
Southwestern Atlantic	6	81	10	0	3	31 (*123*)
West Indies	3	85	5	0	7	100 (*9*)
Vietnam	97	1.5	0	0	1.5	95 (*124*)
New Caledonia	89	8	2	1	0	203 (*121*)
Madagascar	66.5	29	3	1	0.5	225 (*122*)

for the giant clam *Tridacna* sp. (*52*). The photosynthetic activity of these zooxanthellae has been demonstrated to be of paramount importance in the energy balance of a coral reef (*159*). It is in fact the amount of available light that restricts tropical reefs to shallow, clear waters, and it would thus appear that the very existence of coral reefs rests on this association of unicellular algae and coelenterates.

The existence of this symbiotic association greatly obscures the origin of the substances isolated from zooxanthellae-containing coelenterates. Any of the chemicals we consider in this chapter might *a priori* be produced by the symbiotic algae, the animal itself, the association of both, or be of exogenous (e.g., dietary) origin. This is discussed in more detail later (see Section VII).

II. TERPENOIDS ISOLATED FROM ALCYONARIA

This section comprises exclusively the terpenoids isolated from coelenterates. All compounds having a well-established structure are presented in a standardized form, in order to allow easy reference and to avoid unnecessary repetition. Compounds whose occurrence or structure is not yet unambiguously established are grouped in a separate section.

A number of the coelenterate terpenoids were known from other sources. Many of the new structures were solved by X-ray diffraction analysis, and others were determined by classic spectroscopic methods and/or straightforward chemical degradations and correlations. We have therefore elected only to name the method(s) used for identification or structure determination and to provide the reader interested in further details with the original references. A special effort has been made to assess the correct absolute configuration wherever possible. This was in our opinion especially necessary in the case of the sesquiterpenes, where the literature is often confusing because of the repeated use of wrong structures.

The following abbreviations are used: A, Alcyonacea; G, Gorgonacea; P, Pennatulacea; S, Stolonifera; MW, molecular weight; TP, terrestrial plant(s); XR, X-ray diffraction analysis. The abbreviation AC indicates that the absolute configuration is shown. When the compound is also known from nonmarine sources, the physical data are those reported for the sample of marine origin.

A. Sesquiterpenes

1. Marine occurrence: *Plexaurella dichotoma* (G) (*32, 161*), *Plexaurella grisea* (G) (*32*), *Plexaurella fusifera* (G) (*32*).

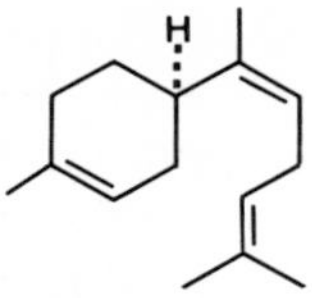

$C_{15}H_{24}$
MW 204
Oil
$[\alpha]_D$ +63.5° (*161*)
n_D^{20} 1.4955 (*161*)
AC

1 (+)-α-Bisabolene

IDENTIFICATION: By comparison of spectral properties with published data (*161*).

ABSOLUTE CONFIGURATION: Established for (−)-α-bisabolene obtained from (+)-nerolidol (*3*).

COMMENT: No (+)-α-bisabolene could be traced in the original reference (*102*) cited by Youngblood (*161*).

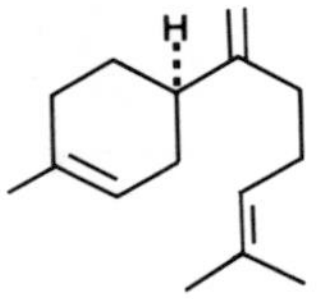

$C_{15}H_{24}$
MW 204
Oil
$[\alpha]_D$ +83.1° (*144*)
n_D^{25} 1.4884 (*144*)
AC

2 (+)-β-Bisabolene

2. MARINE OCCURRENCE: *Plexaurella dichotoma* (G) (*32, 161*), *Plexaurella grisea* (G) (*32*), *Plexaurella fusifera* (G) (*32*), *Plexaurella nutans* (G) (*66*).

IDENTIFICATION: By comparison of its spectral properties (*66, 161*) and gas chromatographic retention time (*66*) with those of an authentic sample.

OTHER SOURCES: Found in *Pinus sibirica* (*100*) and *Chamaecyparis nootkatensis* (*2*); the (−) antipode is very common in TP (*97*).

COMMENTS: The formula depicted in Weinheimer *et al.* (*144*) for (+)-β-bisabolene represents in fact (α)-bisabolene. This error is also found in Premuzic (*104*). A methyl group is lacking in the structure reported in Ciereszko and Karns (*32*). The structure reported in Andersen and Syrdal (*2*) and Jeffs and Lytle (*66*) is (−)-β-bisabolene.

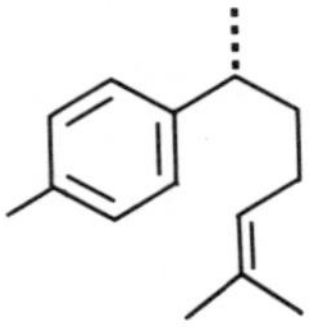

$C_{15}H_{22}$
MW 202
Oil
$[\alpha]_D$ −39.31° (*161*)
n_D^{20} 1.5015 (*161*)
AC

3 (−)-α-Curcumene

3. Marine occurrence: *Plexaurella dichotoma* (G) (*32, 161*), *Plexaurella grisea* (G) (*32*), *Plexaurella fusifera* (G) (*32*), *Plexaurella nutans* (G) (*66*), *Muricea elongata* (G) (*66*).

Identification: By comparison with an authentic sample (*161*).

Absolute configuration: Established by ozonolysis into (−)-4-p-tolylpentanoic acid (*59*).

Other sources: Common in TP (*97*).

Comments: Both antipodes have been isolated from TP, e.g., *Zingiberaceae* (*44, 97*). The absolute configuration of the (+) antipode was independently established (*44*).

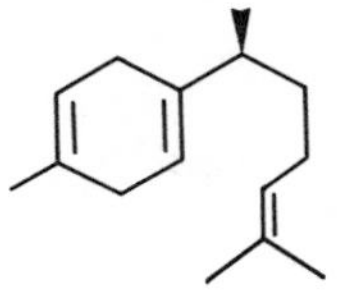

$C_{15}H_{24}$
MW 204
Oil
AC

4 (+)-β-Curcumene

4. Marine occurrence: *Plexaurella dichotoma* (G) (*32, 161*), *Plexaurella grisea* (G) (*32*), *Plexaurella fusifera* (G) (*32*).

Identification: By comparison with an authentic sample prepared by Birch reduction of (−)-α-curcumene (*161*).

Other sources: Found in TP, e.g., *Nectandra elaiophora* (*92*).

Comments: Although the compound was described as (+)-β-curcumene in Ciereszko and Karns (*32*), its optical rotation was not reported in the original work (*161*).

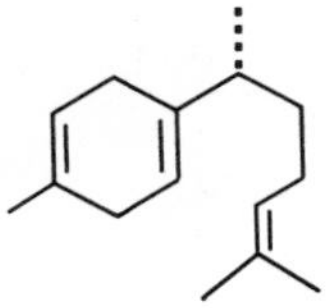

$C_{15}H_{24}$
MW 204
Oil
AC

5 (−)-β-Curcumene

5. Marine occurrence: *Muricea elongata* (G) (*66*), *Plexaurella nutans* (G) (*66*).

Identification: By comparison with an authentic sample (*66*).

Absolute configuration: Established (*66*) by synthesis from (−)-α-curcumene (*59*).

Other sources: Found in TP, e.g., *Curcuma aromatica* (*97*).

6. Marine occurrence: *Eunicea mammosa* (G) (*32, 144, 147, 161*).

Structure and absolute configuration: Derived from spectral data and Cope rearrangement to (+)-β-elemene (**83**) (*147, 161*).

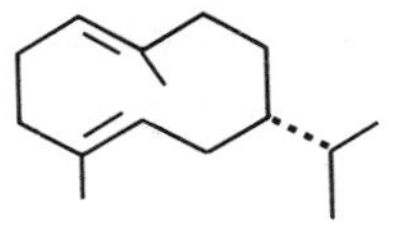

$C_{15}H_{24}$
MW 204
Oil
$[\alpha]_D$ −3.2° (*161*)
n_D^{25} 1.5089 (*161*)
AC

6 (−)-Germacrene

Other sources: Postulated in TP (*11*) and recently shown to be the alarm pheromone of the alfalfa aphid *Therioaphis maculata* (*16*).

Comments: See Section III. This elusive compound could be isolated only by using extreme experimental care (e.g., chromatography on sugar-impregnated Florisil).

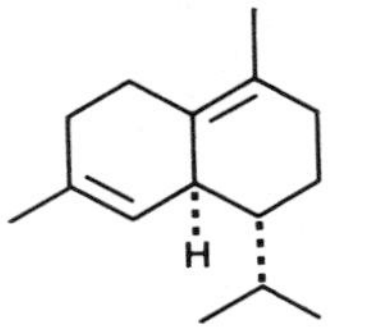

$C_{15}H_{24}$
MW 204
Oil
$[\alpha]_D$ −90.5° (161)
n_D^{25} 1.5070 (*161*)
AC

7 (−)-δ-Cadinene

7. Marine occurrence: *Pseudoplexaura porosa* (G) (*32*, *161*), *Pseudoplexaura wagenaari* (G) (*32*), *Pseudoplexaura flagellosa* (G) (*32*).

Identification: By comparison of its spectral properties with published data (*161*).

Absolute configuration: Established for the (+) antipode (*91*).

Other sources: Both (+) and (−) antipodes occur in TP (*91*, *97*, *153*).

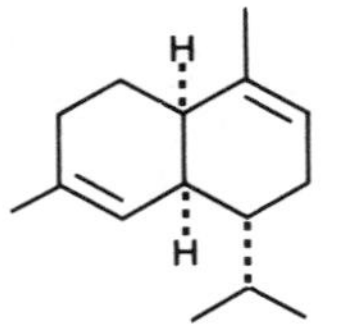

$C_{15}H_{24}$
MW 204
Oil
$[\alpha]_D$ +67.2° (*161*)
n_D^{25} 1.5051 (*161*)
AC

8 (+)-α-Muurolene

8. Marine occurrence: *Eunicea mammosa* (G) (*32*, *144*), *Eunicea palmeri* (G) (*32*, *144*), *Plexaurella dichotoma* (G) (*32*, *144*, *161*), *Plexaurella grisea* (G) (*32*), *Plexaurella fusifera* (G) (*32*), *Pseudoplexaura porosa* (G) (*32*, *144*, *161*), *Pseudoplexaura wagenaari* (G) (*32*), *Pseudoplexaura flagellosa* (G) (*32*).

Identification: By comparison of its spectral properties with those published for the (−) antipode (*161*).

Absolute configuration: Established for the (−) antipode (*154*).

Other sources: None; the (−) antipode is known from TP (*154, 162*).
Comments: The structure given in Scheuer (*109*) is in fact a cadinene. Baker and Murphy (*6*), Faulkner and Andersen (*53*), and Weinheimer *et al.* (*144*) depict the incorrect absolute configuration. This error probably arises from erroneous configurations given in Zabza *et al.* (*162*) and corrected in a rather cryptic note added in proof.

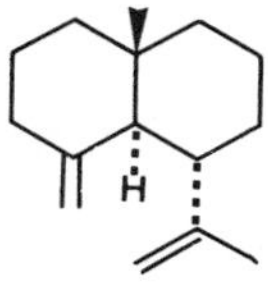

$C_{15}H_{24}$
MW 204
Oil
$[\alpha]_D$ +13.9° (*141*)
n_D^{25} 1.501° (*146*)
AC

9 (+)-β-Gorgonene

9. Marine occurrence: *Pterogorgia americana* (G) (*141, 146*).
Structure and absolute configuration: Determined by XR on the $AgNO_3$ complex (*60, 146*).
Comments: (−)-β-Gorgonene has been synthesized from (+)-maaliol (*98*); (±)-β-gorgonene has been obtained by total synthesis (*14*); *Pterogorgia* has been referred to as *Pseudopterogorgia* (*141*).

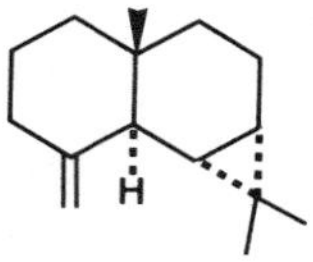

$C_{15}H_{24}$
MW 204
Oil
$[\alpha]_D$ +11.6 (*141*)
n_D^{25} 1.4992 (*141*)
AC

10 (+)-γ-Maaliene

10. Marine occurrence: *Pterogorgia americana* (G) (*141, 146*).
Identification: Based on the spectral properties and conversion to the corresponding 1,2-diol, which was compared to an authentic sample obtained from the (−) antipode (*141, 146*) of known absolute configuration (*8*).
Other sources: None, but the (−) antipode has been obtained by synthesis (*8*).

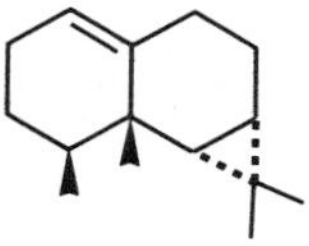

$C_{15}H_{24}$
MW 204
Oil
$[\alpha]_D$ −78.5° (*141*)
n_D^{25} 1.5005 (*141*)
AC

11 (−)-1(10)-Aristolene

11. MARINE OCCURRENCE: *Pterogorgia americana* (G) (*32, 141, 144, 146*).

IDENTIFICATION: By comparison of its spectral properties with those of the (+) antipode (*141, 146*).

ABSOLUTE CONFIGURATION: Established for the (+) antipode (*22*).

OTHER SOURCES: Only the (+) antipode is known from TP (*97*).

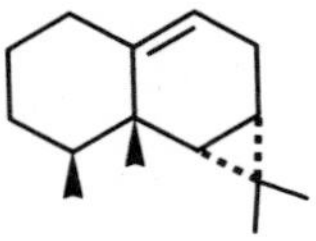

$C_{15}H_{24}$
MW 204
Oil
$[\alpha]_D$ +80.9° (*141*)
n_D^{25} 1.4975 (*141*)
AC

12 (+)-9-Aristolene

12. MARINE OCCURRENCE: *Pterogorgia americana* (G) (*32, 141, 144, 146*).

IDENTIFICATION: By comparison of its spectral properties with published data (*141, 146*).

ABSOLUTE CONFIGURATION: Established by conversion to (+)-aristolone (*26*).

OTHER SOURCES: Found in *Ferula communis* (*26*). The (−) antipode is more common in TP (*138, 139*).

COMMENT: This compound has been referred to as α-ferulene (*26*).

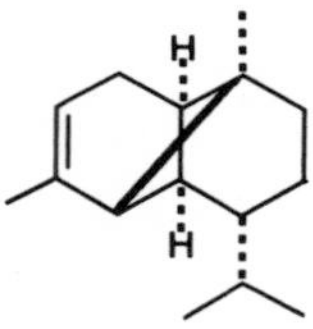

$C_{15}H_{24}$
MW 204
Oil
$[\alpha]_D$ +6.93° (*161*)
n_D^{25} 1.4899 (*161*)
AC

13 (+)-α-Copaene

13. MARINE OCCURRENCE: *Pseudoplexaura porosa* (G) (*32, 161*), *Pseudoplexaura wagenaari* (G) (*32*), *Pseudoplexaura flagellosa* (G) (*32*).

IDENTIFICATION: By comparison of its spectral properties with those published for the (−) antipode (*161*).

ABSOLUTE CONFIGURATION: Established for (−)-α-copaene by transformation into (−)-cadinene dihydrochloride (*46, 69*).

OTHER SOURCES: *Brachylaena hutchinsii* (*79*).

COMMENT: The (−) antipode has been found both in TP (*97*) and in the marine alga *Dictyopteris divaricata* (*64*).

14. MARINE OCCURRENCE: *Sinularia mayi* (A) (*10*), *Eunicea palmeri* (G) (*32*), *Pseudoplexaura porosa* (G) (*32, 161*), *Pseudoplexaura wagenaari* (G) (*32*), *Pseudoplexaura* (G) (*32*).

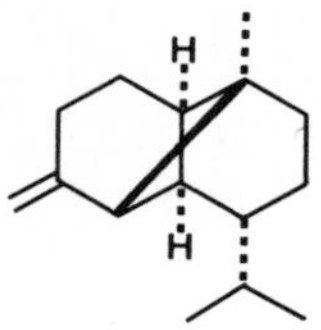

$C_{15}H_{24}$
MW 204
Oil
$[\alpha]_D$ +9.7° (*161*)
n_D^{25} 1.4988 (*161*)
AC

14 (+)-β-Copaene

IDENTIFICATION: By comparison of its physical data and spectral properties with published data (*161*). This was further substantiated by isomerization of (+)-β-copaene to (−)-δ-cadinene and (+)-α-muurolene (*161*).

ABSOLUTE CONFIGURATION: Deduced from that of (−)-β-copaene (*46, 69*).

OTHER SOURCES: None; the (−) antipode is known from TP (*97*).

COMMENTS: One of a few cases reported of a terpene coexisting in gorgonians and alcyonaceans. This compound was erroneously reported as (+)-β-ylangene in Washecheck (*141*) and Weinheimer *et al.* (*144*). Ylangenes are isopropyl isomers of copaenes (*94, 155*). Both are natural compounds and can be easily confused (*107*). Moreover, when this compound was first isolated (*141*), β-copaene was unreported in the literature.

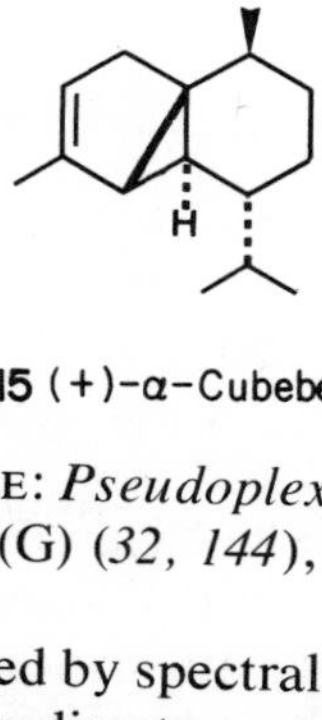

$C_{15}H_{24}$
MW 204
Oil
$[\alpha]_D$ +24.06° (*161*)
n_D^{25} 1.4790 (*161*)
AC

15 (+)-α-Cubebene

15. MARINE OCCURRENCE: *Pseudoplexaura porosa* (G) (*32, 144, 161*), *Pseudoplexaura wagenaari* (G) (*32, 144*), *Pseudoplexaura flagellosa* (G) (*32, 144*).

IDENTIFICATION: Suggested by spectral properties and chemical degradation (*161*) and confirmed by direct comparison with an authentic sample of (−)-α-cubebene (*161*).

ABSOLUTE CONFIGURATION: Established for the (−) antipode by synthesis (*119*).

OTHER SOURCES: Only the (−) antipode is known from TP, e.g., *Piper cubeba* (*95*).

16. MARINE OCCURRENCE: *Pseudoplexaura porosa* (G) (*32, 161*), *Pseudoplexaura wagenaari* (G) (*32*), *Pseudoplexaura flagellosa* (G) (*32*).

IDENTIFICATION: By comparison of its spectral properties with those published for (−)-β-bourbonene and bourbonone (*161*).

$C_{15}H_{24}$
MW 204
Oil
$[\alpha]_D$ +68.27° (*161*)
n_D^{25} 1.4876 (*161*)
AC

16 (+)-β-Epibourbonene

Absolute configuration: Established for the (−) antipode (*81, 83*).

Other sources: Only (−)-β-bourbonene is reported from TP, e.g., *Geranium bourbon* (*82*).

Comments: On the basis of physical and spectral arguments, (+)-β-epibourbonene is probably epimeric with (+)-β-bourbonene at the isopropyl site (*161*). Two total syntheses of α- and β-bourbonenes have been performed (*21, 156*). It is worth noting that α- and β-bourbonenes can be obtained by photoisomerization of germacrene D (*160*). The structures reported in Baker and Murphy (*6*) and Ciereszko and Karns (*32*) are erroneous.

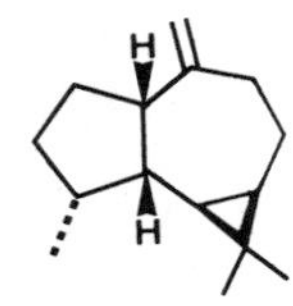

$C_{15}H_{24}$
MW 204
Oil
$[\alpha]_D$ +36.40° (*161*)
n_D^{25} 1.5002 (*161*)

17 Alloaromadendrene

17. Marine occurrence: *Pseudoplexaura porosa* (G) (*32, 161*), *Pseudoplexaura wagenaari* (G) (*32*), *Pseudoplexaura flagellosa* (G) (*32*), *Cespitularia* aff. *subviridis* (A) (*18*).

Identification: On the basis of the spectral properties reported for the (−) antipode (*48, 152*).

Absolute configuration: Significant discrepancy between the optical rotations of the compound from *P. porosa* and (−)-alloaromadendrene ($[\alpha]_D$ −21°) from TP precluded the attribution of the complete relative and absolute configuration (*161*). The optical rotation of the compound from *C.* aff. *subviridis* could not be measured (*18*).

Other sources: The (−) antipode is common in TP (*97*).

Comment: The compound has been synthesized (*24*).

18. Marine occurrence: *Cespitularia* aff. *subviridis* (A) (*18*).

Identification: By comparison of the physical and spectral properties with those published (*47*).

Absolute configuration: Established (*24, 47*).

Other sources: Known from a variety of TP (*47, 97*).

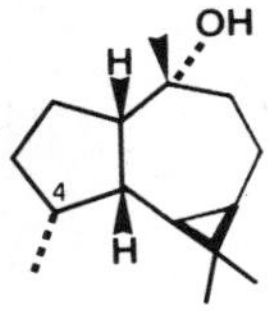

$C_{15}H_{26}O$
MW 222
mp 102°–103° (*20*)
$[\alpha]_D$ +5° (*18*)
AC

18 (+)-Ledol

COMMENT: The original (*47*) stereochemistry at C-4 was later corrected (*23, 24*).

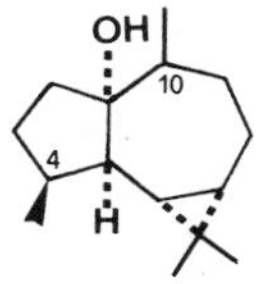

$C_{15}H_{26}O$
MW 222
Oil
$[\alpha]_D$ +14.3° (*28*)
AC

19 (+)-Palustrol

19. MARINE OCCURRENCE: *Cespitularia* aff. *subviridis* (A) (*28*).

IDENTIFICATION: On the basis of spectral data. The structure first selected with the CONGEN computer program (*28*) was further confirmed by comparison with published data for the (−) antipode (*49*).

ABSOLUTE CONFIGURATION: Established for the (−) antipode (*23, 24, 49*).

OTHER SOURCES: Only the (−) antipode is known from TP (*49, 97*).

COMMENTS: The stereochemistry at C-10 is still unknown (*18*). The stereochemistry originally proposed for C-4 (*49*) was later corrected (*23, 24*).

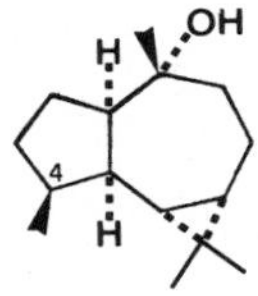

$C_{15}H_{26}O$
MW 222
$[\alpha]_D$ −5° (*18*)
AC

20 (−)-Viridifloral

20. MARINE OCCURRENCE: *Cespitularia* aff. *subviridis* (A) (*18*).

IDENTIFICATION: By comparison of its spectral properties with those of the (+) antipode (*18, 47*).

ABSOLUTE CONFIGURATION: Established for the (+) antipode (*47*).

OTHER SOURCES: Only the (+) antipode is known from TP (*97*).

COMMENT: The original (*47*) stereochemistry at C-4 was later corrected (*23, 24*).

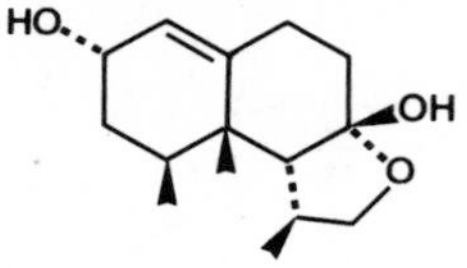

$C_{15}H_{24}O_3$
MW 252
mp 160°–163° (*128*)
AC

21 Lemnacarnol

21. Marine occurrence: *Lemnalia carnosa* (A) (*128*).

Structure and absolute configuration: Determined by XR (*72, 128*).

Comment: See lemnalactone (**24**).

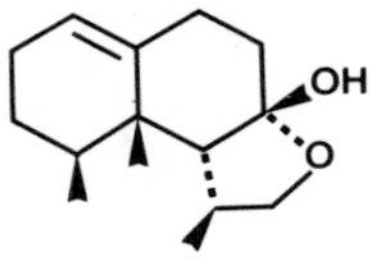

$C_{15}H_{24}O_2$
MW 236
mp 101°–103° (*43*)
$[\alpha]_D$ −165° (*43*)
AC

22 2-Deoxylemnacarnol

22. Marine occurrence: *Lemnalia africana* (A) (*43*), *Lemnalia laevis* (A) (*43*), *Paralemnalia thyroides* (A) (*43*).

Structure and absolute configuration: Derived from chemical correlation with lemnacarnol (**21**) (*43*).

Comment: See lemnalactone (**24**).

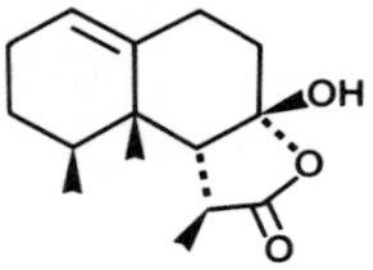

$C_{15}H_{22}O_3$
MW 250
mp 161°–162° (*43*)
$[\alpha]_D$ −20° (*43*)
AC

23 2-Deoxy-12-oxolemnacarnol

23. Marine occurrence: *Paralemnalia thyrsoides* (A) (*43*).

Structure: Determined by chemical correlation with lemnacarnol (**21**) (*43*) and confirmed by XR (*86*).

Absolute configuration: Derived from chemical correlation with lemnacarnol (**21**) (*43*).

Comment: See lemnalactone (**24**).

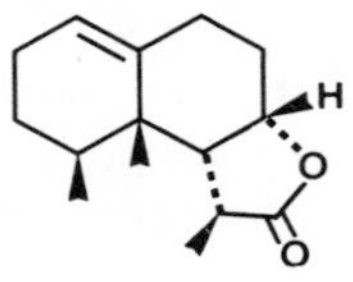

$C_{15}H_{22}O_2$
MW 234
mp 108°–109°
AC

24 Lemnalactone

24. Marine occurrence: *Paralemnalia digitiformis* (A) (*5*).

Structure: Determined by chemical and spectroscopic methods (*5*) and confirmed by chemical correlation with lemnacarnol (**21**) (*43*).

Absolute configuration: Derived from chemical correlation with lemnacarnol (**21**) (*43*).

Comment: The carbon skeleton of lemnalactone, lemnacarnol, and their relatives is antipodal to that of the known nardosinane sesquiterpenes isolated from *Nardostachys chinensis* (*108*).

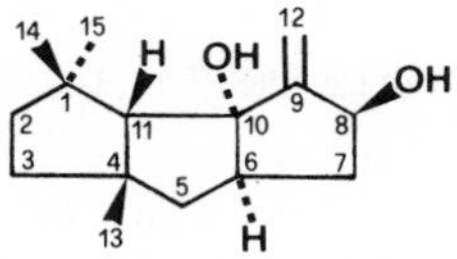

$C_{15}H_{24}O_2$
MW 236
mp 113°–114° (*113*)
$[\alpha]_D$ +41° (*113*)
AC

25 $\Delta^{9(12)}$-Capnellene-8β,10α-diol

25. Marine occurrence: *Capnella imbricata* (A) (*113*).

Structure and absolute configuration: Determined by chemical correlation with $\Delta^{9(12)}$-capnellene-3β,8β,10α-triol (**27**) (*113*).

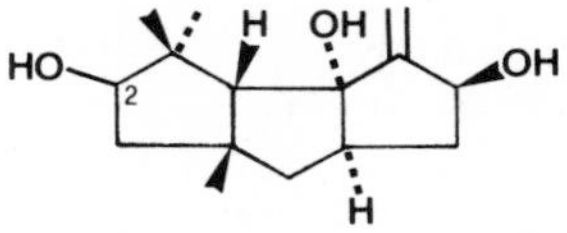

$C_{15}H_{24}O_3$
MW 252
Oil (*113*)
AC

26 $\Delta^{9(12)}$-Capnellene-2ξ,8β,10α-triol

26. Marine occurrence: *Capnella imbricata* (A) (*113*).

Structure and absolute configuration: Determined by chemical correlation with $\Delta^{9(12)}$-capnellene-3β,8β,10α-triol (**27**). The configuration at C-2 is still unknown (*113*).

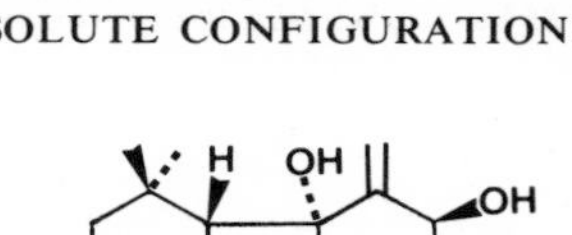

$C_{15}H_{24}O_3$
MW 252
mp 114°–117° (*68*)
$[\alpha]_D$ +2° (*68*)
AC

27 $\Delta^{9(12)}$-Capnellene-3β,8β,10α-triol

27. Marine occurrence: *Capnella imbricata* (A) (*68, 113*).

Structure and absolute configuration: Determined by XR (*68, 71*).

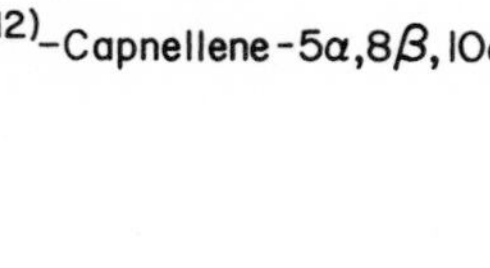

$C_{15}H_{24}O_3$
MW 252
mp 132°–133° (*113*)
$[\alpha]_D$ +34° (*113*)
AC

28 $\Delta^{9(12)}$-Capnellene-5α,8β,10α-triol

28. MARINE OCCURRENCE: *Capnella imbricata* (A) (*113*).

STRUCTURE AND ABSOLUTE CONFIGURATION: Derived from chemical correlation with $\Delta^{9(12)}$-capnellene-3β,8β,10α-triol (**27**).

$C_{15}H_{24}O_4$
MW 268
mp 192°–195° (*114*)
$[\alpha]_D$ +70° (*114*)

29 $\Delta^{9(12)}$-Capnellene-3β, 8β, 10α,14-tetrol

29. MARINE OCCURRENCE: *Capnella imbricata* (A) (*113, 114*).

STRUCTURE: Determined by XR (*114*).

COMMENT: The absolute configuration is assumed to be the same as that of the other capnellene sesquiterpenes (*114*).

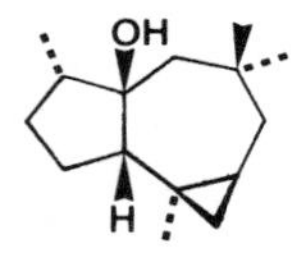

$C_{15}H_{26}O$
MW 222
mp 58°–60° (*126*)
$[\alpha]_{579}$ +59.5° (*126*)
AC

30 Africanol

30. MARINE OCCURRENCE: *Lemnalia africana* (A) (*126*), *Lemnalia nitida* (A) (*43*).

STRUCTURE AND ABSOLUTE CONFIGURATION: Determined by XR (*70, 126*).

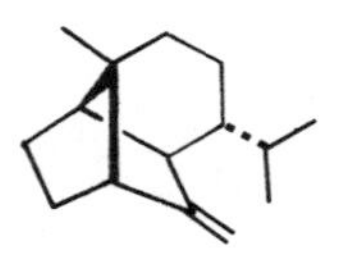

$C_{15}H_{24}$
MW 204
Oil
$[\alpha]_D$ −142° (*10*)
AC

31 Sinularene

31. MARINE OCCURRENCE: *Sinularia mayi* (A) (*10*).

STRUCTURE AND ABSOLUTE CONFIGURATION: Determined by XR on a crystalline derivative (*10*).

$C_{15}H_{18}O$
MW 214
Oil
$[\alpha]_D$ 0° (*141*)

32 Furoventalene

32. MARINE OCCURRENCE: *Gorgonia ventalina* (G) (*141, 143*).
STRUCTURE: Suggested on the basis of its spectral properties and confirmed by total synthesis (*141, 143*).
COMMENTS: It has been suggested (*141*) that the formation of this unusual skeleton could arise from dimethylallylation of a monoterpenoid intermediate (e.g., piperitone) at the γ position, followed by furan development and aromatization. Thus far, no other compound possesses this skeleton. Furoventalene could be an artifact produced during steam distillation of the gorgonian extract (*32*).

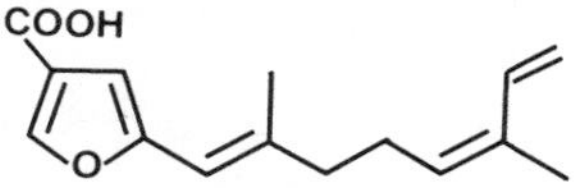

COOH

$C_{15}H_{18}O_3$
MW 246
mp 99°–100° (*37*)
$[\alpha]_D$ 0° (*37*)

33 *Sinularia gonatodes* Compound I

33. MARINE OCCURRENCE: *Sinularia gonatodes* (A) (*37*).
STRUCTURE: Determined by spectroscopic and chemical methods (*37*).

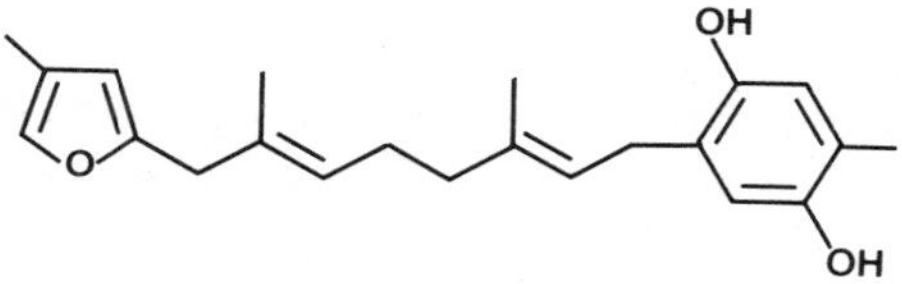

OH OH

$C_{22}H_{28}O_3$
MW 340
mp 88°–89° (*39*)

34 2-[3,7-Dimethyl-8-(4-methyl-2-furyl)-2,6-octadienyl]-5-methylhydroquinone

34. MARINE OCCURRENCE: *Sinularia lochmodes* (A) (*39*).
STRUCTURE: Proposed on the basis of its spectral properties (*39*).

B. Diterpenes

According to the suggestion of Weinheimer *et al.* (*149*), we shall use the following representation and numbering for the cembrane derivatives, the reference unsaturation Δ^7 (or some vestigial indication of its original location) always being situated above the 1,8 axis of skeletal symmetry. Consequently, all compounds discussed here are renumbered according to this convention.

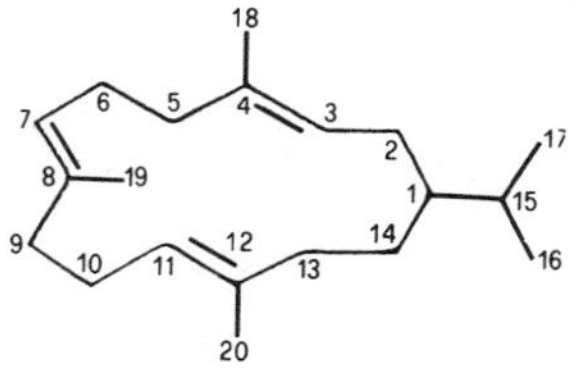

18 7 6 5 4 3 2 17 8 19 1 15 9 10 11 12 13 14 16 20

The geometry of the double bonds (*E* or *Z*) and the stereochemistry of the epoxides (cis or trans), when known are indicated (tr = trans). The lack of an *R* or *S* symbol next to an asymmetric center means that its absolute configuration has not been established.

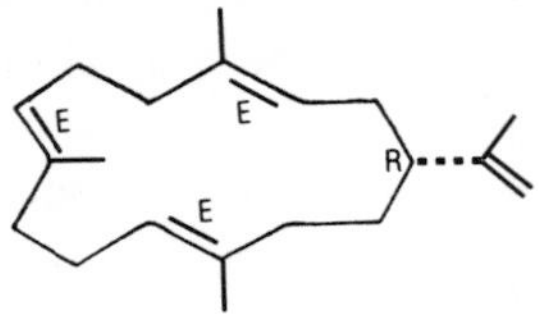

$C_{20}H_{32}$
MW 272
Oil
$[\alpha]_D$ −16° (*57*)
AC

35 *R*-(−)-Cembrene A

35. Marine occurrence: *Sinularia flexibilis* (A) (*57*).

Identification: Suggested by the spectral data and confirmed by chemical correlation with (−)-nephthenol (**37**) (*57*).

Other sources: Known from *Commiphora mukul* (*99*), *Pinus koraiensis* (*115*), and *Picea obovata* (*115*) and as a trail pheromone from the termite *Nasutitermes exitiosus* (*13*). In the last case the name neocembrene A was used, and no rotatory power was reported.

Comment: A total synthesis of (±)-cembrene A has been reported (*80*).

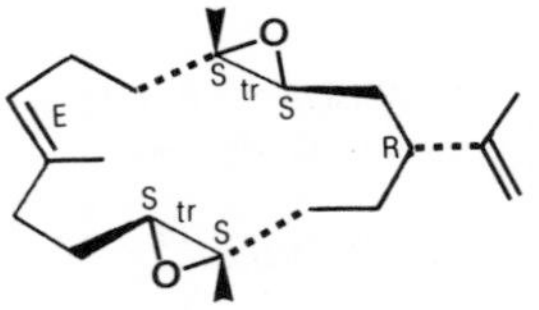

$C_{20}H_{32}O_2$
MW 304
mp 66°–68° (*57*)
$[\alpha]_D$ +63° (*57*)
AC

36 3,4,11,12-Diepoxycembrene A

36. Marine occurrence: *Sinularia flexibilis* (A) (*57*).

Structure and absolute configuration: Determined by chemical correlation with *R*-(−)-cembrene A (**35**) (*57*). The stereochemistry of the oxirane rings rests on chemical arguments and NOE measurements (*56*, *57*).

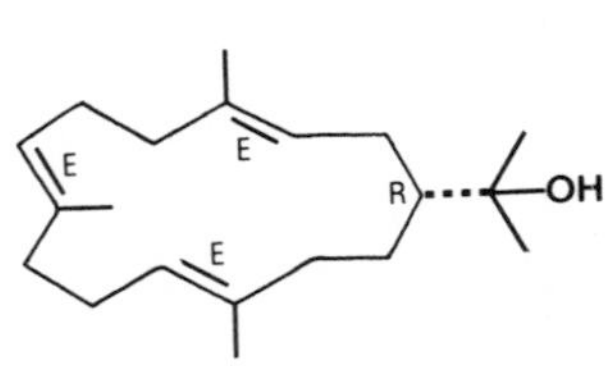

$C_{20}H_{34}O$
MW 290
Oil
$[\alpha]_D$ −36° (*129*)
AC
Nuclear Overhouser Effect

37 Nephthenol

37. MARINE OCCURRENCE: *Nephthea* sp. (A) (*111*), *Litophyton viridis* (A) (*129*), *Litophyton arboreum* (A) (*73*), *Lobophytum pauciflorum* (A) (*73*), *Sarcophyton decaryi* (A) (*73*).

STRUCTURE: Based on chemical degradation and spectral data (*111*). The structure was confirmed by total synthesis (*80*). This established the all-trans geometry of the double bonds.

ABSOLUTE CONFIGURATION: Determined for the sample from *L. viridis* by oxidation to (−)-homoterpenyl methyl ketone of known *R* configuration (*129*). It has not been determined whether nephthenol samples from the other alcyonarians have the same absolute configuration.

OH
OH

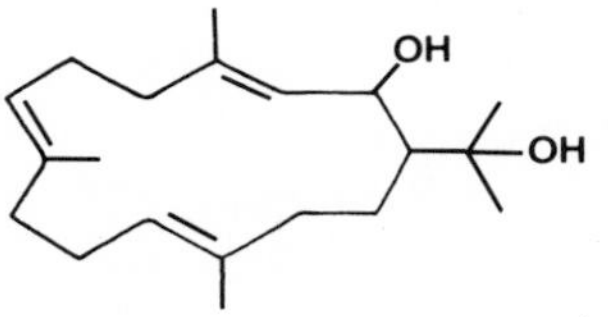

$C_{20}H_{34}O_2$
MW 306
mp 98°–99° (*129*)
$[\alpha]_D$ −104° (*129*)

38 2-Hydroxynephthenol

38. MARINE OCCURRENCE: *Litophyton viridis* (A) (*129*).

STRUCTURE: Based on chemical degradation, spectral data, and correlation with cembrane (*129*). The geometry of the double bonds and the configurations at C-1 and C-2 have not been determined.

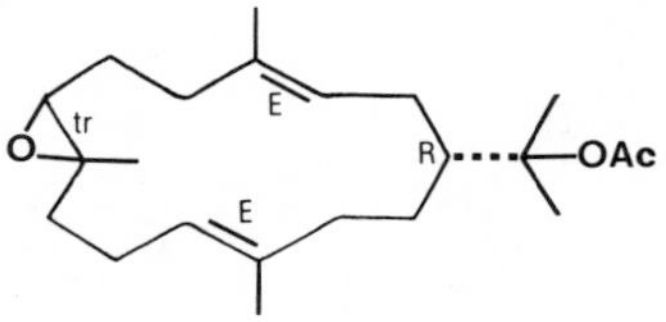

tr
O
E
R
OAc
E

$C_{22}H_{36}O_3$
MW 348
Oil
$[\alpha]_D$ −20.7° (*111*)
AC

39 Epoxynephthenol acetate

39. MARINE OCCURRENCE: *Nephthea* species (A) (*111*).

STRUCTURE: Based on spectral properties and chemical degradation. The geometry of the double bonds and the stereochemistry of the epoxide rest on Nuclear Overhouser Effect (NOE) experiments (*111*).

ABSOLUTE CONFIGURATION: Established by oxidation to (−)-homoterpenyl methyl ketone of known *R* configuration (*111*). The absolute configuration at C-7 and C-8 was not determined.

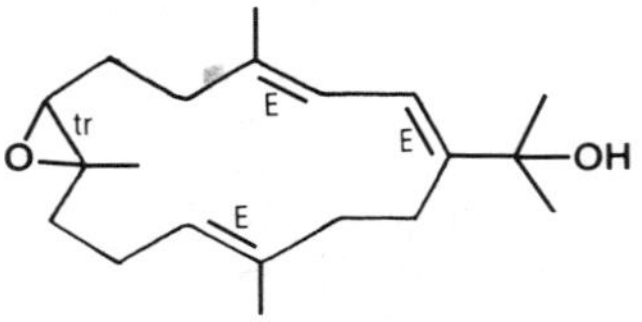

$C_{20}H_{32}O_2$
MW 304
mp 105°–106° (*36*)
$[\alpha]_D$ +61.8° (*36*)

40 2-[(E,E,E)-7',8'-Epoxy-4',8',12'-trimethylcyclotetradeca-1',3',11'-trienyl]propan-2-ol

40. MARINE OCCURRENCE: *Sarcophyton* species (A) (*36*).
STRUCTURE: Deduced from the spectral data and confirmed by XR (*36*).

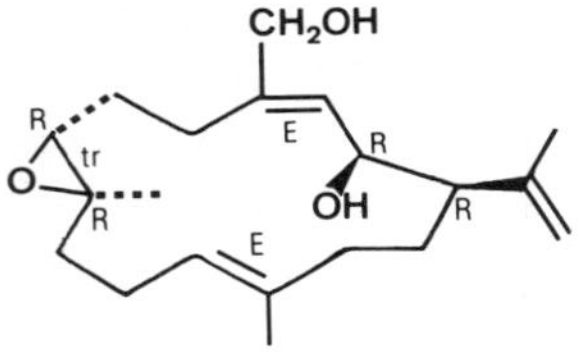

$C_{20}H_{32}O_3$
MW 320
mp 109°–110° (*148*)
$[\alpha]_D$ −87° (*148*)
AC

41 Asperdiol

41. MARINE OCCURRENCE: *Eunicea asperula* (G) (*148*), *Eunicea tourneforti* (G) (*148*).
STRUCTURE AND ABSOLUTE CONFIGURATION: Determined by XR (*148*).

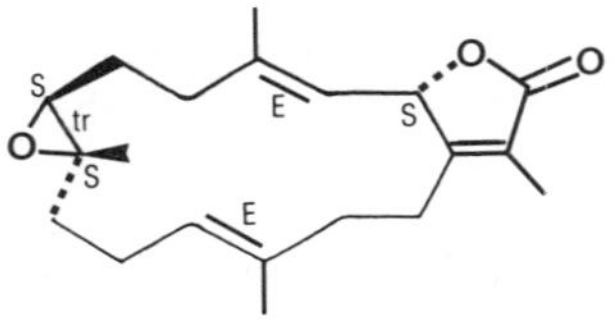

$C_{20}H_{28}O_3$
MW 316
mp 133°–134° (*12*)
$[\alpha]_D$ +92° (*12*)
AC

42 Sarcophine

42. MARINE OCCURRENCE: *Sarcophyton glaucum* (A) (*12, 93*).
STRUCTURE: Deduced from the spectral properties and confirmed by XR (*12*).
ABSOLUTE CONFIGURATION: Determined by CD (*74*).

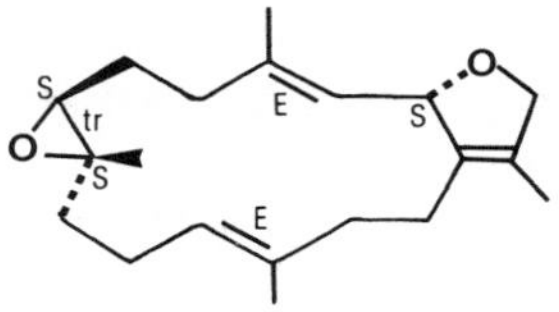

$C_{20}H_{30}O_2$
MW 302
Oil
$[\alpha]_D$ +40° (*76*)
AC

43 *Sarcophyton glaucum* Compound 2

43. MARINE OCCURRENCE: *Sarcophyton glaucum* (A) (*76*), *Sarcophyton* species (A) (*73*), *Lobophytum* species (A) (*73*).

STRUCTURE: Determined by comparison of its spectral properties with those of sarcophine (**42**) (*76*). On the basis of biogenetic assumptions, this compound is considered to be the 16-deoxy derivative of sarcophine (**42**) (*74*). No chemical correlation has been performed to support this assessment.

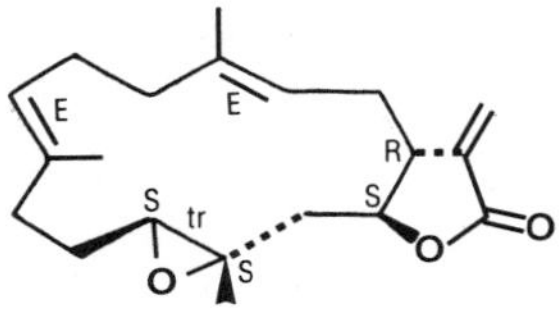

$C_{20}H_{28}O_3$
MW 316
mp 137°–138° (*127*)
$[\alpha]_D$ +7° (*127*)
AC

44 Lobophytolide

44. MARINE OCCURRENCE: *Lobophytum cristagalli* (A) (*127*).

STRUCTURE: Deduced from the spectral properties and confirmed by XR (*71b, 127*).

ABSOLUTE CONFIGURATION: Determined by XR (*71b*).

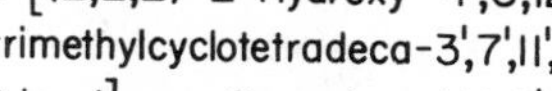

$C_{20}H_{28}O_2$
MW 300
mp 101°–102° (*38*)
$[\alpha]_D$ + 77.9° (*38*)

45 2-[(E,E,E)-2'-Hydroxy-4',8',12'-trimethylcyclotetradeca-3',7',11',-trienyl]prop-2-enoic acid 1,2'-lactone

45. MARINE OCCURRENCE: *Lobophytum michaelae* (A) (*38*).

STRUCTURE: Deduced from the spectral data (*38*). The trans geometry of the double bonds was suggested by ^{13}C nmr spectroscopic arguments, whereas the cis junction of the γ-lactone was derived from the value of the coupling constant between H_1 and H_2 on the assumption that the situation prevalent for γ-lactones fused to six-membered rings is also valid for macrocyclic compounds. This might be questioned.

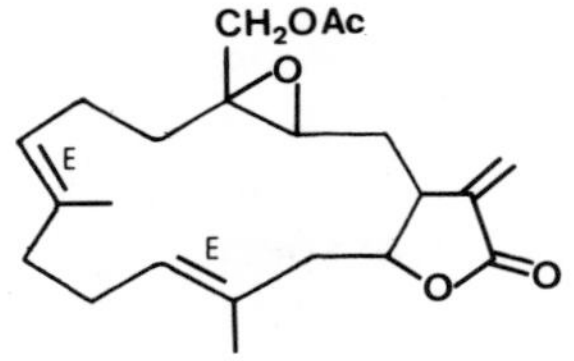

$C_{22}H_{30}O_5$
MW 374
mp 114°–115° (*75*)
$[\alpha]_D$ −58° (*75*)

46 Lobolide

46. MARINE OCCURRENCE: *Lobophytum crassum* (A), (*73, 75*).

STRUCTURE: Based on spectral data and chemical degradation (*75*). The trans geometry of the double bonds is suggested by the nmr data.

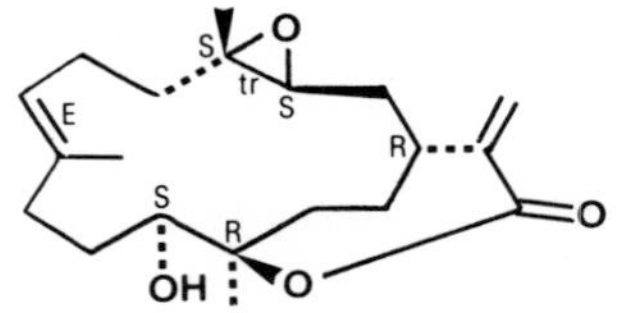

$C_{20}H_{30}O_4$
MW 334
mp 170°–173° (*130*)
176° –177° (*149*)
$[\alpha]_D$ + 76° (CH_3OH) (*130*)
+134° (CH_2Cl_2) (*149*)
AC

47 Sinulariolide

47. MARINE OCCURRENCE: *Sinularia flexibilis* (A) (*130, 149*).

STRUCTURE: Based on chemical degradation and spectral data and confirmed by XR (*71a, 130*).

ABSOLUTE CONFIGURATION: Determined by XR (*71a, 130*).

COMMENT: No direct comparison has been performed between the original sample of sinulariolide (*130*) and that isolated from a specimen of *S. flexibilis* collected on the Great Barrier Reef (*149*).

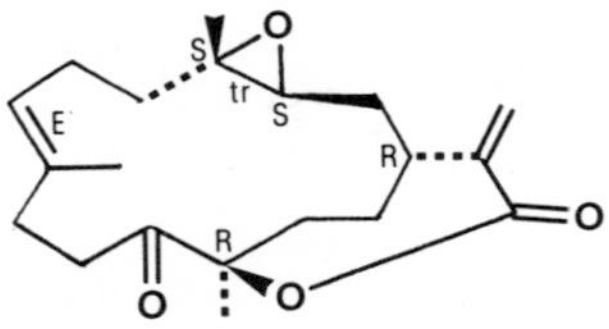

$C_{20}H_{28}O_4$
MW 332
mp 120° (*57*)
$[\alpha]_D$ +87° (*57*)
AC

48 11-Dihydrosinulariolide

48. MARINE OCCURRENCE: *Sinularia flexibilis* (A) (*57*), *Sinularia notanda* (A) (*73*).
STRUCTURE AND ABSOLUTE CONFIGURATION: Established by its spectral properties and by correlation with sinulariolide (**47**) (*57*).

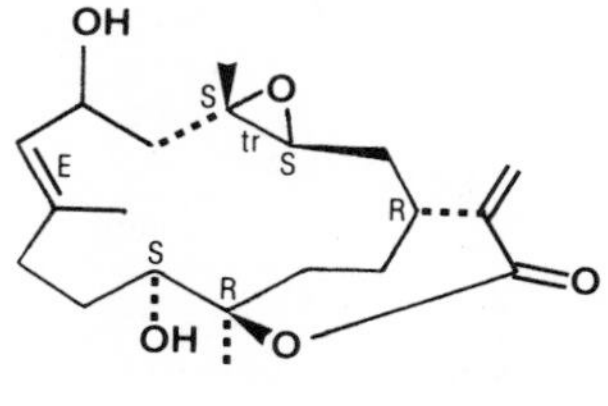

$C_{20}H_{30}O_5$
MW 350
mp 192°–194° (*57*)
$[\alpha]_D$ +54.5° (*57*)
AC

49 6ξ-Hydroxysinulariolide

49. MARINE OCCURRENCE: *Sinularia flexibilis* (A) (*57*).
STRUCTURE AND ABSOLUTE CONFIGURATION: Established by its spectral properties and by chemical correlation with sinulariolide (**47**) (*57*).
COMMENT: This compound might be an artifact since it has been shown to arise from air oxidation of sinulariolide (**47**) (*57*).

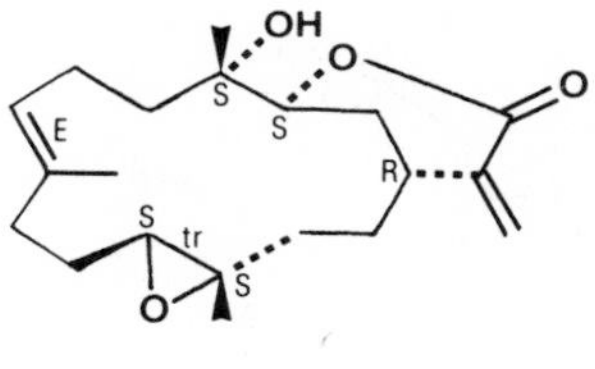

$C_{20}H_{30}O_4$
MW 334
mp 150°–152° (*149*)
$[\alpha]_D$ −127° (*149*)
AC

50 Sinularin

50. MARINE OCCURRENCE: *Sinularia flexibilis* (A) (*149*).
STRUCTURE AND ABSOLUTE CONFIGURATION: Established by XR (*149*).
COMMENT: See dihydrosinularin (**51**).

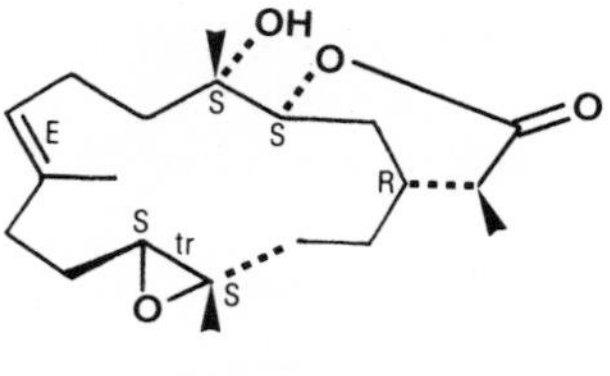

$C_{20}H_{32}O_4$
MW 336
mp 110°–112° (*149*)
$[\alpha]_D$ −45° (*149*)
AC

51 Dihydrosinularin

51. MARINE OCCURRENCE: *Sinularia flexibilis* (A) (*149*).
STRUCTURE AND ABSOLUTE CONFIGURATION: Established by XR (*149*).

COMMENT: Both sinularin and dihydrosinularin have also been isolated by Wells (*151*) from another sample of *S. flexibilis* from the Great Barrier Reef.

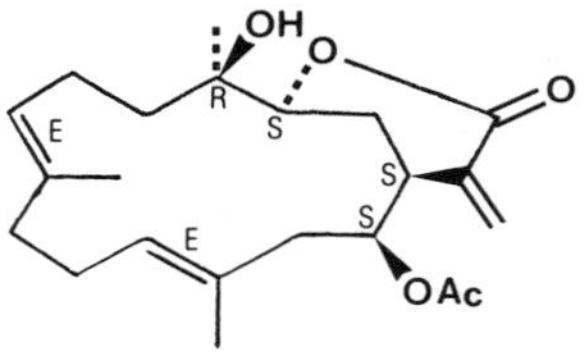

$C_{22}H_{30}O_5$
MW 376
mp 144°–145° (*33*)
$[\alpha]_D$ +70.4° (*33*)
AC

52 Crassin acetate

52. MARINE OCCURRENCE: *Pseudoplexaura porosa* (G) (*144*), *Pseudoplexaura wagenaari* (G) (*144*), *Pseudoplexaura flagellosa* (G) (*144*), *Pseudoplexaura crucis* (G) (*142*).

STRUCTURE AND ABSOLUTE CONFIGURATION: Established by XR on crassin *p*-iodobenzoate (*61*).

COMMENT: In the first isolation reports, *P. porosa* was named by its junior synonym *P. crassa* (*33*).

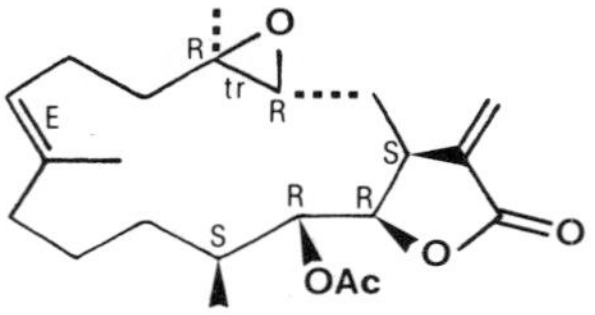

$C_{22}H_{32}O_5$
MW 376
mp 157°–159° (*105*)
$[\alpha]_D$ +8° (*105*)
AC

53 Eupalmerin acetate

53. MARINE OCCURRENCE: *Eunicea palmeri* (G) (*32*), *Eunicea succinea* (G) (*32*), *Eunicea mammosa* (G) (*32*), *Eunicea succinea* var. *plantaginea* (G) (*31*).

STRUCTURE: Based on spectral properties and chemical degradation (*105*) and confirmed by XR on the bromine addition product of eupalmerin acetate (*50, 136*).

ABSOLUTE CONFIGURATION: Determined by XR (*50*).

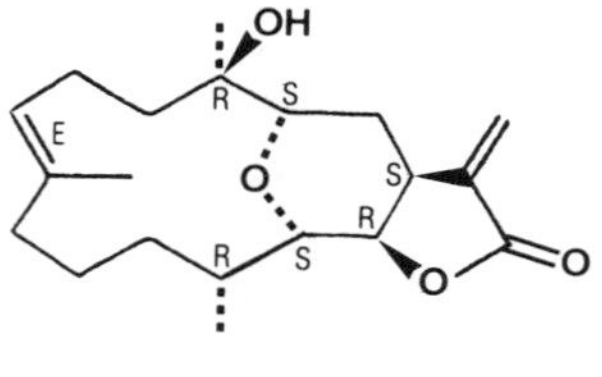

$C_{20}H_{30}O_4$
MW 334
mp 154°–155° (*144*)
$[\alpha]_D$ −89.4° (*144*)
AC

54 Eunicin

54. MARINE OCCURRENCE: *Eunicea mammosa* (G) (*145*), *Eunicea palmeri* (G) (*31*).

STRUCTURE: Based on spectral data and chemical evidences (*87, 145*) and confirmed by XR on the corresponding iodoacetate (*62*).

ABSOLUTE CONFIGURATION: Determined by XR (*62*).

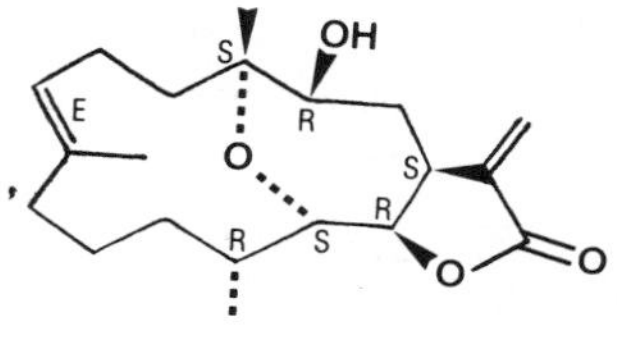

$C_{20}H_{30}O_4$
MW 334
mp 139°–141° (*144*)
$[\alpha]_D$ +2.8° (*144*)
AC

55 Jeunicin

55. MARINE OCCURRENCE: *Eunicea mammosa* (G) (*144*).

STRUCTURE AND ABSOLUTE CONFIGURATION: Determined by XR on the *p*-iodobenzoate derivative (*137*).

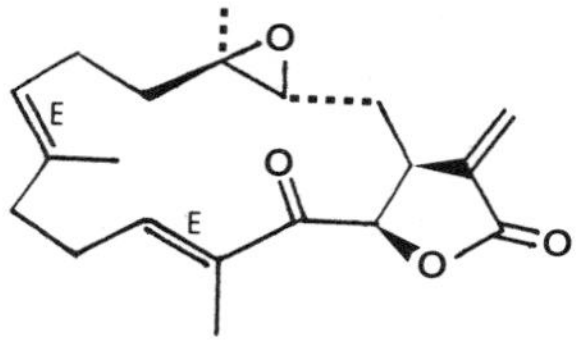

$C_{20}H_{26}O_4$
MW 330
mp 175°–176° (*27*)

56 Peunicin

56. MARINE OCCURRENCE: *Eunicea succinea* var. *plantaginea* (G) (*31*).

STRUCTURE: Determined by XR (*27*).

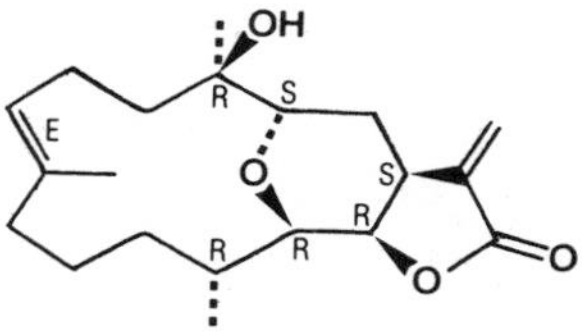

$C_{20}H_{30}O_4$
MW 334
Oil
$[\alpha]_D$ −147° (*55*)
AC

57 Cueunicin

57. MARINE OCCURRENCE: *Eunicea mammosa* (G) (*55*).

STRUCTURE: Based on spectral properties and chemical degradation (*55*).

ABSOLUTE CONFIGURATION: Assumed to be the same as that of eunicin except at C-13 (*55*).

COMMENT: The structure reported in Ciereszko and Karns (*32*) is erroneous and has been corrected (*55*).

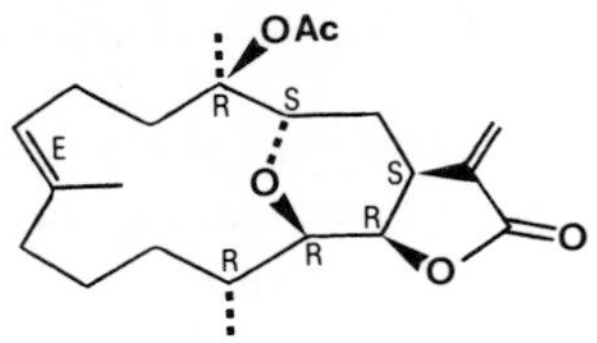

$C_{22}H_{32}O_5$
MW 376
mp 141°–142° (55)
$[\alpha]_D$ −152° (55)
AC

58 Cueunicin acetate

58. MARINE OCCURRENCE: *Eunicea mammosa* (G) (*55*).
STRUCTURE AND ABSOLUTE CONFIGURATION: Determined by correlation with cueunicin (**57**) (*55*).

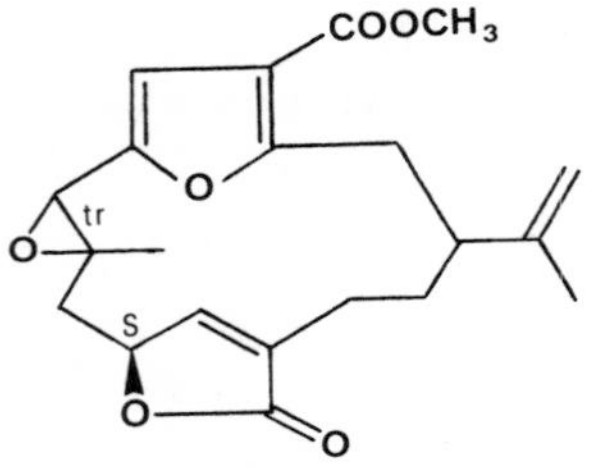

$C_{21}H_{24}O_6$
MW 372
mp 204°–206° (88)
$[\alpha]_D$ +44° (88)

59 Pukalide

59. MARINE OCCURRENCE: *Sinularia abrupta* (A) (*88*).
STRUCTURE: Based on spectral properties, mainly nmr decoupling (*88*). The trans stereochemistry of the oxirane ring was established by NOE experiments (*88*).
ABSOLUTE CONFIGURATION: Obtained at C-10 by CD (*88*).

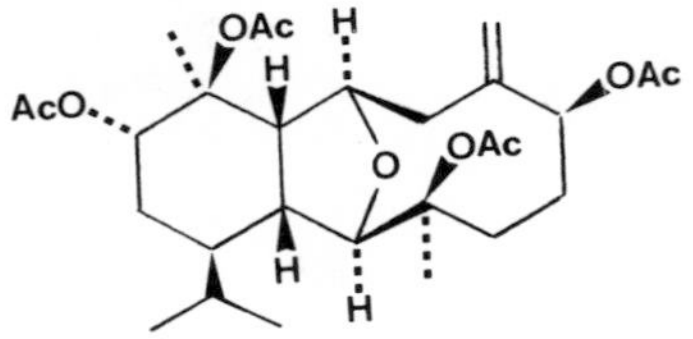

$C_{28}H_{42}O_9$
MW 522
mp 186°–188° (77)
$[\alpha]_D$ −36° (77)

60 Eunicellin

60. MARINE OCCURRENCE: *Eunicella stricta* (G) (*77*).
STRUCTURE: Determined by XR on the dibromo derivative (*78*).

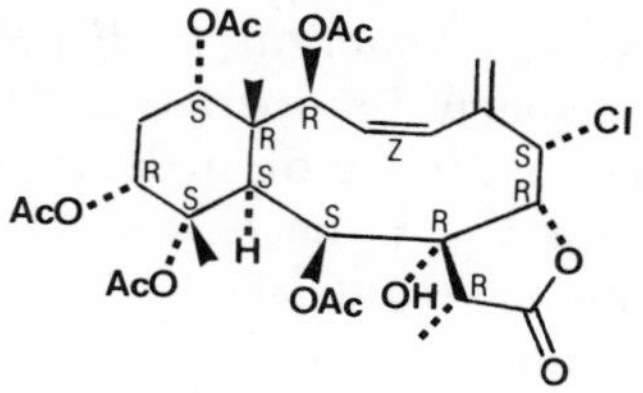

$C_{30}H_{39}O_{13}Cl$
MW 642
mp 245°–250° (*7*)
$[\alpha]_D$ −90° (*7*)
AC

61 Briarein A

61. MARINE OCCURRENCE: *Briareum asbestinum* (G) (*7, 25, 63*).
STRUCTURE AND ABSOLUTE CONFIGURATION: Determined by XR (*25*).

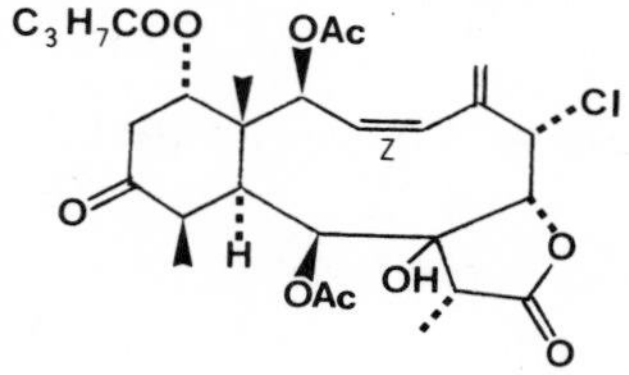

$C_{28}H_{37}O_{10}Cl$
MW 568

62 Ptilosarcone

62. MARINE OCCURRENCE: *Ptilosarcus guyneyi* (P) (*150, 158*).
STRUCTURE: Suggested by comparison of its nmr spectrum with that of briarein A (**61**) (*158*).

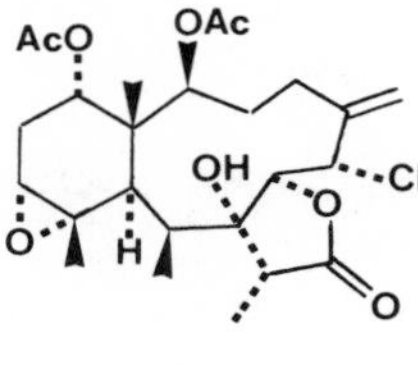

$C_{26}H_{35}O_{10}Cl$
MW 542
mp 179°–181° (*157*)
$[\alpha]_D$ +65° (*157*)

63 Stylatulide

63. MARINE OCCURRENCE: *Stylatula* species (P) (*157*).
STRUCTURE: Determined by XR (*157*).

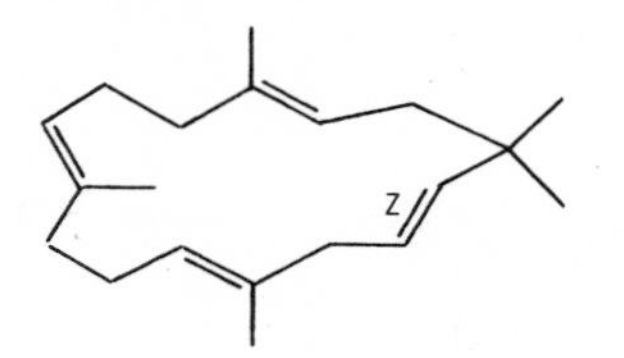

$C_{20}H_{32}$
MW 272
Oil
$[\alpha]_D$ 0° (*58*)

64 Flexibilene

64. MARINE OCCURRENCE: *Sinularia flexibilis* (A) (*58*).

STRUCTURE: Based on spectral properties (*58*). The trans geometry of the disubstituted double bond rests on ir data.

COMMENTS: This compound is an isoprenolog of the sesquiterpene humulene. It is interesting to note its similarity with casbene, a diterpene isolated from *Ricinus communis* (*116*).

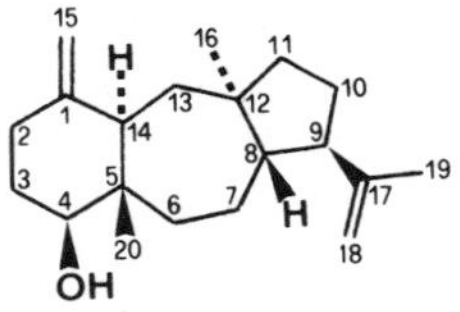

$C_{20}H_{32}O$
MW 288
mp 107°–110° (*19*)
$[\alpha]_D$ +51° (*19*)

65 4β-Hydroxyclavulara-1(15),17-diene

65. MARINE OCCURRENCE: *Clavularia inflata* (S) (*19*).

STRUCTURE: Determined by chemical correlation with 1α,4β-dihydroxyclavular-17-ene (**66**) (*19*).

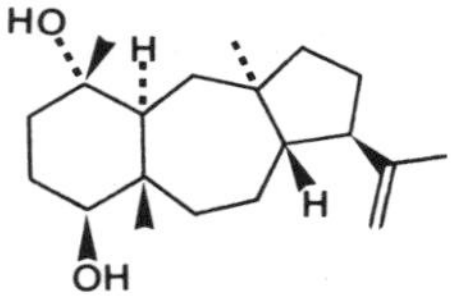

$C_{20}H_{34}O_2$
MW 306
mp 165°–167° (*19*)
$[\alpha]_D$ +9° (*19*)

66 1α,4β-Dihydroxyclavular-17-ene

66. MARINE OCCURRENCE: *Clavularia inflata* (S) (*19*).

STRUCTURE: Determined by XR (*19*).

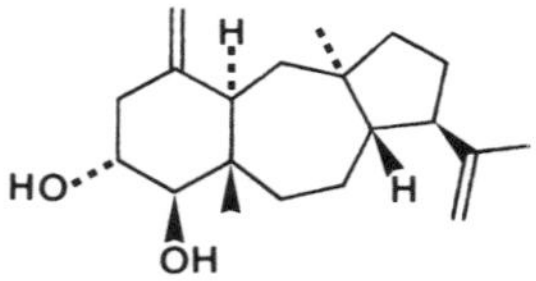

$C_{20}H_{32}O_2$
MW 304
mp 78°–80° (*19*)
$[\alpha]_D$ +51° (*19*)

67 3α,4β-Dihydroxyclavulara-1(15),17-diene

67. MARINE OCCURRENCE: *Clavularia inflata* (S) (*19*).

STRUCTURE: Determined by correlation with 1α,4β-dihydroxyclavular-17-ene (**66**) (*19*).

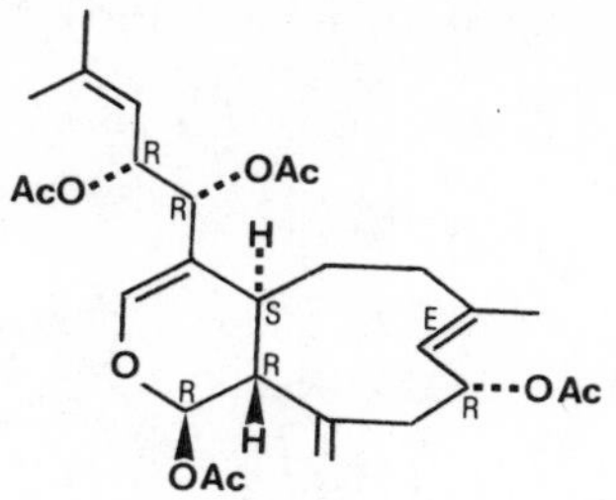

$C_{28}H_{38}O_9$
MW 518
mp 141°–142° (*135*)
$[\alpha]_D$ −36.7° (*135*)
AC

68 Xenicin

68. MARINE OCCURRENCE: *Xenia elongata* (A) (*135*).
STRUCTURE AND ABSOLUTE CONFIGURATION: Determined by XR (*110, 135*).

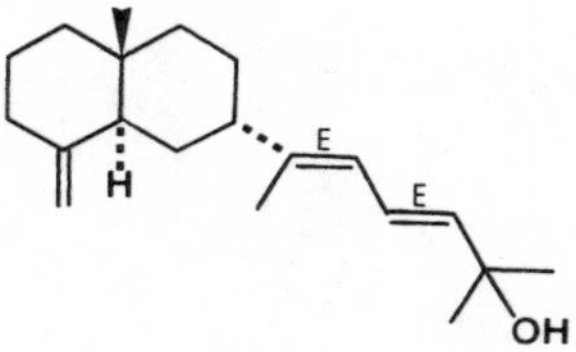

$C_{20}H_{32}O$
MW 288
Oil
$[\alpha]_D$ +83° (*40*)

69 Unnamed

69. MARINE OCCURRENCE: *Lobophytum hedleyi* (A) (*40*).
STRUCTURE: Based on spectral properties and chemical degradations (*40*).

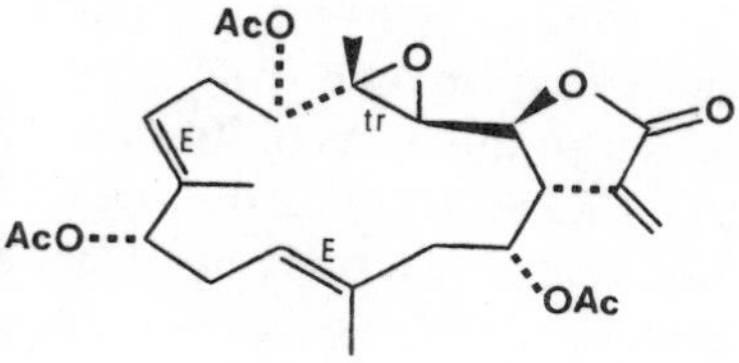

$C_{26}H_{34}O_9$
MW 490
Amorphous
$[\alpha]_D$ − 16° (*132*)

70 Crassolide

70. MARINE OCCURRENCE: *Lobophytum crassum* (A) (*17, 125*).
STRUCTURE: Determined by XR (*132*).
COMMENT: The structure reported in Braekman (*17*) and Tursch (*125*), based on chemical degradations and ^{1}H nmr decoupling experiments, is erroneous and has been corrected (*132*).

C. Terpenoids Whose Final Structures Have Not Been Determined

1. Sesquiterpenes

$C_{15}H_{24}$
MW 204
bp 71°/2.3 torr
$[\alpha]_D$ +24.5° (*33*)

71. MARINE OCCURRENCE: *Pseudoplexaura porosa* (G) (*33*).
COMMENTS: This compound was the first sesquiterpene reported from an animal (*33*). It yields (+)-cadinene dihydrochloride on HCl treatment. Since it was later shown (*161*) that *P. porosa* contains several sesquiterpene hydrocarbons that afford the same hydrochloride, no definite structure can be assigned to this compound. However, it was reported as (+)-cadinene by Premuzic (*104*).

2. Diterpenes

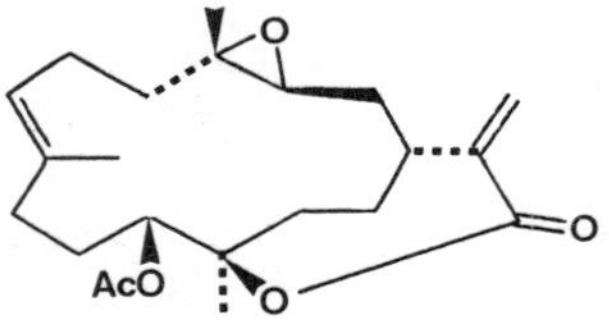

$C_{22}H_{32}O_5$
MW 376

AcO O O O

72 11-Episinulariolide acetate

72. MARINE OCCURRENCE: *Sinularia flexibilis* (A) (*56, 57*), *Sinularia notanda* (A) (*73*), *Sinularia querciformis* (A) (*73*).
COMMENTS: The compound isolated from *S. flexibilis* is supposed to be 11-episinulariolide acetate on the basis of its spectral properties only (*56*). A derivative having the same spectral properties has been independently isolated from *S. notanda* and *S. querciformis* (*73*), but no direct comparison has been performed between the two samples.
73. Several cembrane derivatives having the same empirical formula, $C_{20}H_{30}O_2$ (MW 302), as *Sarcophyton glaucum* compound 2 (**43**) have been isolated from different species of *Sarcophyton* (see Table 2). On the basis of their spectral properties, compounds **73a** and **73b** are supposed to be stereoisomers of **43**, whereas **73c** is probably an isomer having the epoxide function located on carbon atoms C-11 and C-12. A number of derivatives with related spectral properties have been isolated independently by Coll and Faulkner from *Sarcophyton* and *Lobophytum* species (*35*).

TABLE 2

Cembrane Derivatives, $C_{20}H_{30}O_2$

Compound	Species		$[\alpha]_D$	Ref.
Sarcophytoxide (**73a**)	*Sarcophyton trocheliophorum* *S. glaucum*	Oil	−137°	*125, 134*
S. glaucum compound 3 (**73b**)	*S. glaucum*	Oil	—	*76*
Isosarcophytoxide (**73c**)	*S. trocheliophorum*	Oil	—	*125, 134*

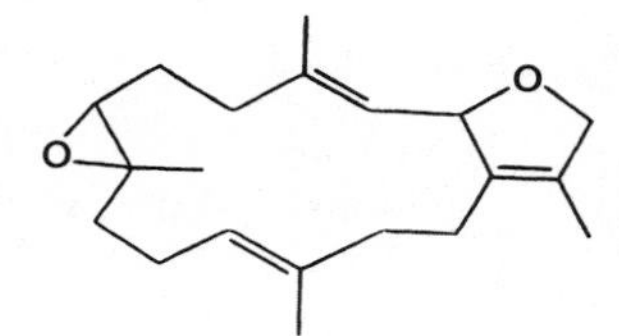

73a and **73b** (stereoisomers of **43**)

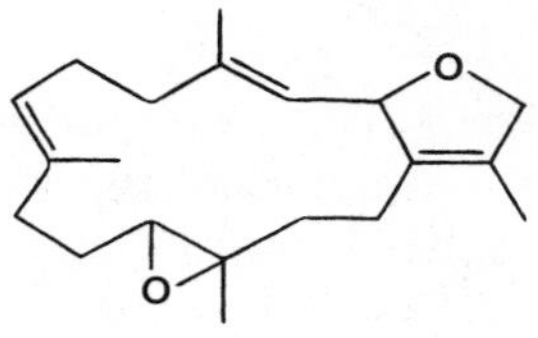

73c

74. The isolation from *Sarcophyton glaucum* of two stereoisomers (*S. glaucum* compounds 4 and 5) of sarcophine (**42**) was reported by Kashman *et al.* (*76*). It is claimed that these two cembranolides can be interconverted in a chloroform solution (*76*). Their stereochemistry has not yet been established (*74*).

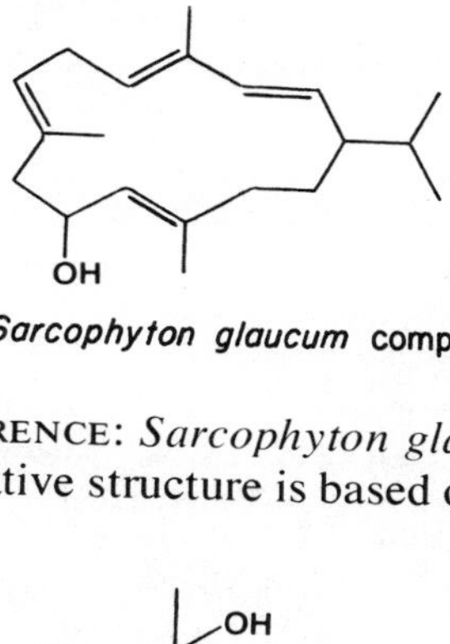

$C_{20}H_{32}O$
MW 288
mp 143°–145° (*76*)

75 *Sarcophyton glaucum* compound 6

75. Marine occurrence: *Sarcophyton glaucum* (A) (*76*).
Comment: This tentative structure is based only on spectral properties (*76*).

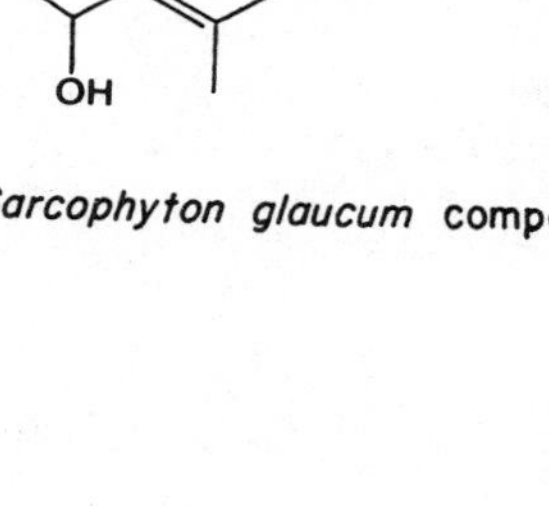

$C_{20}H_{34}O_2$
MW 306
Oil

76 *Sarcophyton glaucum* compound 7

76. Marine occurrence: *Sarcophyton glaucum* (A) (*76*).
Comment: This tentative structure is based only on spectral data (*76*).

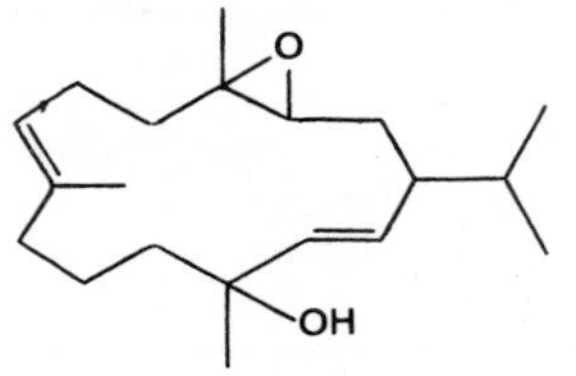

$C_{20}H_{34}O_2$
MW 306
Oil

77 Trocheliophorol

77. Marine occurrence: *Sarcophyton trocheliophorum* (A) (*125, 134*).
Comment: This tentative structure is based on spectral data and preliminary chemical degradation (*134*).

	78a Briarein B	**78b** Briarein C	**78c** Briarein D
Formula	$C_{32}H_{43}O_{13}Cl$	$C_{28}H_{39}O_{11}Cl$	$C_{30}H_{41}O_{11}Cl$
MW	670	584	612
mp	230°–233°	248°–253°	153°–160°
$[\alpha]_D$	−95°	−96°	−99°

78. Marine occurrence: *Briareum asbestinum* (G) (*7*).
Comments: The structure of briarein B is supposed to be the same as that of briarein A (**61**) except that one of the acetate groups is replaced by a butyrate group (*7*). Spectroscopic and chemical data on briareins C and D point to a close structural relationship with briareins A and B (*7*).

79. The isolation of three new cembrane derivatives from a *Pseudoplexaura* species (G) has been reported by Gross (*55*). Their structures are still to be determined.

	Compound B1	Compound B2	Compound B3
Formula	$C_{20}H_{34}O_3$	$C_{20}H_{32}O_3$	$C_{20}H_{36}O_3$
MW	322	320	324
mp	111°–112°	86°–87°	—
$[\alpha]_D$	−7.8°	+61°	—

80. Two different extracts of the soft coral *Sclerophyton capitalis* (A) yielded two, still undetermined diterpenes. The empirical formulas were reported to be $C_{23}H_{34}O_7$ (*32*) and $C_{22}H_{34}O_4$ (mp 111°–112°) (*55*), respectively.

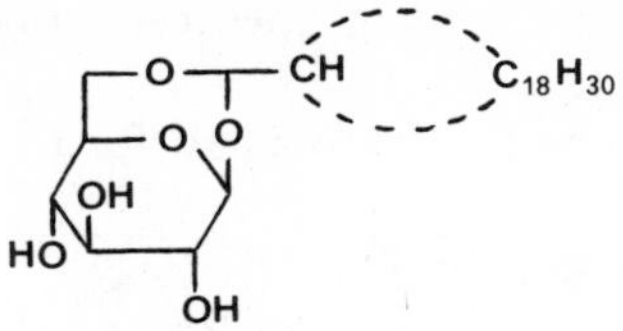

$C_{26}H_{42}O_6$
MW 450
Amorphous

81 Lemnalialoside

81. MARINE OCCURRENCE: *Lemnalia digitata* (A) (*125, 131*).
COMMENT: Preliminary results indicate that this compound is a bicyclic diterpene linked to a D-glucose through an unusual ketal linkage (*131*).

III. ARTIFACTS

Only one of the diterpenes so far reported from coelenterates is suspected to be an artifact. It is 6ξ-hydroxysinulariolide (**49**) (*57*), which was isolated from an extract of *Sinularia flexibilis* but was observed to be generated by air oxidation of sinulariolide (**47**).

The situation is more complex in the case of coelenterate sesquiterpenes. In contrast to the diterpenes many of the volatile sesquiterpenes have been subjected to rather drastic treatment such as distillation or steam distillation. Furoventalene (**32**) (*143*), for instance, could be an artifact generated during steam distillation. (+)-Calamenene (**82**), arising during distillation of the hydrocarbon fraction of *Pseudoplexaura porosa*, is obviously lacking in the fresh hydrocarbon extract (*161*).

It should be kept in mind that some of the reported structures could easily originate from a more reactive natural precursor. This is illustrated by the many compounds that can arise from a germacrene precursor (*65, 90, 118, 147, 160*).

The following three sesquiterpenes were isolated from gorgonians and have been shown to be artifacts.

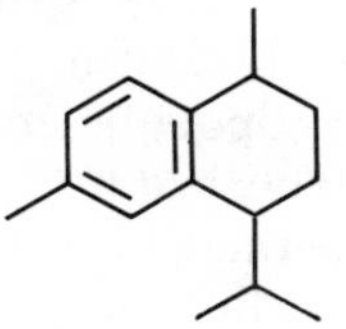

$C_{15}H_{22}$
MW 202
Oil
$[\alpha]_D$ +66.4° (*161*)
n_D^{25} 1.5164 (*161*)

82 (+)-Calamenene

82. MARINE OCCURRENCE: *Pseudoplexaura porosa* (G) (*144, 161*).

IDENTIFICATION: On the basis of spectral properties and comparison with published data (*161*).

ABSOLUTE CONFIGURATION: Established for (+)-calamenene from TP (*84*).

OTHER SOURCES: Common in TP (*97*).

COMMENT: Calamenene is an artifact formed during distillation of the hydrocarbon fraction from *Pseudoplexaura porosa* (*161*).

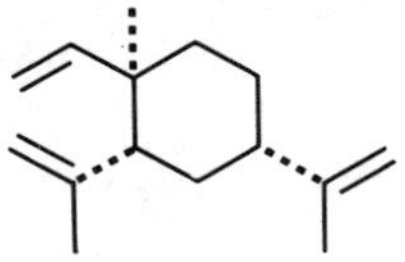

$C_{15}H_{24}$
MW 204
Oil
$[\alpha]_D$ +15.1° (*161*)
n_D^{25} 1.4910 (*161*)
AC

83 (+)-β-Elemene

83. MARINE OCCURRENCE: *Eunicea mammosa* (G) (*141, 144, 147, 161*), *Dictyopteris divaricata* (alga) (*64*).

IDENTIFICATION: On the basis of spectral properties and comparison with published data (*141, 147*).

ABSOLUTE CONFIGURATION: Established for the (−) antipode (*117*).

OTHER SOURCES: Both (+) and (−) antipodes are common in TP (*97*).

COMMENTS: See (−)-β-selinene (**84**). The structure reported in Scheuer (*109*) is erroneous.

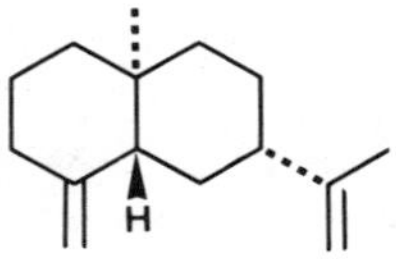

$C_{15}H_{24}$
MW 204
Oil
$[\alpha]_D$ −20.4° (*141*)
n_D^{25} 1.5035 (*141*)
A$\overline{C}$

84 (−)-β-Selinene

84. MARINE OCCURRENCE: *Eunicea mammosa* (G) (*141, 147, 161*).

IDENTIFICATION: On the basis of its spectral properties (*141, 147*).

ABSOLUTE CONFIGURATION: Established for the (+) antipode (*34*).

OTHER SOURCES: The (+) antipode is common in TP (*97*).

COMMENTS: It has been established (*141, 147, 161*) that (+)-β-elemene (**83**) readily arises from (−)-germacrene A (**6**) through a thermal Cope rearrangement (*118*). Moreover, in contact with silica gel, (+)-β-elemene (**83**) yields a mixture of (−)-β-selinene and (+)-α-selinene (*141*). The latter compound was not reported in a subsequent study (*161*). For rearrangements involving germacrene sesquiterpenes, see Jain and McCloskey (*65*) and Takeda (*118*).

IV. DISTRIBUTION OF TERPENOIDS IN ALCYONARIA

Table 3 gives the distribution of the terpenoids in the different genera and orders of Alcyonaria. In our opinion, the general paucity of sesquiterpene hydrocarbons in Alcyonacea simply reflects the fact that the presence of these compounds has not been systematically checked. In contrast it might be significant that, although sesquiterpene alcohols are widespread in Alcyonacea, they have so far not been reported from Gorgonacea.

The earlier observation (*125*) that in Alcyonacea sesquiterpenes and diterpenes are not found together in the same species has been upheld so far. This does not hold at the generic level (e.g., in *Sinularia*). Both groups of compounds coexist in the same species in the order Gorgonacea.

TABLE 3

Distribution of Alcyonaria Terpenoids

Animal	Sesquiterpene hydrocarbons	Oxygenated sesqui-terpenes	Cembrane and cyclized cembrane diterpenes	Noncembrane diterpenes
Alcyonacea				
Capnella		+		
Cespitularia	+	+		
Lemnalia		+		
Litophyton			+	
Lobophytum			+	+
Nephthea			+	
Paralemnalia		+		
Sarcophyton			+	
Sinularia		+	+	
Xenia				+
Gorgonacea				
Briareum			+	
Eunicea	+		+	
Eunicella			+	
Gorgonia		+		
Muricea	+			
Plexaurella	+			
Pseudoplexaura	+		+	
Pterogorgia	+			
Stolonifera				
Clavularia				+
Pennatulacea				
Ptilosarcus			+	
Stylatula			+	

Although cembrane-related diterpenes are dominant in Alcyonacea, Gorgonacea, and Pennatulacea, it is striking that the first diterpenes isolated from Stolonifera clearly depart from the familiar cembrane skeleton. The order Telestacea has not been examined so far.

V. ABSOLUTE CONFIGURATION

A. Sesquiterpenes

Table 4 compares the signs of rotation of sesquiterpenes from coelenterates and from other sources. In their 1973 review (*32*) Ciereszko and Karns noted that "each of the sesquiterpenes isolated from the marine coelenterates is the optical antipode of the form found, where known, in terrestrial plants." Further examples of this antipodal relationship have been reported (*18, 28, 43*), but as can be seen from Table 4 the situation is not so clear-cut today. In several instances [e.g., 9-aristolene (**12**), α-copaene (**13**), δ-cadinene (**7**)] both antipodes are produced by terrestrial plants, but

TABLE 4

Sign of Rotation of Marine and Nonmarine Sesquiterpenes

Compound	Marine	Nonmarine
β-Bourbonene (**16**)[a]	(+)	(−)
α-Cubebene (**15**)	(+)	(−)
α-Copaene (**13**)	(+)	(+) and (−)[b]
β-Copaene (**14**)	(+)	(−)
1(10)-Aristolene (**11**)	(−)	(+)
9-Aristolene (**12**)	(+)	(+) and (−)[b]
Palustrol (**19**)	(+)	(−)
Viridiflorol (**20**)	(−)	(+)
Ledol (**18**)	(+)	(+)
γ-Maaliene (**10**)	(+)	(−)
δ-Cadinene (**7**)	(−)	(+)[b] and (−)
α-Muurolene (**8**)	(+)	(−)
α-Curcumene (**3**)	(−)	(+) and (−)
β-Curcumene (**4**)	(−)	(+) and (−)
β-Bisabolene (**2**)	(+)	(+) and (−)[b]
Germacrene A (**6**)	(−)	(−)

[a] The (+)-β-bourbonene of marine origin is in fact (+)-β-epibourbonene (**16**).

[b] This indicates the more frequently encountered antipode.

it is true that the "marine antipode" is often much rarer in terrestrial plants. Ledol (**18**) is an obvious exception. The same antipode of germacrene A (**6**) has been isolated from aphids and coelenterates. Moreover, the marine derivatives **21** to **24** of the nardosinane series (*43, 128*) have a skeleton that is antipodal to that of their terrestrial relatives (*54, 108, 140*).

A similar antipodal relationship seems to exist in the case of liverworts (*89, 96*) when compared to higher plants. It is to be expected that the finding of further "unconventional" antipodes from terrestrial sources will blur even more the distinction between marine and nonmarine sesquiterpenes.

B. Diterpenes

With a few exceptions all coelenterate diterpenes reported so far are cembrane or cyclized cembrane derivatives. It is generally admitted that the cembrane skeleton originates from a cyclization of geranylgeranyl pyrophosphate, a hypothesis supported by the observation that in all derivatives the geometry of the double bonds, when known, is always *E*, as in geranylgeraniol. The initial cyclization can a priori yield the two antipodal ions I and II.

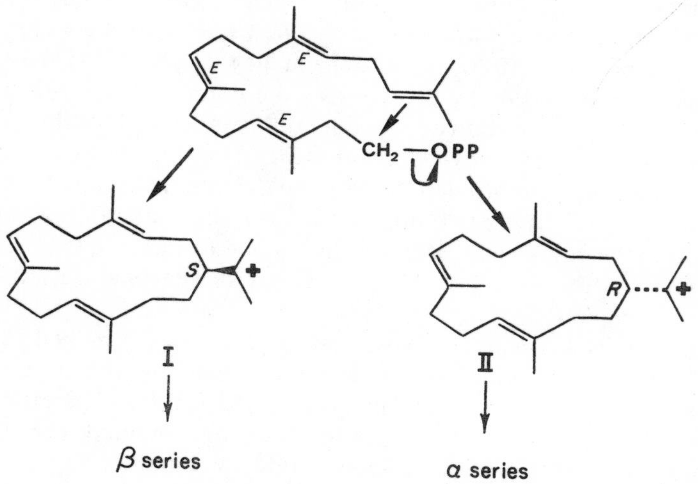

E E E CH_2—OPP

I → β series

II → α series

The *S* antipode (I) leads to a series of compounds of corresponding configuration in which the isopropyl group is β (pointing upward) when the double bonds are located as shown here. This is called the β series. In a similar fashion, the *R* antipode (II) leads to an α series in which the isopropyl is α (pointing downward). Obviously, one could convert an α

compound into a β compound either by inverting the configuration of the carbon atom C-1 bearing the isopropyl or by shifting the double bonds by moving each to the other side of its methyl group.

All cembrane diterpenes of known absolute configuration at C-1 that have so far been reported from the order Alcyonacea (**35**, **36**, **37**, **39**, **44**, **47**, **48**, **49**, **50**, **51**) belong to the α series. This is also the case for all cembrane derivatives isolated from terrestrial plants, e.g., cembrene (*45*), mukulol (*103*), and tobacco thunberganoids (*1*). In contrast, all cembrane diterpenes isolated from the order Gorgonacea belong to the β series (**41**, **52**, **53**, **54**, **55**, **57**, **58**).

The existence of the two series could be explained by the presence of two different enzymatic systems for geranylgeranyl pyrophosphate cyclization, a mechanism allowing for epimerization at C-1, or a rather improbable mechanism inducing the shift of all double bonds.

VI. VARIATION OF TERPENOID CONTENT

Although no systematic research has yet been reported on this topic, it might be worthwhile to assemble a number of scattered observations on the subject. *Capnella imbricata* occurs in small colonies of about 10 cm in height. Comparison of pooled colonies of *C. imbricata* collected at five localities in the South Moluccas and one in New Guinea (*113*) showed that all samples possessed diol (**25**), whereas the content in the related triols (**26–28**) and tetrol (**29**) showed considerable variation. Monthly observations at Laing Island (New Guinea) over a period of about one year indicated a marked variation in the terpenoid content of colonies collected simultaneously in the same biotope. This variation affects the pattern of oxidation of the capnellane skeleton. Initial results indicate that individual variation is more important than the effect of seasonal variation on a statistical sample (*67*).

The gorgonians *Pseudoplexaura porosa*, *P. wagenaari*, and *P. flagellosa* collected over a wide geographical range exhibited a marked consistency in their content of crassin acetate (**52**) and sesquiterpenes. In contrast, *Eunicea mammosa* yields the diterpenes eunicin (**54**), jeunicin (**55**), cueunicin (**57**), cueunicin acetate (**58**), or eupalmerin acetate (**53**) depending on the location of collection. Analogous variation has also been observed in the case of *Eunicea palmeri* and *E. succinea*. *Eunicea mammosa* also shows a geographic variation in its sesquiterpene hydrocarbon content (*29, 31, 32*).

It is interesting that in the same location the terpenoid content seems to vary over long periods of time. For instance, it was reported that *E.*

mammosa from Jamaica yielded pure jeunicin (**55**) in 1960 and that later collections yielded both jeunicin (**55**) and eunicin (**54**), with eunicin more common in recent collections (*29, 31*). Samples of *Briareum asbestinum* collected on a small rock off St. Thomas, Virgin Islands, yielded briareins A (**61**) and B (**78a**). Specimens collected one year later in the same month, on the same rock, yielded only briareins C (**78b**) and D (**78c**). Samples of *B. asbestinum* collected at Grand Bahama yielded only briareins C and D (*7*).

Sinulariolide (**47**), accompanied by several closely related derivatives (**36**, **48**, **49**, **72**), was the major diterpene in a sample of *Sinularia flexibilis* from the South Moluccas (*56, 57*). In a sample of *S. flexibilis* collected on the Great Barrier Reef, sinulariolide was isolated together with sinularin (**50**) and dihydrosinularin (**51**) (*149*); these were not detected in the sample of the South Moluccas, in which even trace compounds were characterized. In a third sample of *S. flexibilis*, also from the Great Barrier Reef, only sinularin (**50**) and its dihydro derivative (**51**) were found (*151*). Like the gorgonian cembranolides **53**, **54**, and **55** (*50*), sinulariolide and sinularin could originate from a common precursor (*56*).

Comparison of a number of samples of *Sarcophyton trocheliophorum* also shows a variation in terpenoid content. This variation is not only geographic, since two neighboring specimens collected at Anse Gaulette, La Digue, Seychelles, differed in that one contained mostly trocheliophorol (**77**) whereas the other yielded sarcophytoxide (**73a**) (*134*). *Sarcophyton glaucum* collected in the Red Sea afforded sarcophine (**42**) and a number of structurally related derivatives (*76*). It was reported that "the relative amounts of the various diterpenes in diverse batches of *S. glaucum* changed remarkably as a function of the collection location and period of the year" (*76*). We observed similar variations in specimens collected in the Seychelles and various localities in the South Moluccas. Although we isolated none of the compounds reported from the Red Sea samples, we did find very closely related substances (*134*).

It is noteworthy that in a given species the variation in terpenoid content arises only from the functionalization and not from the skeleton. The interpretation of these variations is not straightforward. It is difficult at this stage to separate the various factors that might be involved, for instance, geographic, individual, or seasonal variations. Other factors could be involved, such as age and sex of the colonies (there are female and male colonies) or changes in the symbiotic association between polyps and xanthellae (see Section VII).

In addition to these natural factors, one should keep in mind the possibility of identification errors. The systematics of Octocórallia are sometimes delicate (especially the identification of young colonies). When it comes to the symbiotic zooxanthellae, systematics are still a matter of controversy (*112*). Furthermore, our chemical data do not necessarily reflect the situation in live colonies. Many of the coelenterate terpenoids are notoriously unstable, and it is possible that in some instances our view of these mixtures simply reflects the way they have decomposed during handling.

VII. ORIGIN OF COELENTERATE TERPENOIDS

Since most of the coelenterates examined so far live in symbiosis with zooxanthellae, what is the origin of the terpenoids? Three obvious solutions might be considered: production by the animal polyps alone, by the algae alone, or by the association of both symbionts.

The isolation of a compound from one of the symbionts is not really proof of its origin, since it is conceivable that compounds produced by one of the partners are transferred to the other. Very little work has been published on this subject. The only incorporation experiments so far reported deal with *Pseudoplexaura porosa*, in which crystals of crassin acetate (**52**) are found in the tissues (*106*). Early experiments established that the gorgonian–zooxanthellae complex has the capability of *de novo* synthesis of crassin acetate (*106*). More recently it was reported (*4*) that intact zooxanthellae did not incorporate labeled precursors, whereas a cell-free preparation of these organisms enzymatically incorporated ^{14}C-labeled mevalonate into crassin acetate in the absence of any enzymatic system from the animal part. In surprising contrast, the sesquiterpenes, although present in the isolated algae (*141*), were reported to be synthesized by the animal tissues alone (*4*). It would thus seem that in the case of *Pseudoplexaura porosa* the sesquiterpenes are produced by the animal polyps and that the diterpenes are produced by the algae. On the other hand, in the case of the biosynthesis of prostaglandins by *Pseudoplexaura homomalla*, it was suggested (*41, 42*) that both symbionts are involved.

In our opinion it is too early to generalize from these results, and many more data will be needed to clarify the situation. However, the role of the zooxanthellae in the production of terpenoids is supported by the fact that in our laboratory terpenoids were detected only in animals in which xanthellae were present. Species devoid of algae did not yield any of these compounds. For instance, eunicellin (**60**) has been found in the gorgonian *Eunicella stricta* but has not been detected in the deeper-water form, *E. stricta* var. *aphyta*, which is devoid of zooxanthellae.

It has been suggested (*125*) that, if the zooxanthellae are solely responsible for the elaboration of the coelenterate terpenoids, the great diversity of chemical structures would then imply the existence of distinct forms (species? varieties? strains?) of these algae. The variations observed in the terpenoid content of a given coelenterate species could then be at least partially explained by variations in its zooxanthellae population. The data accumulated so far, however, indicate a great specificity of some types of terpenes in certain genera or species of animals (e.g., capnellane derivatives found only in *Capnella imbricata*, the presence of cembrane diterpenes in most of the *Lobophytum* and *Sarcophyton* species, and nardosinane sesquiterpenes found only in *Lemnalia* and *Paralemnalia* species). This would be expected if the symbiotic associations between polyps and zooxanthellae do not occur at random but, on the contrary, exhibit a marked degree of specificity.

These views recently found experimental support when it was shown (*112*) that the common zooxanthella *Symbiodinium microadriaticum* exists in a variety of genetically distinct strains, which are recognized by protein electrophoresis, isozyme pattern, cell size, and infective variation. These strains have specific host distribution. There is no one-to-one correspondence between strains and host species, since five strains were isolated from more than one host species and different individuals of the same host species harbored two distinct strains.

VIII. PHYSIOLOGICAL ACTIVITY

It is well known that a number of coelenterate terpenoids exhibit interesting biodynamic properties. Besides other activities, ptilosarcone (**62**) is toxic to mice by injection (*150*). This is also the case for sarcophine (**42**) (*93*). Many compounds are toxic to fish, as illustrated by lobolide (**46**) (*75*), sarcophine (**42**) (*93*), lobophytolide (**44**) (*125*), crassolide (**70**) (*125*), and africanol (**30**) (*125*). However, it should not be inferred that all coelenterate terpenoids are endowed with this property; several are tolerated by fish in quite high concentrations, e.g., sinulariolide (**47**).

Antineoplastic properties, among others, have been reported for sinulariolide (**47**), sinularin (**50**) and its dihydro derivative (**51**) (*149*), crassin acetate (**52**) (*29, 105, 142*), eupalmerin acetate (**53**) (*105*), jeunicin (**55**) (*105*), eunicin (**54**) (*105*), and asperdiol (**41**) (*148*).

Crassin acetate also affects growth and motility of the ciliate *Tetrahymena pyriformis* (*101*). Activity against algae is quite widespread. Sinulariolide (**47**), africanol (**30**), lobophytolide (**44**), lemnalialoside (**81**), and the capnellane derivatives **25** and **28**, when present in minute concentrations in culture media, markedly inhibit the growth of the marine unicellular algae *Chaetoceros septentrionalis*, *Astrionella japonica*, *Thal-*

lasioscira excentricus, *Protocentrum micans*, and *Amphidinium carterae*, whereas lemnacarnol (**21**) is active only against the two latter species (*125*).

Stylatulide (**63**) is toxic to larvae of the copepod *Tisbes furcata* (*157*). Sarcophine (**42**) inhibits the enzyme phosphofructokinase (*51*) and shows a strong antiacetylcholine action on the isolated guinea pig ileum (*93*).

IX. ECOLOGICAL SIGNIFICANCE

It is quite plausible that some of the compounds reviewed here might have a pheromonal activity. No evidence for such action has yet been presented, but it is interesting that two of these terpenoids have a well-established function as chemical messengers in the terrestrial environment. Cembrene A (**35**) is the trail pheromone of the termite *Nasutitermes exitiosus* (*13*), and (−)-germacrene A (**6**) has been shown to be an alarm pheromone of aphids (*16*).

There is quite ample evidence, however, to indicate that many of the coelenterate terpenoids possess a protective function for the colonies. Sessile reef coelenterates must fend off attack by large predators and also ensure that they are not invaded by microorganisms, larvae, and/or algae. Field observations demonstrate that these defensive functions are indeed performed in a most efficient way.

As indicated above, many of the terpenoids reviewed here are toxic to fish at low concentrations. It was demonstrated that these compounds could in fact ward off predation during many experiments in which aquarium fish were fed either with their usual food flakes or with the same flakes containing small measured amounts of given toxic terpenoids. The "protected" flakes were subjected to many fewer bites and, when seized by the fish, were quite systematically and immediately rejected (*67*). In some cases, these toxins can be emitted by the colonies in the event of attack. The xeniid *Cespitularia* sp. aff. *subviridis*, when even slightly molested, was observed to produce an abundant mucus that was lethal to a variety of small marine crustaceans. This mucus contained the toxic terpene palustrol (**19**) in substantial amounts (*133*). Since feeding deterrent action (unpleasant taste?) would probably take place well below lethal concentration, many coelenterates could be effectively protected by terpenoids occurring at concentrations below 0.001% and thus escaping routine isolation techniques.

In the case of the abundant reef octocorals, these feeding deterrent terpenoids might have a considerable ecological significance by protecting the major contributors to the biomass against grazing macropredators. By participating in the control of the amount and direction of predation in the reef biotope, these substances could be compared to "switches" directing the food chain (and hence the energy flow) in one given path rather than another.

From the information summarized in Section VIII one can surmise that many terpenoids would be equally effective in the protection of the coelenterate colonies against invasion by microorganisms and algae. It is of special interest that some terpenes are active against dinoflagellates, the group to which the zooxanthellae belong. If these terpenes are indeed produced by the symbiotic algae, it would be tempting to speculate that such compounds could be used to ensure the suggested specificity of the coelenterate–zooxanthellae associations.

APPENDIX

TABLE A1

Known Terpenoids Isolated from Coelenterates

Africanol (**30**)
(−)-1(10)-Aristolene (**11**)
(+)-9-Aristolene (**12**)
Asperdiol (**41**)
Alloaromadendrene (**17**)
(+)-α-Bisabolene (**1**)
(+)-β-Bisabolene (**2**)
Briarein A (**61**)
Briarein B (**78a**)
Briarein C (**78b**)
Briarein D (**78c**)
(−)-δ-Cadinene (**7**)
(+)-Calamenene (**82**)
$\Delta^{9(12)}$-Capnellene-8β,10α-diol (**25**)
$\Delta^{9(12)}$-Capnellene-3β,8β,10α,14-tetrol (**29**)
$\Delta^{9(12)}$-Capnellene-2ξ,8β,10α-triol (**26**)
$\Delta^{9(12)}$-Capnellene-3β,8β,10α-triol (**27**)
$\Delta^{9(12)}$-Capnellene-5α,8β,10α-triol (**28**)
R-(−)-Cembrene A (**35**)
(+)-α-Copaene (**13**)
(+)-β-Copaene (**14**)
Crassin acetate (**52**)
Crassolide (**70**)
(+)-α-Cubebene (**15**)
Cueunicin (**57**)
Cueunicin acetate (**58**)
(−)-α-Curcumene (**3**)
(+)-β-Curcumene (**4**)
(−)-β-Curcumene (**5**)
11-Dehydrosinulariolide (**48**)
2-Deoxylemnacarnol (**22**)
2-Deoxy-12-oxolemnacarnol (**23**)
3,4,11,12-Diepoxycembrene A (**36**)
Dihydrosinularin (**51**)
3α,4β-Dihydroxyclavulara-1(15),17-diene (**67**)
1α,4β-Dihydroxyclavular-17-ene (**66**)
2-[3,7-Dimethyl-8-(4-methyl-2-furyl)-2,6-octadienyl]-5-methylhydroquinone (**34**)
(+)-β-Elemene (**83**)
(+)-β-Epibourbonene (**16**)
11-Episinulariolide acetate (**72**)
Epoxynephthenol acetate (**39**)
2-[(*E,E,E*)-7′,8′-epoxy-4′,8′,12′-trimethylcyclotetradeca-1′,3′,11′-trienyl]propan-2-ol (**40**)
Eupalmerin acetate (**53**)
Eunicellin (**60**)
Eunicin (**54**)
Flexibilene (**64**)

(*Continued*)

TABLE A1 (*Continued*)

Furoventalene (**32**)	*Pseudoplexaura* compound B1 (**79a**)
(−)-Germacrene A (**6**)	*Pseudoplexaura* compound B2 (**79b**)
(+)-β-Gorgonene (**9**)	*Pseudoplexaura* compound B3 (**79c**)
4β-Hydroxyclavulara-1(15),17-diene (**65**)	Ptilosarcone (**62**)
2-Hydroxynephthenol (**38**)	Pukalide (**59**)
6ξ-Hydroxysinulariolide (**49**)	Sarcophine (**42**)
2-[(*E*,*E*,*E*)-2′-Hydroxy-4′,8′,12′-trimethylcyclotetradeca-3′,7′,11′-trienyl]prop-2-enoic acid-1,2′-lactone (**45**)	Sarcophytoxide (**73a**)
	Sarcophyton glaucum compound 2 (**43**)
	Sarcophyton glaucum compound 3 (**73b**)
	Sarcophyton glaucum compound 4 (**74a**)
Isosarcophytoxide (**73c**)	*Sarcophyton glaucum* compound 5 (**74b**)
Jeunicin (**55**)	*Sarcophyton glaucum* compound 6 (**75**)
(+)-Ledol (**18**)	*Sarcophyton glaucum* compound 7 (**76**)
Lemnacarnol (**21**)	*Sclerophyton* compounds (**80**)
Lemnalactone (**24**)	(−)-β-Selinene (**84**)
Lemnalialoside (**81**)	Sinularene (**31**)
Lobolide (**46**)	*Sinularia gonatodes* compound 1 (**33**)
Lobophytolide (**44**)	Sinularin (**50**)
(+)-γ-Maaliene (**10**)	Sinulariolide (**47**)
(+)-α-Muurolene (**8**)	Stylatulide (**63**)
Nephthenol (**37**)	Trocheliophorol (**77**)
(+)-Palustrol (**19**)	Unnamed (**69**)
Peunicin (**56**)	(−)-Viridiflorol (**20**)
Pseudoplexaura porosa compound $C_{15}H_{24}$ (**71**)	Xenicin (**68**)

TABLE A2

Animal Species and Their Terpenoids

Alcyonacea	*Paralemnalia digitiformis* (**24**)
Capnella imbricata (**25, 26, 27, 28, 29**)	*P. thyrsoides* (**22, 23**)
Cespitularia aff. *subviridis* (**17, 18, 19, 20**)	*Sarcophyton* (**40, 43**)
Lemnalia africana (**22, 30**)	*Sarcophyton glaucum* (**42, 43, 73a, 73b, 74a, 74b, 75, 76**)
L. carnosa (**21**)	*S. decaryi* (**37**)
L. digitata (**81**)	*S. trocheliophorum* (**73a, 73c, 77**)
L. laevis (**22**)	*Sclerophyton capitalis* (**80a, 80b**)
L. nitida (**30**)	*Sinularia abrupta* (**59**)
Litophyton arboreum (**37**)	*S. flexibilis* (**35, 36, 47, 48, 49, 50, 51, 64, 72**)
L. viridis (**37, 38**)	*S. gonatodes* (**33**)
Lobophytum (**43**)	*S. lochmodes* (**34**)
Lobophytum crassum (**46, 70**)	*S. mayi* (**14, 31**)
L. cristagalli (**44**)	*S. notanda* (**48, 72**)
L. hedleyi (**69**)	*S. querciformis* (**72**)
L. michaelae (**45**)	*Xenia elongata* (**68**)
L. pauciflorum (**37**)	
Nephthea (**37, 39**)	

(*Continued*)

TABLE A1 (*Continued*)

Gorgonacea	*P. nutans* (**2**, **3**, **5**)
Briareum asbestinum (**61**, **78a**, **78b**, **78c**)	*Pseudoplexaura* (**79a**, **79b**, **79c**)
Eunicea asperula (**41**)	*Pseudoplexaura crucis* (**52**)
E. mammosa (**6**, **8**, **53**, **54**, **55**, **57**, **58**, **83**, **84**)	*P. flagellosa* (**7**, **8**, **13**, **14**, **15**, **16**, **17**, **52**)
E. palmeri (**8**, **14**, **53**, **54**)	*P. porosa* (**7**, **8**, **13**, **14**, **15**, **16**, **17**, **52**, **71**, **82**)
E. succinea (**53**)	*P. wagenaari* (**7**, **8**, **13**, **14**, **15**, **16**, **17**, **52**)
E. succinea var. *plantaginea* (**53**, **56**)	*Pterogorgia americana* (**9**, **10**, **11**, **12**)
E. tourneforti (**41**)	
Eunicella stricta (**60**)	Pennatulacea
Gorgonia ventalina (**32**)	*Ptilosarcus gurneyi* (**62**)
Muricea elongata (**2**, **3**, **5**)	*Stylatula* (**63**)
Plexaurella dichotoma (**1**, **2**, **3**, **4**, **8**)	
P. fusifera (**1**, **2**, **3**, **4**, **8**)	Stolonifera
P. grisea (**1**, **2**, **3**, **4**, **8**)	*Clavularia inflata* (**65**, **66**, **67**)

REFERENCES

1. Aasen, A. J., Junker, N., Enzell, C. R., Berg, J. E., and Pilotti, A. M., *Tetrahedron Lett.* p. 2607 (1975).
2. Andersen, N. H., and Syrdal, D. D., *Phytochemistry* **9**, 1325 (1970).
3. Andersen, N. H., and Syrdal, D. D., *Tetrahedron Lett.* p. 2455 (1972).
4. Anderson, D. G., *Mar. Nat. Prod. Symp., Aberdeen*, personal communication (1975).
5. Baker, J. T., *Pure Appl. Chem.* **48**, 35 (1976).
6. Baker, J. T., and Murphy, V., "Handbook of Marine Science: Compounds from Marine Organisms," Vol. 1. CRC Press, Cleveland, Ohio, 1976.
7. Bartholomé, C., Ph.D. thesis, Univ. of Brussels, 1974.
8. Bates, R. B., Büchi, G., Matsuura, T., and Shaffer, R. R., *J. Am. Chem. Soc.* **82**, 2327 (1960).
9. Bayer, F. M., "The Shallow-water Octocorallia of the West Indian Region." Nijhoff, The Hague, 1961.
10. Beechan, C. M., Djerassi, C., Finer, J. S., and Clardy, J., *Tetrahedron Lett.* p. 2395 (1977).
11. Bellesia, F., Pagnoni, U. M., and Trave, R., *Tetrahedron Lett.* p. 1245 (1974).
12. Bernstein, J., Schmeuli, U., Zadock, E., Kashman, Y., and Néeman, I., *Tetrahedron* **30**, 2817 (1974).
13. Birch, A. J., Brown, W. V., Corrie, J. E. T., and Moore, B. P., *J. Chem. Soc., Perkin Trans. 1*, p. 2653 (1972).
14. Boekman, R. E., Jr., and Stealey, M. A., *J. Org. Chem.* **40**, 1755 (1975).
15. Bouillon, J., *in* "Chemical Zoology" (M. Florkin and B. T. Scheer, eds.), Vol. 2, p. 81. Academic Press, New York, 1968.
16. Bowers, W. S., Nishino, C., Montgomery, M. E., Nault, L. R., and Nielson, M. W., *Science* **196**, 680 (1977).
17. Braekman, J. C., *in* "Marine Natural Products Chemistry" (D. J. Faulkner and W. H. Fenical, eds.), p. 5. Plenum, New York, 1977.

18. Braekman, J. C., Daloze, D., Ottinger, R., and Tursch, B., *Experientia* **33**, 993 (1977).
19. Braekman, J. C., Daloze, D., Schubert, R., Albericci, M., Tursch, B., and Karlsson, R., *Tetrahedron* **34**, 1551 (1978).
20. Braekman, J. C., Daloze, D., and Tursch, B., unpublished observations (1975).
21. Brown, M., *J. Org. Chem.* **33**, 162 (1968).
22. Büchi, G., Greuter, F., and Tokoroyama, T., *Tetrahedron Lett.* p. 827 (1962).
23. Büchi, G., Hofheinz, W., and Paukstelis, J. V., *J. Am. Chem. Soc.* **88**, 4113 (1966).
24. Büchi, G., Hofheinz, W., and Paukstelis, J. V., *J. Am. Chem. Soc.* **91**, 6473 (1969).
25. Burks, J. E., van der Helm, D., Chang, C. Y., and Ciereszko, L. S., *Acta Crystallogr., Sect. B* **33**, 704 (1977).
26. Carboni, S., De Settimo, A., Malaguzzi, V., Marsili, A., and Pacini, P. L., *Tetrahedron Lett.* p. 3017 (1965).
27. Chang, C. Y., Ciereszko, L. S., van der Helm, D., and Wu, R. K. K., personal communication (1977).
28. Cheer, C. J., Smith, D. H., Djerassi, C., Tursch, B., Braekman, J. C., and Daloze, D., *Tetrahedron* **32**, 1807 (1976).
29. Ciereszko, L. S., *Trans. N.Y. Acad. Sci.* **24**, 502 (1962).
30. Ciereszko, L. S., *Proc. Food Drugs Sea Symp., Mar. Technol. Soc., Washington, D.C.* p. 297 (1974).
31. Ciereszko, L. S., *Int. Symp. Ecol. Manage. Trop. Shallow Water Communities, Indonesia,* 1976.
32. Ciereszko, L. S., and Karns, T. K. B., *in* "Biology and Geology of Coral Reefs" (O. A. Jones and R. Endean, eds.), p. 183. Academic Press, New York, 1973.
33. Ciereszko, L. S., Sifford, D. H., and Weinheimer, A. J., *Ann. N.Y. Acad. Sci.* **90**, 917 (1960).
34. Cocker, W., and McMurry, T. B. H., *Tetrahedron* **8**, 181 (1960).
35. Coll, J. C., personal communication (1977).
36. Coll, J. C., Harves, G. B., Liyanage, N., Obershänsli, W., and Wells, R. J., *Aust. J. Chem.* **30**, 1305 (1977).
37. Coll, J. C., Mitchell, S. J., and Stockie, G. J., *Tetrahedron Lett.* p. 1539 (1977).
38. Coll, J. C., Mitchell, S. J., and Stockie, G. J., *Aust. J. Chem.* **30**, 1859 (1977).
39. Coll, J. C., Liyanage, N., Stockie, G. J., Van Altena, I., Nemorin, J. N. E., Sternhell, S., and Kazlauskas, R., Austr. J. Chem., **31**, 157 (1978).
40. Coll, J. C., Liyanage, N., Stockie, G. J., and Van Altena, I., Austr. J. Chem., **31**, 163 (1978).
41. Corey, E. J., and Washburn, W. N., *J. Am. Chem. Soc.* **96**, 934 (1974).
42. Corey, E. J., Washburn, W. N., and Chen, J. C., *J. Am. Chem. Soc.* **95**, 2054 (1973).
43. Daloze, D., Braekman, J. C., Georget, P., and Tursch, B., *Bull. Soc. Chim. Belg.* **86**, 47 (1977).
44. Damodaran, N. P., and Dev, S., *Tetrahedron* **24**, 4113 (1968).
45. Dauben, W. G., Thiessen, W. E., and Resnick, P. R., *J. Org. Chem.* **30**, 1693 (1965).
46. De Mayo, P., Williams, R. E., Büchi, G., and Feairheller, S. H., *Tetrahedron* **21**, 619 (1965).
47. Dolejs, L., and Šorm, F., *Collect. Czech. Chem. Commun.* **25**, 1837 (1960).
48. Dolejs, L., Motl, O., Soucek, M., Herout, V., and Šorm, F., *Collect. Czech. Chem. Commun.* **25**, 1483 (1960).
49. Dolejs, L., Herout V., and Šorm, F., *Collect. Czech. Chem. Commun.* **26**, 811 (1961).
50. Ealick, S. E., van der Helm, D., and Weinheimer, A. J., *Acta Crystallogr., Sect. B* **31**, 1618 (1975).
51. Erman, A., and Néeman, I., *Toxicon* **15**, 207 (1977).

52. Fankboner, P. V., *Biol. Bull.* (*Woods Hole, Mass.*) **141**, 222 (1971).
53. Faulkner, D. J., and Andersen, R. J., *in* "The Sea" (E. D. Goldberg, ed.), p. 679. Wiley (Interscience), New York, 1973.
54. Grandi, R., Marchesini, A., Pagnoni, U. M., and Trave, R., *Tetrahedron Lett.* p. 1765 (1973).
55. Gross, R. A., Ph.D. thesis, Univ. of Oklahoma, Norman, 1974.
56. Herin, M., Ph.D. thesis, Univ. of Brussels, 1977.
57. Herin, M., and Tursch, B., *Bull. Soc. Chim. Belg.* **85**, 707 (1976).
58. Herin, M., Colin, M., and Tursch, B., *Bull. Soc. Chim. Belg.* **85**, 801 (1976).
59. Honwad, V. K., and Rao, A. S., *Tetrahedron* **21**, 2593 (1965).
60. Hossain, M. B., and van der Helm, D., *J. Am. Chem. Soc.* **90**, 6607 (1968).
61. Hossain, M. B., and van der Helm, D., *Rec. Trav. Chim. Pays-Bas* **88**, 1413 (1969).
62. Hossain, M. B., Nicholas, A. F., and van der Helm, D., *Chem. Commun.* p. 385 (1968).
63. Hyde, R. W., Ph.D. thesis, Univ. of Oklahoma, Norman, 1966.
64. Irie, T., Yamamoto, K., and Masamune, T., *Bull. Chem. Soc. Jpn.* **37**, 1053 (1964).
65. Jain, T. C., and McCloskey, J. E., *Tetrahedron Lett.* p. 5139 (1972).
66. Jeffs, P. W., and Lytle, L. T., *Lloydia* **37**, 315 (1974).
67. Kaisin, M., and Tursch, B., unpublished observations (1977).
68. Kaisin, M., Sheikh, Y. M., Durham, L. J., Djerassi, C., Tursch, B., Daloze, D., Braekman, J. C., Losman, D., and Karlsson, R., *Tetrahedron Lett.* p. 2239 (1974).
69. Kapadia, V. H., Nagasampagi, B. A., Naik, V. G., and Dev, S., *Tetrahedron* **21**, 607 (1965).
70. Karlsson. R., *Acta Crystallogr., Sect. B* **32**, 2609 (1976).
71. Karlsson, R., *Acta Crystallogr., Sect. B* **33**, 1143 (1977).
71a. Karlsson, R., *Acta Crystallogr., Sect. B* **33**, 2027 (1977).
71b. Karlsson, R., *Acta Crystallogr., Sect. B* **33**, 2032 (1977).
72. Karlsson, R., and Losman, D., *Acta Crystallogr., Sect. B* **32**, 1614 (1976).
73. Kashman, Y., M. Bodner, Y. Loya, Y. Bernayahu, Israel J. Chem., **16**, 1 (1977).
74. Kashman, Y., *in* "Marine Natural Products Chemistry" (D. J. Faulkner and W. H. Fenical, eds.), p. 17. Plenum, New York, 1977.
75. Kashman, Y., and Groweiss, A., *Tetrahedron Lett.* p. 1159 (1977).
76. Kashman, Y., Zadock, E., and Néeman, I., *Tetrahedron* **30**, 3615 (1974).
77. Kennard, O., Watson, D. G., Riva Di Sanseverino, L., Tursch, B., Bosmans, R., and Djerassi, C., *Tetrahedron Lett.* p. 2879 (1968).
78. Kennard, O., Watson, D. G., and Riva Di Sanseverino, L., *Acta Crystallogr., Sect. B* **26**, 1038 (1970).
79. Klein, E., and Schmidt, W., *J. Agric. Food Chem.* **19**, 1115 (1971).
80. Kodama, M., Matsuki, Y., and Itô, S., *Tetrahedron Lett.* p. 3065 (1975).
81. Krepinsky, J., Samek, Z., and Šorm, F., *Tetrahedron Lett.* p. 3209 (1966).
82. Krepinsky, J., Samek, Z., Šorm, F., and Lamparsky, D., *Tetrahedron Lett.* p. 359 (1966).
83. Krepinsky, J., Samek, Z., Šorm, F., Lamparsky, D., Ochsner, P., and Naves, Y. R., *Tetrahedron, Suppl.* **8**, 53 (1966).
84. Lawda, P. H., Joshi, G. D., and Kulkarni, S. N., *Chem. Ind.* (*London*) p. 1601 (1968).
85. Lenhoff, H. M., *in* "Chemical Zoology" (M. Florkin and B. T. Scheer, eds.), Vol. 2, p. 157. Academic Press, New York, 1968.
86. Losman, D., Karlsson, R., and Sheldrick, G. M., *Acta Crystallogr., Sect. B* **33**, 1959 (1977).
87. Middlebrook, R. E., Ph.D. Thesis, Univ. of Oklahoma, Norman, 1966.

88. Missakian, M. G., Burreson, B. J., and Scheuer, P. J., *Tetrahedron* **31**, 2513 (1975).
89. Money, T., *in* "Terpenoids and Steroids," Specialist Periodical Reports, (K. H. Overton, ed.) Vol. 5, p. 52. Chem. Soc., London, 1975.
90. Morikawa, K., and Hirose, Y., *Tetrahedron Lett.* p. 1799 (1969).
91. Nagasampagi, B. A., Yankov, L., and Dev, S., *Tetrahedron Lett.* p. 1913 (1968).
92. Naves, Y. R., *Bull. Soc. Chim. Fr.* p. 987 (1951).
93. Néeman, I., Fishelson, L., and Kashman, Y., *Toxicon* **12**, 593 (1974).
94. Ohta, Y., and Hirose, Y., *Tetrahedron Lett.* p. 1601 (1969).
95. Ohta, Y., Sakai, T., and Hirose, Y., *Tetrahedron Lett.* p. 6365 (1966).
96. Ohta, Y., Andersen, N. H., and Liu, C. B., *Tetrahedron* **33**, 617 (1977).
97. Ourisson, G., Munavalli, S., and Ehret, C., "Selected Constants: Sesquiterpenoids." Pergamon, Oxford, 1966.
98. Paknikar, S. K., and Sood, V. K., *Tetrahedron Lett.* p. 4853 (1973).
99. Patil, V. D., Nayak, U. R., and Dev, S., *Tetrahedron* **29**, 341 (1973).
100. Pentegova, V. A., Motl, O., and Herout, V., *Collect. Czech. Chem. Commun.* **26**, 1362 (1961).
101. Perkins, D. L., and Ciereszko, L. S., *Hydrobiologia* **42**, 77 (1973).
102. Pliva, J., Herout, V., and Šorm, F., *Collect. Czech. Chem. Commun.* **16**, 158 (1951).
103. Prasad, R. S., and Dev, S., *Tetrahedron* **32**, 1437 (1976).
104. Premuzic, E., *Fortschr. Chem. Org. Naturst.* **29**, 417 (1971).
105. Rehm, S. J., Ph.D. thesis, Univ. of Oklahoma, Norman, 1971.
106. Rice, J. R., Papastephanou, C., and Anderson, D. G., *Biol. Bull.* (*Woods Hole, Mass.*) **138**, 334 (1970).
107. Roberts, J. S., *in* "Chemistry of Terpenes and Terpenoids" (A. A. Newman, ed.), p. 99. Academic Press, New York, 1972.
108. Rucker, G., *Justus Liebigs Ann. Chem.* p. 311 (1975).
109. Scheuer, P. J., "Chemistry of Marine Natural Products." Academic Press, New York, 1973.
110. Schmitz, F. J., *in* "Marine Natural Products Chemistry" (D. J. Faulkner and W. Fenical, eds.), p. 293. Plenum, New York, 1977.
111. Schmitz, F. J., Vanderah, D. J., and Ciereszko, L. S., *Chem. Commun.* p. 407 (1974).
112. Schoenberg, D. A., and Trench, R. K., *in* "Coelenterate Ecology and Behavior" (G. O. Mackie, ed.), p. 423. Plenum, New York, 1976.
113. Sheikh, Y. M., Singy, G., Kaisin, M., Eggert, H., Djerassi, C., Tursch, B., Daloze, D., and Braekman, J. C., *Tetrahedron* **32**, 1171 (1976).
114. Sheikh, Y. M., Djerassi, C., Braekman, J. C., Daloze, D., Kaisin, M., Tursch, B., and Karlsson, R., *Tetrahedron* **33**, 2115 (1977).
115. Shmidt, E. N., Kashtanova, N. K., and Pentegova, V. A., *Khim. Prir. Soedin.* p. 694 (1970); *Chem. Abstr.* **74**, 112243 (1971).
116. Sitton, D., and West, C. A., *Phytochemistry* **14**, 1921 (1975).
117. Sykora, V., Herout, V., and Šorm, F., *Collect. Czech. Chem. Commun.* **21**, 267 (1956).
118. Takeda, K., *Tetrahedron* **30**, 1525 (1974).
119. Tanaka, A., Tanaka, R., Uda, H., and Yoshikoshi, A., *J. Chem. Soc., Perkin Trans.*, p. 1721 (1972).
120. Taylor, D. L., *Adv. Mar. Biol.* **11**, 1 (1973).
121. Tixier-Durivault, A., "Les Octocoralliaires de Nouvelle-Calédonie." Mus. Natl. Hist. Nat., Paris, 1966.
122. Tixier-Durivault, A., "Faune de Madagascar, XXI, Octocoralliaires." Orstom, Paris, 1966.

123. Tixier-Durivault, A., "Résultats Scientifiques des Campagnes de la Calypso," Vol. I, p. 24. Masson, Paris, 1969.
124. Tixier-Durivault, A., *Cah. Pac.* p. 14 (1970).
125. Tursch, B., *Pure Appl. Chem.* **48**, 1 (1976).
126. Tursch, B., Braekman, J. C., Daloze, D., Fritz, P., Kelecom, A., Karlsson, R., and Losman, D., *Tetrahedron Lett.* p. 747 (1974).
127. Tursch, B., Braekman, J. C., Daloze, D., Herin, M., and Karlsson, R., *Tetrahedron Lett.* p. 3769 (1974).
128. Tursch, B., Colin, M., Daloze, D., Losman, D., and Karlsson, R., *Bull. Soc. Chim. Belg.* **84**, 81 (1975).
129. Tursch, B., Braekman, J. C., and Daloze, D., *Bull. Soc. Chim. Belg.* **84**, 767 (1975).
130. Tursch, B., Braekman, J. C., Daloze, D., Hěrin, M., Karlsson, R., and Losman, D., *Tetrahedron* **31**, 129 (1975).
131. Tursch, B., Braekman, J. C., Charles, C., Daloze, D., Herin, M., Kelecom, A., and Van Haelen, M., unpublished observations (1975).
132. Tursch, B., Braekman, J. C., Daloze, D., Dedeurwaerder, H., and Karlsson, R., *Bull. Soc. Chim. Belg.* **87**, 75 (1978).
133. Tursch, B., Braekman, J. C., and Kelecom, A., unpublished observations (1974)
134. Tursch, B., Cornet, P., Braekman, J. C., and Daloze, D., unpublished observations (1975).
135. Vanderah, D. J., Steudler, P. A., Ciereszko, L. S., Schmitz, F. J., Ekstrand, J. D., and van der Helm, D., *J. Am. Chem. Soc.* **99**, 5780 (1977).
136. van der Helm, D., Ealick, S. E., and Weinheimer, A. J., *Cryst. Struct. Commun.* **3**, 167 (1974).
137. van der Helm, D., Enwall, E. L., Weinheimer, A. J., Karns, T. K. B., and Ciereszko, L. S., *Acta Crystallogr., Sect. B* **32**, 1558 (1976).
138. Vrkoc, J., Krepinsky, J., Herout, V., and Šorm, F., *Tetrahedron Lett.* pp. 225, 735 (1963).
139. Vrkoc, J., Krepinsky, J., Herout, V., and Šorm, F., *Collect. Czech. Chem. Commun.* **29**, 795 (1964).
140. Von Rudloff, E., and Nair, G. V., *Can. J. Chem.* **42**, 421 (1964).
141. Washecheck, P. H., Ph.D. thesis, Univ. of Oklahoma, Norman, 1967.
142. Weinheimer, A. J., and Matson, J. A., *Lloydia* **38**, 378 (1975).
143. Weinheimer, A. J., and Washecheck, P. H., *Tetrahedron Lett.* p. 3315 (1969).
144. Weinheimer, A. J., Schmitz, F. J., and Ciereszko, L. S., *Drugs Sea, Trans. Drugs Sea Symp., Mar. Technol. Soc., Washington, D.C.* p. 135 (1967).
145. Weinheimer, A. J., Middlebrook, R. E., Bledsoe, J. O., Marsico, W. E., and Karns, T. K. B., *Chem. Commun.* p. 384 (1968).
146. Weinheimer, A. J., Washecheck, P. H., van der Helm, D., and Hossain, M. B., *Chem. Commun.* p. 1070 (1968).
147. Weinheimer, A. J., Youngblood, W. W., Washecheck, P. H., Karns, T. K. B., and Ciereszko, L. S., *Tetrahedron Lett.* p. 497 (1970).
148. Weinheimer, A. J., Matson, J. A., van der Helm, D., and Poling, M., *Tetrahedron Lett.* p. 1295 (1977).
149. Weinheimer, A. J., Matson, J. A., Hossain, M. B., and van der Helm, D., *Tetrahedron Lett.* p. 2923 (1977).
150. Wekel, J. C., *Proc. Food Drugs Sea Symp., Mar. Technol. Soc., Washington, D.C.* p. 324 (1974).
151. Wells, R. J., personal communication, 1975.
152. Wenninger, J. A., and Yates, R. L., *J. Am. Oil Chem. Soc.* **53**, 949 (1970).

153. Westfelt, L., *Acta Chem. Scand.* **20**, 2841 (1966).
154. Westfelt, L., *Acta Chem. Scand.* **20**, 2852 (1966).
155. Westfelt, L., *Acta Chem. Scand.* **21**, 152 (1967).
156. White, J. D., and Gupta, D. N., *J. Am. Chem. Soc.* **90**, 6171 (1968).
157. Wratten, S. J., Faulkner, D. J., Hirotsu, K., and Clardy, J., *J. Am. Chem. Soc.* **99**, 2824 (1977).
158. Wratten, S. J., Fenical, W., Faulkner, D. J., and Wekell, J. C., *Tetrahedron Lett.* p. 1559 (1977).
159. Yonge, C. M., *Adv. Mar. Biol.* **1**, 209 (1963).
160. Yoshihara, Y., Ohta, Y., Sakai, T., and Hirose, Y., *Tetrahedron Lett.* p. 2263 (1969).
161. Youngblood, W. W., Ph.D. thesis, Univ. of Oklahoma, Norman, 1969.
162. Zabza, A., Romanuk, M., and Herout, V., *Collect. Czech. Chem. Commun.* **31**, 3373 (1966).

Chapter 5

Applications of ^{13}C nmr to Marine Natural Products

JAMES J. SIMS, ALLAN F. ROSE, and RICHARD R. IZAC

I. INTRODUCTION

A. Background

The natural products chemist faces a challenge with each new compound he isolates. He often must determine the structure of a completely unknown compound. Years ago this meant extensive degradation and comparison with known substances. In the modern era, structure investigation is for the most part carried out with various energy probes known collectively as spectroscopy. With these powerful methods one can, in favorable cases, solve a structure using only a few milligrams of a compound.

MARINE NATURAL PRODUCTS

ISBN 0-12-624002-7

Structure determination by spectroscopy depends on empirical correlation of spectral information with organic structures. The larger the data bank of empirical facts the better the correlation should be. However, the history of natural products shows that for any new molecular feature spectral correlations are usually lacking. Sometimes two techniques give conflicting results. Thus, the scientist working in this area often uncovers gaps in existing data and must seek new ways to achieve his goal of proving the existence of a particular arrangement of atoms.

In general there are two ways to approach this problem. One method employs an existing technique by developing new correlations with model compounds or by improving the technique. The second method uses other techniques in a complementary manner. The most common instrumental methods used in structure determination are mass spectrometry, proton nuclear magnetic resonance (^{1}H nmr), infrared spectroscopy, and ultraviolet spectroscopy. It is important to note that any one of these methods alone seldom succeeds in suggesting a structure. In our laboratory the usual order of determining spectra is ^{1}H nmr, ^{13}C nmr, and then the others.

The newest technique, ^{13}C nmr, has in a short time become as important in structure work as any of the others. Many problems previously unsolvable have yielded to the additional information provided by ^{13}C nmr. An avalanche of publications in this field has resulted from a need to build up the data banks of empirical correlations, to find new applications, and to point out complementary uses.

One of the features of marine natural products work is the large number of new organic structures, which often do not fit easily into the well-known groups of natural products. For instance, there is a high occurrence of organohalogen compounds, which because of their rarity often do not fit the existing spectral correlations well.

Our objective in this chapter is to provide information on ^{13}C nmr, specifically of compounds isolated from marine organisms. We have omitted certain widely distributed compounds such as sterols. This chapter has two major aims: first to compile pertinent chemical shift data and second to provide a record of assignment methods that we have found valuable in our own work. We have determined the ^{13}C nmr spectra of a number of the compounds that are included.

We hope that as a result of this chapter researchers will publish complete ^{13}C nmr data for each new compound, even if every resonance cannot be assigned. There is a growing tendency to publish only resonances that are pertinent to a particular structural argument. It is frustrating to see a compound in the literature with incomplete ^{13}C nmr data when one has a related structure and wishes to compare the two.

In ^{13}C nmr spectroscopy there is a great difference between assigning resonances to known structures and deriving a structure from resonance data. The latter is extremely difficult, if not impossible. Often, only a few unambiguous structure assignments can be made in a spectrum. It is, of course, relatively easy to count the number of hydrogens on each carbon. The difficulty arises when one wishes to connect the carbons. This is fortunately one of the strong points of ^1H nmr. Spin–spin coupling in ^{13}C nmr spectral interpretation has not progressed to the point where a great amount of connectivity information can be obtained. However, rapid progess is being made in this area.

Probably the most dramatic argument for the use of ^{13}C nmr spectroscopy is that the structures of chondriol (**197**), violacene (**28**), nidifocene (**92**), and chondrocole A (**45**) were revised on the basis of ^{13}C nmr data (*vide infra*). Three of the compounds were shown to have halogens on carbons different from those originally assigned; the fourth contained an unrecognized tetrasubstituted double bond. Demonstrating a tetrasubstituted double bond is difficult by most methods but becomes trivial with ^{13}C nmr.

B. Instrumental Methods

Although it is not our intent to delve into the theory of nuclear magnetic resonance, we believe it would be appropriate to relate some of our experiences with the various instrumental techniques presently available. There are a number of excellent texts that detail carbon assignment techniques. We have found the monographs by Stothers (1972), Gray (1975), Breitmaier and Voelter (1974), and Wehrli and Wirthlin (1976) to be most useful. The latter two are especially useful for instrumental methods and parameters.

One common difficulty in structure determination arises when the amount of compound available for analysis is small. Typical quantities vary from a few milligrams to several grams. The instrumental techniques to be used are circumscribed not only by the amount of material available but also by the number of carbon atoms in the compound. If a few milligrams of compound are available, it is possible to obtain an excellent proton noise decoupled spectrum, but off-resonance spectra require much longer instrument time.

The four techniques that we have found to be of greatest use are described in the following paragraphs.

1. Proton Noise Decoupling

The proton noise decoupled spectrum is the "normal" carbon spectrum. The ^{13}C–^1H multiplets are collapsed to singlets by a second decoupling field. The second field is usually a radiofrequency (rf) oscillator capable of generating a wide range of frequencies. We center the broad-band proton decoupler, operating near maximal power, in the center of the proton spectrum (5δ, 450 Hz at 90 MHz) with high-frequency modulation. Under these conditions the frequency bandwidth is at least 1200 Hz. Carbon nuclei are excited by a short, intense rf pulse from the observed channel transmitter. The duration of the pulse is adjusted to obtain a tip angle of 30°–45°. Under these conditions even quaternary carbons, which have long relaxation times, are observed.

The advantage of this technique is that full, or nearly full, nuclear Overhauser enhancement (NOE) of carbon signals is observed so that a spectrum can be obtained with a very small sample. The spectrum is not complicated by ^{13}C–^{1}H couplings. The number of carbon atoms in the molecule can usually be ascertained by counting the number of "lines" in the spectrum. However, because of variations in the NOE and differences in relaxation times of the carbon atoms, the spectrum cannot be integrated. If carbon resonances overlap, it is difficult to obtain an accurate carbon count. The disadvantage of this technique is that broad-band ^{13}C–^{1}H decoupled spectra do not contain information provided by fully or partially coupled spectra.

2. *Single-Frequency Off-Resonance*

Carbon-13 nuclei are spin–spin coupled to protons and show not only large one-bond couplings ($^{1}J_{CH}$) but also smaller two- ($^{2}J_{CCH}$), three- ($^{3}J_{CCCH}$), and four-bond ($^{4}J_{CCCCH}$) couplings. We will ignore couplings to other nuclei (^{13}C, ^{19}F, ^{31}P, ^{15}N). In a fully coupled spectrum, NOE is lost and the time (number of acquisitions) required to obtain a spectrum is long.

Much of the coupling information can be obtained in a coherent single-frequency off-resonance decoupling (SFORD) experiment. In this technique the proton decoupler power is reduced and the decoupling frequency is moved "off" the proton region (off-resonance). The net effect is reduced coupling constants. Methyl groups appear as compressed quartets, methylenes as compressed triplets, and methine carbons as compressed doublets. True coupling constant information is lost. Nuclear Overhauser enhancement is retained so that off-resonance spectra can be obtained in a reasonable time period. Off-resonance spectra require four to eight times the number of acquisitions of proton noise decoupled spectra.

SFORD spectra are usually uncomplicated by long-range couplings. One exception occurs when the protons attached to a carbon are magnetically nonequivalent. In this case the observed carbon pattern deviates from that of an isolated A_nX pattern. Hagaman (1976) discussed the significance of geminal proton nonequivalence in structure elucidation.

We use a decoupler power of 3350 Hz and a decoupler offset of -4δ (in the proton region). Under these conditions, the methyl reduced coupling constants are of the order of 20 Hz, and methine reduced couplings are of the order of 35 Hz. To determine the decoupler power level, we use methanol, which has $^1J_{CH} = 141$ Hz. The relationship between reduced coupling constants, $^rJ_{CH}$, and one-bond couplings, $^1J_{CH}$, is given by

$$^rJ_{CH} = {}^1J_{CH}\Delta\nu(\gamma H_0/2\pi)^{-1} \quad (1)$$

where $\Delta\nu$ is the frequency difference, in hertz, between the resonance position of a given proton signal and the decoupler offset, and $\gamma H_0/2\pi$ is the power setting, in hertz, of a decoupler (Archer *et al.*, 1970; Wehrli and Wirthlin, 1976, pp. 66–69). Using $^1J_{CH}$ of methanol (141 Hz) and $\Delta\nu = 66.06$ Hz (at 90 MHz and -4δ offset) allowed us calibrate the power and modulation controls of a Bruker WH-90D spectrometer.

When we use $^rJ_{CH}$ as a diagnostic indicator, the power setting is increased to 5200 Hz and the decoupler offset is changed to 0δ.

3. Specific Proton—Carbon Decoupling

Proton magnetic resonance offers the advantage of proton–proton couplings. By multiplet pattern analysis and spin–spin decoupling experiments, it is often possible to connect a chain of protons. Once this proton chain is constructed, the next problem is the placement of electronegative atoms. The proton chemical shifts of protons in the environment of an electronegative atom tend to overlap. This is especially true for chloro- and bromomethylenes and chloro- and bromomethines. This overlap of chemical shifts frequently prohibits absolute placement of chlorine, bromine, or oxygen atoms onto the carbon backbone of a new molecule.

Additional structure information about the bonded electronegative atom can be obtained by direct correlation of the carbon atom and its attached proton. This can be done by a technique called specific proton–carbon decoupling. Specific proton–carbon decoupling experiments, also referred to as single-frequency on-resonance decoupling, are instrumentally similar to sford experiments. The decoupler offset is centered on the proton resonance, and the spectrum is recorded at low decoupler power (and low modulation). The power level is reduced to a level just sufficient to collapse the ^{13}C–^1H spin–spin coupling. The result is a singlet for the carbon bonded to the irradiated proton while the remainder of the spectrum appears as if it were a single-frequency off-resonance spectrum.

In this experiment the ^{12}C–^1H signal is irradiated when in fact the ^{13}C–^1H signal should be brought into resonance. However, ^{13}C–^1H resonances are not greatly shifted from $^{12}C-^{1}H$ signals, and this irradiation frequency error is rarely a complication.

Specific decoupling experiments require not only that the proton chemical shift be known but also that the exact position of the proton decoupler be known. We have found it advantageous to use the TMS quartet as an internal standard. The position of the proton decoupler is adjusted so that this signal collapses to a singlet and this frequency is recorded. The chemical shift of the proton signal to be decoupled is then added to this value to obtain the exact irradiation frequency.

4. Gated Decoupling

Additional information can be obtained from a fully coupled spectrum. The simplest way to obtain such a spectrum is to turn off the decoupler and record the spectrum. In this manner true $^{x}J_{CH}$ coupling constants can be determined. However, a significant loss of sensitivity, resulting from loss of NOE, requires extended periods of data acquisition. Some of the NOE can be regained if decoupling is permitted.

In a gated decoupling experiment the decoupler is turned (gated) off before the pulse and turned (gated) on after data acquisition. The resulting spectrum is identical to one obtained if the decoupler were off during the entire experiment.

Such fully coupled spectra allow one to determine $^{1}J_{CH}$ as well as $^{2}J_{CCH}$ and $^{3}J_{CCCH}$. The sign of $^{1}J_{CH}$ is positive, whereas $^{2}J_{CCH}$ is both positive and negative; $^{3}J_{CCCH}$, which is often equal to or greater than $^{2}J_{CCH}$, is usually positive (Stothers, 1972, pp. 331–348). Because of the uncertainty of the sign of two-bond coupling constants, computer simulation of the multiplet pattern is necessary if coupling constants are to be determined.

In the fully coupled spectrum of acetone the methyl group appears as a quartet (J = 125.5 Hz) and the carbonyl as a septet (J = 5.5 Hz). The complexity of these spectra increases drastically as the number of carbon atoms increases. We have found it virtually impossible to completely interpret sesquiterpene gated decoupled spectra. However, certain carbons sometimes stand out because of the number of hydrogen atoms two, three, or four bonds away.

II. MONOTERPENES

The first report of polyhalogenated monoterpene constituents of a marine organism implicated the digestive gland of the sea hare *Aplysia californica* (Faulkner and Stallard, 1973). Sea hares are macrophagous herbivores, and hence it was soon afterward that reports appeared of the isolation of these and other monoterpenes from their algal diet, in this case the red alga *Plocamium cartilagineum* (Stallard and Faulkner, 1974). In addition to *P. cartilagineum*, *P. violaceum*, *P. costatum*, *P. coccineum*, *P. mertensi*, and *Chondroccus hornemanni* have been shown to contain monoterpenes.

For the sake of simplicity, we have divided the monoterpenes into two broad categories: linear and cyclic. The linear monoterpenes are arranged by increasing unsaturation, whereas the cyclic compounds are further subdivided into monocarbocyclic, monoheterocyclic, and bicyclic.

A. Linear

The halogen content of the monoterpenes can usually be determined by analysis of the mass spectrum. For chlorine- and bromine-containing compounds, one major difficulty in structure analysis is the correct placement of the halogens. The location of the halogens in crystalline compounds has most frequently been ascertained by single-crystal X-ray diffraction studies. However, careful analysis of the ^{13}C nmr spectrum coupled with ^{1}H nmr studies has often resulted in unambiguous structure assignments. Empirically derived substituent effects in linear and branched alkanes, summarized in Table 1, demonstrated a small but consistent carbon chemical shift difference for chlorine and bromine (Wehrli and Wirthlin, 1976, pp. 36–39). Examination of Table 1 reveals that primary chlorines resonate approximately 11 ppm downfield from the corresponding bromine carbon. Likewise, secondary carbons bearing

TABLE 1
Substituent Parameters for Replacement of a Hydrogen by X in Linear and Branched Alkanes[a]

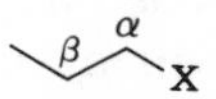

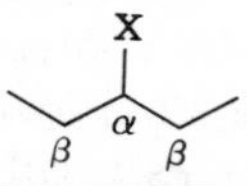

X	α		β	
	n	Iso	*n*	Iso
F	+68	+63	+9	+6
Cl	+31	+32	+11	+10
Br	+20	+25	+11	+10
I	−6	+4	+11	+12

[a] Wehrli and Wirthlin (1976, pp. 36–39).

chlorine resonate approximately 7 ppm toward lower field than the corresponding brominated carbon.

Further support for an approximately 11 ppm difference between primary carbons bearing chlorine or bromine can be seen by the C-1 chemical shifts of olefins **1–4** (Crews and Kho-Wiseman, 1977). In the 2-butenes **1** and **2**, the chlorine carbon resonates 12.2 ppm downfield from the corresponding bromine carbon whereas, in the 3-methyl-2-butenes **3** and **4**, the chemical shift difference is 11.6 ppm. In both sets of compounds (**1** and **2** or **3** and **4**) the β carbon chemical shifts are nearly identical and indicate that the β substituent effects of these electronegative elements are equal. Thus, chemical shift and multiplicity data of the α carbon should serve to distinguish between these halogens.

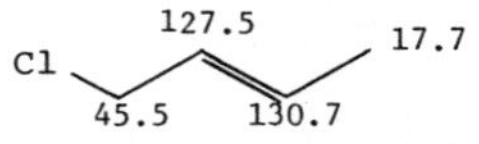

127.5 17.7
Cl
45.5 130.7

1

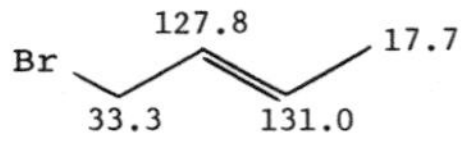

127.8 17.7
Br
33.3 131.0

2

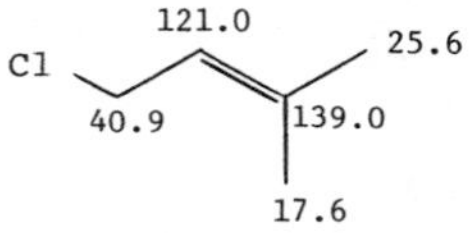

121.0 25.6
Cl
40.9 139.0
17.6

3

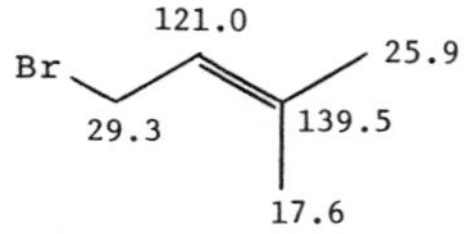

121.0 25.9
Br
29.3 139.5
17.6

4

However, carbon chemical shifts are only numbers and are nearly meaningless without the chemical shifts of known halogenated model compounds with which to compare them. Although close model compounds may be lacking, it is still possible to calculate, with reasonable accuracy, the desired carbon chemical shift using a suitable nonhalogenated system. The model compound chosen need not contain the entire carbon skeleton of the unknown compound but merely the portion of interest. Several examples involving nonhalogenated model compounds to calculate carbon–halogen chemicals shifts will be given. Table 2 lists the carbon chemical shift assignments of 18 halogenated monoterpenes (**5–22**) obtained from marine organisms.

The use of substituent constants to calculate carbon chemical shifts for compounds **5** and **6** will be illustrated. Monoterpene **23** serves as a suitable model system, having the unsaturation correctly placed and electronegative groups at C-2 and C-6. The β- and γ-substituent effects are nearly equal for electronegative groups, and small differences between them will be ignored. Thus, given the C-3 chemical shift of compound **23**

TABLE 2

Carbon Chemical Shifts of Linear Monoterpenes[a]

Compound	C-1	C-2	C-3	C-4	C-5	C-6	C-7	C-8	C-9	C-10	Reference[b]
5		67.3	71.1			74.0				39.0	Burreson *et al.* (1975b)
6			64.5								Burreson *et al.*)1975b)
7		71.8	64.9			71.9				40.2	Burreson *et al.* (1975b)
8		68.0	65.8								Burreson *et al.* (1975b)
9		71.6	70.7								Burreson *et al.* (1975b)
10		68.0	71.4								Burreson *et al.* (1975b)
11							62.7	60.2			Burreson *et al.* (1975b)
12	27.6	66.9	138.0	127.5	67.9	70.4	135.0	122.5	41.5	25.7	Stierle and Sims (1978)
13	49.7	68.8	133.5	130.5	67.4	70.2	134.8	122.6	37.3	25.4	Stierle and Sims (1978)
14	49.6	68.9	133.5	130.3	67.7	71.4	138.6	110.3	37.3	25.3	Crews (1977)
15	27.6	66.9	138.0	127.4	67.6	71.6	138.6	110.1	41.5	25.5	Stierle and Sims (1978)
16	27.8		137.1	127.6	67.8		137.9	110.6	41.5	28.0	Crews (1977)
17	115.6	129.9	56.4	41.4	134.0[c]	136.0[c]	71.3	37.0	16.2	14.4	Kazlauskas *et al.* (1976b)
18	108.2	135.4	133.0	127.9	69.0	71.6	138.7	110.0	19.5	25.4	Crews (1977)
19	108.2		133.2	128.5	71.9		137.6	110.5	19.1	28.4	Crews (1977)
20	143.9	137.2	122.5	134.0	69.5	71.5	139.5	116.3	189.3	24.6	Crews and Kho (1974)
21	21.8[c]	140.0[d]	128.1[e]	31.5[f]	34.7[f]	142.5[d]	126.8[e]	117.6[g]	20.2[c]	112.8[g]	Burreson *et al.* (1975b)
22	130.8	141.6	191.2	129.3	133.0	68.8	61.9	46.1	16.2	26.7	Stierle and Sims (1978)

[a] The δ values are in parts per million downfield from TMS in $CDCl_3$ solution.
[b] References for carbon data ^{13}C nmr.
[c–g] Values within a row may be interchanged.

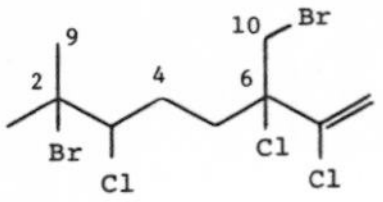

5

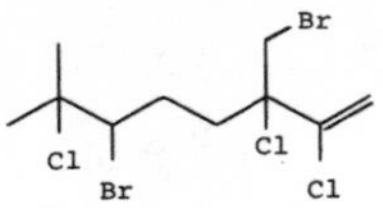

6

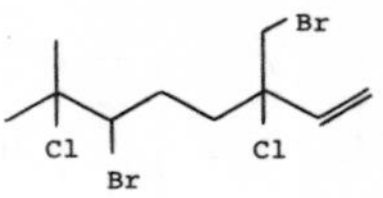

7

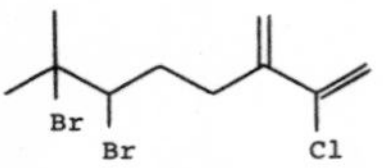

8

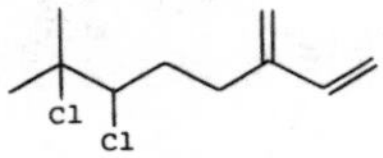

9

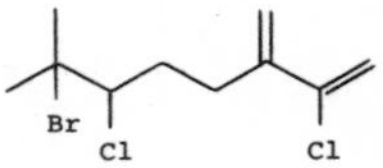

10

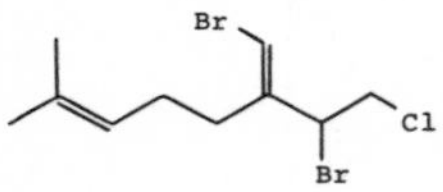

11

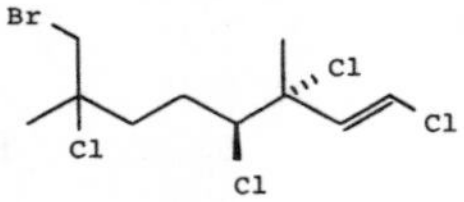

12

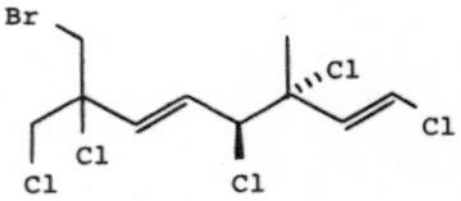

13

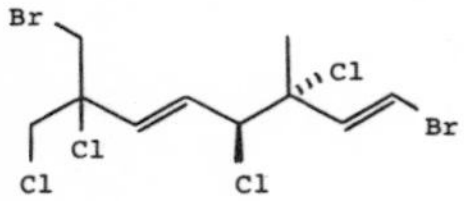

14

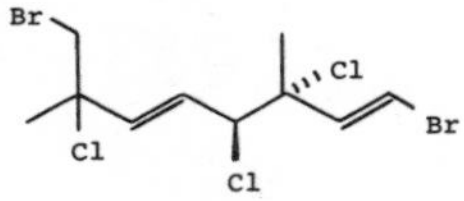

15

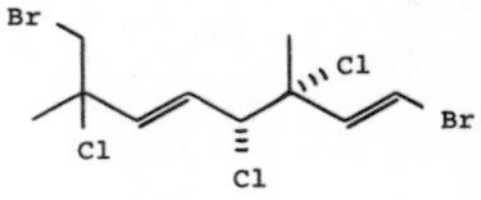

16

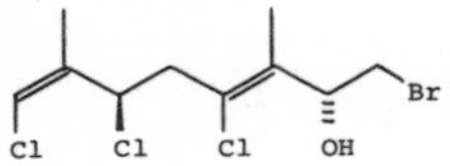

17

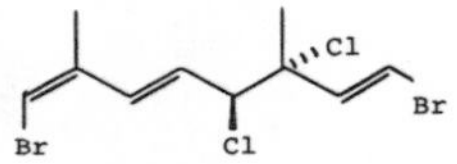

18

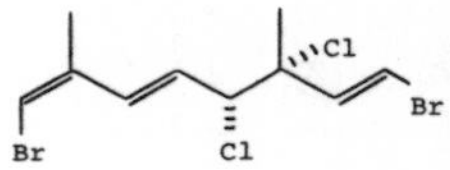

19

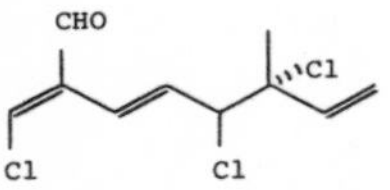

20

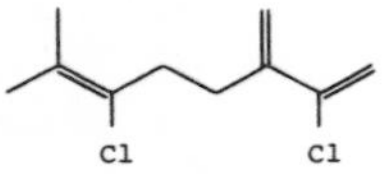

21

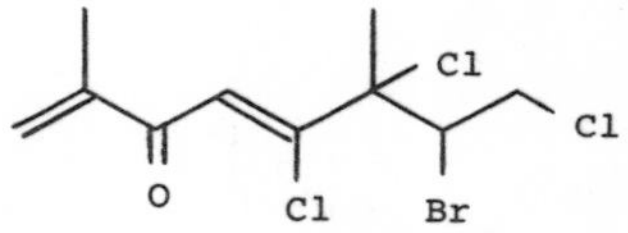

22

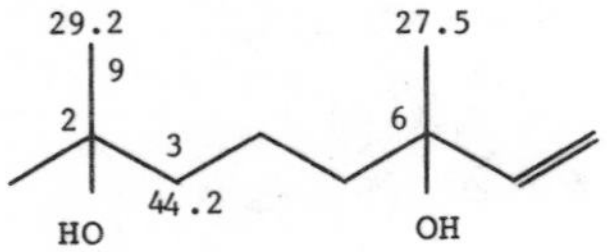

23

(44.2 ppm) and the halogen substituent parameters given in Table 1, the predicted chemical shift of C-3 of the chlorinated carbon in compound **5** is 44.2 + 32 ppm, or 76.2 ppm, structure **5a**. This value differs from the experimentally determined value by 5.1 ppm. Similar calculations predict a C-3 chemical shift of 44.2 + 25 ppm, or 69.2 ppm, for the bromine-bearing carbon in compound **6**. The calculated chemical shift deviates from the experimentally determined value by 4.3 ppm, as depicted on structure **6a**.

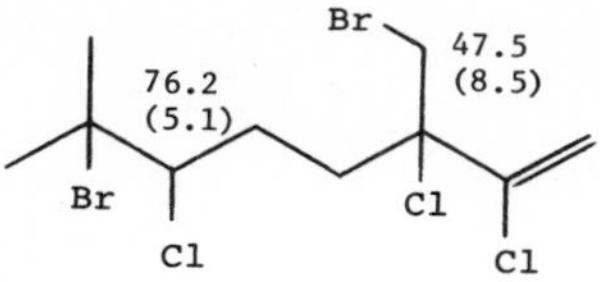

Br 47.5 (8.5)
76.2 (5.1)
Br Cl Cl Cl

5a

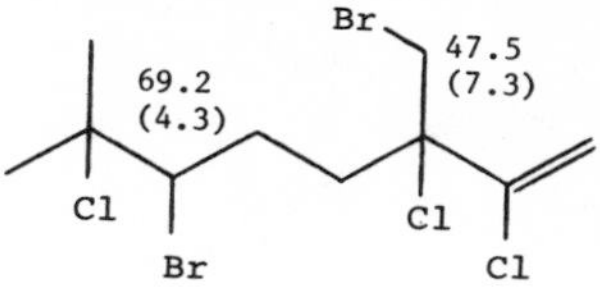

Br 47.5 (7.3)
69.2 (4.3)
Cl Cl Br Cl

6a

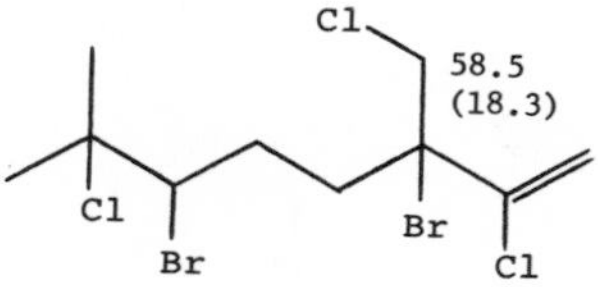

Cl 58.5 (18.3)
Cl Br Br Cl

6b

The chemical shift of C-10 in monoterpene **23** (27.5 ppm) serves as a template to calculate the chemical shift of C-10 in compound **6**. Calculations predict a chemical shift of 27.5 + 20 ppm = 47.5 ppm for a bromo substituent, structure **6a**, and a value of 27.5 + 31 ppm = 58.5 ppm if C-10 bears a chlorine atom, structure **6b**. The experimental value of 40.2 ppm clearly excludes a chlorine substituent at C-10 in compound **6**.

The substituent constants in Table 1 were not determined using vicinal electronegative groups. The average value of the deviation between experimental and calculated chemical shifts is 6 ppm. It is reasonable that bulky vicinal groups are mutually shielding, and a correlation term of −6 ppm should be added to such calculations. Similar but smaller correction terms are required for carbon chemical shift calculations in 1,2-dimethylcyclohexane systems.

By the use of the substituent constants in Table 1, model compounds (compound **24** for C-5, compound **23** for C-9) and the vicinal correction term of −6 ppm, the chemical shift values for C-5 and C-9 of compound **12** were determined. The results are shown on structure **12a**. Calculations for the alternate bromine and chlorine substitution pattern at C-5 and C-9 are shown on structure **12b**. The values in parentheses are the deviation from the experimental value. These calculations eliminate structure **12b** and support the proposed structure **12**.

Compounds **12** and **13** contain a terminal vinyl chloride with a chemical shift for the terminal olefinic carbon of 122.5 ppm. Compounds **14–16**, **18**, and **19** contain a terminal vinyl bromide with an average chemical shift of 110.3 ppm. This difference is large and should serve to distinguish between the two functional groups.

Compounds **12**, **13**, **14**, **15**, and **18** have the same relative stereochemistry at C-5 and C-6, which is opposite to that of compounds **16** and **19**.

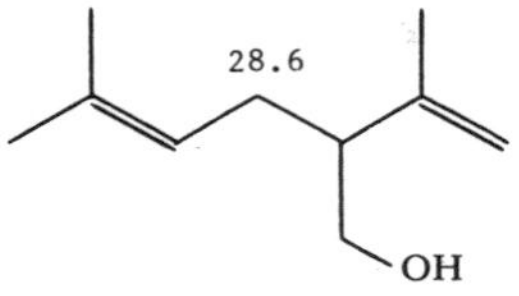

Chiral carbons C-5 and C-6 do not differ sufficiently in chemical shift to permit distinctive assignment (see Table 2). The C-10 methyl group, however, is shifted in the two sets of compounds. The cause of this steric compression shift can be explained by examination of the molecular models of compounds **15** and **16**. Newman projections down the C-5–C-6 bond are shown in Fig. 1. The conformation in which the two bulky R groups are trans [conformer (a)] should be of lower energy. In compound **15**, conformer (a) has the methyl (C-10) in a sterically congested environment. However, in the lowest-energy conformer (a) of compound **16**, the methyl group (C-10) is in a less hindered environment and hence, relative to compound **15**, the methyl group (C-10) of compound **16** should resonate at lower field. In compound **15** C-10 resonates at 25.5 ppm and is shifted downfield (+2.5 ppm) to 28.0 ppm in compound **16**.

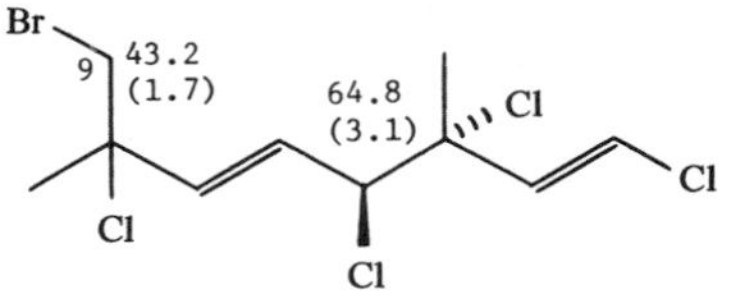

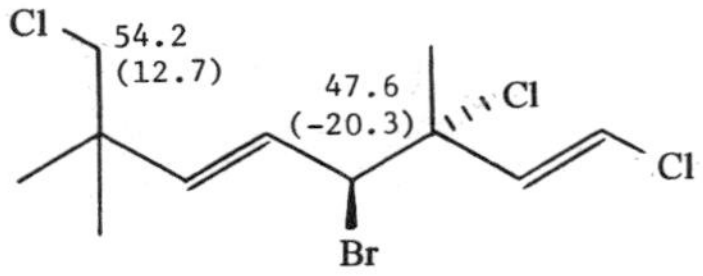

The coupling constants for costatol (**17**), determined in a gated decoupling experiment, are indicated on structure **17a**. The two methyl groups, C-9 and C-10, have identical one-bond ^{13}C–^{1}H coupling constants of 129 Hz. The highest-field off-resonance quartet centered at 14.4 ppm shows long range ($^{3}J_{CCCH}$) coupling to the proton on C-7. Each line of the quartets is split into a doublet (J = 4 Hz). This signal therefore originates from C-10. The C-9 methyl is a quartet that shows long-range coupling to two protons and appears as a quartet of triplets (J = 129, 6, 6 Hz). Carbon C-9 is coupled to the proton on C-1 and to the proton on C-3. The gated decoupled spectrum contains two methylene triplets. The resonance at 41.4 ppm shows one-bond ^{13}C–^{1}H coupling of 131 Hz, whereas the signal of 37.0 ppm shows one-bond coupling of 152 Hz. This increased coupling indicates the attachment of an electronegative element at this center. Thus, C-8, a bromomethylene, is assigned to the resonance at 37.0 ppm whereas C-8, a methylene bonded to two carbon atoms, is assigned to the 41.4 ppm signal.

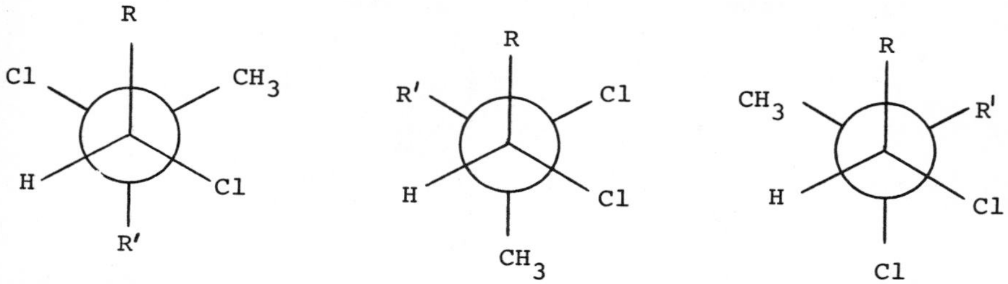

15 $R = C_5H_7BrCl$
$R' = C_2H_2Br$

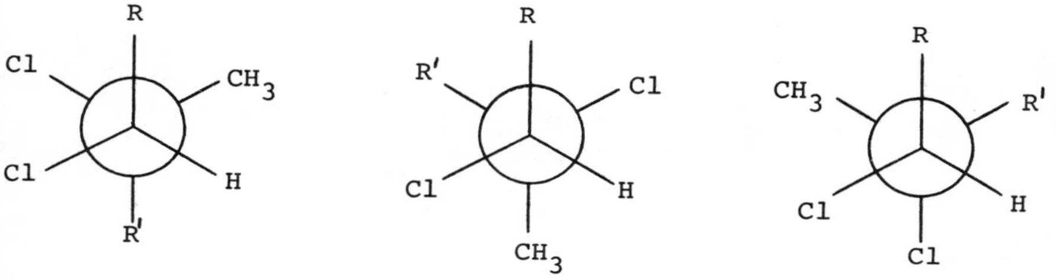

16

Fig. 1. Newman projections along the C-5–C-6 bond for conformers of compounds **15** and **16**.

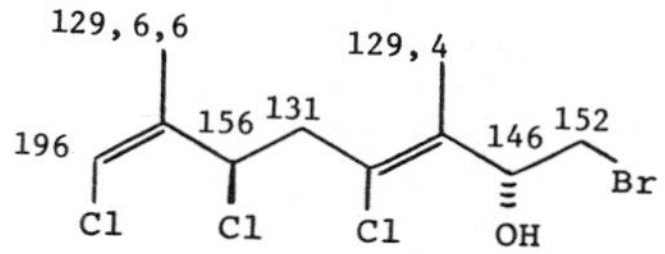

17a

B. Cyclic

1. Monocarbocyclic

The monocyclic terpenes are subdivided into those with normal head-to-tail linkage and those with a rearranged carbon skeleton. Table 3 lists the ^{13}C nmr chemical shifts of the unrearranged polyhalogenated cyclic monoterpenes (**25–35**).

a. *Normal Head-to-Tail.* The proposed structure of the first reported halogenated cyclic monoterpene, violacene (**28**), isolated from *Plocamium violaceum*, later proved to be incorrect. Mynderse and

TABLE 3

^{13}C nmr Chemical Shifts of Monocyclic Monoterpenes[a]

Compound	C-1	C-2	C-3	C-4	C-5	C-6	C-7	C-8	C-9	C-10	Reference
25	44.8	56.4	41.4	66.8	70.5	55.0	29.8	26.1	139.4	108.0	González *et al*. (1977)
26	41.4	56.1	39.3	60.5	70.1	47.9	33.4	30.4	138.2	116.8	Higgs *et al*. (1977)
27	43.5	67.1	40.6	55.2	70.6	52.7	26.1	20.1	140.6	119.1	Norton *et al*. (1977)
28	41.6	63.9	37.9[b]	58.9	71.0	48.3	38.7[b]	27.0	135.1	118.7	Higgs *et al*. (1977)
28	42.0	64.1	38.3[b]	59.0	71.3	48.8	38.8[b]	27.4	135.4	119.5	Crews and Kho (1975)
29	42.0	64.6	39.1[b]	51.1	70.8	48.5	40.3[b]	27.4	135.3	119.1	Higgs *et al*. (1977)
29	42.1	64.6	40.4[b]	51.2	70.9	48.6	39.2[b]	27.5	135.4	119.2	Stierle and Sims (1978)
30	40.7	64.2	33.7[b]	124.4	132.4	36.7[b]	48.9	25.6	135.3	119.1	Stierle and Sims (1978)
31	40.7	64.1	33.9[b]	124.7	133.0	36.7[b]	37.2	25.6	135.3	119.1	Stierle and Sims (1978)
32	43.4	67.9	32.7[b]	32.9[b]	143.2	44.6	111.4	26.1	136.3	119.1	Stierle and Sims (1978)
33	43.5	63.2	41.5[b]	61.5	142.3	41.1[b]	115.3	26.5	134.1	120.3	Stierle and Sims (1978)
34	43.3	65.1	43.3[b]	58.2	140.9	45.5[b]	113.6	26.2	113.3	120.7	Stierle and Sims (1978)
35	43.3	65.5	44.1[b]	49.3	140.7	45.2[b]	116.1	26.2	133.5	120.6	Stierle and Sims (1978)

[a] The δ values are in parts per million downfield from TMS in $CDCl_3$ solution.
[b] Signals within a row may be reversed.

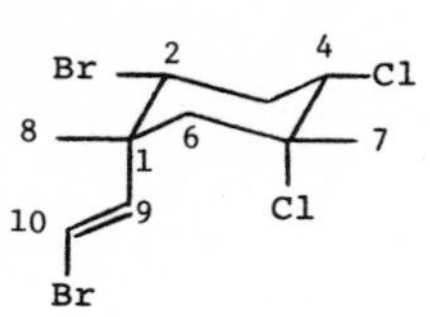

25 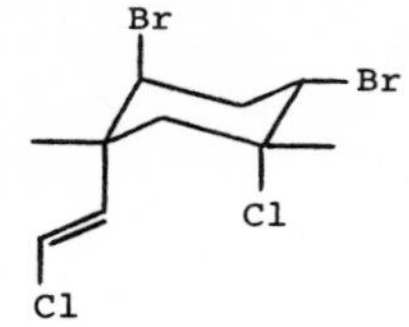 26 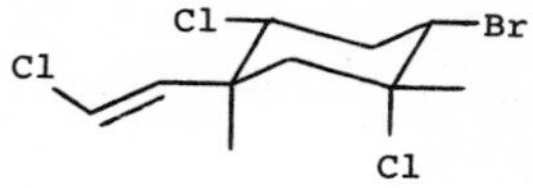 27

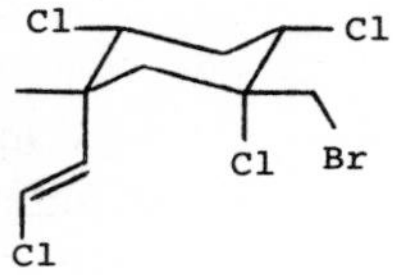

28 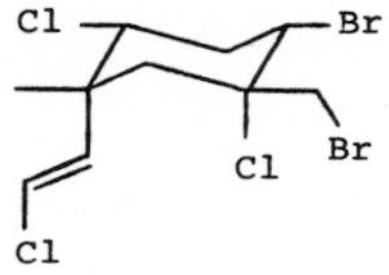 29 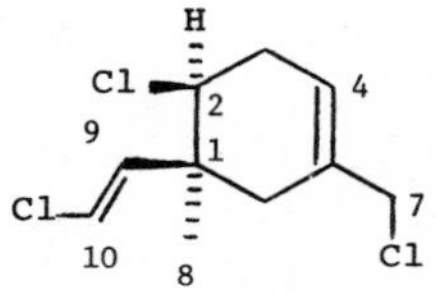 30

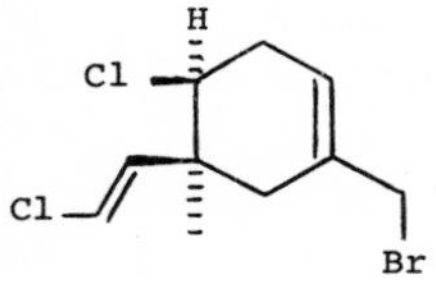

31 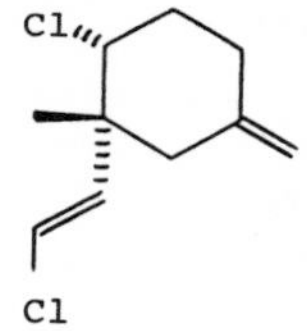 32 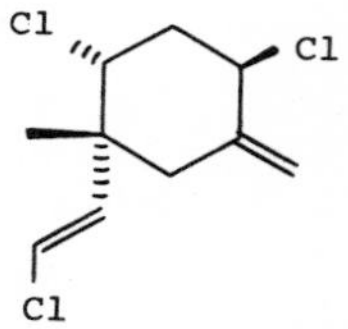 33

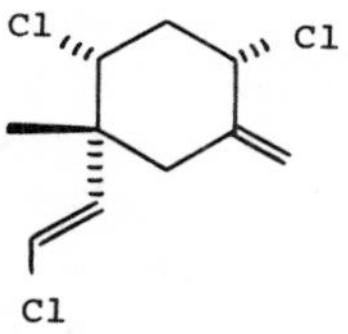

34 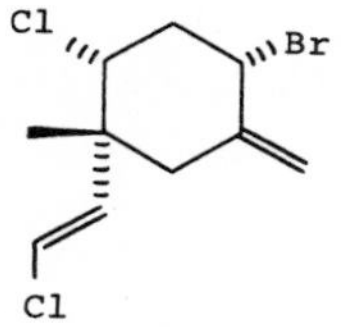 35

Faulkner (1974) proposed structure **28a** for violacene on the basis of a chromous sulfate dehalogenation reaction, ozonolysis, and detailed analysis of the 220 MHz ^{1}H nmr spectra. Support for the structure was obtained by mass spectral fragmentation analysis, which displayed an $M^+ - CH_2Cl$ ion cluster but no cluster for $M^+ - CH_2Br$. This seemed to confirm the presence of a chlorine atom at C-7. However, during the three years since the structure of violacene was proposed, a great deal of knowledge about carbon, and especially carbon–halogen, chemical shifts has been gained.

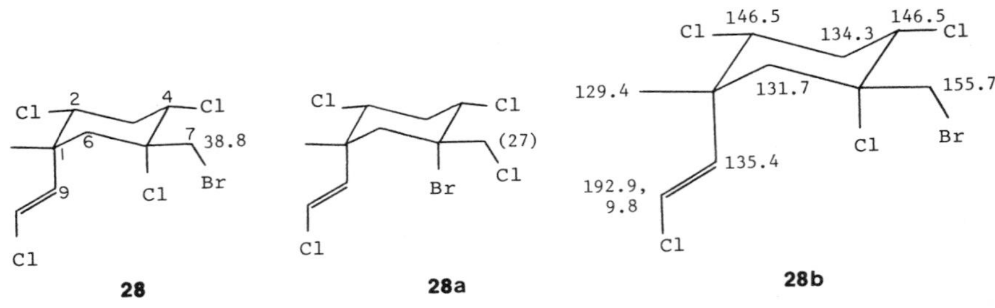

The ^{13}C nmr spectrum of violacene exhibits three off-resonance methylene triplets at 38.3, 38.8, and 48.8 ppm. Carbon C-6, which is adjacent to two quarternary centers, should resonate at lowest field and is assigned the 48.8 ppm resonance. Support for this assignment is found in compounds **25–27**, which do not contain a halomethylene group but show a similarly placed low-field methylene resonance. The one-bond coupling constant of this carbon is 131.7 Hz, a value that is consistent with attachment of two carbon atoms to this methylene center. Thus, one of the resonances near 38 ppm is the halomethylene signal for C-7. This resonance occurs at too high a field for a chloromethylene but is reasonable for a bromomethylene.

In 1-methylcyclohexanol (**36**), the methyl group absorbs at 29.8 ppm. Given this methyl resonance as a model and the substituent parameters from Table 1, the calculated chemical shift for the chloromethylene carbon of structure **36a** is 54.8 ppm (29.8 + 31.6 ppm). For the bromomethylene carbon of structure **36b** a resonance of 43.8 ppm (29.8 + 20.6 ppm) is predicted. The equatorial halogen at C-4 of violacene (**28**) will result in a steric compression shift (high field) of C-7. This implies that the calculated chemical shifts are maximal values; this clearly excludes a chloromethylene part structure in violacene. Partly on the basis of ^{13}C nmr data, the structure of violacene was revised to **28** (van Engen *et al.*, 1977). One-bond ^{13}C–^{1}H coupling constants (hertz) for violacene are shown in **28b** (Crews and Kho, 1975). Carbon C-10 displays two-bond ^{13}C–^{1}H coupling to the protons on C-9.

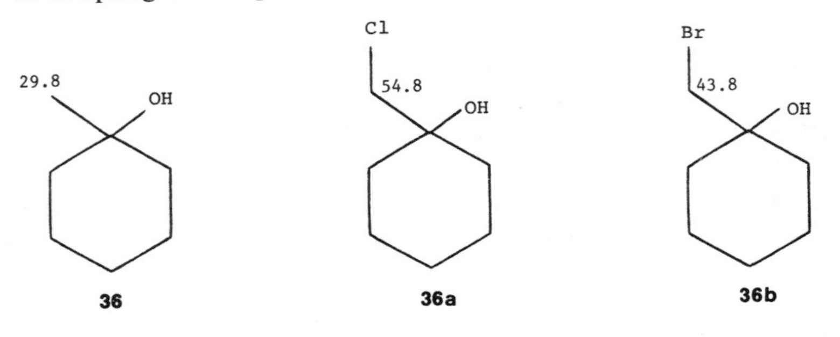

A similar bromomethylene from chloromethylene chemical shift differentiation is seen in compounds **30** and **31** (Table 3). The bromomethylene carbon, C-7, of compound **31** resonates at 37.2 ppm, whereas the environmentally comparable chloromethylene carbon, C-7, of compound **30** resonates at 48.9 ppm, a chemical shift difference of 11.7 ppm.

Compounds **25** and **26** are epimeric at C-2. In compound **25** the C-2 bromine atom and C-8 (methyl) are *trans* (diequatorial), resulting in a steric compression shift to higher field of C-8. In compound **26**, the C-2 bromine is axial, which relieves the steric compression of C-8 (methyl). The C-8 methyl group of compound **26** (30.4 ppm) therefore resonates at lower (4.3 ppm) field than C-8 of compound **25** (26.1 ppm).

The C-8 methyl group of violacene (**28**) resonates at 27.4 ppm. Epimerization of C-1 leads to compound **27**. In this compound the C-2 chlorine is equatorial and the methyl (C-8) is axial, but the axial methyl (C-8) now experiences 1,3 interactions with the hydrogens on C-3 and the chlorine on C-5 and is shifted (−7.3 ppm) to higher field, resonating at 20.1 ppm.

b. *Rearranged.* The 1,5-dimethyl-2-ethylidenecyclohexane ring system probably results from vinyl migration in the head-to-tail monocyclic monoterpenes. The ^{13}C nmr chemical shifts for the rearranged monocyclic compounds **37–42** are shown in Table 4.

The normal monocyclic monoterpenes can be separated from the rearranged monocyclic monoterpenes by a ^{13}C nmr off-resonance singlet at 42 ± 3 ppm. This signal, assigned to C-1 of the head-to-tail monocyclic series, is absent in the rearranged compounds.

Compounds **37** and **38** differ by a chlorine or bromine atom at C-1, but the carbon chemical shift difference at this center is only 0.3 ppm. The chemical shift of quaternary carbons is an unreliable indicator of the nature of the attached electronegative group. However, carbon chemical shifts can be used to differentiate bromine from chlorine on tertiary carbons (compare C-4 of compounds **37** or **38** versus **39** or C-4 of compound **40** versus **41**).

Structure **40b** lists carbon–proton coupling constants in hertz for plocamene-B (**40**). Increased one-bond coupling constants were used to assign halogen-bearing carbons (Crews and Kho, 1975). The olefinic carbons C-9 and C-10 show two-bond coupling. Carbon C-9 is coupled to the proton on C-10 by 13 Hz, whereas C-10 shows two-bond coupling to the C-9 proton of 9 Hz.

TABLE 4

^{13}C nmr Chemical Shifts of Rearranged Monocyclic Monoterpenes[a]

Compound	C-1	C-2	C-3	C-4	C-5	C-6	C-7	C-8	C-9	C-10	Reference
37	67.8	52.1	36.4	58.6	62.5	53.1	34.0	32.7	131.8	120.4	Norton *et al*. (1977)
38	67.5	53.7	36.4	58.0	71.2	57.2	33.7	28.1	131.3	121.1	Higgs *et al*. (1977)
38	67.4	53.5	36.3	57.1	71.1	57.1	33.6	28.0	131.2	120.9	González *et al*. (1977b)
39	67.4	52.7	35.3	65.2	71.2	57.4	32.1	28.0	131.4	120.8	Higgs *et al*. (1977)
40	123.8	129.8	34.5	64.1	69.3	48.5	30.3	18.4	130.3	117.7	Crews and Kho (1975)
41	124.5	130.3	35.8	57.4	69.2	48.2	32.1	18.6	130.2	117.7	Higgs *et al*. (1977)
42	142.6	45.2	40.1	58.5	71.9	49.6	32.5	112.7	132.8	119.2	Higgs *et al*. (1977)

[a] The δ values are in parts per million downfield from TMS in $CDCl_3$ solution.

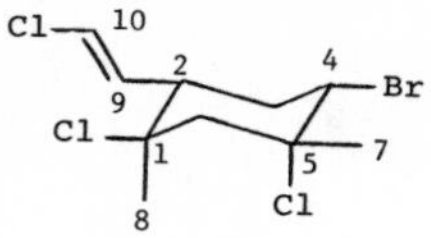

Cl 10 2 4 Br 9 Cl 1 5 7 8 Cl

37

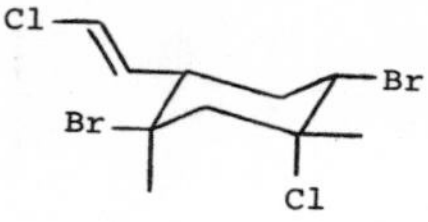

Cl Br Br Cl

38

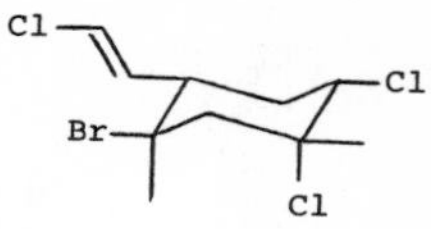

Cl Cl Br Cl

39

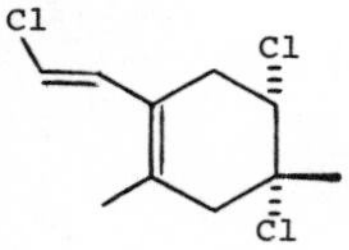

Cl Cl Cl

40

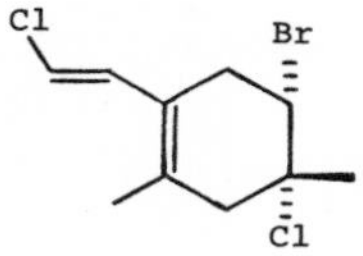

Cl Br Cl

41

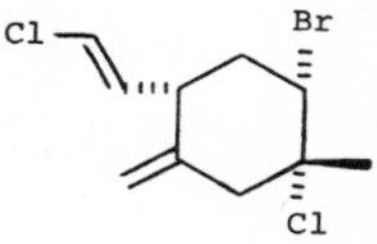

Cl Br Cl

42

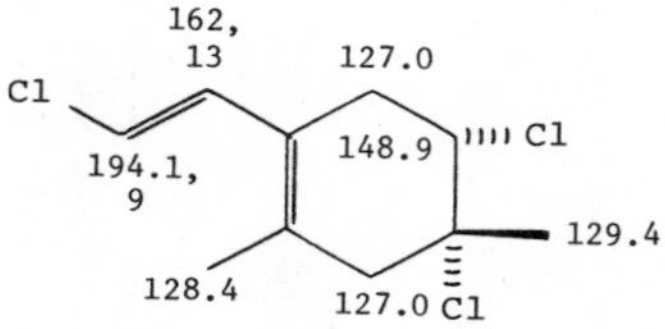

162, 13 127.0 Cl 148.9 Cl 194.1, 9 129.4 128.4 127.0 Cl

40b

2. *Monoheterocyclic*

Costatone (**43**), a linear monoterpene cyclized to a stable hemiketal, was isolated from *Plocamium costatum* collected at Robe, South Australia, by Stierle *et al.* (1976), who also isolated costatolide (**44**), a lactone, as a natural product and as a degradation product of costatone. The ^{13}C nmr chemical shift assignments are shown in Table 5.

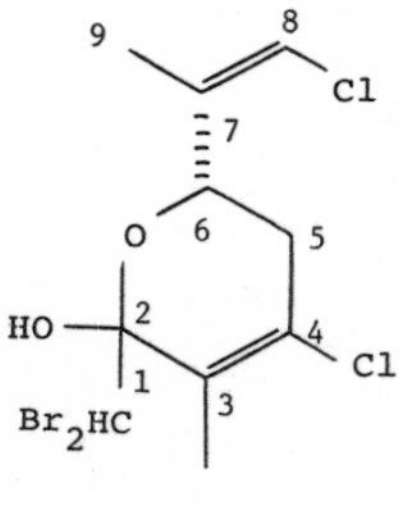

9 8 Cl 7 O 6 5 HO 2 4 Cl 1 3 Br_2HC

43

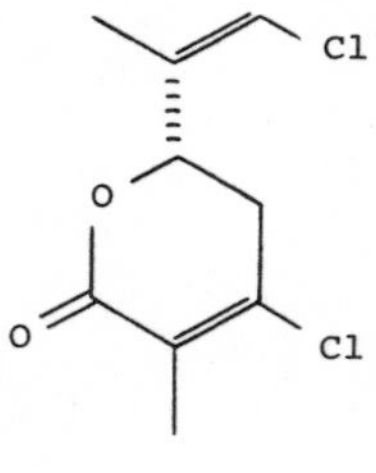

Cl O O Cl

44

The direct attachment of a polar substituent has a profound and predictable effect on ^{13}C–^{1}H one-bond coupling constants, $^{1}J_{CH}$. Some $^{1}J_{CH}$

TABLE 5

^{13}C nmr Chemical Shifts of Monoheterocyclic Monoterpenes[a]

Compound	C-1	C-2	C-3	C-4	C-5	C-6	C-7	C-8	C-9	C-10	Reference
43	51.6	97.0	132.0	137.2	35.0	67.6	125.7	113.4	16.0	13.6	Stierle *et al.* (1976)
43	51.6	97.1	132.0	137.1	35.1	67.5	125.6	113.4	16.0	13.8	Kazlauskas *et al.* (1976a)
44	—	163.4	134.5	145.6	34.9	73.6	125.0	115.3	16.0	13.8	Stierle *et al.* (1976)

[a] The δ values are in parts per million downfield from TMS in $CDCl_3$ solution.

values for vinyl systems are listed in Table 6. Although it is not possible to differentiate among the halogens by $^{1}J_{CH}$ values, it is possible to assign halogen-bearing olefins on the basis of an increased $^{1}J_{CH}$ value. The $^{1}J_{CH}$ values obtained for costatone (**43**) are listed on structure **43a**. The coupling constant for C-8 of 195 Hz and its chemical shift of 113.4 ppm are in agreement for attachment of a polar (Cl) substituent to this carbon.

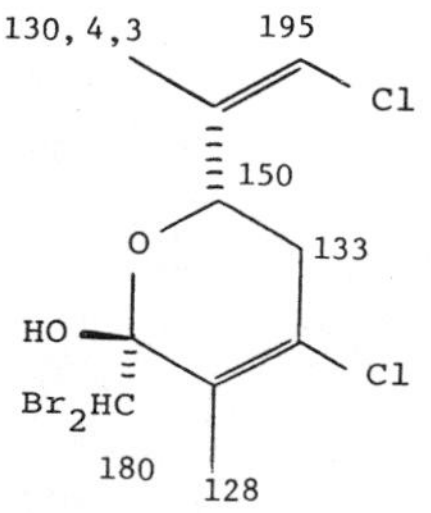

43a

The same effect of $^{1}J_{CH}$ in substituted methane systems has been observed; Table 7 lists these values. The observed $^{1}J_{CH}$ for C-1, 180 Hz, is in agreement with structure **43**, in which this carbon atom contains two bromine substituents. The one-bond coupling constants for C-5 (133 Hz), C-9 (128 Hz), and C-10 (130 Hz) are near the sp^3 theoretical value of 125 Hz, indicating that these carbon atoms are not substituted with electronegative atoms. Carbon C-6, which has an increased $^{1}J_{CH}$ constant of 150 Hz, must have an attached electronegative element.

The gated decoupled spectrum of costatone (**43**) also reveals that the signal at 16.0 ppm is a quartet ($^{1}J_{CH}$ = 130 Hz) of doublets ($^{x}J_{CH}$ = 4 Hz) of doublets ($^{x}J_{CH}$ = 3 Hz). This signal, by virtue of its multiplicity and high-field chemical shift, can be assigned to either C-9 or C-10. If this

TABLE 6

One-Bond ^{13}C–^{1}H Coupling Constants in Vinyl Systems[a]

X	$^{1}J_{CH}$ (Hz)
H	156.2
C	157.4
Cl	194.9
Br	196.6
I	190.9

[a] Stothers (1972, pp. 332–348).

TABLE 7

One-Bond ^{13}C–^{1}H Coupling Constants of Substituted Methanes[a]

X	CH_3X	CH_2X_2
CH_3	125.0	—
Cl	148.6	176.5
Br	150.5	177.7
CH_3O	139.6	161.8
$CH_2 = C(CH_3)$	126	—

[a] Stothers (1972, pp. 332–348).

resonance is assigned to C-10, the small coupling is ${}^{4}J_{CCCCH}$. On the assumption that ${}^{4}J_{CCCCH}$ values are smaller than ${}^{3}J_{CCCH}$ values, which in aliphatic systems are typically 3–6 Hz (Stothers, 1972, pp. 348–362), the observed coupling is too large to be accounted for by four-bond coupling between C-10 and protons on C-1. The observed doublet of doublet pattern can, however, be explained by ${}^{3}J_{CCCH}$ between C-9 and the proton on C-6 and the proton on C-8.

3. Bicyclic

Bicyclic monoterpenes have been isolated from *Chondroccus hornemanni*. Chondrocole A (**45**) is a major metabolite of *C. hornemanni* collected at Halona blowhole on Oahu, Hawaii (Burreson *et al.*, 1975a). An incorrect structure of chondrocole A (**45a**) was determined by analysis of the ${}^{1}H$ nmr spectrum and mass spectral fragments corresponding to M^{+} Br and $-$ M^{+} $-$ Br $-$ HCl. The bromine atom was placed at C-4, where its facile loss in the mass spectrometer would generate an allylic radical. Placement of the chlorine at C-6 rather than at C-4 would accommodate the loss of HCl. The published ${}^{13}C$ nmr data were in agreement with the proposed structure displaying doublets at 63.8 and 54.4 ppm that were assignable to halomethine carbons.

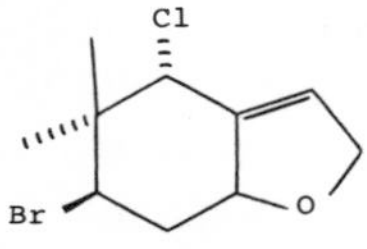

45 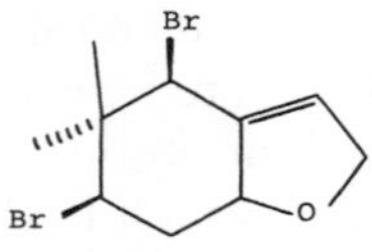 **46** 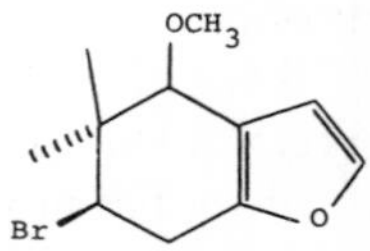 **47**

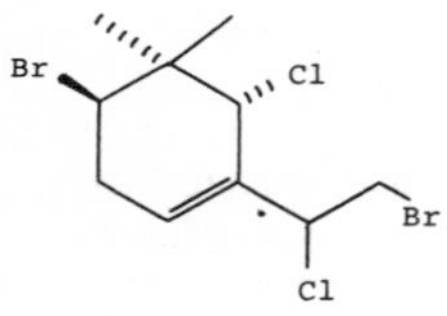

48 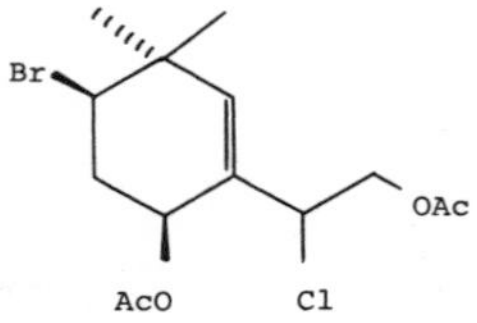 **49** 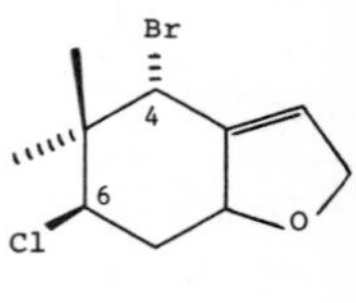 **45a**

The structure of chondrocole A (**45a**), however, proved to be incorrect (Moore, 1977). In the ${}^{1}H$ nmr spectrum the C-4 proton was a singlet at $\delta 4.64$, whereas the C-6 methine appeared as a doublet of doublets (J = 4, 13 Hz) at $\delta 4.45$. Single-frequency on-resonance decoupling of the $\delta 4.64$ proton collapsed the ${}^{13}C$ nmr signal at 63.8 ppm. This chemical shift was in agreement for a chlorine substituent. Likewise, irradiation of the δ4.45 proton signal collapsed the ^{13}C nmr signal at 54.4 ppm, which corresponded to a bromomethine. These new data led to a revision of the structure of chondrocole A (**45**) as shown. The ^{13}C nmr chemical shifts for the bicyclic monoterpenes chondrocole A (**45**), chondrocole C (**46**), and compound **47** along with two related monocyclic monoterpenes (**48** and **49**) are listed in Table 8.

III. SESQUITERPENES

A. Aromatic

Brominated aromatic sesquiterpenes of the cuparene class were first isolated from gastropod mollusks of the genus *Aplysia* by Yamamura and Hirata (1963). The same compounds were then isolated by Irie *et al.* (1966) from the red alga *Laurencia glandulifera*.

1. Cyclopentane Class

Table 9 lists the ^{13}C nmr chemical shift assignments of laurene (**50**), allolaurinterol (**51**), laurenisol acetate (**53**), bromolaurenisol acetate (**55**), and isolaurinterol (**56**). Table 10 lists the ^{13}C nmr chemical shift assignments of laurene (**50**), 3α-bromocuparene (**58**), and 3α-isobromocuparene (**59**).

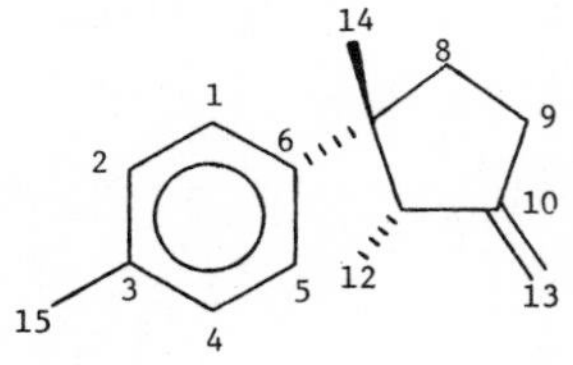

14 8 1 9 6 2 10 3 5 12 13 15 4

50

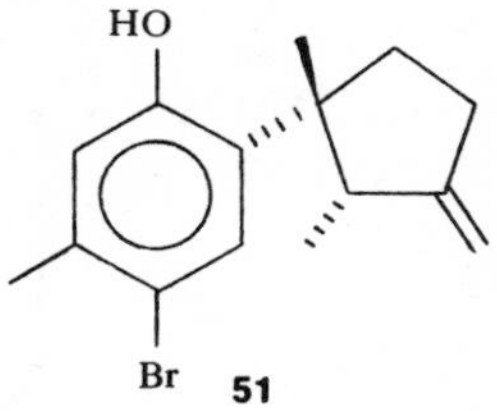

HO

Br **51**

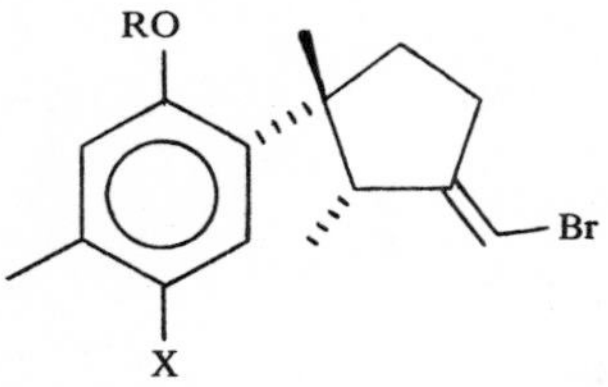

RO

Br

X

53 X = H
55 X = Br

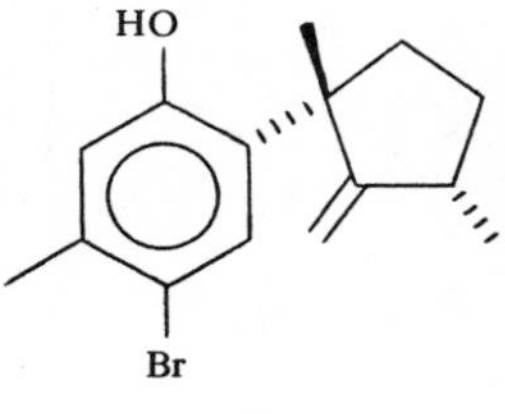

HO

Br

56

TABLE 8

^{13}C nmr Chemical Shifts of Bicyclic Monoterpenes[a]

Compound	C-1	C-2	C-3	C-4	C-5	C-6	C-7	C-8	C-9	C-10	Reference
(**45**)	75.4	122.3	137.6	63.8	41.7	54.4	41.7	80.7	27.6	21.0	Burreson *et al.* (1975b)
(**46**)	75.3	124.8	138.3	55.7	43.6	54.8	41.4	82.6	29.1	16.0	Woolard *et al.* (1978)
(**47**)[b]	141.3	110.3	117.0	80.0	40.9	56.4	33.1	148.6	24.7	20.4	Woolard *et al.* (1978)
(**48**)	34.5[d]	60.0	134.7	61.2	40.2	56.2	30.0[d]	127.2	28.2	19.5	Woolard *et al.* (1978)
(**49**)[c]	65.6	55.7	138.3	131.5	37.6	54.9	36.2	68.9	28.4	23.9	Woolard *et al.* (1978)

[a] The δ values are in parts per million downfield from TMS in $CDCl_3$ solution.
[b] Methyl ether, 57.8 ppm.
[c] Acetate signals: 20.7 (q), 20.9 (q), 169.8 (s), 170.1 (s) ppm.
[d] Signals may be reversed.

TABLE 9

Carbon Chemical Shifts of Aromatic Sesquiterpenes[a]

Carbon	Compound				
	50[b]	**51**[b]	**53**[d]	**55**[d]	**56**[e]
1	126.9 (125.5)	152.5 (150.8)	154.1 (148.5)	153.7 (146.9)	153.0 (151.6)
2	128.6 (129.3)	118.2 (118.3)	123.8 (123.3)	125.4 (125.0)	120.2 (118.3)
3	134.8 (135.0)	136.2 (139.8)	137.0 (142.1)	137.7 (143.4)	136.8 (139.8)
4	128.6 (129.3)	115.5 (116.5)	126.6 (126.5)	121.7 (121.8)	115.4 (114.9)
5	126.9 (125.5)	131.8 (130.3)	128.1 (127.3)	132.1 (129.9)	131.6 (134.2)
6	144.4 (148.3)	134.0 (137.4)	134.9 (136.0)	136.7 (139.4)	132.9 (137.1)
7	49.0	48.4	49.3	49.4	49.9
8	34.7	34.7	33.3	33.4	39.1
9	29.2	27.9	28.3	28.7	31.4
10	157.4	157.8	148.1	147.2	37.9
11	50.5	48.4	49.4	49.4	164.9
12	17.3[c]	19.5[c]	18.8[c]	18.8[c]	106.9
13	105.6	106.5	98.6	99.0	20.9[c]
14	29.7[c]	25.8[c]	26.9[c]	26.5[c]	27.7[c]
15	20.9	22.2[c]	21.2[c]	22.3[c]	22.2[c]
Other			169.2	169.8	
			20.6	21.5	

[a] The δ values are in parts per million downfield from TMS in $CDCl_3$ solution. The values in parentheses are calculated chemical shifts.
[b] Kazlauskas *et al.* (1976b).
[c] Values within a column may be interchanged.
[d] Rose and Sims (1977b).
[e] Izac and Sims (1977).

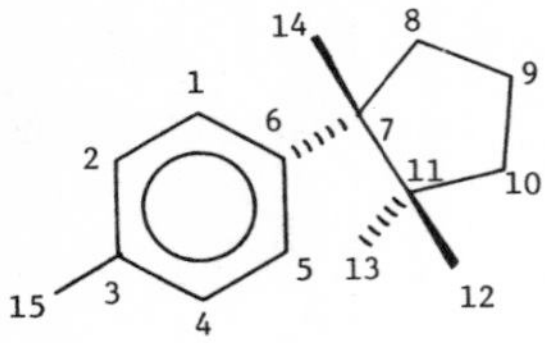

57 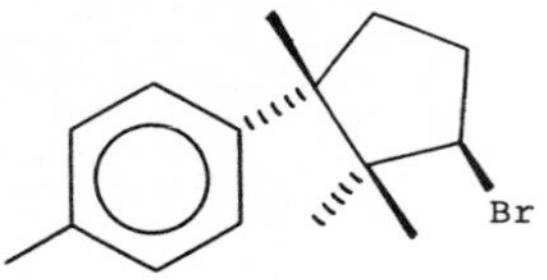 **58** 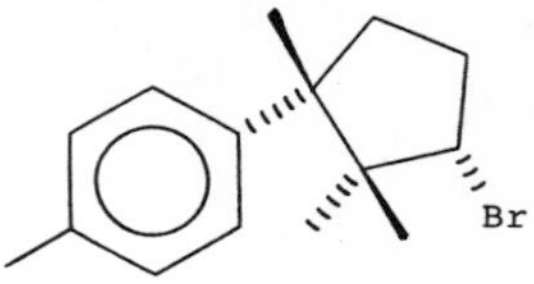 **59**

The aromatic carbon resonances of the cyclopentane-type aromatic sesquiterpenes are assigned by calculating their chemical shifts. Substituent effects in benzene systems have been extensively studied (Stothers, 1972, pp. 196–205). It has been found that the substituent effects in benzene systems are additive. The chemical shift, δ_C, of an aromatic carbon, j, can be calculated by Eq. (2), where A is the empirically derived substituent constant for the ipso, ortho, meta, and para positions. Some relevant substituent constants are listed in Table 11.

TABLE 10

Carbon Chemical Shifts of Bromocuparenes[a]

	Compound		
Carbon	**50**[b]	**58**[e]	**59**[e]
1	126.9 (125.5)	127.3	126.6
2	128.6 (129.3)	128.2	128.6
3	134.8 (135.0)	135.4	135.5
4	128.6 (129.3)	128.2	128.6
5	126.9 (125.5)	127.3	126.6
6	144.4 (148.3)	144.0	142.7
7		47.4	47.8
8		36.6	33.7[c]
9		33.2	31.4[c]
10		62.0	63.6
11		48.5	48.6
12		22.6[c]	21.0[d]
13		21.0[c]	22.2[d]
14		25.1[c]	25.5[d]
15		20.8	20.7

[a] The δ values are in parts per million downfield from TMS in $CDCl_3$ solution. The values in parentheses are calculated chemical shifts.

[b] Kazlauskas *et al.* (1976b).

[c,d] Assignments in a column may be reversed.

[e] Izac and Sims (1977).

$$\delta_{Cj}\ (\text{ppm}) = 128.5 + \Sigma\, A_{i,o,m,p} \qquad (2)$$

Laurene (**50**), isolated from *Laurencia glandulifera, L. pacifica,* and *L. filiformis*, was characterized by Irie *et al.* (1965). The ^{13}C nmr spectrum of laurene (**50**) showed only 13 signals: four aromatic, two olefinic, and seven high-field sp^3 carbons. The presence of four aromatic resonances is typical of symmetrically substituted systems. Equation (2) and the substituent constants in Table 11 correctly predict a para substitution pattern in laurene (**50**). The calculated chemical shift values, indicated in parentheses in Table 9, are used to assign the aromatic carbon resonances. Thus, the lowest-field aromatic singlet (144.4 ppm; calculated 148.3 ppm) is assigned to C-6, whereas the higher-field singlet (134.8 ppm; calculated 135.0) is assigned to C-3. A methyl group on a benzene ring results in a slight (+0.8 ppm) downfield shift of the ortho carbon resonances, whereas a *tert*-butyl group with its greater size causes a larger (−3.1 ppm) shift to higher field of ortho carbons. Hence, the off-resonance doublet at 128.6 ppm is assigned to C-2 and C-4, whereas the higher-field doublet at 125.9 ppm represents C-1 and C-5. The two remaining low-field signals at 157.4 and 105.6 ppm are assigned to C-10 and C-13, respectively, values typical of methylenecyclopentanes (Stothers, 1972, pp. 71–76).

The seven high-field signals of laurene (**50**) can be assigned by consideration of off-resonance multiplicities and deshielding effects of α and β substituents. The off-resonance singlet nature of the signal at 49.0 ppm assigns it to C-7, whereas the doublet at 50.5 ppm must originate from C-11. Of the two off-resonance triplets, C-8, which is adjacent to a quaternary center, should resonate at lower field than C-9, which should be slightly deshielded by the exocyclic methylene. Thus, C-8 is assigned to the resonance at 34.7 ppm and C-9 to the resonance at 29.2 ppm. The

TABLE 11

Substituent Constants for Substituted Benzenes[a]

	A			
R	Ipso	Ortho	Meta	Para
CH_3	+9.3	+0.8	0	−2.9
$C(CH_3)_3$	+22.4	−3.1	−0.1	−2.9
OH	+26.9	−12.7	+1.4	−7.3
$OCOCH_3$	+23	−6	+1	−2
OCH_3	+31.4	−14.4	+1.0	−7.7
Cl	+6.2	+0.4	+1.3	−1.9
Br	−5.5	+3.4	+1.7	−1.6

[a] Wehrli and Wirthlin (1976, p. 45).

three methyl groups resonate at 29.7, 20.9, and 17.3 ppm. The aromatic methyl of *p*-xylene resonates at 20.9 ppm (Stothers, 1972, p. 98). An identically situated signal in laurene (**50**) permits its assignment to the aromatic methyl at C-15. The remaining two methyl groups can be distinguished by consideration of steric compression shifts. When carbon atoms are in a sterically crowded environment they resonate at higher field than do similar carbons in an uncrowded environment. Thus, C-12, which is cis to the bulky phenyl ring, should resonate at higher field than C-14, which is cis to a hydrogen and therefore is in a less crowded environment. The signal at 17.3 ppm is therefore the sterically congested C-12 methyl, whereas C-14 resonates at 29.7 ppm.

Allolaurinterol (**51**) was isolated from the petroleum ether extract of freeze-dried *Laurencia filiformis* f. *heteroclada* collected on the coast of South Australia (Kazlauskas *et al.*, 1976b). The methylenecyclopentane ring of allolaurinterol (**51**) was identical to that of laurene (**50**). The chemical shift assignments paralleled those of laurene (**50**). The aromatic ring carbon chemical shifts were indicative of a substituted bromophenol, and assignments were made from the calculated shifts (Table 9).

Laurenisol (**52**) (Irie *et al.*, 1969) is unstable but forms a stable acetate (**53**). The introduction of an acetate on the aromatic ring shifts the ipso carbon downfield by about 23 ppm (Table 9) while causing a 6 ppm upfield shift of the ortho carbons. The assignments of the aromatic signals along with the calculated chemical shift values are given in Table 9.

The methylenecyclopentane ring shows the effect of the bromine at C-13. The heavy-atom effect of the bromine results in shielding of the attached carbon so that C-13 now resonates at 98.6 ppm. Carbon C-10, which is adjacent to the carbon bearing bromine, is shielded and resonates at slightly higher field (148.1 ppm) than in laurene (**50**) (157.4 ppm). The remaining high-field signals (Table 9) are assigned by comparison of multiplicities and chemical shifts with laurene (**50**).

Like laurenisol (**52**), bromolaurenisol (**54**) is unstable at room temperature but forms a stable acetate (**55**). The ^{13}C nmr chemical shifts of bromolaurenisol acetate (**55**) (Table 9) parallel those of laurenisol acetate (**53**). The introduction of the bromine on the aromatic ring shifts C-4 downfield by about 5 ppm.

Isolaurinterol (**56**) was isolated by Irie *et al.* (1970) as a minor component of *Laurencia intermedia* Yamada. Carbon-13 nmr data for isolaurinterol (**56**) became available when Izac and Sims (1977) examined the lipid extract of *L. pacifica*. The substitution pattern on the aromatic ring of isolaurinterol (**56**) was identical to that of allolaurinterol (**51**). The aromatic carbon shift assignments followed from those of allolaurinterol (**51**). Off-resonance carbon multiplicities and chemical shift considerations permitted assignments of the methylenecyclopentane ring. The methyl groups could not be unambiguously assigned.

Although cuparene (**57**) (Enzell and Erdtman, 1958), a common heartwood constituent of conifers in the family Cupressaceae, is unknown in the marine environment, the cuparene skeleton represented by 3α-bromocuparene (**58**) and 3α-isobromocuparene (**59**) was reported by Suzuki *et al.* (1975). The bromocuparenes are present in *Laurencia glandulifera* and *L. pacifica*. The aromatic signals of the bromocuparenes (Table 10) can be assigned by comparison to laurene (**50**). The 1.3 ppm upfield shift of C-6 in 3α-isobromocuparene (**59**) probably reflects the cis orientation of the C-10 bromine atom and phenyl ring.

The off-resonance ^{13}C nmr spectrum of 3α-bromocuparene (**58**) displays two high-field singlets at 47.4 and 48.5 ppm. In laurene (**50**), C-7 resonates at 49.0 ppm. The introduction of a second methyl group at C-11 of laurene (**50**) should cause an upfield shift of about 1.2 ppm at C-7 (Levy and Nelson, 1972). Thus, C-7 of 3α-bromocuparene (**58**) should resonate near 47.8 ppm. Assignment of the higher-field (47.4 ppm) signal of 3α-bromocuparene (**58**) to C-7 and the signal of 48.5 ppm to C-11 is consistent not only with the expected shift based on the laurene (**50**) model, but also with the anticipated downfield shift of C-11 caused by the halogen at C-10. The chemical shift of 62.0 ppm assigns this signal to the bromomethine carbon, C-10.

Assignments of C-8 and C-9 can be made by consideration of bromine β- and γ-shielding parameters (Wehrli and Wirthlin, 1976, pp. 36–39) and of 1,1-dimethylcyclohexane (**63**) as a model compound. Using the average (axial, equatorial) bromine values, the calculated chemical shift values for 4-bromo-1,1-dimethylcyclohexane (**64**) are determined. In this compound C-2, which is adjacent to a quaternary center, resonates at lower field strength than C-3, which is alpha to the bromomethine center. The same trends are expected in the analogous cyclopentane system so that C-8, which is adjacent to the quaternary center, is the lower-field signal at 36.6 ppm, whereas C-9, which is alpha to the halomethine, resonates at 33.2 ppm. The signal at 20.8 ppm is the aromatic methyl C-15 by analogy with *p*-xylene (20.9 ppm). The remaining methyl signals cannot be unambiguously assigned.

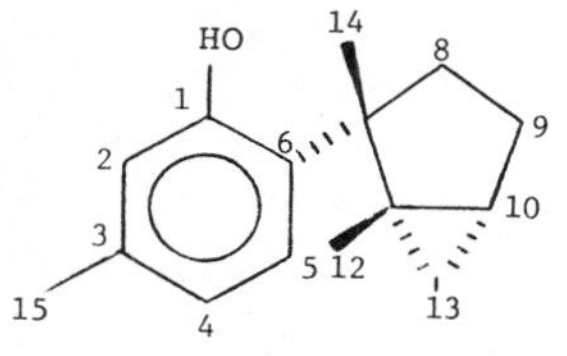

HO 1 2 3 4 5 6 8 9 10 12 13 14 15

60

HO X_2 X_1

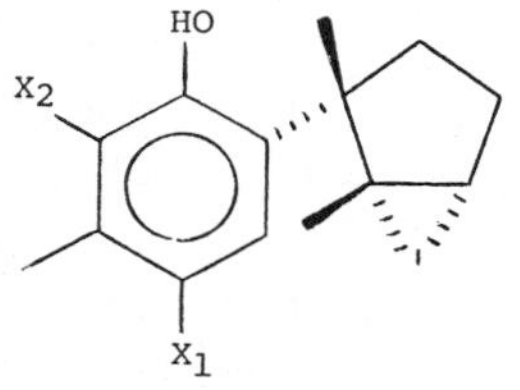

61 $X_1 = H$; $X_2 = Br$
62 $X_1 = X_2 = Br$

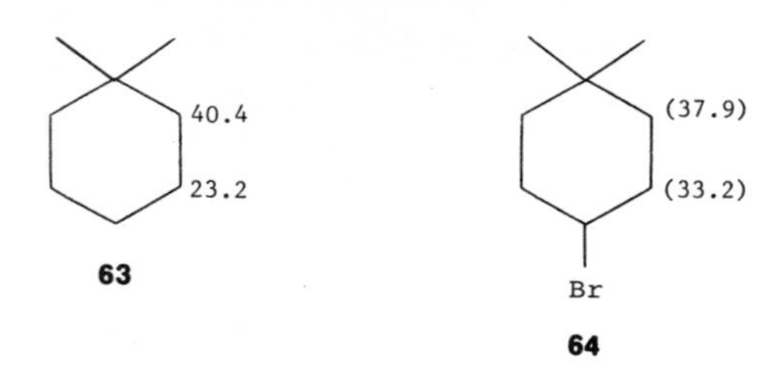

The ^{13}C nmr spectrum of 3α-isobromocuparene (**59**) is assigned using 3α-bromocuparene (**58**) as a model. The bromomethine carbon, C-10, of 3α-isobromocuparene (**59**) is shifted to lower field by 1.5 ppm and probably reflects relief of steric compression due to less 1,3 interaction between the C-10 bromine atom and the rotationally restricted phenyl group. The shifts observed (Table 10) for C-8 and C-9 are somewhat unexpected. The upfield shift of these carbons could indicate that the cyclopentane ring is not planar. In 3α-bromocuparene (**58**) the C-10 bromine atom is pseudoequatorial. Epimerization of C-10 would generate 3α-isobromocuparene (**59**) with the bromine atom in a pseudoaxial conformation. The pseudoaxial C-10 bromine atom, because of 1,3 interactions, would experience a shift to higher field with respect to C-8. A similar but smaller upfield shift of C-9 (β effect) should be observed going from the pseudoequatorial to pseudoaxial conformation.

2. [*3.1.0*] *Bicyclic Class*

The ^{13}C nmr chemical shift assignments of debromolaurinterol (**60**), laurinterol (**61**), and bromolaurinterol (**62**) are listed in Table 12.

Debromolaurinterol (**60**) (Irie *et al.*, 1970) differs from laurene (**50**) in that it contains a hydroxyl at C-1 and the exocyclic methylene is closed at C-11 to form a cyclopropane. The symmetry of the aromatic region observed in the ^{13}C nmr spectrum of laurene (**50**) is absent in debromolaurinterol (**60**), which displays six aromatic signals. From the calculated chemical shift values indicated in Table 12 it can be seen that the phenolic carbon, C-1, should resonate at lowest field and can be assigned to the carbon signal at 153.8 ppm. By similar consideration of the substituent constant effects at the ortho, meta, and para positions, C-6 is represented by the resonance at 131.3 ppm, C-5 resonates at 128.6 ppm, C-4 gives rise to the signal at 120.7 ppm, C-3 to the signal at 136.5 ppm, and C-2 to the signal at 117.1 ppm.

The strain of the cyclopropane ring of debromolaurinterol (**60**) can be seen by the high-field resonance of C-11 at 29.6 ppm. Similarly, C-10, an off-resonance doublet, resonates at 18.7 ppm, whereas the C-13 methylene absorbs at 16.4 ppm. Carbon C-7, a singlet at 47.8 ppm in the

TABLE 12

Carbon Chemical Shifts of Aromatic Sesquiterpenes[a,b]

Carbon	Compound 60	Compound 61	Compound 62
1	153.8 (151.6)	153.8 (150.8)	150.1 (154.2)
2	117.1 (116.6)	118.8 (118.3)	114.3 (112.8)
3	136.5 (136.4)	135.9 (139.8)	135.0 (143.2)
4	120.7 (122.0)	114.9 (116.5)	115.0 (118.2)
5	128.6 (126.9)	132.3 (130.3)	131.7 (128.7)
6	131.3 (135.4)	134.1 (137.1)	134.2 (139.1)
7	47.8	48.0	48.9
8	35.9	35.9	34.9
9	25.2	25.3	25.2
10	18.7 [7.1]	18.6 [7.0]	18.6
11	29.6	29.6	29.4
12	24.2[c] [7.4][d]	24.4[c] [7.5]	24.1[c]
13	16.2	16.2	16.2
14	23.4[c] [7.4]	23.5[c] [7.5]	22.3[c]
15	20.5 [8.5]	22.2 [8.5]	23.9

[a] The δ values are in parts per million downfield from TMS in $CDCl_3$ solution. Data from Izac and Sims (1977).
[b] Values in parentheses are calculated chemical shifts.
[c] Assignments may be reversed.
[d] Values in brackets are residual coupling constants.

single-frequency off-resonance spectrum, can be assigned by comparison to laurene (**50**). Hence, the highest-field methylene signal at 35.9 ppm is assigned to C-8, which is adjacent to a quaternary center. The signal at 25.2 is assigned to C-9 by default.

The methyl groups can be differentiated by consideration of reduced coupling constants, ^{r}J, obtained from the single-frequency off-resonance decoupled spectrum. Two factors affect the one-bond coupling constant, $^{1}J_{CH}$. First, $^{1}J_{CH}$ is proportional to the *s* character of the carbon atom and increases depending on hybridization in the order $sp^3 < sp^2 < sp$. Second, $^{1}J_{CH}$ depends on the electronegativity of groups bonded to carbon. Electronegative substituents increase the effective nuclear charge at the carbon atom, which increases $^{1}J_{CH}$. However useful $^{1}J_{CH}$ values may be, fully coupled spectra are rarely determined. The measurement of one-bond coupling constants is hampered by loss of NOE, which requires prolonged instrument time. In addition the spectrum is more complex due to multiplet overlap and second-order effects ($^{2}J_{CH}$, $^{3}J_{CH}$). Although most of the NOE can be regained using gated decoupling techniques (Gray, 1975), single-frequency off-resonance spectra, which display reduced coupling constants, ${}^{r}J_{CH}$, yield similar information.

The relationship between one-bond, ${}^{1}J_{CH}$, and reduced one-bond carbon–hydrogen, ${}^{r}J_{CH}$, coupling constants is given by Eq. (1). Two assumptions of this equation are that the decoupler power is greater than both $\Delta\nu$ and ${}^{1}J_{CH}$ (Wehrli and Wirthlin, 1976, pp. 66–69).

The ${}^{1}J_{CH}$ values of methyl groups bonded to carbon atoms are approximately equal (Stothers, 1972, p. 337). If $\gamma H_0/2\pi$ is held constant, Eq. (1) reduces to

$$ {}^{r}J_{CH} \sim \Delta\nu \qquad (2)$$

The reduced coupling constants should increase as the distance between the decoupler offset frequency and the proton chemical shift increases. Thus, if the decoupler offset is at higher field in the ${}^{1}H$ region than TMS, ${}^{r}J_{CH}$ will be proportional to the proton chemical shift and independent of the carbon chemical shift. The ${}^{r}J_{CH}$ of a methyl group whose protons resonate at $\delta 1.8$ will be larger than that of a methyl group that resonates at $\delta 1.5$.

In debromolaurinterol (**60**), the methyl quartet at 20.5 ppm has a larger ${}^{r}J$ than the signals at 23.4 or 24.2 ppm, and hence it is assigned to the aromatic methyl, C-15. This assignment was confirmed by selective proton decoupling experiments. The C-12 and C-14 methyl groups cannot be unambiguously assigned.

The carbon chemical shift assignments of laurinterol (**61**) and bromolaurinterol (**62**) can be made by comparison with debromolaurinterol (**60**) (Table 12). The aliphatic ring carbons are essentially unaltered by the introduction of bromine atoms to the aromatic ring, and thus chemical shift assignments parallel those of debromolaurinterol (**60**). Again, the aromatic ring of laurinterol (**61**) and bromolaurinterol (**62**) shows the chemical shifts expected from the addition of one and two bromine atoms onto the debromolaurinterol (**60**) skeleton.

3. *Cyclic Ether Class*

The ${}^{13}C$ nmr assignments of two five-membered ring ethers, aplysin (**65**) and debromoaplysin (**66**), are indicated on the structures. Table 13 lists the ${}^{13}C$ nmr chemical shift assignments of four six-membered ring ethers, filiformin (**67**), filiforminol (**68**), filiformic acid (**69**), and methyl filiformate (**70**).

The ${}^{13}C$ nmr spectrum of aplysin (**65**) can be assigned using benzene substituent constants (Table 11), off-resonance multiplicities and specific proton decoupling. The aromatic resonances, C-1 through C-6, are as-

TABLE 13

Carbon Chemical Shifts of Aromatic Sesquiterpenes[a]

Carbon	Calculated[b]	Compound			
		67	**68**	**69**	**70**
1	155.2	152.3	151.6	150.1	150.6
2	116.6	117.4	117.5	118.0	118.2
3	139.3	136.4	136.5	137.3	137.0
4	116.0	114.4	114.9	116.1	115.6
5	129.8	128.4	128.6	128.6	128.4
6	135.3	130.1	130.3	129.3	129.5
7		44.9	44.9	44.9	44.7
8		42.2	41.8	42.6	42.7
9		37.3	33.0	35.6	35.7
10		85.3	87.4	87.5	87.7
11		46.4	43.1	45.8	45.7
12		7.4	7.9	7.7	7.8
13		23.0	66.0	176.2	171.9
14		20.4	20.4	19.6	19.6
15		22.5	22.5	22.5	22.5

[a] The δ values are in parts per million downfield from TMS in $CDCl_3$ solution. From Kazlauskas *et al.* (1976b).
[b] See Table 11 for substituent parameters.

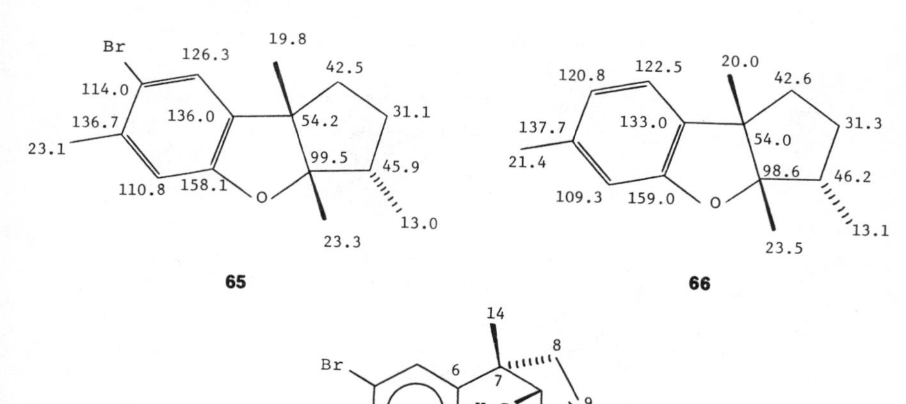

67 R = CH_3
68 R = CH_2OH
69 R = COOH
70 R = $COOCH_3$

signed on the basis of calculated shift values. Although the calculated values are derived from mono- and disubstituted aromatic systems, reasonable agreement is obtained with polysubstituted benzenes. The calculated shift values for C-4 (116.0 ppm) and C-2 (116.6 ppm) are too close to permit rigorous assignment. However, these carbons are readily distinguished by off-resonance multiplicity; C-2 is a doublet, whereas C-4 is a singlet. A gated decoupled spectrum shows C-2 and C-5 to have identical coupling constants of 162 Hz, but whereas C-5 is a doublet, C-2 shows three-bond coupling ($^3J_{CCCH}$ = 4 Hz) to the aromatic methyl and appears as a doublet of quartets.

The aliphatic carbons of aplysin (**65**) are assigned as follows. Chemical shift considerations assign the resonance at 99.5 ppm to C-11 and the singlet at 54.2 ppm to C-7. The only high-field doublet in the ^{13}C nmr spectrum appears at 45.9 ppm and is ascribed to C-10. The higher-field triplet at 42.5 ppm is assigned to C-8 because it is adjacent to the quaternary center. The methylene signal at 31.1 ppm represents C-9. The aromatic methyl group of aplysin (**65**) resonates at 2.3δ. If the proton decoupler is centered at this resonance and the spectrum is recorded at low decoupler power, the carbon signal at 23.1 ppm collapses to a singlet and is the tallest peak in the spectrum. This allows unambiguous assignment of the 23.1 ppm resonance to the aromatic methyl, C-15. The resonance at 19.8 ppm is assigned to C-14 by comparison to debromolaurinterol (**60**). The high-field resonance of C-13, 13.0 ppm, probably results from interaction of this methyl group with the nonbonded electrons on the ether oxygen.

Additional support for the methyl group assignments in aplysin (**65**) is found in the reduced coupling constants. In the ^1H nmr spectrum, the aromatic methyl, C-15, resonates at δ2.3; C-12 at δ1.33; the benzylic methyl, C-14, at δ1.25; and the tertiary methyl, C-13, at δ1.10. The reduced coupling constants for the four methyl quartets decrease in the order 23.1 ppm (^rJ = 7.2 Hz), 23.3 ppm (^rJ = 6.0 Hz), 19.8 ppm (^rJ = 5.5 Hz), 13.0 ppm (^rJ = 5.0 Hz). The largest ^rJ is assigned to the carbon whose protons resonate at lowest field, whereas the smallest ^rJ is assigned to the carbon whose protons resonate at highest field. Thus, the aromatic methyl, C-15, is the 23.1 ppm signal and C-12, the methyl group on an oxygenated carbon, is the 23.3 ppm resonance. The benzylic methyl, C-14, is the signal at 19.8 ppm, and the tertiary methyl, C-13, is the signal at 13.0 ppm.

Treatment of aplysin (**65**) with lithium aluminum hydride in diethyl ether yields debromoaplysin (**66**). The lower-field carbon chemical shifts reflect the removal of the bromine atom from the aromatic system. With the exception of C-15, the aliphatic carbon resonances are very similar to those of aplysin (**65**). Carbon C-15, which now absorbs at 21.4 ppm, is shifted upfield by 1.7 ppm from its position in aplysin (**65**). The 21.4 ppm chemical shift assignment of C-15 in debromoaplysin (**66**) was confirmed by specific proton–carbon decoupling of the ^1H nmr signal at δ2.27. The remaining shift assignments follow from aplysin (**65**).

Kazlauskas *et al.* (1976b) reported the structural elucidation of a series of six-membered ring ether compounds from *Laurencia filiformis* and reported their ^{13}C nmr spectra. Structurally related compounds were isolated by Irie *et al.* (1969) as degradation products in the structural elucidation of laurenisol (**52**). Assignment of the aromatic signals (Table 13) in filiformin (**67**), filiforminol (**68**), and filiformic acid (**69**) and its methyl ester (**70**) followed from aplysin (**65**). The agreement between the calculated value of 116.6 ppm and experimental values of 117.4 ppm for C-2 reflects the decreased strain of the six-membered ring ether relative to the five-membered ring in aplysin (**65**). In aplysin, C-2 resonates at 110.8 ppm.

Assignments of the carbocyclic ring of **67–70** again parallel those found in aplysin (**65**) if substituent effects are included. The off-resonance singlet near 87 ppm is the signal for C-10, whereas the higher-field singlet near 44.9 ppm corresponds to C-7. The off-resonance doublet found near 45 ppm is designated as C-11. The β effect of the ether linkage at C-10 results in approximately a 5 ppm upfield shift of C-9 relative to this carbon in aplysin (**65**); C-9 now resonates at 35–37 ppm. In filiforminol (**68**) the alcohol at C-13 exerts a shielding β-substituent effect on C-9, resulting in a 4.3 ppm upfield shift of this center, which now resonates at 33.0 ppm. By analogy with aplysin (**65**) the aromatic methyl group resonates at 22.5 ppm. The rigid skeleton of these compounds holds the C-12 methyl group above the plane of the aromatic ring, where it is strongly shielded by the aromatic ring current. In this configuration, C-12 resonates at 7.4–7.9 ppm. The signal at 23.0 ppm of filiformin (**67**), which is absent in filiforminol (**68**), is assigned to the C-13 methyl. In filiformic acid (**69**), where C-13 is a carboxyl group, the C-14 methyl is shifted, relative to filiformin (**67**), by 0.8 ppm. This shift is typical of γ-shielding effects.

B. Chamigranes

Sesquiterpenes possessing the chamigrane carbon skeleton are produced by several species of marine algae of the genus *Laurencia*. It has been suggested (Kato *et al.*, 1970; González *et al.*, 1976) that the biosynthesis of these metabolites involves cyclization of a farnesyl pyrophosphate to yield either a γ-bisabolene or a bromocyclofarnesyl pyrophosphate intermediate. Bromonium ion-catalyzed cyclization of the γ-bisabolene or proton-catalyzed cyclization of the bromofarnesyl pyrophosphate could generate the bromochamigranes.

For carbon chemical shift evaluation, the *Laurencia* chamigranes seem to fall into three structure types: those with an A-ring endocyclic double bond, those with an A-ring exocyclic double bond, and those with additional cyclization to yield five-membered ring ethers.

1. Endocyclic A-Ring Olefins

The carbon chemical shift assignments for compounds **71–79** are colligated in Table 14. Examination of the carbon multiplicities of **71** allows the assignment of only 2 of the 15 carbons: C-7 to the 139.6 ppm singlet and C-8 to the 122.9 ppm doublet. Chemical shift considerations assign the 71.0 ppm signal to the chlorine-bearing quaternary center, C-3.

Assignments of the remaining carbon signals of **71** require cross-correlation of the chemical shifts with those of structurally related compounds. The signals at 60.8 and 62.9 ppm of compound **71** could be

TABLE 14

Carbon Chemical Shifts of A-Ring Endocyclic Chamigrenes[a]

	Compound								
Carbon	**71**[g]	**72**[g]	**73**[g]	**74**[h]	**75**[g]	**76**[i]	**77**[h]	**78**[g]	**79**[j]
1	31.6	31.4	31.2	35.9[d]	25.6	25.8	26.5[d]	29.8[e]	70.6
2	39.5[d]	39.1	39.0	36.1[d]	39.6	39.0	27.8[d]	28.8[e]	35.2
3	71.0	70.6	70.5	71.9	70.8	71.1	55.5	134.0	n.o.[c]
4	62.9	62.4	62.9	66.1	62.3[d]	63.1	58.7	121.5[d]	62.2
5	40.3[d]	40.1	40.0	39.1	40.2	39.5	29.4[d]	31.1[e]	47.2[d]
6	47.7	47.9	47.7	48.1	46.2	49.1	43.0	44.0	n.o.
7	139.6	142.7	144.4	139.9	62.9[d]	58.1	141.6	140.8	n.o.
8	122.6	124.3	121.6	122.0	65.4	56.2	119.6	122.6[d]	34.9[d]
9	36.3	73.0	75.1	36.2	72.2	124.1	35.7	36.6	122.4
10	60.8	70.7	62.2	61.2	67.2	143.6	61.9	63.3	139.2
11	42.8	45.5	45.8	42.7	43.9	45.8	42.0	41.8	n.o.
12	17.1[f]	17.9[d]	18.0[d]	17.3[e]	18.3	23.8[d]	17.6	17.4	23.8[e]
13	24.6[f]	25.1[d]	24.9[d]	22.1[e]	24.9[e]	23.5[d]	20.6[e]	24.9	26.3[e]
14	26.0[e]	25.7	25.6	25.8[e]	25.4	24.3[d]	25.4	23.4	112.7
15	24.1[e]	24.1	24.0	24.7[e]	24.5[e]	24.7[d]	22.3[e]	23.4	28.2[e]
Other			170.7						
			20.9						

[a] The δ values are in parts per million downfield from TMS in $CDCl_3$ solution.
[b] In C_6D_6 solution.
[c] n.o., not observed.
[d–f] Signals within a column may be reversed.
[g] Izac and Sims (1977).
[h] Kurosawa (1977).
[i] Rose and Sims (1977a).
[j] Erickson (1977).

assigned to either C-4 or C-10. Introduction of a hydroxyl at C-9 should shift the C-10 signal downfield, whereas C-4, which is removed from the influence of the hydroxyl, should remain unshifted. In compound **72**, which differs from compound **71** by a hydroxyl at C-9, the bromomethine signals are at 62.4 and 70.7 ppm. Hence, the unshifted 62.9 ppm resonance of compound **71** is assigned to C-4, whereas the 60.8 ppm signal, which

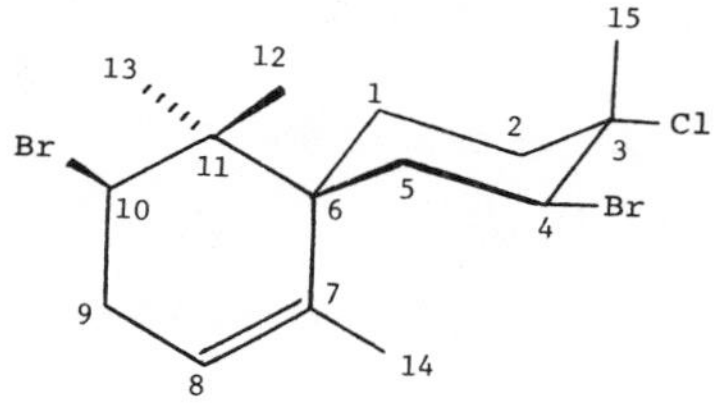

71

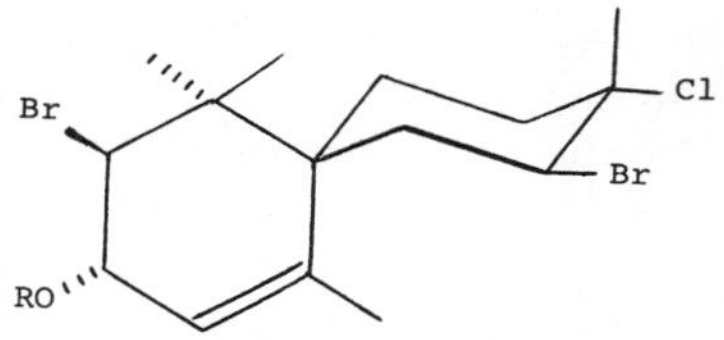

72 R = H
73 R = Ac

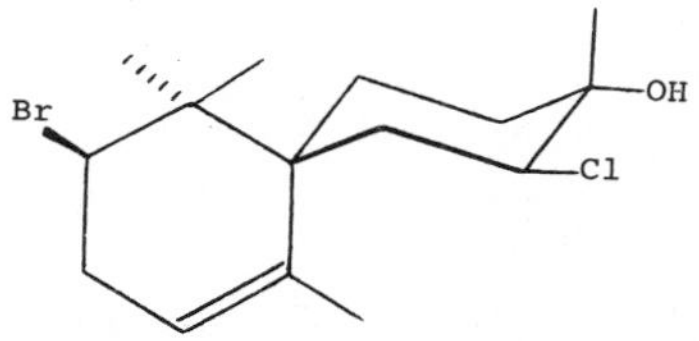

74

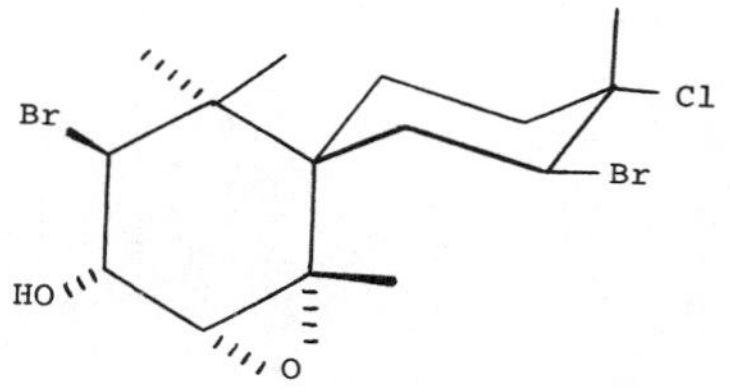

75

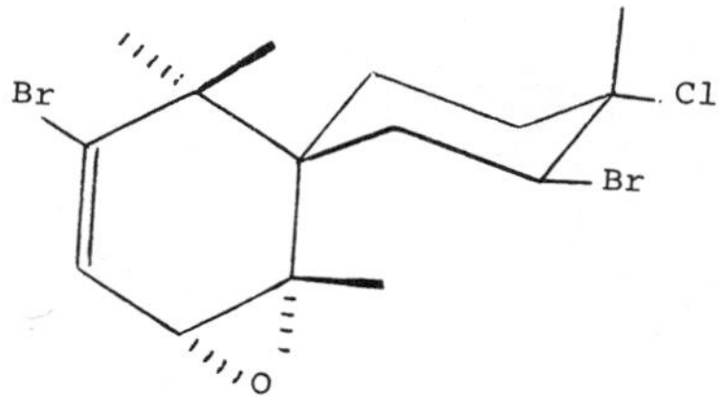

76

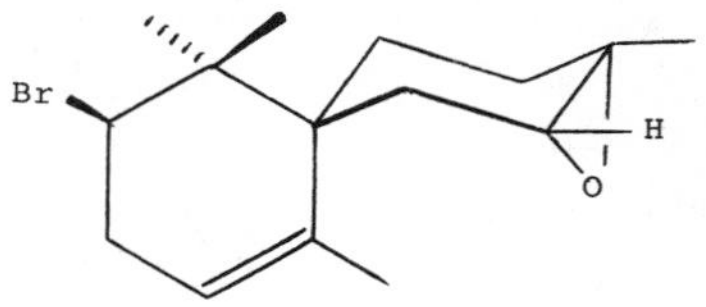

77

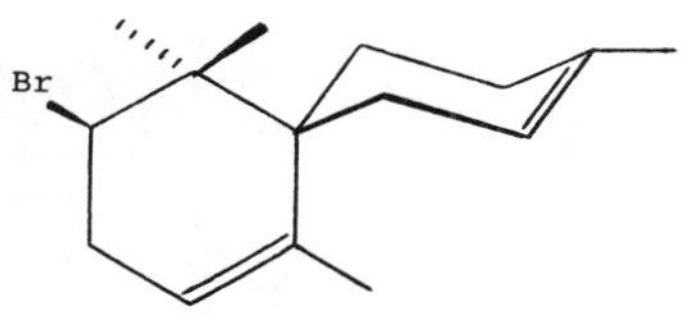

78

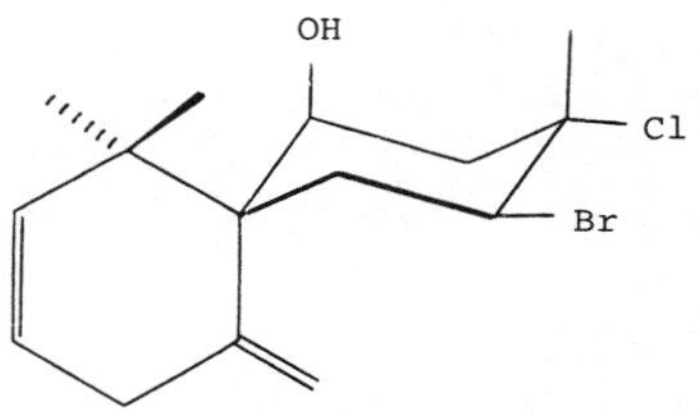

79

shows an anticipated 9.9 ppm downfield shift, is assigned to C-10. The resonances at 62.4 and 70.7 ppm of compound **72** are then assigned to C-4 and C-10, respectively.

Of the four methylene carbons in **71**, C-9 can be assigned to the signal at 36.3 ppm. This signal disappears in **72**, where C-9 is a hydroxymethine carbon. Carbon C-5, which is alpha to both a quaternary center and an equatorial bromine atom, should resonate at lower field than C-2 and hence is assigned to the signal at 40.3 ppm. Carbon C-2 is assigned to the resonance at 39.5 ppm. Carbon C-1 is the 31.6 ppm resonance by default.

On the basis of residual coupling constants, the four methyl resonances of **71** can be divded into two groups. The resonances at 24.1 and 26.0 ppm have larger ^{r}J values than the signals at 17.1 and 24.6 ppm and can therefore be assigned to C-14 or C-15. The chemical shift of C-14, unlike that of C-15, should be sensitive to substitution at C-9. In **72**, where C-9 contains a hydroxyl substituent, C-14 resonates at 25.7 ppm, an upfield shift of 0.3 ppm from its position in **71**, whereas the 24.1 ppm signal is unchanged. Thus, in **71** the signal at 26.0 ppm is assigned to C-14, and the 24.1 ppm signal is assigned to C-15. Dreiding molecular models suggest that C-13 experiences shielding γ-gauche interactions with the protons on C-2 and should therefore resonate at higher field than C-12. Thus, C-13 is assigned the resonance at 17.1 ppm, whereas C-12 resonates at 24.6 ppm.

The assignments of the chemical shifts of compounds **72** and **73** (Table 14) follow from those of **71**. In cyclohexanes the introduction of an equatorial hydroxyl deshields the β carbon by 8 ppm while shielding the γ carbon by 3 ppm (Wehrli and Wirthlin, 1976, p. 45). Similar but smaller shifts (5 and 3 ppm, respectively) are observed for an equatorial acetoxy substituent. The introduction of the hydroxyl at C-9 (**72**) shifts the neighboring carbon bearing bromine (C-10) downfield by an anticipated 10 ppm. However, when C-9 is substituted with an acetoxy group, C-10 shifts downfield by only 1.4 ppm. Thus, acetylation of bromohydrin (**72**) yields **73**, and this results in an upfield shift of C-10 by 8.5 ppm.

Specific proton decoupling of the vinyl methyl (δ2.02) in compound **72** collapses the multiplet at 25.7 ppm to a singlet. This assigns the 25.7 ppm resonance to C-14. The quaternary methyl (δ1.70) is partially decoupled in this experiment, and the ^{13}C nmr resonance at 24.1 ppm is observed as a broadened singlet. Thus, C-15 is the resonance at 24.1 ppm.

Glanduliferol (**74**) (Suzuki *et al.*, 1974b) differs from **71** by the substituents on the B ring containing an axial hydroxyl at C-3 and an equatorial chlorine at C-4. The A-ring carbons of **74** (Table 14) are unchanged ($\pm$0.4 ppm) from those of **71**. It is interesting to note the chemical shifts of C-3 in **71** and glanduliferol (**74**). In **71** this quaternary center bears a chlorine and in glanduliferol (**74**) it bears an oxygen, but the chemical shift difference of these carbon atoms is only 0.9 ppm. Again, carbon chemical shifts of quaternary centers cannot be used as a reliable indicator of substituents.

Cyclohexane substituent parameters predict a net (Cl to OH and Br to Cl) downfield shift of 5 ppm for the chlorine-bearing carbon, C-4, of **74** relative to the shift of C-4 in **71**. This is in reasonable agreement with the 3.2 ppm downfield shift observed for C-4 in glanduliferol (**74**) relative to the same carbon in **71**. In contrast to quaternary centers, the chemical shift of tertiary carbons is an accurate indicator of the attached atoms and can be used to distinguish between halogens.

Compound **75** has the same B-ring substitution pattern as compounds **71–73**. The chemical shift assignments for these compounds coincide except for C-1. In compound **75** this center is shifted upfield (~6 ppm) and resonates at 25.6 ppm. It is possible that C-1 is shielded by the nonbonded electrons on the epoxide oxygen. The epoxide carbons, C-7 and C-8, are assigned to the resonances at 62.9 and 65.4 ppm, respectively. The tertiary epoxide carbon, C-8, resonates at lower field than the quaternary epoxide carbon, C-7, because of the deshielding β effect of the hydroxide at C-9.

Deoxyprepacifenol (**76**), like compound **75**, exhibits a high-field triplet (25.8 ppm) assigned to C-1. In deoxyprepacifenol (**76**) the quaternary epoxide carbon, C-7 (58.1 ppm), resonates at lower field than the tertiary epoxide carbon, C-8 (56.2 ppm).

The A-ring carbon shifts of compound **77** are very similar to those of compound **71** and glanduliferol (**74**). The B-ring epoxide carbons of compound **77** show a reversal of chemical shift, with the tertiary center resonating at lower field than the quaternary epoxide carbon.

Carbon chemical shift assignments of the A-ring carbons of compound **78** are made by reference to compound **71**. Carbon C-5 of compound **78** is shifted to higher field relative to its assignment in compound **71**, reflecting both the absence of the C-4 halogen and ring carbon shielding due to the double bond.

The chemical shift assignments of nidifidienol **79** are made by chemical shift and substituent parameter considerations. The three methylene signals of nidifidienol (**79**) are assigned as follows. In monoterpene **80** (Bohlmann *et al.*, 1975) the methylene carbon between the olefins resonated at 29.6 ppm. Thus, the highest-field resonance, 34.9 ppm, of nidifidienol (**79**) is assigned to C-8. The remaining methylene signals were distinguished by substituent chemical shift considerations. Carbons C-2 and C-5 are alpha to a quaternary center. Substituent parameters predict that the β carbon atom should be shifted downfield 5 ppm by an axial hydroxyl but 12 ppm by an equatorial bromine (Wehrli and Wirthlin, 1976,

p. 45). Thus, C-5, which is beta to an equatorial bromide, is assigned to the higher field signal at 47.2 ppm while C-2, which is beta to an axial hydroxyl, is assigned the resonance at 35.2 ppm.

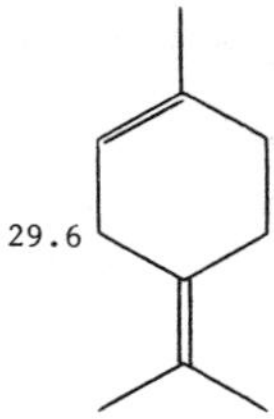

29.6

80

2. *Exocyclic A-Ring Olefins*

The carbon chemical shift assignments for elatol (**81**), elatol acetate (**82**), debromoelatol (**83**) and its acetate (**84**), 10-bromo-β-chamigrene (**85**), and nidificene (**86**) are listed in Table 15.

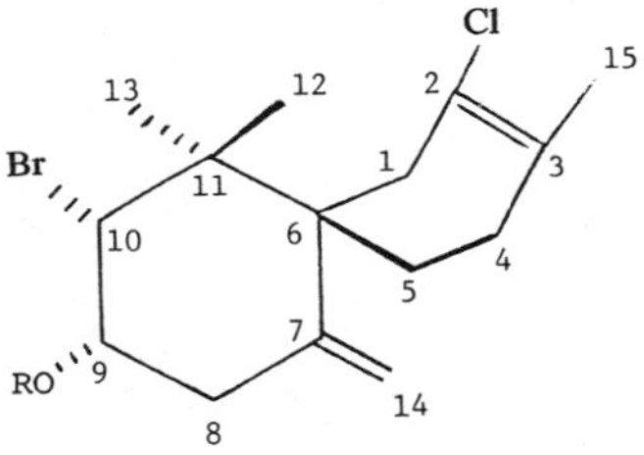

81 R = H
82 R = Ac

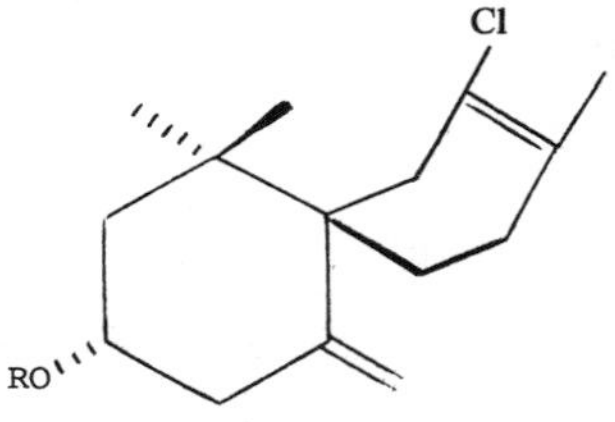

83 R = H
84 R = Ac

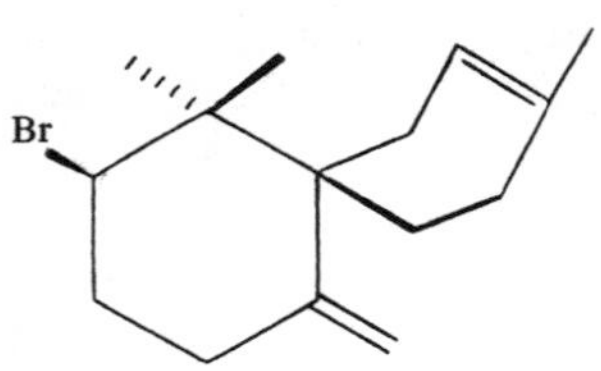

85

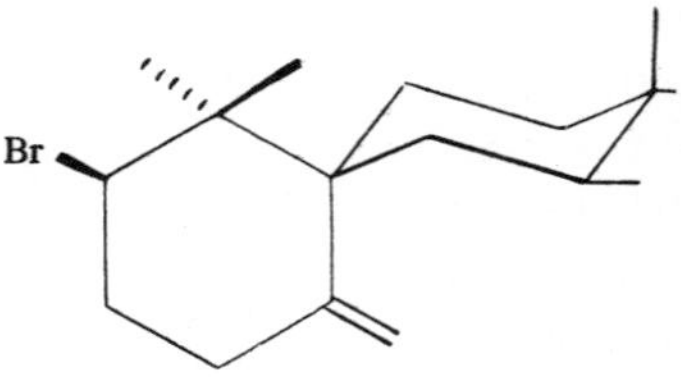

86

The second series of spirobicyclic sesquiterpenes is typified by elatol (**81**). Elatol (**81**) contains the (−)-β-chamigrane skeleton with an exocyclic methylene in the A ring. Carbon chemical shift assignments require the use of residual coupling constants, specific proton decoupling, and a series of related compounds. Off-resonance multiplicities allow the as-

TABLE 15

Carbon Chemical Shifts of A-Ring Exocyclic Chamigrenes[a]

Carbon	Compound					
	81[g]	**82**[g]	**83**[h]	**84**[h]	**85**[h]	**85**[b,h]
1	38.5[c]	38.4[c]	39.8[c]	40.0[c]	30.4[c]	25.5[c]
2	124.0[d]	123.8[d]	124.3[d]	124.5[d]	119.7	37.2[c]
3	127.8[d]	127.7[d]	127.8[d]	127.8[d]	132.7	68.3
4	25.5[e]	25.4[e]	25.6[e]	25.7[e]	25.6[d]	63.5
5	29.2[e]	29.2[e]	29.7[e]	29.7[e]	26.6[d]	40.4[c]
6	49.0	48.8	47.1	47.1	50.8	50.4
7	140.6	140.4	143.8	143.2	145.6	145.7
8	37.9[c]	36.5[c]	37.3[c]	36.3[c]	35.7	35.8[c]
9	72.0	73.3	68.0	70.7	33.1[c]	33.5[c]
10	70.7	62.9	43.5	37.4	66.1	67.9
11	42.9	43.1	37.1	37.1	42.8	43.9
12	20.6[f]	19.9[f]	25.1[f]	25.0[f]	17.5[e]	17.4[d]
13	24.1[f]	24.0[f]	25.9[f]	25.2[f]	23.9[e]	23.6[d]
14	115.6	115.4	114.0	113.9	112.6	114.7
15	19.3	19.2	19.3	19.4	23.1[e]	23.9
Other		169.5		170.2		
		20.8		21.3		

[a] The δ values are in parts per million downfield from TMS in $CDCl_3$ solution.
[b] This compound may not be nidificene (**86**). See text.
[c-f] Signals within a column may be reversed.
[g] Rose and Sims (1977b).
[h] Izac and Sims (1977).

signment of the signals at 115.6 ppm (dd) and 140.6 (s) to the exocyclic methylene carbons C-14 and C-7. In the proton spectrum of elatol (**81**) the exocyclic methylenes are nonequivalent and appear as broadened singlets (δ5.10 and 4.80). This difference in chemical shift of the protons attached to C-14 is reflected in the off-resonance spectrum. Carbon C-14 appears as a doublet of doublets rather than the expected triplet (Hagaman, 1976).

The C-14 shift of 115.6 ppm is at lower field (10.8 ppm) than is reported for the *exo*-methylene (104.8 ppm) of 2,2-dimethylmethylenecyclohexane (Grover and Stothers, 1975). Likewise, the ring carbon, C-7, is shifted to higher field. In 2,2-dimethylmethylenecyclohexane the C-1 olefin resonates at 156.5 ppm, whereas in elatol a similar carbon, C-7, appears at 140.6 ppm. In linear 1-halo-1-alkene systems the halogen-bearing carbon resonates at higher field than its olefinic partner. If this is also the case in tetrasubstituted systems, the olefinic carbons in the B ring, C-2 and C-3, can be assigned to the carbon signals at 124.0 and 127.8 ppm, respectively.

The two low-field methine signals of elatol (**81**), which are assignable to either the hydroxymethine carbon, C-9, or the bromomethine carbon, C-10, are assigned using two complementary methods. In the proton spectrum, H-9 absorbs at δ4.1 (m), whereas H-10 resonates at δ4.62 (d, J = 3 Hz). When the decoupler offset is centered at δ1.05 (in the proton region), the reduced coupling constant for the doublet centered at 70.7 ppm is larger than that of the doublet centered at 72.0 ppm. Since the reduced coupling constants are proportional to the frequency difference of the decoupler offset and the proton absorption, this allows C-10, whose protons resonate at lower field and hence have a larger reduced coupling constant, to be assigned to the carbon resonance at 70.7 ppm. Therefore, C-9 is represented by the signal at 72.0 ppm.

The second method of assigning the methine carbons employs selective proton decoupling experiments. When the decoupler offset (at low power level) is centered on the proton signal at δ4.62 (CHBr), the carbon signal at 70.7 ppm collapses to a singlet while the 72.0 ppm signal remains a doublet. Likewise, selective decoupling of the δ4.01 (CHOH) proton resonance causes the 72.0 ppm signal to collapse to a singlet while the signal centered at 70.7 ppm remains a doublet. This confirms the assignment of C-9 to 72.0 ppm and C-10 to the resonance at 70.7 ppm.

The three methyl groups of elatol (**81**), which appear as quartets centered at 24.1, 20.6, and 19.3 ppm, are differentiated as follows. Selective on-resonance decoupling experiments allow the unambiguous assignment of C-15. In the ^{1}H nmr spectrum of elatol (**81**), the vinyl methyl (C-15) resonates at δ1.70, whereas the *gem*-dimethyl protons are equivalent and are absorbing at δ1.05. Irradiation of H-15 collapses the ^{13}C nmr signal at 19.3 ppm, thus confirming its assignment to C-15. Likewise, irradiation of the six-proton singlet at δ1.05 (C-12 and C-13) collapses the ^{13}C nmr signals at 24.1 and 20.6 ppm. The *gem*-dimethyl signals are distinguished by steric compression shift arguments.

In general, axial methyl groups resonate at higher field than do equatorial methyl groups. This high-field shift of axial methyls is attributed to shielding γ-gauche interactions. γ-Gauche interactions perturb the C–H bond with a concomitant drift of electron density toward the carbon atom, which results in increased shielding (Dalling and Grant, 1972). Examination of Dreiding molecular models suggests that the axial methyl (C-12) experiences more severe γ-gauche interactions with H-9 and H-5. The equatorial methyl (C-13) also experiences γ-gauche interactions with H-1 and H-5, but the average C-12–H-9 and C-12–H-5 distance is smaller than the average C-13–H-1 and C-13–H-5 distance. Thus, C-12 is assigned to the higher-field resonance at 20.6 ppm. Carbon C-13 is then the 24.1 ppm signal.

The four methylene carbons in elatol (**81**) are assigned by residual coupling and calculated chemical shift considerations. The residual coupling constants of the methylene signals at 38.5 and 37.9 ppm are larger than the signals at 29.2 or 25.4 ppm. The low-field signals could be assigned to either C-8 or C-1. The protons attached to these two centers resonate at lower field than do those attached to C-4 or C-5. The 37.9 ppm signal is assigned to C-8 by virtue of its upfield shift in elatol acetate (**82**). Cyclohexane ring substituent parameters predict that acetylation of an equatorial alcohol would result in a net 3 ppm upfield shift of the β carbon (Wehrli and Wirthlin, 1976, p. 45). In elatol acetate (**82**), C-8 is shifted upfield but only by 0.6 ppm.

The remaining three methylene signals are assigned by calculated chemical shifts. In 1-methylcyclohexene (**88**), C-3 resonates at 26.7 ppm, C-4 and C-5 at 24.4 ppm, and C-6 at 31.5 ppm. By the use of methyl substituent parameters and geminal correction terms (Wehrli and Wirthlin, 1976, p. 45), the calculated values for 1,4,4-trimethylcyclohexene (**88a**) are obtained. From model compound **88a**, C-4 is assigned to the resonance at 25.5 ppm and C-5 is assigned to the resonance at 29.2 ppm. Carbon C-1, adjacent to a fully substituted olefinic carbon and a quaternary center, is assigned to the lowest field signal at 38.5 ppm.

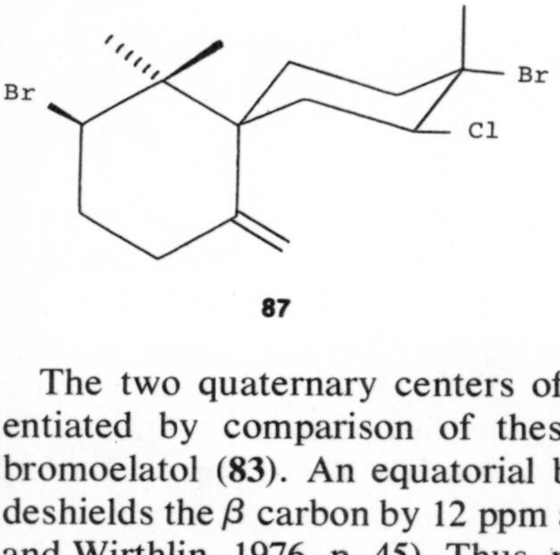

87 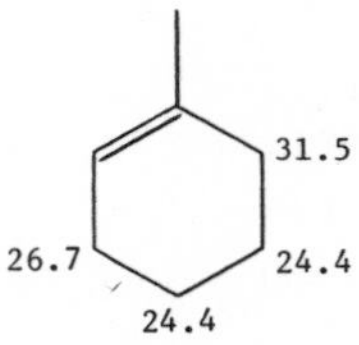 **88** 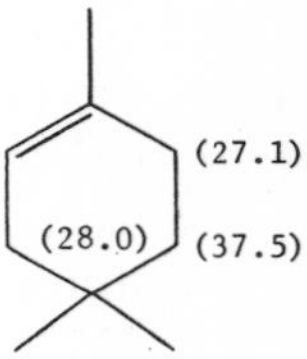 **88a**

The two quaternary centers of elatol (**81**), C-6 and C-11, are differentiated by comparison of these chemical shifts with those of debromoelatol (**83**). An equatorial bromine atom on a six-membered ring deshields the β carbon by 12 ppm and the γ carbon by only 1 ppm (Wehrli and Wirthlin, 1976, p. 45). Thus, removal of the bromine atom at C-10 of elatol should result in a 12 ppm upfield shift of the β C-11 quaternary center, whereas C-6, which is gamma to the halogenated center, should be only slightly affected. Thus, the C-11 of elatol (**81**) is assigned to the resonance at 42.9 ppm, which shifts upfield (5.8 ppm) to 37.1 ppm in debromoelatol (**83**). Therefore, C-6 is assigned to the 49.0 ppm signal, which also shifts upfield (1.9 ppm) in debromoelatol (**83**) to 37.1 ppm. The small upfield shift (5.8 ppm) of the quaternary center upon removal of the bromine atom is noteworthy. The substituent parameters quoted (Wehrli and Wirthlin, 1976, p. 45) for bromine were determined in systems where the β and γ carbons were methylenes. Quaternary carbons, which are less polarizable, should be less affected by inductive effects of substituents at adjacent centers. Shifts of quaternary centers will therefore be smaller than shifts of methylene centers.

The ^{13}C nmr chemical shifts of elatol acetate (**82**) (Table 15) follow from the assignments of elatol (**81**). Acetylation of an alcohol should shift the α carbon to lower field. This trend is observed in elatol acetate (**82**) when C-9 resonates at 73.3 ppm, a downfield shift of 1.3 ppm from elatol (**81**). The C-10 bromomethine carbon of elatol acetate (**82**) resonates at 62.9 ppm, an upfield shift of 7.8 ppm from elatol (**81**) (C-10, 70.7 ppm). A similar shift of the bromine carbon is observed upon acetylation of bromohydrins and probably reflects the greater strain repulsion of the acetate. The acetate methyl, which resonates at $\delta 2.08$, is assigned to the ^{13}C nmr signal at 20.8 ppm. This assignment is confirmed by selective on-resonance decoupling experiments.

Debromoelatol (**83**) is prepared by treatment of elatol (**81**) with lithium aluminum hydride in diethyl ether. For the most part, the ^{13}C nmr chemical shifts can be interpreted by reference to elatol (**81**). Several signals, however, are worthy of attention. The appearance of a new off-resonance triplet centered at 43.5 ppm is assigned to the C-10 methylene. Carbon C-1 of debromoelatol (**83**) resonates at 39.8 ppm, 1.3 ppm downfield from elatol (**81**) (38.5 ppm). The *gem*-dimethyl carbons, C-12 and C-13, are nearly equivalent and resonate at 25.9 and 25.1 ppm, respectively.

The structure of 10-bromo-β-chamigrene (**85**), a metabolite of a common California red alga (*Laurencia pacifica*), was elucidated by Izac and Sims (1977). The A-ring carbons are assigned using elatol (**81**) as a model system. Removal of the C-10 bromine atom of elatol (**81**) shifts the C-9 hydroxymethine upfield by 4 ppm. It is anticipated that removal of the C-9 hydroxide of elatol (**81**) would therefore shift the C-10 bromomethine carbon upfield by 4 ppm. This is, in fact, observed, and the C-10 bromomethine of 10-bromochamigrene resonates at 66.1 ppm. B-Ring assignments follow from those of elatol (**81**).

Compounds **71**, **72**, **73**, **75**, and **76** contain diequatorial bromine and chlorine substituents at C-4 and C-3. In these compounds the tertiary carbon bearing bromine resonates at 62.7 ± 0.4 ppm. We have isolated a dibromochlorochamigrene compound from *Laurencia obtusa*, green variety, which by ^{1}H nmr (60 MHz) and infrared spectral comparison is nidificene (**86**). The ^{13}C nmr spectrum of this compound displays a singlet at 68.3 ppm and a doublet at 63.5 ppm. Although the chemical shifts of these carbon atoms indicate the attachment of halogens, we are not convinced that the 68.3 ppm signal is a quaternary carbon bearing chlorine. It is possible that our compound has chlorine and bromine interchanged and should be represented by structure **87**.

Some tentative support for structure **87** is found in the ^{13}C nmr chemical shift of the C-15 methyl group. In compounds **71**, **72**, **73**, **75**, and **76** C-15 resonates at 24.3 ± 0.4 ppm. In our compound, C-15 resonates at 23.6 ppm. Although it is difficult to use the chemical shift of quaternary carbons as a diagnostic tool for differentiation of attached electronegative atoms, in a homologous series the ^{13}C nmr chemical shift might be useful for excluding a substituent. In compounds **71**, **72**, **73**, **75**, **76**, and **118**, the quaternary chlorine-bearing carbon adjacent to an equatorial bromine atom resonates at 70.9 ± 0.6 ppm. In our compound the quaternary carbon bearing halogen resonates at 68.3 ppm. Although the chemical shift difference between the numbers cited is very small, together they prompt a closer look at the structure of the *L. obtusa* metabolite.

3. Cyclic Ethers

The chemical shift assignments of pacifenol (**89**), johnstonol (**90**), and johnstonol acetate (**91**) (Table 16) are determined by off-resonance and selective proton decoupling, gated decoupling experiments, and correlation to previously assigned compounds. Chemical shift assignments for nidifocene (**92**) (Waraszkiewicz *et al.*, 1977) are determined by comparison to those of compounds **79**, **86**, and **89**.

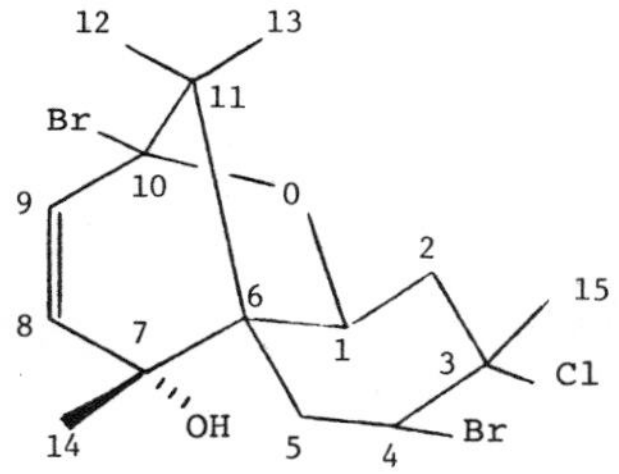

89

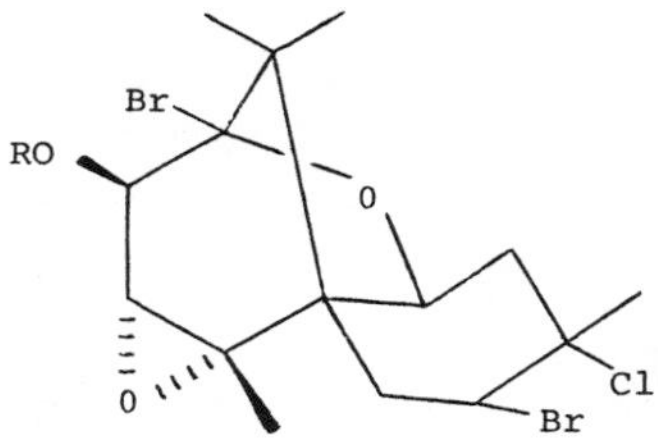

90 R=H
91 R=Ac

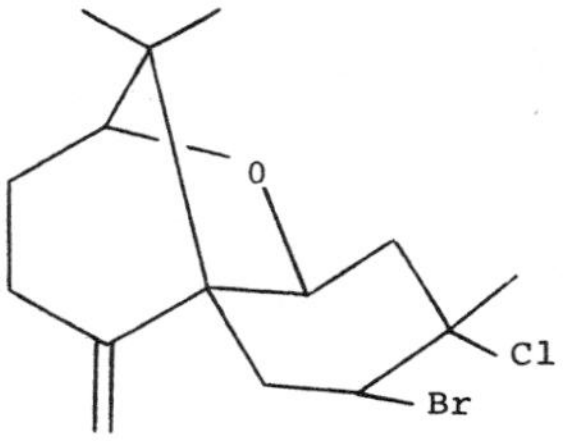

92

TABLE 16

13C nmr Chemical Shifts of Chamigrene Cyclic Ethers[a]

Carbon	Compound			
	89[f]	**90**[g]	**91**[g]	**92**[b,h]
1	74.1	74.8	74.8	78.5
2	34.3	33.9	33.9	32.0
3	69.0	68.1	68.1	n.o.[c]
4	59.4	58.1	58.0	60.8
5	46.1	45.4	45.4	47.5
6	53.4[d]	50.7[d]	50.8[d]	n.o.
7	77.0	61.4	61.4	n.o.
8	134.3	60.0	59.1	31.7[d]
9	132.4	74.8	73.1	28.4[d]
10	99.8	113.6	106.8	85.4
11	52.0[d]	49.6[d]	49.8[d]	n.o.
12	23.5[e]	18.4[e]	18.4[e]	22.2[e]
13	24.7[e]	21.7[e]	21.6[e]	24.3[e]
14	33.6	31.3	31.3	106.1
15	25.2	24.8	24.6	28.1
Other			168.5	
			21.0	

[a] The δ values are in parts per million downfield from TMS in $CDCl_3$ solution.
[b] In C_6D_6 solution.
[c] n.o., not observed.
[d,e] Signals within a column may be reversed.
[f] Rose and Sims (1977b).
[g] Izac and Sims (1977).
[h] Waraszkiewicz *et al.* (1977).

The chemical shift of C-10 in pacifenol (**89**), which bears an oxygen and a bromine, is reminiscent of the shift of ketals. The olefinic carbons of pacifenol (**89**) were differentiated by a gated decoupling experiment, which showed C-9 (132.4 ppm) to be a doublet and C-8 (134.3 ppm) to be a doublet of quartets resulting from three-bond coupling to the C-14 methyl group. Two quaternary centers, C-6 and C-11, cannot be distinguished, and their assignments may be reversed. In the A-ring endocyclic and the A-ring exocyclic chamigrenes, C-6 resonates at lower field than does C-11. The same chemical shift trend is assumed to hold for the cyclic ether series, and C-6 is assigned to the lower-field signal.

The two methylene signals of pacifenol (**89**), 46.1 ppm and 34.3 ppm, are assigned to C-5 and C-2, respectively. Both centers are adjacent to quaternary centers. The C-2 methylene is adjacent to an axial ether, which should deshield this center by only 2 ppm. On the other hand, C-5 is adjacent to an equatorial bromine atom, which should deshield it by 12 ppm. Thus, the lower-field signal at 46.1 ppm is assigned to C-5.

The assignments of C-1 through C-5 of johnstonol (**90**) and its acetate (**91**) follow from those of pacifenol (**89**). The epoxide carbons, C-7 and C-8, are assigned to the resonances at 61.4 ppm and 60.0 ppm, respectively, by their chemical shift and carbon multiplicities. In johnstonol acetate (**91**) the C-9 acetate and the C-10 bromine atom are diequatorial. Steric repulsion results in an upfield shift of both centers, C-10 by 6.8 ppm and C-9 by 1.7 ppm.

In nidifocene (**92**), the assignments of C-1 through C-5 are made by reference to pacifenol (**89**) (Table 16). Carbon C-14 is assigned by its characteristic shift of 106.1 ppm. Carbon C-10, which lacks the bonded bromine atom of pacifenol (**89**), is shifted upfield by 14.4 ppm, resonating at 85.4 ppm.

C. Miscellaneous Sesquiterpenes

A number of additional sesquiterpenes have been isolated from marine organisms. Oppositol (**93**) (Hall *et al.*, 1973) was isolated from the red alga *Laurencia subopposita*, but its ^{13}C nmr spectrum was not published until the alga was reinvestigated (Wratten and Faulkner, 1977b). The ^{13}C nmr chemical shift assignments, shown on the formula (**93**), were determined by chemical shift values, carbon multiplicities, and model compounds. The side chain signals were assigned by reference to olefin **94** (Couperus *et al.*, 1976). Dalling *et al.* (1973) showed that the bridgehead methyl group of *trans*-decalin is strongly shielded because of 1,3 interactions, resulting in a high-field resonance of 15.8 ppm. Molecular models suggested that the bridgehead methyl, C-13, of trans-fused indane systems should experience similar 1,3 interactions, causing a high-field absorption of this group. Thus, the bridgehead axial methyl group, C-13, was assigned to the

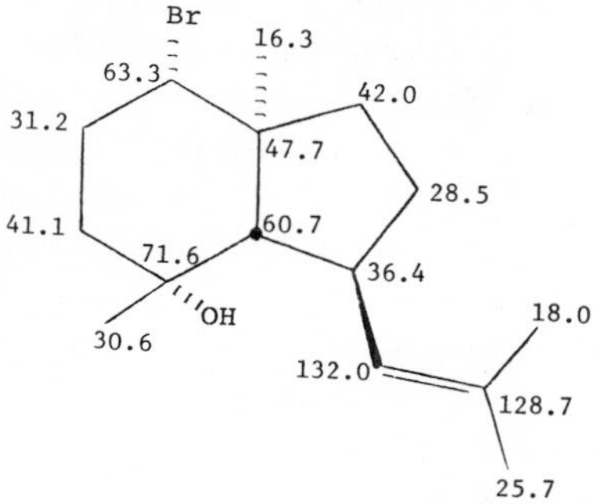

93

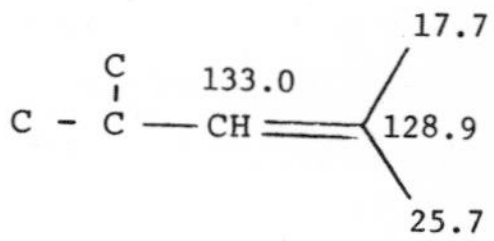

94

highest-field quartet at 16.3 ppm. The steric compression (high-field) shift of the methyl group on *cis*-olefins (Levy and Nelson, 1972) permitted assignment of the resonance at 18.0 ppm to the C-11 methyl group.

Cafieri *et al*. (1973) and Iengo *et al*. (1977) reported compounds related to oppositol (**93**) from a marine sponge (*Axinella cannabina*) and proposed that the carbon skeleton be referred to as the axane skeleton. These compounds contain additional isonitrile, isothiocyanate, and formamido functionality at C-8. Unfortunately, no carbon data have yet been reported.

In addition to oppositol (**93**), Wratten and Faulkner (1977b) obtained a new algal metabolite having the germacrane skeleton (**95**). The proposed structure, 7β,10α-dihydroxy-3β-isopropyl-10β-methyl-6-methylene-*trans*-cyclodecene (**96**), was obtained by analysis of the 220 MHz ^{1}H nmr spectrum and its rearrangement to 8(15)-dehydrooplopanone (**97**).

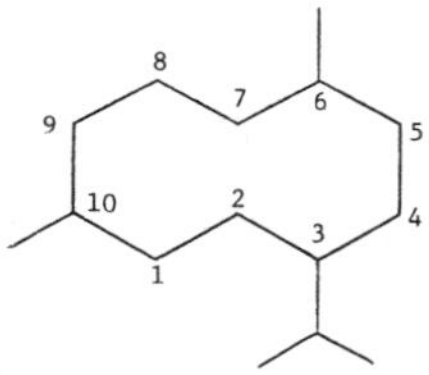

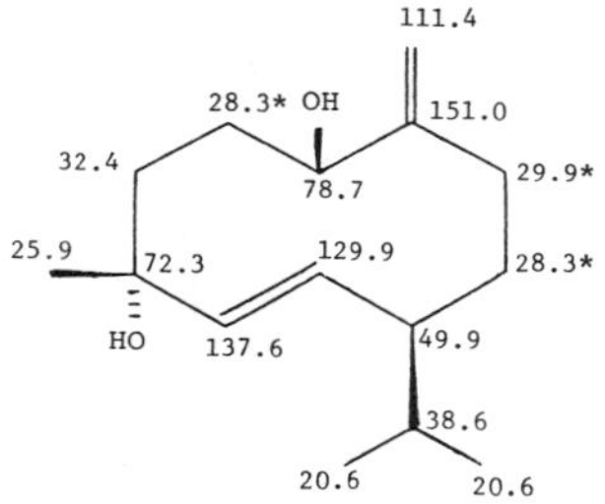

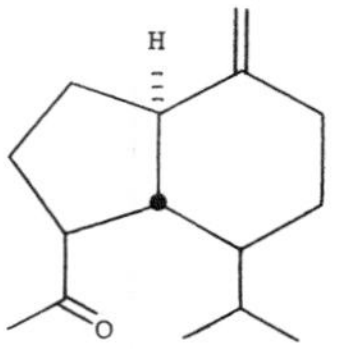

A novel sesquiterpenoid isonitrile dichloride possessing the axinyssane ring skeleton (**98**) was reported by Wratten and Faulkner (1977a). The structure of the isonitrile dichloride (**99**) was determined by a combination of chemical and spectroscopic evidence. The isonitrile dichloride carbon, which resonates at 127.1 ppm, generates a very low intensity signal. Confirmation of this chemical shift was obtained by the synthesis of cyclohexylisonitrile dichloride (**100**). In this compound the isonitrile dichloride carbon resonated at 122.0 ppm. The side chain carbons (C-7 through C-12) were assigned by comparison to model olefin **101** (Couperus *et al*., 1976). The rather high field shift of C-8 (25.9 ppm in model **101** versus 21.7 ppm in **99**) suggests strong 1,3 interactions between C-8 and C-13.

Fenical and Howard (1978) obtained 1*S*-bromo-4*R*-hydroxyselin-7-ene (**102**) as a metabolite of a *Laurencia* species collected in the Gulf of California and determined the structure by chemical methods. Compound **102** was also obtained by Rose and Sims (1977a) as a minor metabolite of the same genus collected in Australia. Compound **103**, obtained by dissolving metal reduction of compound **102**, was used to assign C-2, C-3,

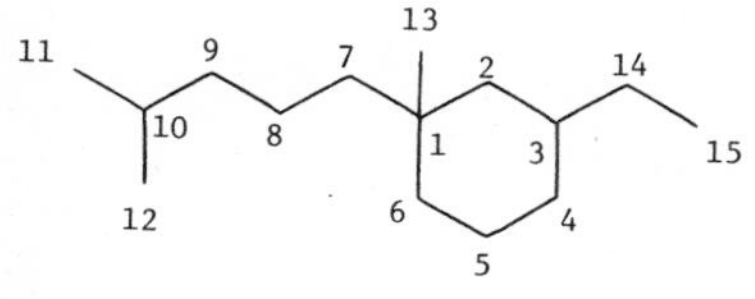

98

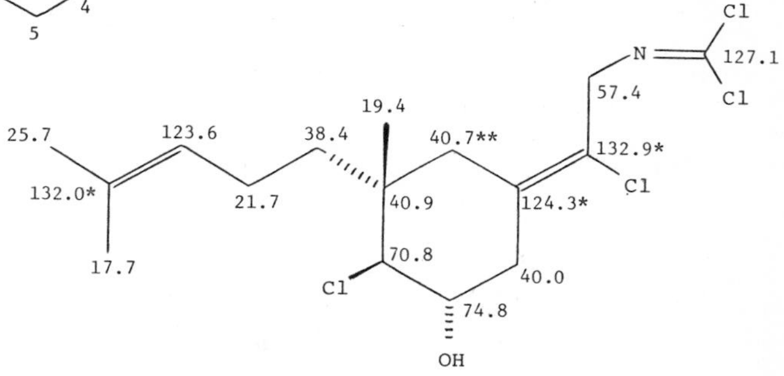

99

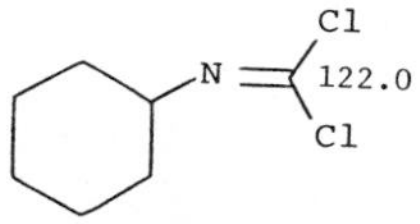

100

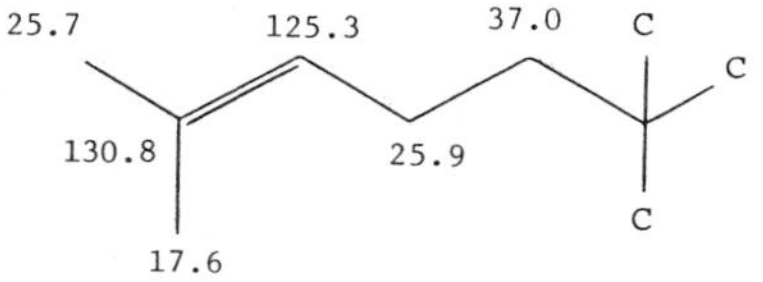

101

C-6, and C-9; the results are given in Table 17. A related metabolite (**104**) along with compound **102** was isolated from a *Laurencia* species also collected in Australia. Compound **104** was crystalline, and its structure, including absolute stereochemistry, was determined by X-ray analysis (Rose *et al.*, 1978).

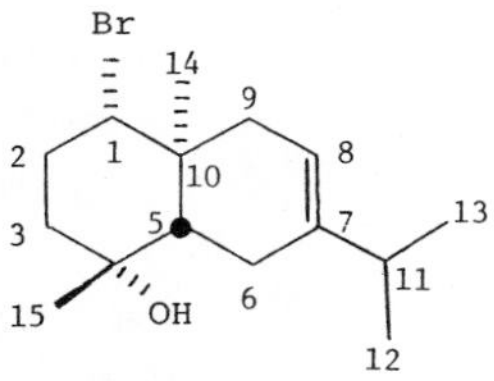

102

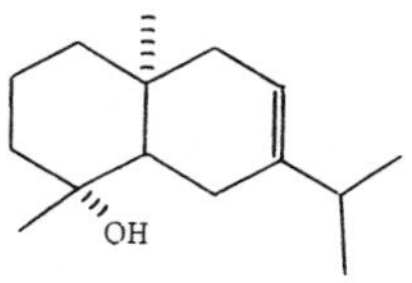

103

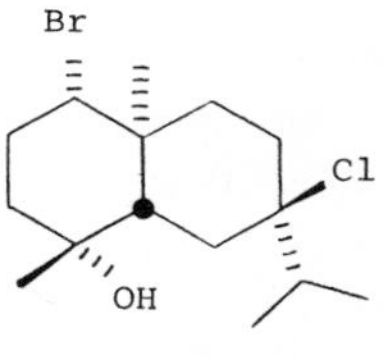

104

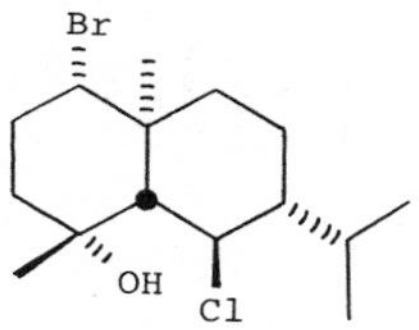

105

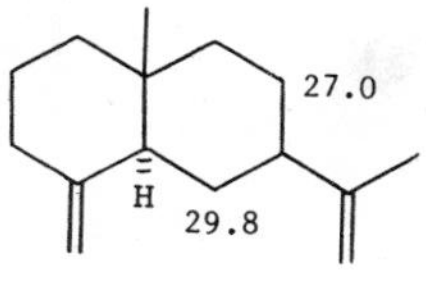

106

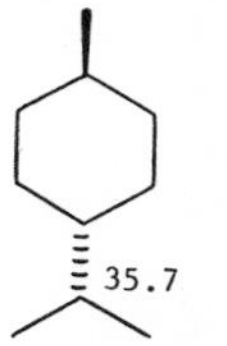

107

TABLE 17

Carbon Chemical Shifts of Selinane Derivatives[a]

Carbon	Compound				
	102[b,e]	**102**[f]	**103**[f]	**104**[f]	**105**[g]
1	68.5	68.4	41.3	67.5	67.5
2	24.6	24.3	18.8	30.3	30.4
3	42.5[c]	42.4[c]	41.6[c]	42.7[c]	44.4[c]
4	70.4	71.0	71.6	71.4	72.9
5	48.1	48.2	47.3	47.4	58.4
6	30.5	30.2	30.2	37.3	63.6
7	142.0	141.9	142.5	81.0	52.3
8	116.4	116.2	116.6	33.8[c]	19.3
9	43.1[c]	42.9[c]	44.6[c]	33.0[c]	39.9[c]
10	38.6	38.4	32.4	39.4	42.4
11	34.9	34.8	35.1	40.9	27.2[d]
12	21.9	21.8	21.9	17.7	21.2
13	21.3	21.2	21.3	17.6	15.2
14	14.2	14.0	18.0	14.3	14.9
15	29.6	29.8	23.3	29.8	35.7[d]

[a] The δ values are in parts per million downfield from TMS in $CDCl_3$ solution.
[b] In C_6D_6 solution.
[c] Signals within a column may be reversed.
[d] See text.
[e] Fenical and Howard (1978).
[f] Rose and Sims (1977a).
[g] Kazlauskas *et al.* (1977c).

Laurencia filiformis grows in three distinct forms: *heteroclada, dendritica,* and *filiformis*. Kazlauskas *et al.* (1977c) collected *L. filiformis* f. *heteroclada* at a depth of 2 m near Cape Jervis, South Australia, and isolated an additional selinane derivative, heterocladol (**105**). As in the case of compound **104**, the structure and absolute configuration of heterocladol (**105**) were determined by single-crystal X-ray diffraction studies.

When compared to sesquiterpene **106** (Wenkert *et al.*, 1976), the chemical shift of C-8 (19.3 ppm) of heterocladol (**105**) seems slightly high field. The equatorial chlorine atom at C-6 could inhibit free rotation of the isopropyl group, which would increase the steric environment about C-8. This increased steric crowding would result in a shift of C-8 to higher field than that observed in sesquiterpene **106**.

The heterocladol (**105**) carbon resonances at 27.2 and 35.7 ppm were assigned to C-11 and C-15 on the basis of the reported doublet and quartet multiplicities, respectively, of these signals. However, reference to monoterpene **107** (Bohlmann *et al.*, 1975) in which the *gem*-dimethyl carbon resonates at 35.7 ppm, and oppositol (**93**), in which the hydroxyl methyl resonates at 30.6 ppm, suggests that these multiplicity assignments should be reinvestigated.

Spirolaurenone (**108**) (Suzuki *et al.*, 1970) probably results from ring contraction of the chamigrene epoxide (**109**) since both compounds are present in *L. glandulifera* (Suzuki *et al.*, 1974a). The ^{13}C nmr chemical shifts (Kurosawa, 1977) for spirolaurenone (**108**) are given on the formula.

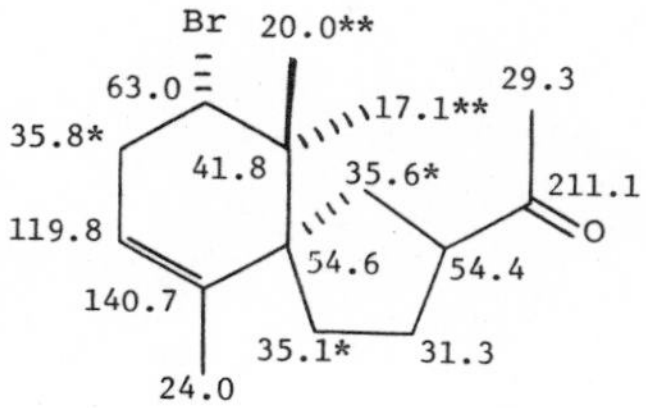

108

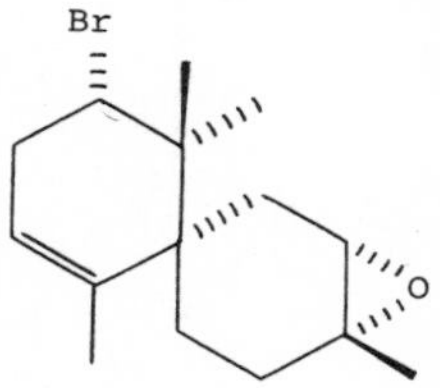

109

Two pathways have been proposed for the biosynthesis of the *Laurencia* chamigrene metabolites. Howard and Fenical (1976a) isolated two isomeric alcohols, α-snyderol (**110**) and β-snyderol (**111**), from *L. snyderiae*. The presence of β-snyderol (**111**) in *Laurencia* was used to support the proposed monocyclofarnesol biosynthetic pathway (Kato *et al.*, 1970). The ^{13}C nmr chemical shift assignments, indicated on the formulas, were aided by the use of monoterpene **112** (Bohlmann *et al.*, 1975) as a model.

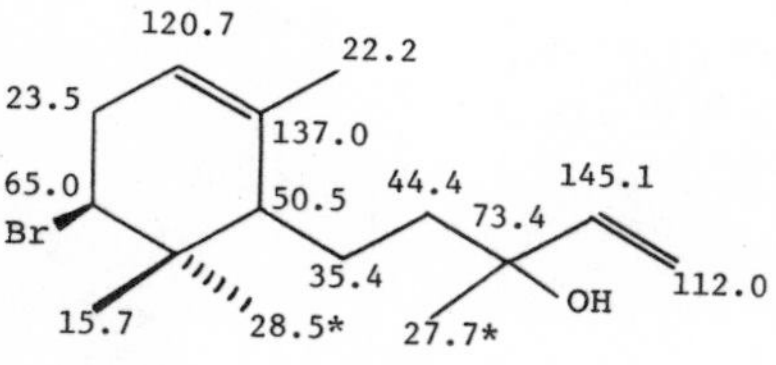

110

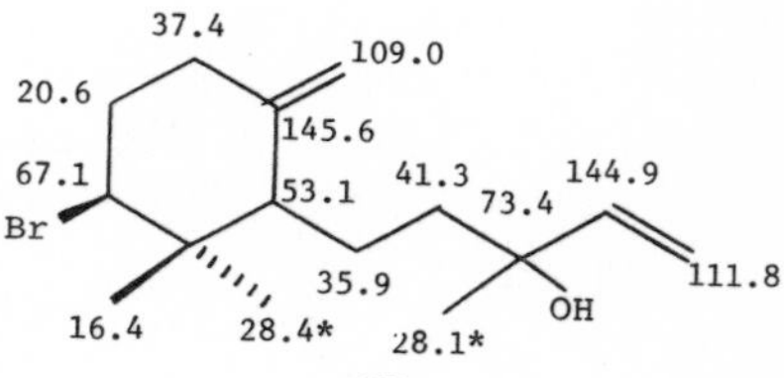

111

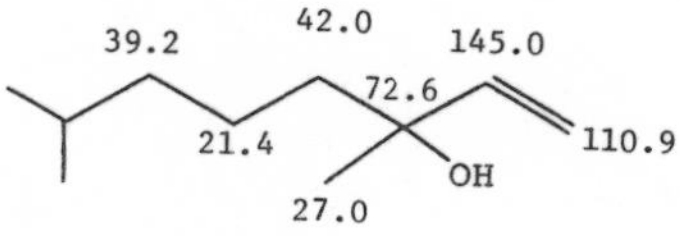

112

Cyclization of α- or β-snyderol to generate a six-membered cyclic ether yields 3β-bromo-8-epicaparrapi oxide (**113**) (Faulkner, 1976), a metabolite of *L. obtusa* collected from the English Channel. The *gem*-dimethyl groups on C-4 were assigned by comparison to compound **114** (Wehrli and Wirthlin, 1976, p. 44). The highest-field methylene (22.4 ppm) was assigned to C-2, again using compound **114** as a model. The remaining methylene signals could not be unambiguously assigned. Likewise, the two quaternary ether carbons could not be distinguished.

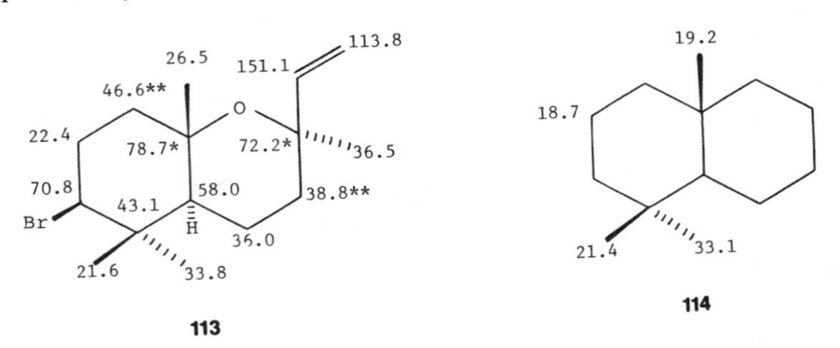

113 **114**

A second proposed biosynthetic pathway leading to the *Laurencia* chamigrenes required an intermediate bisabolene (González *et al.*, 1976). Support for the bisabolene route was found in the isolation of isocaespitol (**115**) (González *et al.*, 1975) and caespitol (**117**) (González *et al.*, 1974) from *L. caespitosa*. The ^{13}C nmr chemical shift assignments for isocaespitol acetate (**116**) and caespitol acetate (**118**) are indicated on the formulas (González *et al.*, 1977b).

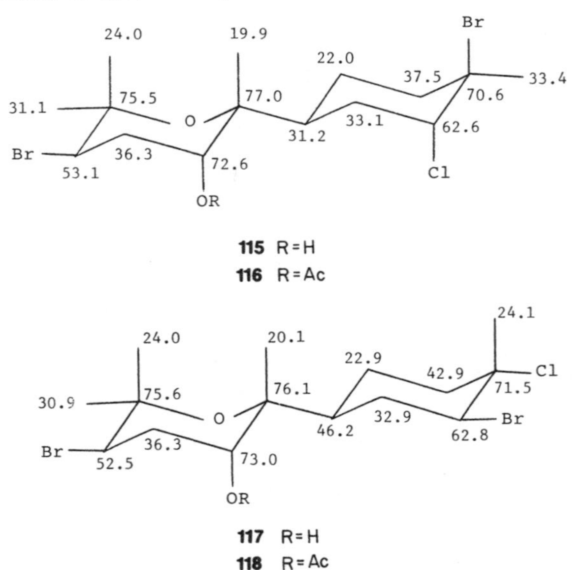

115 R=H
116 R=Ac

117 R=H
118 R=Ac

Two related furanosesquiterpenes, isofuranocaespitane (**119**) and furanocaespitane (**120**), were elucidated by González *et al.* (1977b).

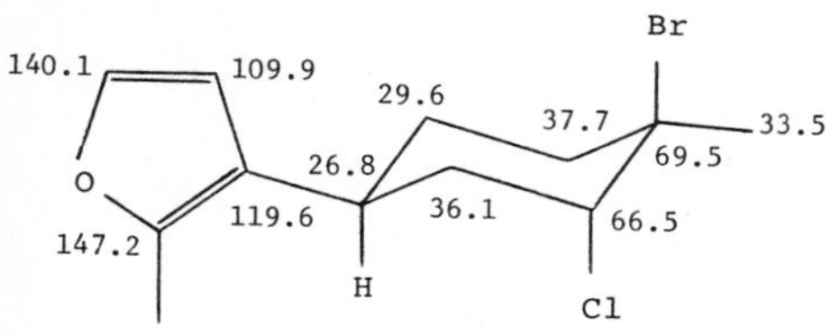

119

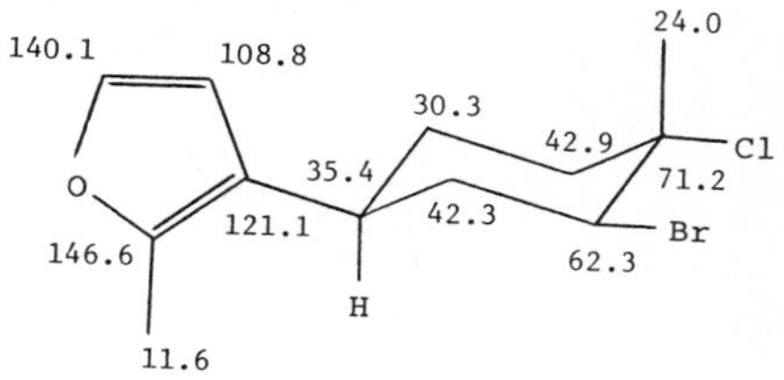

120

Carbon-13 nmr data are available for two nonhalogenated sesquiterpenoid compounds related to snyderol. Solvolysis of the bromine atom of α-snyderol (**110**) followed by methyl migration and elimination of a proton leads to compound **121** (Sun *et al.*, 1976b), a metabolite of *L. nidifica*. The ring carbon signals are assigned by comparison with compound **122** (Bohlmann *et al.*, 1975), and the side chain resonances are assigned by reference to α-snyderol (**110**). The side chain methylene carbons of **121** are shifted to higher field when compared to those of α-snyderol (**110**). Molecular models suggest that ring flattening caused by the introduction of a second double bond into the ring results in increased steric congestion about both chain methylenes. Steric confinement of the methylenes results in their shift to higher-field resonance.

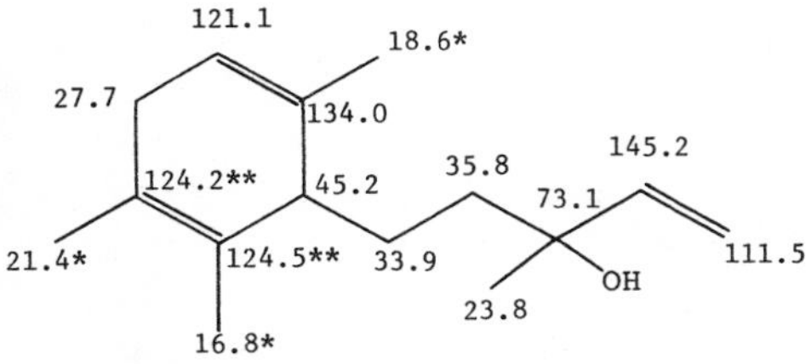

121

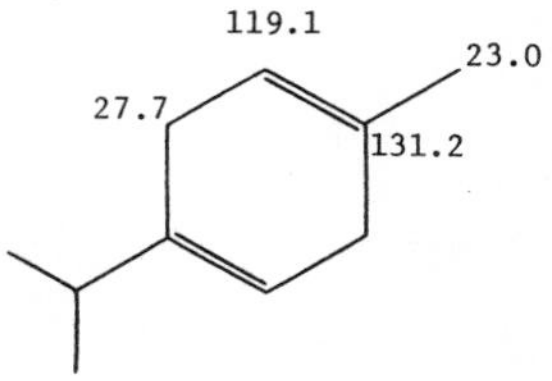

122

Dactyloxene B (**123**) (Schmitz and McDonald, 1974), a metabolite of the sea hare *Aplysia dactylomela*, is also related to α-snyderol (**110**). As probably happens with compound **121**, solvolysis of the bromine atom followed by methyl migration, charge migration, and cyclization yields dactyloxene B (**123**).

Suzuki *et al.* (1978) reported the structural elucidation of bromo ether **124**, a metabolite of the red alga *Laurencia nipponica* Yamada. The structure, including absolute configuration, was determined by X-ray diffraction analysis. The cmr chemical shifts are given on the formula.

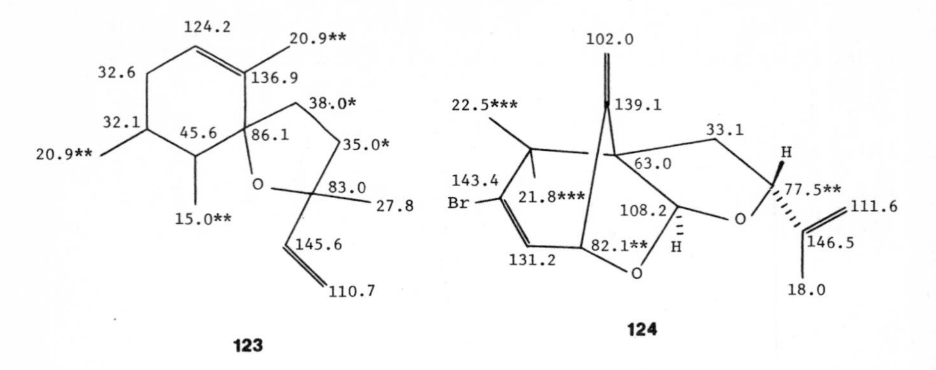

123 **124**

Although a number of furanosesquiterpenes have been isolated from sponges, few have been isolated from other marine sources. Coll *et al.* (1977) reported the isolation of the furanosesquiterpene **125** from the soft coral *Sinularia gonatodes* (phylum Coelenterata).

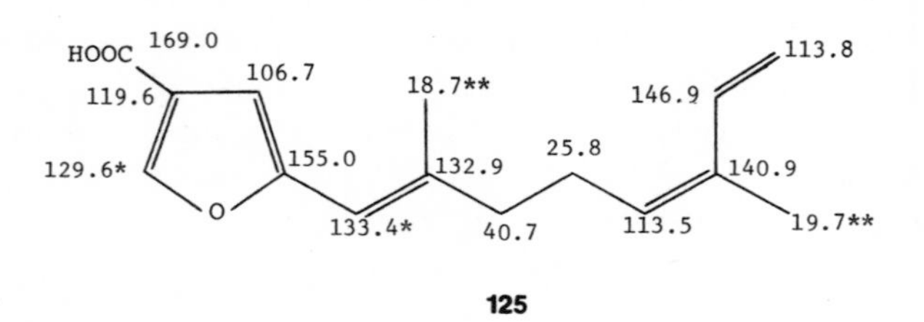

125

The brown alga *Dictyopteris undulata* (= *zonarioides*) has yielded zonarol (**126**) (Fenical and Sims, 1973), a drimane sesquiterpene substituted with a hydroquinone. A related compound, *ent*-chromazonarol (**127**) (Cimino *et al.*, 1975b), has been reported as a metabolite of the sponge *Disidea pallescens*. Migration of one of the *gem*-dimethyls of *ent*-chromazonarol (**127**), analogous to the β-amyrin to friedelin rearrangement (Corey and Ursprung, 1956), could generate avarol (**128**), a metabolite of the related sponge *D. avara* (de Rosa *et al.*, 1976). Catalytic hydrogenation of avarol dimethyl ether (**129**) yields the dihydro compound **130**.

Carbon chemical shift assignments for avarol dimethyl ether (**129**) and dihydroavarol dimethyl ether (**130**) were made by comparison to *trans*-clerodane (**131**) (de Rosa *et al.*, 1976); the results are summarized in Table 18. The ^{13}C nmr data for zonarol (**126**) are also listed in Table 18; assignment of the ring carbons follows from model compound **132** (Wehrli and Wirthlin, 1976, p. 44), whereas assignment of the aromatic ring was determined by calculation (Wehrli and Wirthlin, 1976, p. 47).

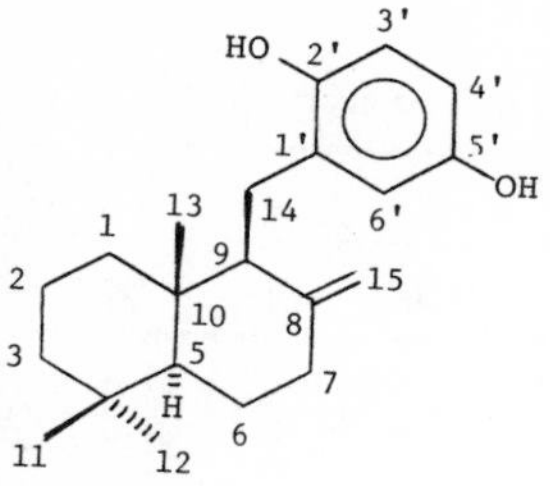

126

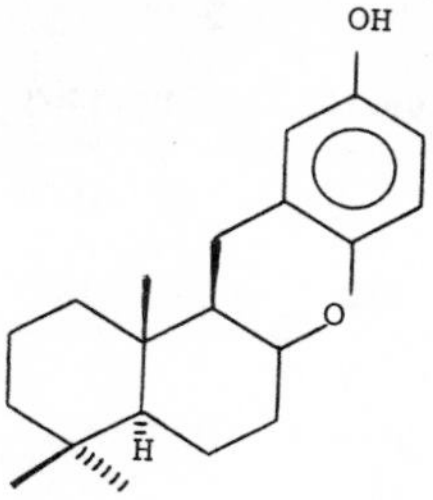

127

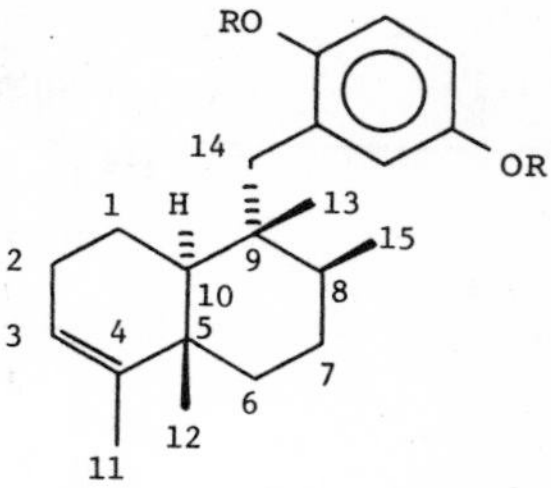

128 R = H
129 R = CH_3

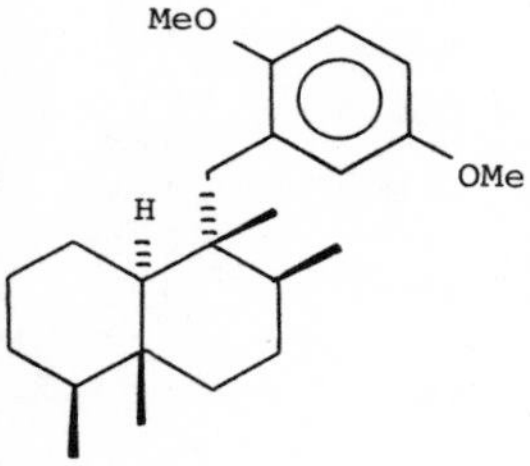

130

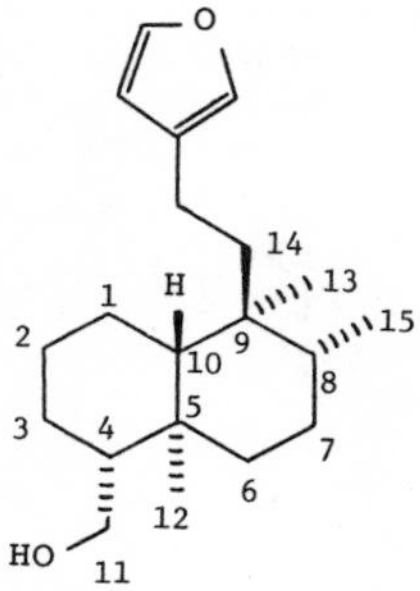

131

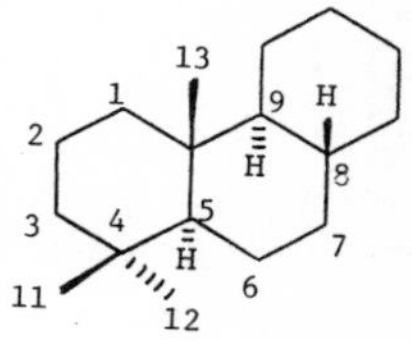

132

Soft corals belong to the phylum Coelenterata, class Anthozoa, in the order Alcyonacea. Sinularene (**133**) (Beechan *et al.*, 1977), a major metabolite of the soft coral *Sinularia mayi*, is a rearranged sesquiterpene based on a novel carbon skeleton. Treatment of sinularene (**133**) with osmium tetroxide gave an oily diol that formed a crystalline mono-*p*-bromobenzoate derivative (**134**). Single-crystal X-ray diffraction studies on the crystalline derivative confirmed the structure of sinularene (**133**).

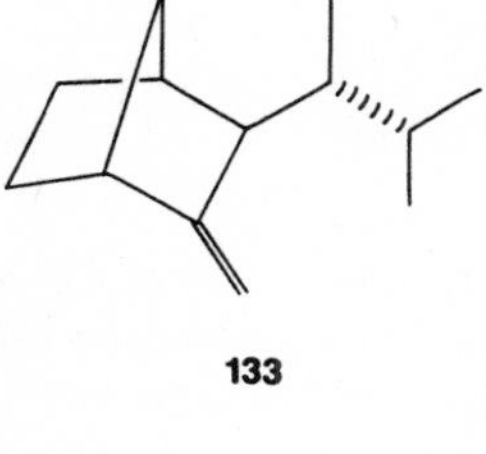

133

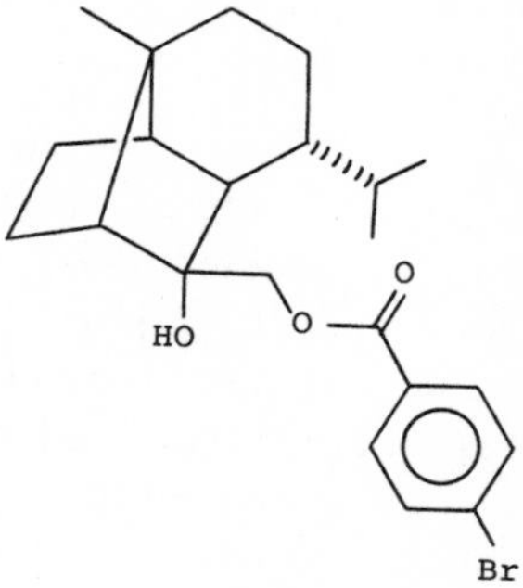

134

TABLE 18

13C nmr Shifts of Zonarol, ent-Chromazonarol, and Avarol[a]

	Compound				
Carbon	**126**[b,g,h]	**132**[i]	**129**[j]	**130**[j]	**131**[j]
1	37.4	39.1	19.5	22.3	21.6
2	19.0	19.0	26.4[c]	26.4[c]	27.3[c]
3	41.8	42.2	120.5	30.4	25.5
4	33.2	33.3	144.5	45.1	54.5
5	55.0[c]	55.5	41.8	41.3	37.0
6	22.8[d]	21.8	37.0[d]	38.0	38.6[d]
7	23.9[d]	35.9	27.6[c]	27.1[c]	27.0[c]
8	147.7	36.8	36.0	36.7[d]	36.5
9	55.7[c]	56.3	38.2	38.7	38.8
10	33.2	36.9	45.8	48.9	49.8
11	21.5	22.0	19.8[e]	12.6[e]	63.6
12	33.4	33.6	17.8[e]	14.4[e]	15.2
13	14.3	14.3	17.1[e]	16.9[e]	16.1
14	41.8	25.1	35.7[d]	35.8[d]	39.5[d]
15		[f]	17.5[e]	17.1[e]	18.2
1′	132.9 (128.7)		129.1	129.4	
2′	147.8 (147.3)		153.3	153.4	
3′	117.3 (115.9)		111.1	111.4	
4′	114.7 (112.2)		111.0	111.0	
5′	149.4 (148.2)		153.1	153.6	
6′	116.9 (115.1)		119.4	119.2	
OCH_3			55.4	55.3	
OCH_3			55.3	55.1	

[a] The δ values are in parts per million downfield from TMS in $CDCl_3$ solution.
[b] In DMSO-d_6 solution.
[c-e] Signals within a column may be reversed.
[f] Obscured by solvent.
[g] Numbers in parentheses are calculated chemical shifts.
[h] Rose and Sims (1977a).
[i] Wehrli and Wirthlin (1976, pp. 36–39).
[j] de Rosa *et al.* (1976).

Carbon-13 nmr studies of sinularene (**133**) revealed the presence of three quartets (20.7, 20.7, and 21.4 ppm), five triplets (22.4, 26.0, 30.2, 31.7, and 101.1 ppm), five doublets (29.1, 42.9, 49.2, 50.2, and 53.3 ppm), and two singlets (46.9 and 162.9 ppm). Additional studies will be required for chemical shift assignments to this hydrocarbon.

Cimino *et al.* (1975a) divided their collection of the sponge *Pleraplysilla spinifera*, which grows on the coelenterate *Paramuricea camaleon* in the Bay of Naples, into two morphologically distinct but taxonomically identical forms. Spiniferin-1 (**135**), the major furanoid component (0.32% of dry animal) of this sponge, was incorrectly assigned to one of two equally plausible alternative structures, **136** or **137**. The ^{1}H nmr signals at $\delta 0.75$ (d, $J = 10$ Hz) and 3.62 (d, $J = 10$ Hz) were attributed to cyclopropane proton resonances. A hint that the proposed structure was incorrect could be seen in the ^{13}C nmr data, about which the authors stated that "there is one carbon missing" and that "the accidental coincidence of the carbon shift of two methine carbons in a molecule without any local symmetry is unexpected."

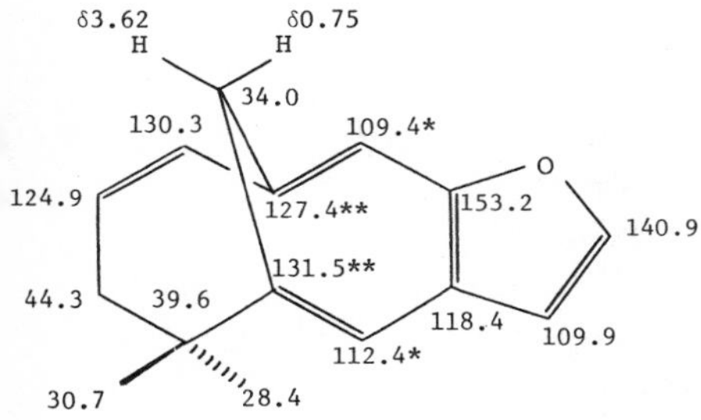

135 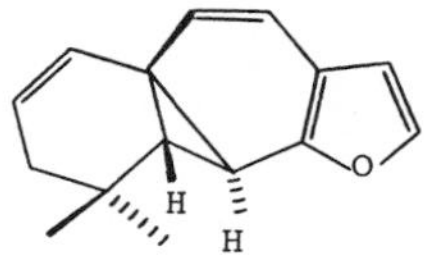 **136** 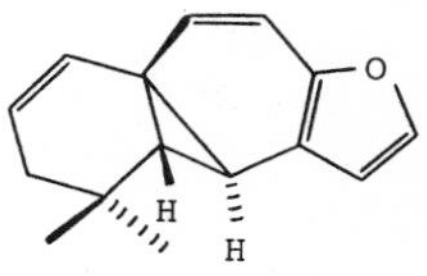 **137**

Careful examination of new ^{13}C nmr data along with some additional chemical proof and biogenetic considerations led to the revised structure for spiniferin-1 (**135**) (Cimino, 1977). Some ^{1}H nmr decoupling experiments demonstrated that the signals at $\delta 0.75$ and 3.62 were mutually coupled and could therefore be assigned to either vicinal cyclopropane protons or strongly nonequivalent methylene protons (Günther and Jikeli, 1977). The correctness of the latter alternative was seen in the ^{13}C nmr off-resonance spectrum. The signal centered at 34.0 ppm appeared as a doublet of doublets, which could then be assigned to the C-6 methylene. Observations of similar "deviant" ^{13}C–^{1}H multiplets have been reported (Hagaman, 1976). Carbon chemical shifts and assignments are indicated on the formula.

IV. DITERPENES

The discovery of a new algal species, *Laurencia irieii*, led to the isolation and characterization of two novel crystalline diterpenes, iriediol (**139**) and irieol A (**141**). The structures of these key compounds were determined by X-ray crystallography (Fenical *et al.*, 1975).

Subsequently, Howard and Fenical (1978) isolated seven related compounds whose structures were confirmed by chemical correlation to iriediol (**139**) or irieol A (**141**). Treatment of iriediol (**139**) with acetic anhydride in pyridine and osmylation of the C-10–C-11 double bond followed by β elimination of the C-7 acetate generated aldehyde **147** and ketone **149**. Similar oxidative degradation of irieol D acetate (**144**) or irieol F (**143**) yielded aldehyde **148** and ketone **149**. The epoxide of irieol A (**141**) was opened and the intermediate diol was cleaved with periodic acid to yield aldehyde **148**. Aldehyde **148** was also obtained by oxidative cleavage of the double bond of irieol (**138**) or the diol in irieol E (**142**). These chemical interconversions confirmed the stereochemistry and substitution pattern of the perhydroindane system. The structures of irieol C (**145**) and irieol B acetate (**146**) were determined by spectral methods.

Using model compounds **147–149** and oppositol (**93**), Howard and Fenical (1978) were able to assign the ^{13}C nmr chemical shifts of all compounds with the exception of irieol A (**141**). The absence of off-resonance data for irieol A (**141**) precluded definite carbon assignments, although some signals were assigned by chemical comparison to irieol (**138**). The ^{13}C nmr shift assignments for irieol (**138**), iriediol (**139**), iriediol acetate (**140**), irieol A (**141**), irieol E (**142**), irieol F (**143**), irieol D acetate (**144**), irieol C (**145**), and irieol B acetate (**146**) are listed in Table 19.

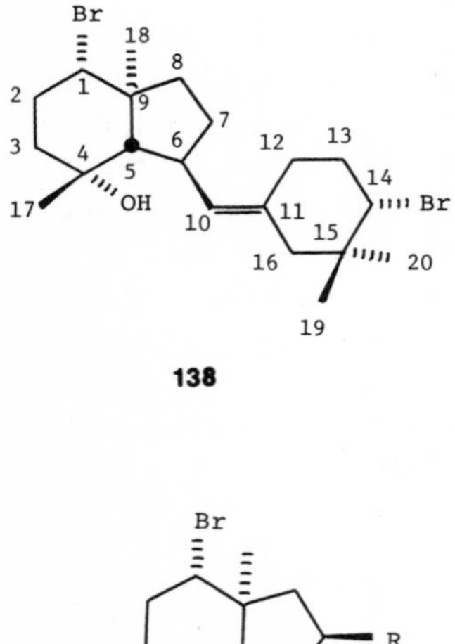

138

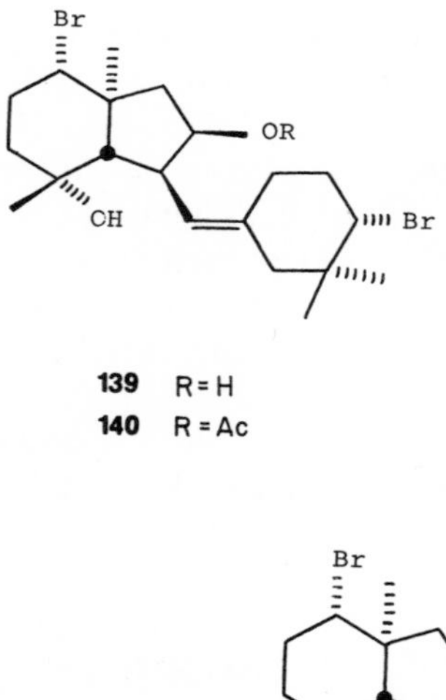

139 R = H
140 R = Ac

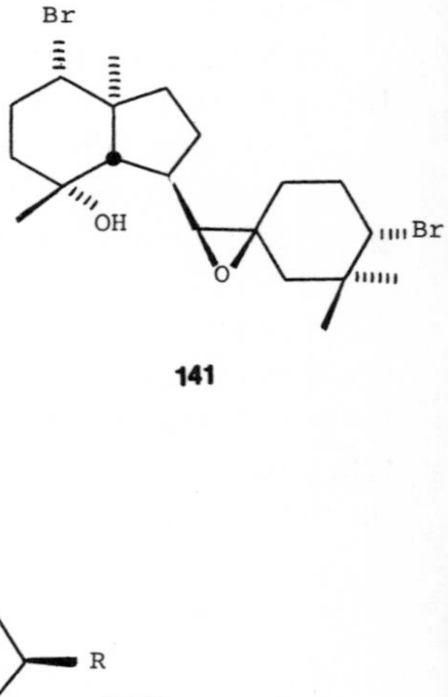

141

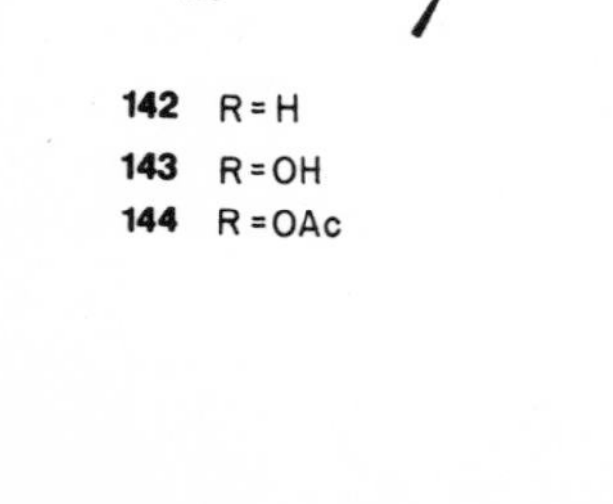

142 R = H
143 R = OH
144 R = OAc

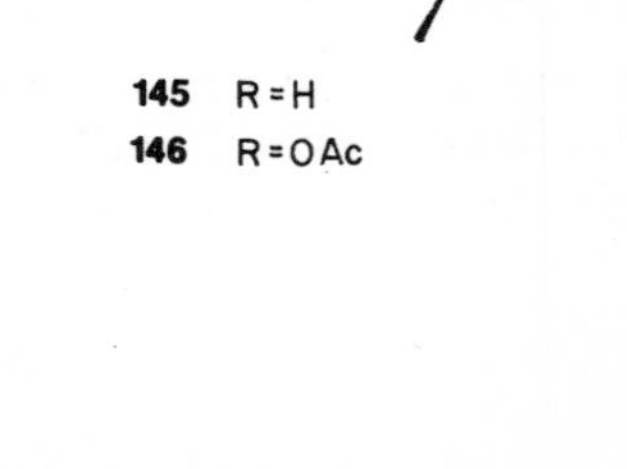

145 R = H
146 R = OAc

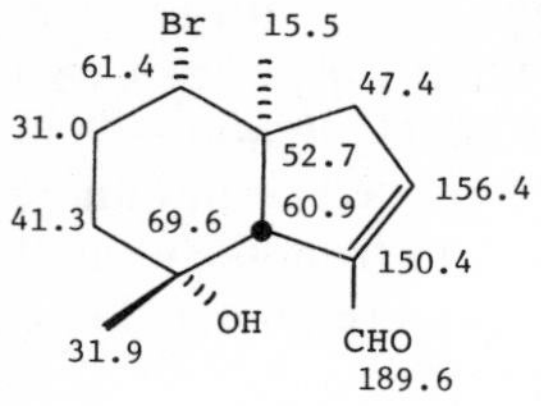

147 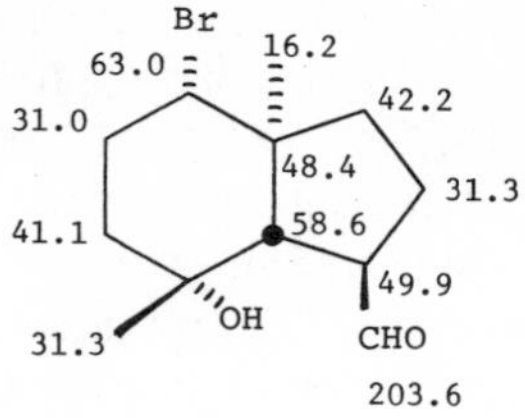 **148** 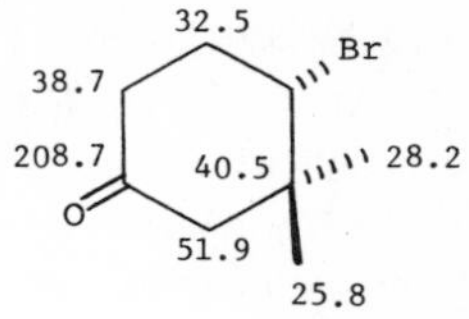 **149**

The first report of a diterpene possessing the perhydroazulene ring skeleton was that of pachydictyol A (**150**) (Hirschfeld *et al.*, 1973) isolated from the brown alga *Pachydictyon coriaceum*. In addition to pachydictyol A (**150**), two related compounds, dictyol A (**151**) and dictyol B (**152**), were

TABLE 19

13C nmr Shifts of Irieols and Iriediols[a,b]

	Compound								
Carbon	**138**	**139**	**140**	**141**[c]	**142**	**143**	**144**	**145**	**146**
1[e]	66.1	65.8	65.9	66.1	65.2	65.1	65.2	65.8	65.3
2	31.2	30.8	30.8	30.9	31.0	30.6	30.6	31.0	30.8
3	40.6	42.9	42.9	41.1	41.5	44.0	44.0	40.7	39.7
4	71.7	71.5	71.2	71.3	72.1	71.7	71.7	72.1	71.7
5	60.9	59.8	59.8	59.7	56.3	53.9	54.1	60.2	57.7
6	35.7	47.2	43.7	d	37.9	42.1	41.1	32.5	36.4
7	28.0	78.2	79.7	d	20.6	75.7[f]	75.6	28.0	73.5
8	42.9	50.6	48.2	43.0	43.6	46.7	44.9	43.1	43.3
9	48.0	46.9	46.9	d	47.0	45.3	47.4	47.4	45.4
10	132.4	129.4	128.3	64.3	78.3	80.3[f]	78.6	52.0	49.1
11	131.9	134.7	135.0	67.9	75.0	75.7	75.3	73.1	72.6
12	33.6	33.8	33.5	d	36.4	38.3	35.4	38.3	38.4
13	29.2	28.4	27.7	d	30.1	30.2	30.2	30.4	30.3
14[e]	64.9	64.5	63.9	64.1	64.4	64.0	62.2	64.8	64.3
15	38.3	38.4	33.8	d	37.1	36.5	36.6	36.6	36.6
16	48.7	48.9	48.2	d	45.8	50.9	50.2	51.3	51.3
17	31.2	31.2	31.0	31.8	32.8	32.8	32.7	31.8	32.4
18	16.4	17.9	17.3	16.2	16.7	18.0	18.1	16.4	17.7
19	23.0	23.1	23.8	24.5	22.9	23.0	22.9	22.8	23.0
20	29.6	29.6	29.1	30.8	31.7	31.5	31.4	31.8	32.1

[a] Carbon data and assignments, Howard and Fenical (1978).
[b] The δ values are in parts per million downfield from TMS in $CDCl_3$ solution.
[c] No off-resonance data.
[d] Other signals at 40.0, 37.1, 36.1, 34.9, 29.4, 25.7, and 24.8 ppm.
[e] Assignments may be reversed.
[f] Signals may be reversed.

isolated from the digestive gland of the sea hare *Aplysia depilans* (Minale and Riccio, 1976) and from the sea hare's food source, the brown alga *Dictyota dichotoma* (Fattorusso *et al.*, 1976). From *D. ligulatus* and *Aplysia depilans*, Danise *et al.* (1977) reported three related compounds, dictyols C (**153**), D (**154**), and E (**155**). *Dictyota flabellata* from the Gulf of California produced pachydictyol A epoxide (**156**) (Robertson and Fenical, 1977), the probable precursor of dictyol C (**153**).

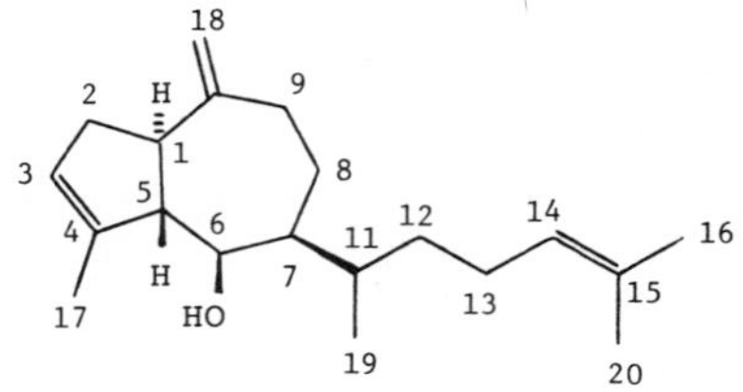

150

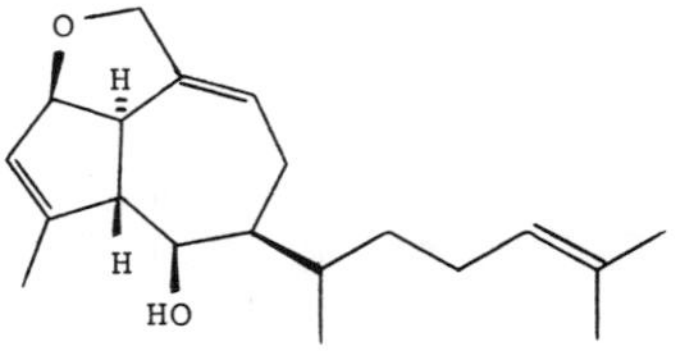

151

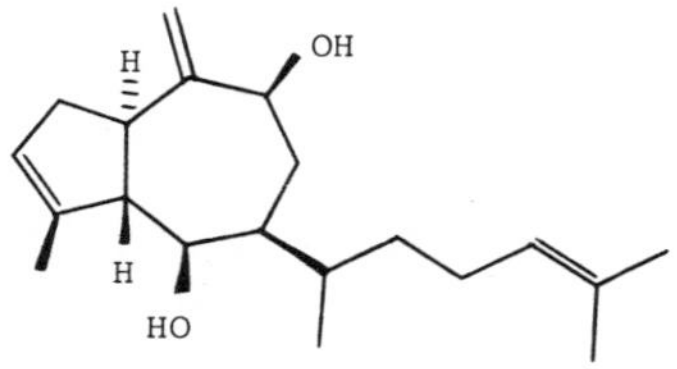

152

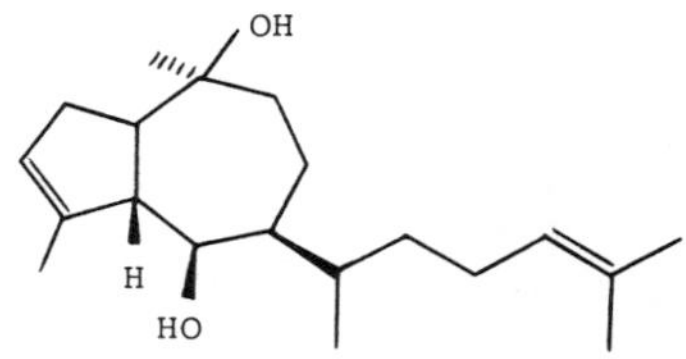

153

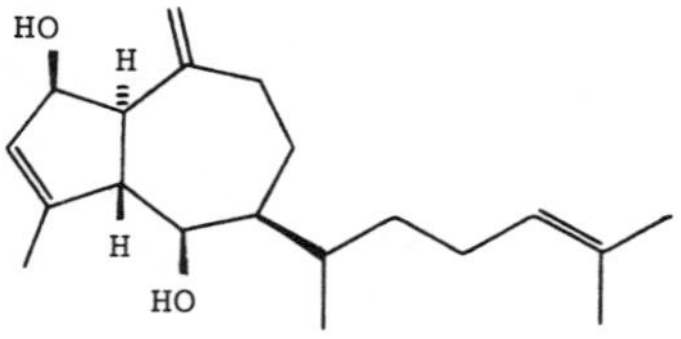

154

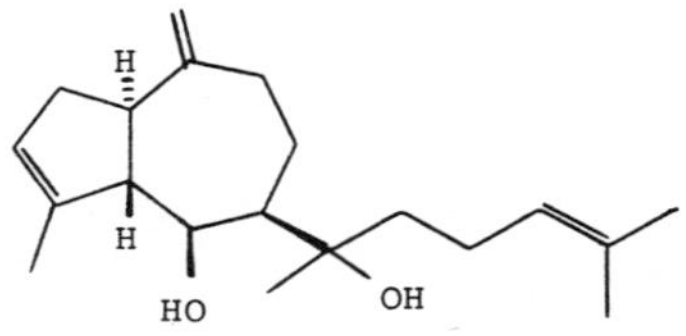

155

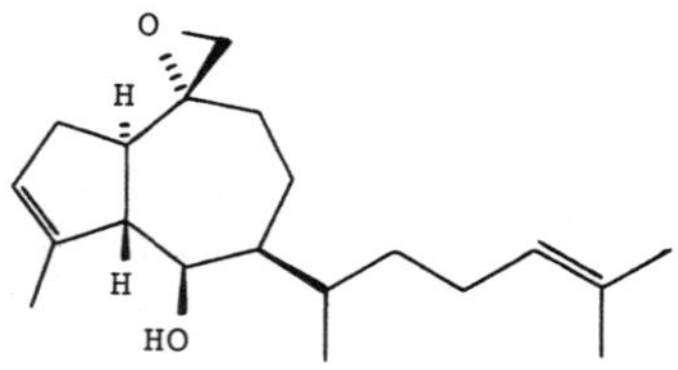

156

TABLE 20

13C nmr Chemical Shifts of Pachydictyol A, Dictyols, and Pachydictyol A Epoxide[a]

	Compound							
Carbon	**150**[f]	**150**[g]	**155**[g]	**152**[h]	**154**[g]	**156**[i]	**153**[g]	**151**[h]
1	46.1	46.5	46.2	43.0[b]	52.6[b]	43.9[e]	50.0[b]	48.4
2	33.9	34.3	33.8	33.8[c]	74.5[c]	31.2[e]	33.0	85.6
3	123.9	124.0	124.4[b]	123.8	125.6	124.2	123.3	123.9
4	141.4	141.5	140.9	140.9	147.8	141.2	142.7	142.8
5	60.4	60.8	60.3	61.2	52.4[b]	50.6[e]	52.8[b]	61.3
6	75.1	75.4	74.4	74.9	75.1[c]	74.6	74.5	74.6
7	47.7	47.9	48.7	43.8[b]	44.4	48.8[e]	49.1	45.0
8	23.5	23.8	21.7	33.4[c]	22.7	20.6[e]	19.7	25.7
9	40.6	40.5	40.5[c]	76.4	35.1[d]	39.7[e]	46.6	121.4
10	152.5	152.6	155.9	154.5	150.3	62.2	72.5	151.2
11	35.1	35.2	76.2	35.0	34.3	35.4[e]	34.5	35.3
12	34.8	35.4	41.0[c]	35.1	35.6[d]	34.9[e]	34.8	34.2
13	25.7	25.9	23.3	25.7	25.5	25.9[e]	25.6	26.4
14	124.7	124.9	124.2[b]	124.6	124.6	125.0	124.7	124.6
15	131.4	131.3	131.7	131.5	131.4	131.3	131.3	132.0
16	25.7	25.6	25.7	25.7	25.7	25.9[e]	25.7	25.7
17	15.9	15.7	15.8	15.6	15.8	15.7[e]	16.3	15.8
18	107.1	107.1	107.1	104.0	111.8	58.0[e]	30.0	75.0
19	17.5	17.6	25.4	17.5	17.3	17.5[e]	17.5	17.4
20	17.7	17.6	17.7	17.7	17.7	17.7[e]	17.7	17.7

[a] The δ values are in parts per million downfield from TMS in $CDCl_3$ solution.
[b-d] Signals within a column may be reversed.
[e] No off-resonance data.
[f] Rose and Sims (1977a).
[g] Danise *et al.* (1977).
[h] Minale and Riccio (1976).
[i] Robertson and Fenical (1977).

The ^{13}C nmr chemical shifts for compounds **150–156** are listed in Table 20. The side chain carbons C-12 through C-16 and C-20 were assigned by comparison with model compounds. Monoterpene **157** was used as a template for compounds **150–154** and **156**, and monoterpene **158** was used as a guide for carbon assignments in compound **155** (Bohlmann *et al.*, 1975).

HO 29.4 37.4 25.6 125.0 130.7 25.7 19.6 17.6

157

HO 72.8 41.4 22.9 125.1 131.0 25.7 26.3 17.6

158

Substitution of a hydroxyl at C-2 (**150** versus **154**) or C-11 (**150** versus **155**) resulted in an expected 41 ppm downfield shift of this carbon. Similar downfield shifts have been observed in cyclohexane and linear systems (Wehrli and Wirthlin, 1976, p. 45). In contrast, substitution of a hydroxyl at C-9 (**150** versus **152**) resulted in a downfield shift of the α carbon by only 35.9 ppm. From this limited amount of data it would appear that, in methylene cycloalkane systems, hydroxide substituent parameters of +36 ppm (C_α) and +10 ppm (C_β) would be more reliable.

The chemical shift of the quaternary epoxide carbon C-10 (compound **156**), 62.2 ppm, was not exceptional and indicated that the epoxide was not excessively strained by the seven-membered ring. However, a rather large γ effect of the expoxide on C-5 and C-8 (compare **156** and **150**), resulting in downfield shifts of these atoms by 9.8 and 2.9 ppm, respectively, was somewhat surprising.

Sun *et al.* (1977) reported the structure of dictyoxepin (**159**), a metabolite of the brown alga *Dictyota acutiloba*. The ^{13}C nmr chemical shift assignments are indicated on the formula (Erickson, 1977).

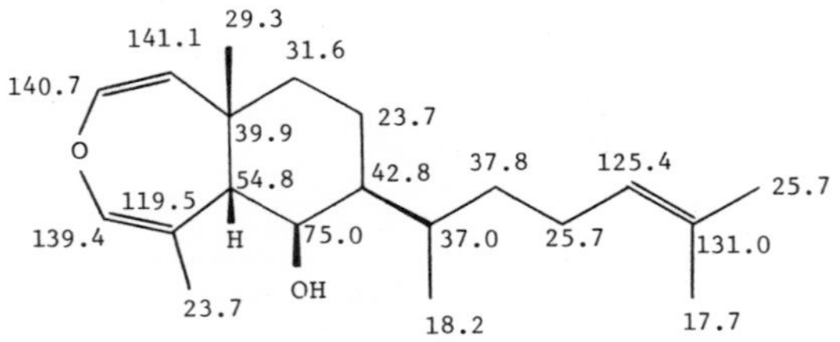

159

Only a few natural products have been reported from green algae (Chlorophyta). Caulerpol (**160**) (Blackman and Wells, 1976), a diterpene alcohol, was isolated from the alga *Caulerpa brownii* growing on the east coast of Tasmania. The structure of caulerpol (**160**) was determined by comparison of ^{1}H nmr chemical shift values with those of retinol and by ozonolysis, which gave levulinic acid. Some of the ^{13}C nmr signals were assigned by reference to the monoterpene alcohol **161** (Bohlmann *et al.*, 1975), but additional signals at 49.1, 40.6, 32.7, 31.8, 30.0, and 27.5 ppm, some of which must represent more than one carbon atom, could not be assigned.

For the most part, brominated diterpenes elaborated by *Laurencia* species possess unrearranged carbon skeletons. Fenical *et al.* (1976) reported the occurrence of a rearranged diterpene, sphaerococcenol A (**162**), in the red alga *Sphaerococcus coronopifolius*. X-ray diffraction

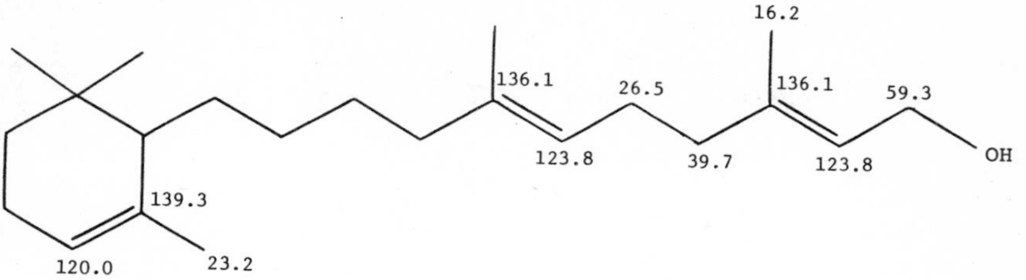

160

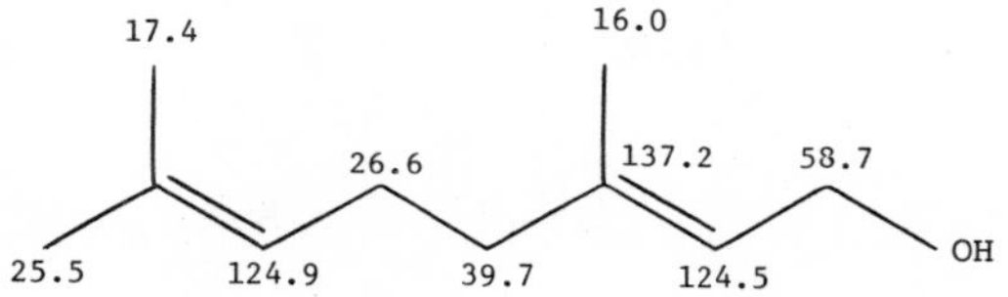

161

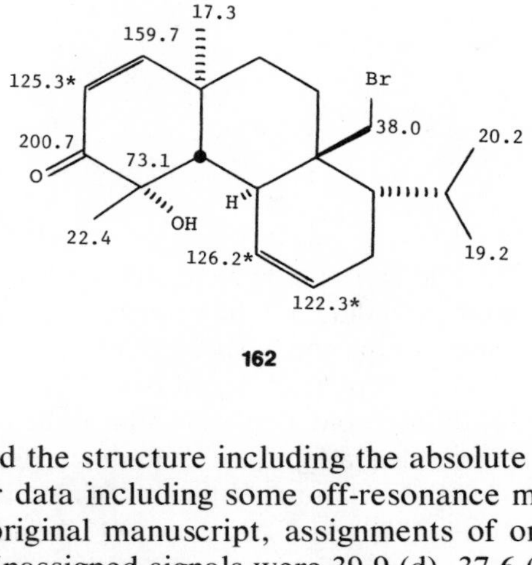

162

studies determined the structure including the absolute stereochemistry. Although ^{13}C nmr data including some off-resonance multiplicities were provided in the original manuscript, assignments of only a few signals could be made. Unassigned signals were 39.9 (d), 37.6 (d), 33.4 (t), 30.9, 29.1, 23.7, and 23.6 ppm.

From the digestive gland of the nocturnal sea hare *Dolabella californica*, Ireland and Faulkner (1976) isolated a crystalline hydroxyacetate, (1S*,2*E*,4*R**,7*E*,10*S**,11*R**,12*R**)-10-acetoxy-18-hydroxy-2,7-dolabelladiene (**164**). This diterpene possesses a new carbon skeleton, which Ireland and Faulkner (1976) called the dolabellane skeleton (**163**). Subsequently, the same authors reported the structure of 12 additional dolabellane-type diterpenes, including ^{13}C nmr data for six compounds (**164–169**) (Ireland

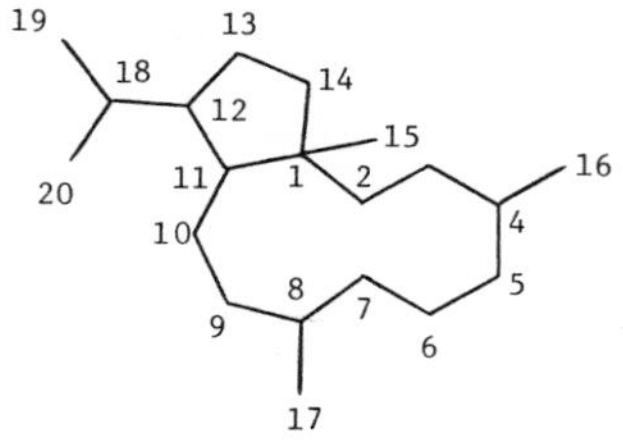

163

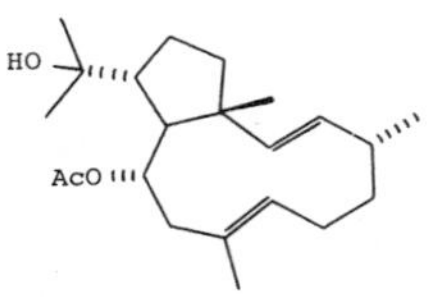

164

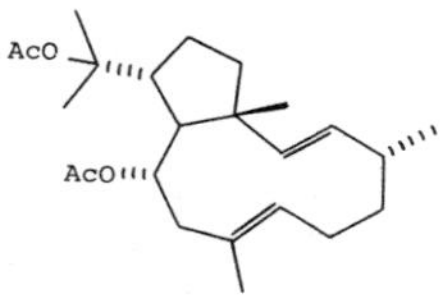

165

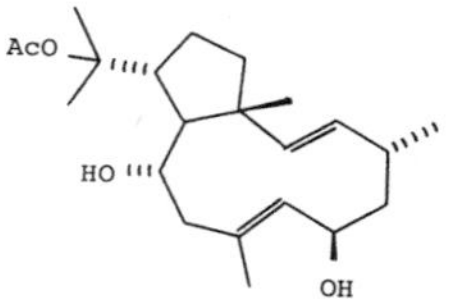

166

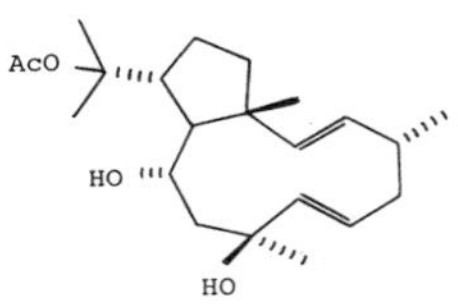

167

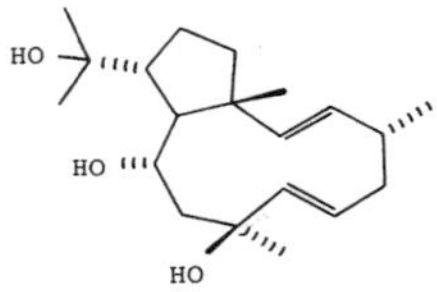

168

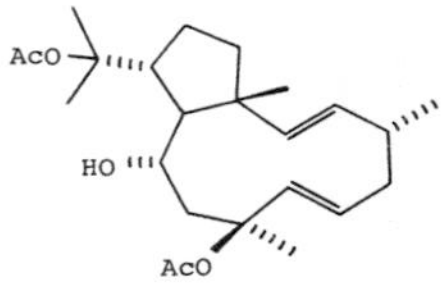

169

and Faulkner, 1977). The carbon chemical shifts are summarized in Table 21. Additional instrumental data will be required for carbon assignments.

Collections of sponges of the genus *Spongia* spp. from different locations on the Great Barrier Reef from near Heron Island to Lizard Island yielded complex mixtures of diterpenoid alcohols and acetates. A combination of preparative thick-layer chromatography, column chromatography, and recrystallization allowed the isolation of four sets of alcohols and acetates. In addition to spongiadiol (**170**) and its diacetate (**171**) and spongiatriol (**172**) and its triacetate (**173**), Kazlauskas *et al.* (1978a) isolated the C-3 epimeric compounds epispongiadiol (**174**), diacetylepispongiadiol (**175**), and epispongiatriol (**176**) and its triacetate (**177**).

The ^{13}C nmr chemical shift data on compounds **170–177** are reported in Table 22. These data suggested the presence of a ketone in all compounds by a signal near 205 ppm. The ^{13}C nmr chemical shifts of compounds **170** and **172**, **171** and **173**, **174** and **176**, and **175** and **177** were quite similar. Compounds **170**, **171**, **174**, and **175** displayed a methyl quartet near 25.5 ppm, which was replaced by a new signal near 60.8 ppm (or 64 ppm) in

TABLE 21

13C nmr Chemical Shifts of Dolabellanes[a,b]

Compound					
164[c]	**165**	**166**	**167**	**168**	**169**
	170.1				170.5
169.8	169.0	169.0	169.5		169.9
135.8	135.4	135.6	142.1	143.1	139.5
131.2	134.4	134.1	138.6	137.4	138.4
128.3	130.5	133.0	132.8	132.4	131.9
127.4	127.3	130.2	126.4	125.5	125.5
73.0	84.9	86.8	86.6	75.0	86.1
71.9	70.7	71.5	74.1	72.7	83.5
		65.3	66.8	67.0	65.8
56.1	53.3	54.2	58.0	57.1	58.6
50.0	47.2	47.5	56.5	56.8	53.7
47.7	45.0	46.2	49.5	49.4	49.4
46.1	38.9	43.1	39.2	47.1	46.4
40.0	37.3	39.8	37.3	38.2	38.9
38.3	36.0	31.2	35.1	37.9	38.0
36.4	26.8	26.0	30.1	34.1	34.6
32.5	26.3	21.9	26.1	31.5	25.8
27.9	26.1	20.0	25.9	29.3	25.4
27.3	25.7	18.9	23.2	26.0	23.0
23.9	25.5	16.9	20.1	23.2	22.6
22.0	23.4	16.0	18.0	16.6	20.3
20.0	22.8			16.4	17.4
18.8	21.0				17.3
	19.1				16.9
	18.0				

[a] Ireland and Faulkner (1977).
[b] Chemical Shifts are in parts per million downfield from TMS in $CDCl_3$ solution.
[c] In C_6D_6 solution.

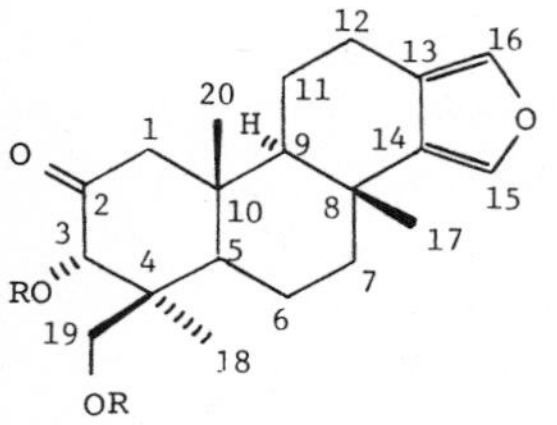

170 R = H
171 R = Ac

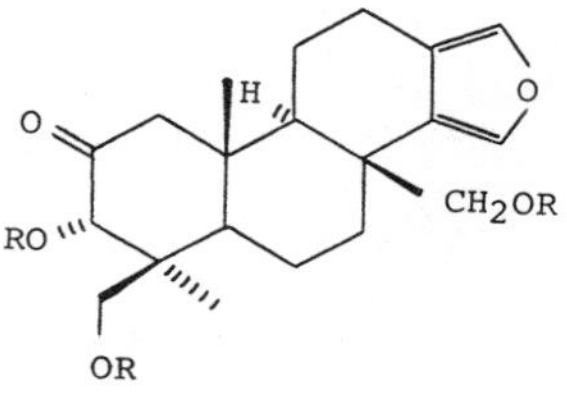

172 R = H
173 R = Ac

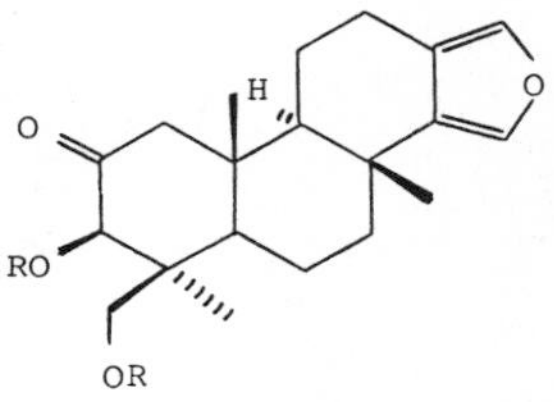

174 R = H
175 R = Ac

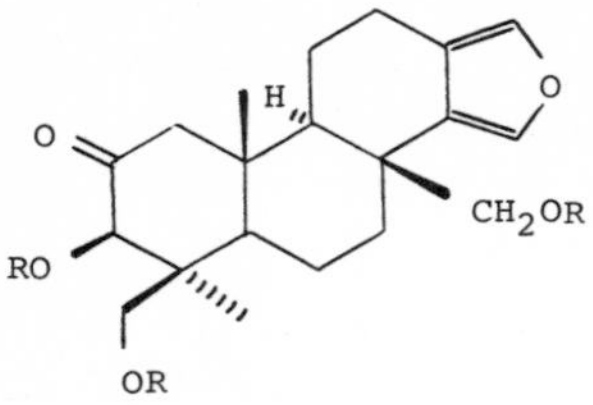

176 R = H
177 R = Ac

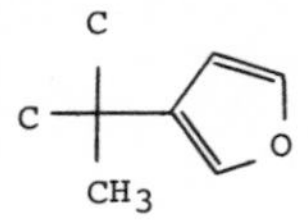

178

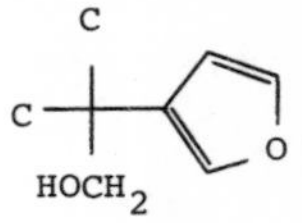

179

TABLE 22

13C nmr Chemical Shifts of Spongiadiols and Spongiatriols[a,b]

Carbon	Compound 170	171	172	173	174	175	176	177
1	53.9	54.0	52.1	53.9	52.9	54.3	53.4	53.8
2	212.8	205.5	212.5	205.0	209.9	202.3	209.9	201.7
3	76.0	77.4	76.1	77.1	84.2	83.0	82.5	82.5
4	46.2	43.8	46.4	43.7	46.4	48.4	45.9	49.3
5	54.4	53.7	51.6	53.7	55.3	56.2	54.7	55.7
6[c]	18.7	[d]	17.7	[e]	18.8	18.8	17.4	17.9
7[c]	20.6	[d]	20.1	[e]	20.5	20.9	20.0	19.9
8	34.0	34.2	39.6	38.1	34.5	34.5	39.6	38.1
9	56.4	56.5	56.4	56.9	55.8	56.2	55.8	56.3
10	38.6	39.9	40.3	39.4	43.1	42.7	42.2	42.1
11[c]	19.2	19.2	18.6	[e]	18.8	19.9	19.2	19.1
12	40.3	40.3	34.3	35.1	40.8	41.0	35.1	35.8
13	119.1	119.1	119.7	119.2	119.1	119.2	119.5	118.9
14	136.2	136.2	130.4	129.1	136.5	136.5	130.4	128.9
15	135.2	135.2	136.2	137.2	135.2	135.2	136.2	137.0
16	136.9	137.0	137.7	137.9	137.0	137.1	137.5	137.5
17	25.3	25.5	60.8	63.9	25.7	25.7	60.8	63.7
18	19.9	19.3	20.9	[e]	24.0	23.8	23.1	23.4
19	66.6	65.9	62.7	65.6	64.0	64.5	62.8	64.2
20	19.9	[d]	18.6	[e]	17.4	16.6	17.1	16.6
Acetate		170.4		170.5		170.4		170.2
Acetate		169.7		170.4		169.4		170.2
Acetate				169.7				169.7
CH_3		20.5		[e]		20.5		20.5
CH_3		[d]		[e]		20.5		20.5
CH_3				[e]				

[a] Carbon data and partial assignments, Kazlauskas *et al*. (1978a).
[b] The δ values are in parts per million downfield from TMS in $CDCl_3$ solution.
[c] These assignments may be reversed.
[d] Unassigned signals at 21.2, 20.8, 20.7, and 19.8 ppm.
[e] Unassigned signals at 21.2, 20.9, 20.5, 19.9, 19.3, and 18.8 ppm.

compounds **171**, **173**, **176**, and **177**. This downfield shift of approximately 35 ppm is indicative of substitution of a hydroxyl at this center. This methyl to hydroxymethylene difference caused a shift in one of the β-furan carbons (C-14). Similarly, a singlet near 34 ppm (**170**, **171**, **174**, and **175**) shifts downfield (~ + 5 ppm) to nearly 39 ppm (**172**, **173**, **176**, and **177**). These data suggested part structure **178** in compounds **170**, **171**, **174**, and **175** and part structure **179** in compounds **172**, **173**, **176**, and **177**.

Acetylation of the alcohols at C-3 and C-19 shifted C-4 to higher field. This shift distinguished C-4 from the quaternary centers C-8 and C-10. The hydroxy substituent at C-17 resulted in a downfield shift of C-8 (compare C-8 of compound **170** and **172**) and permitted its assignment. Carbon C-10 was assigned by default.

Epimerization at C-3 affected the chemical shifts of C-18 (methyl) and, to a lesser extent, C-19 (hydroxymethylene). In compound **170**, in which the C-3 hydroxyl was axial, C-18 (methyl) resonated at 19.9 ppm. Epimerization of C-4 yielded compound **174**. In this compound the C-3 hydroxyl and the C-18 methyl were trans-diequatorial, but the C-18 signal shifted downfield, resonating at 24.0 ppm. Support for this downfield shift was found in the shifts of *cis*- and *trans*-2-methylcyclohexanol. In *cis*-2-methylcyclohexanol the methyl group resonated at 16.5 ppm. Epimerization led to *trans*-2-methylcyclohexanol and a concomitant downfield methyl shift to 19.1 ppm (Stothers, 1972).

Acetoxycrenulatin (**180**) (McEnroe *et al.*, 1977), a diterpene γ-lactone, is a metabolite of the brown algae *Pachydictyon coriaceum* and *Dictyota crenulata*. Carbon chemical shift considerations and comparison shifts with the side chain assignments of pachydictyol A (**150**) allowed the probable assignment of 14 of the 22 observed signals in the ¹³C nmr spectrum. The absence of off-resonance data made these assignments tenuous. Additional carbon signals were observed at 10.3, 17.1, 17.7, 21.3, 23.4, 25.5, 25.7, 25.9, 29.2, 32.8, and 35.7 ppm.

A number of cembrane diterpenes have been reported, but ¹³C nmr

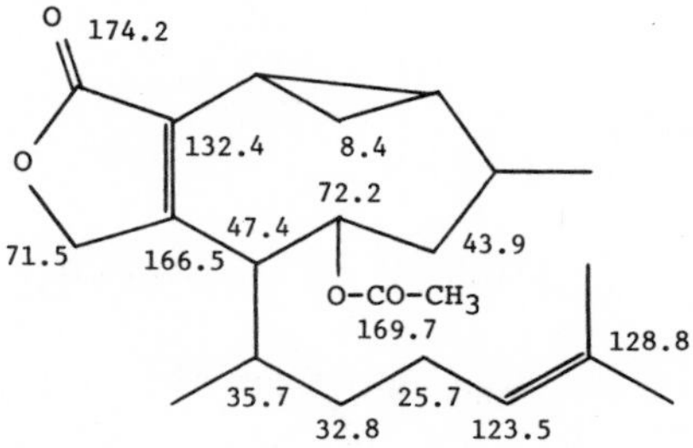

O 174.2
O
132.4 8.4
72.2
47.4
71.5 166.5 43.9
O-CO-CH_3
169.7
128.8
35.7 25.7
32.8 123.5

180

data are available for only a few compounds. A unique furanocembranolide, pukalide (**181**), was isolated as a major metabolite of the octocoral *Sinularia abrupta* Tixier-Durivault found off the southeast coast of Oahu, Hawaii (Missakian *et al.*, 1975). Although the reported ^{13}C nmr data are consistent with the proposed structure, it is somewhat surprising that C-2, adjacent to an isopropenyl and a furan, and C-14, adjacent to an isopropenyl and a carbon-substituted methylene, should both resonate at 32.5 ppm.

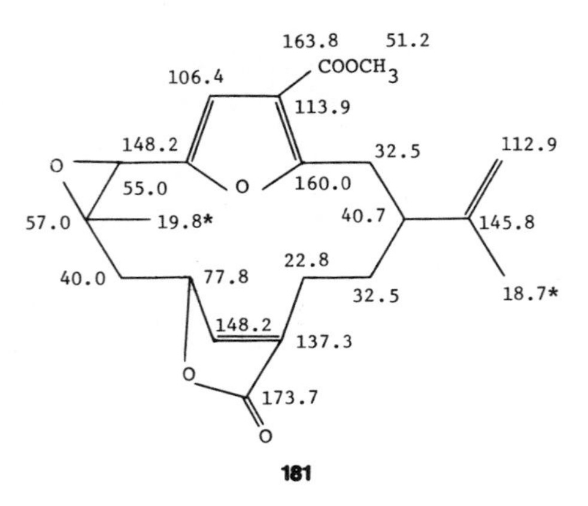

181

The icthyotoxic substance from the soft coral *Lobophytum* species was shown by Kashman and Groweiss (1977) to be lobolide (**182**). The structure of lobolide acetate (**183**) was determined by chemical and spectral methods.

182 R = H **183** R = $COCH_3$

Sarophine (**184**), a cembranolide from the soft coral *Sarcophytum glaucum*, crystallized from the hexane extract of the animal. The stereochemical features, excluding the absolute configuration, of sar-

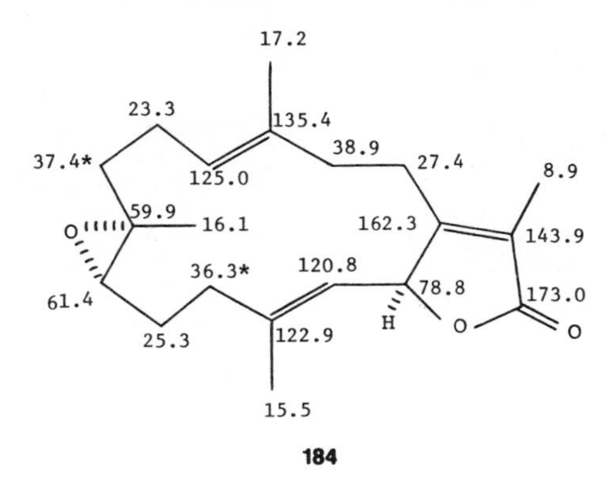

184

cophine (**184**) were determined by X-ray crystallography (Bernstein *et al.*, 1974). Carbon-13 nmr data (Kashman, 1977) were recently published.

Kazlauskas *et al.* (1978b) reported the structural elucidation of two cembranoid lactones, flexibilide (**185**) and dihydroflexibilide (**186**), from the soft coral *Sinularia flexibilis*. Hydrogenation of flexibilide (**185**) gave epidihydroflexibilide (**187**). In addition, the Roche group (Kazlauskas *et al.*, 1978b) isolated a novel 15-membered ring macrocyclic diterpene (**188**). Only partial chemical shift assignments for these compounds were possible. Flexibilide (**185**) also had ^{13}C nmr signals at 38.9(t), 34.7, 33.8, 32.7(t), 27.8(t), and 25.3 ppm. The ^{13}C nmr spectrum of dihydroflexibilide (**186**) contained additional signals at 38.4 (t), 36.8, 34.7 (t), 31.0 (t), and 27.6 ppm. Unassigned signals in epidihydroflexibilide (**187**) were at 38.9, 37.7, 36.2, 34.3, 31.0, 28.6, 27.1, 25.6, 25.1, and 22.6 ppm.

The double-bond shift assignments of compound **188** were made by calculations (Wehrli and Wirthlin, 1976, pp. 41–43). The calculated values are indicated on formula **188a**. The disubstituted olefinic carbons C-5, C-7, and C-13 agreed extremely well with the calculated values. The

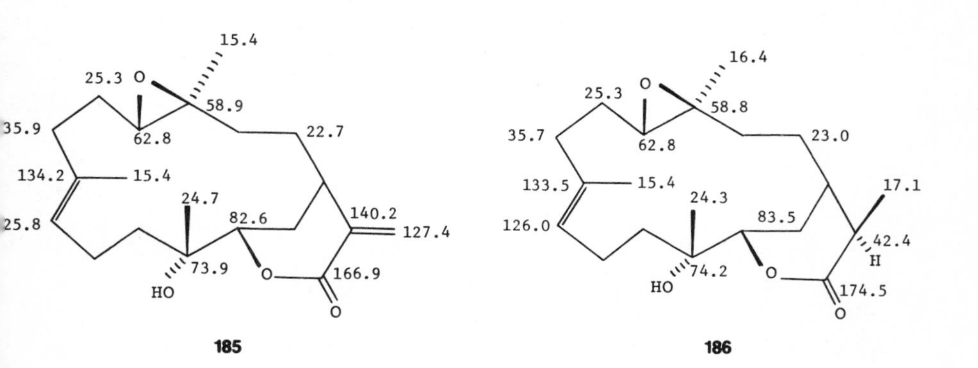

185 **186**

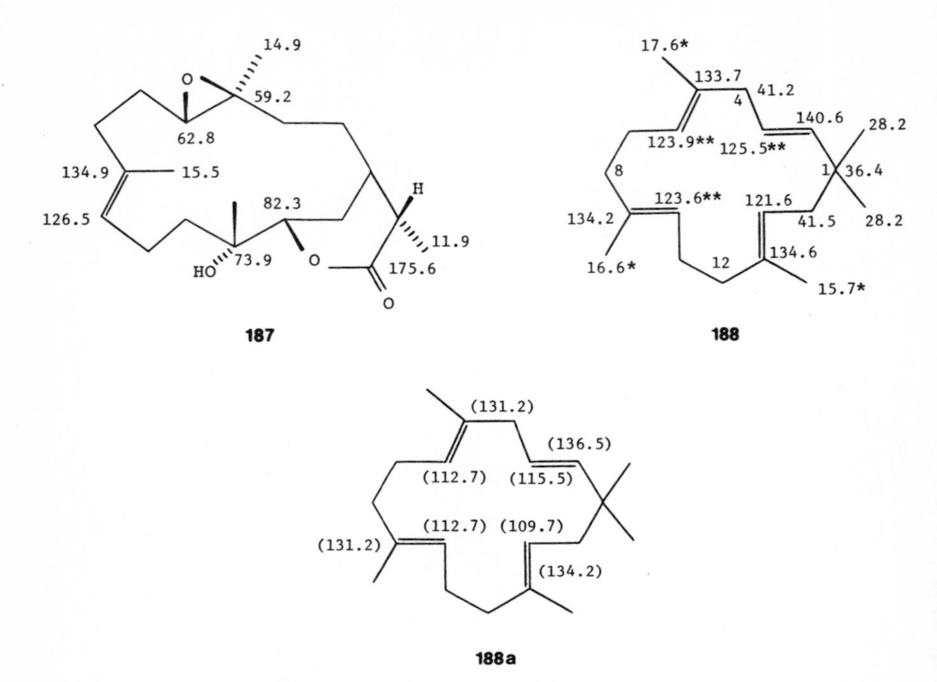

calculated values were ±2 ppm of experimental values. Their monosubstituted partners, C-6, C-10, and C-14, however, did not correlate as well, showing an average ±11ppm deviation. The remaining double bond, C-2–C-3, correlated somewhat better. Despite the deviation between experimental and calculated values, the chemical shift trend was assumed to hold. Thus, the calculated highest-field off-resonance doublet was assigned to C-6 (or C-8). These signals had the highest-field signals in the olefinic region of the spectrum. Unassigned signals in compound **188** were observed at 39.1 (t), 28.8 (t), 24.7 (t), and 24.4 (t) ppm.

V. NONTERPENOID COMPOUNDS

A. Linear C_{15} Compounds

Laurencia nidifica has been an exceedingly rich source of halogenated terpenoid and nonterpenoid sesquiterpenes. By visual inspection *L. nidifica* collected off the south coast of Oahu, Hawaii, was divided into several varieties. These varieties were characterized by markedly different sesquiterpene content. A clumpy pink variety contained predominantly aromatic sesquiterpenes, a nonclumpy pink variety contained chamigrene sesequiterpenes, and the green variety contained nonterpenoid C_{15} compounds.

There are two types of ring systems among the nonterpenoid C_{15} metabolites of *L. nidifica*: the maneonenes and the isomaneonenes. To date, the structures of four maneonenes, *cis*-maneonene A (**189**), *cis*-maneonene B (**190**), *trans*-maneonene B (**191**), (Waraszkiewicz *et al*., 1976), and *cis*-maneonene C (**192**) (Sun *et al*., 1976a), and two isomaneonenes, isomaneonene A (**193**) and isomaneonene B (**194**) (Sun *et al*., 1976a), have been elaborated. The ^{13}C nmr shift assignments for compounds **189–194** are tabulated in Table 23.

The ^{13}C nmr chemical shift assignments in the maneonenes required a recognition of substituent shift effects, off-resonance multiplicities, and specific carbon–proton decoupling experiments. The multiplicities of the protons on C-7, C-9, and C-10 in *cis*-maneonene A (**189**) permitted their assignment. Specific proton–carbon decoupling allowed unambiguous assignment of these carbon atoms.

A chloromethine carbon, C-5, of *cis*-maneonene B (**190**) was assigned to the carbon resonance at 58.3 ppm. Carbons alpha to a cis double bond show a high-field shift relative to their trans counterparts. Thus, isomerization of the C-3–C-4 double bond resulted in a downfield shift of C-5, which in *trans*-maneonene B (**191**) resonated at 63.8 ppm.

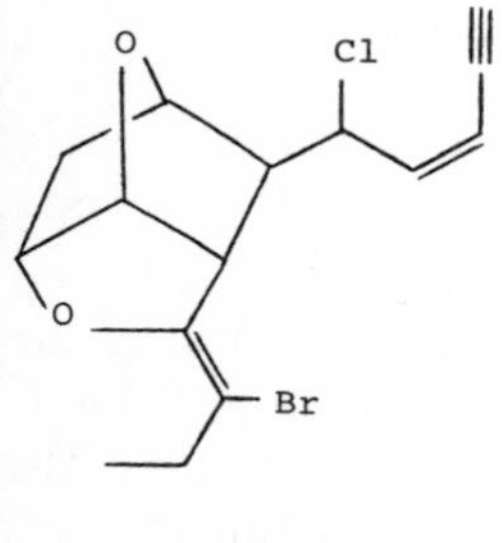

189

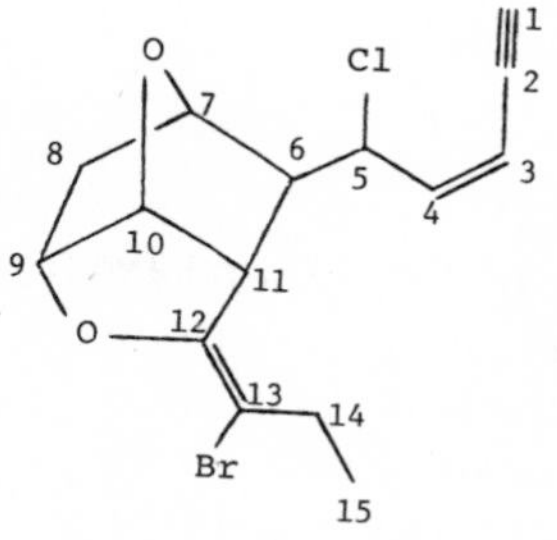

190

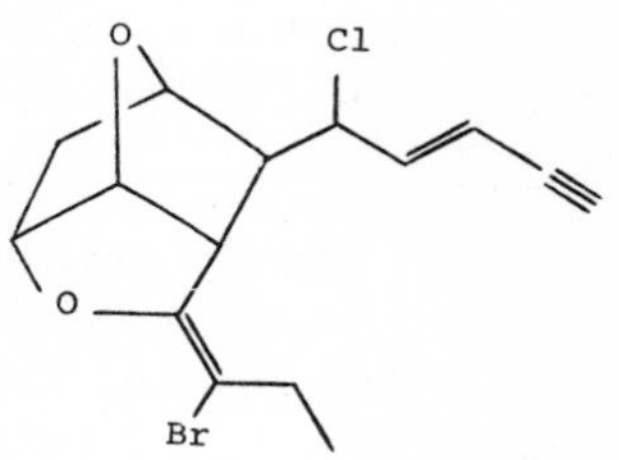

191

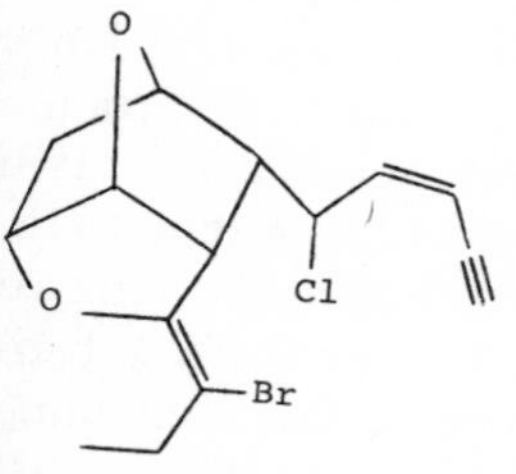

192

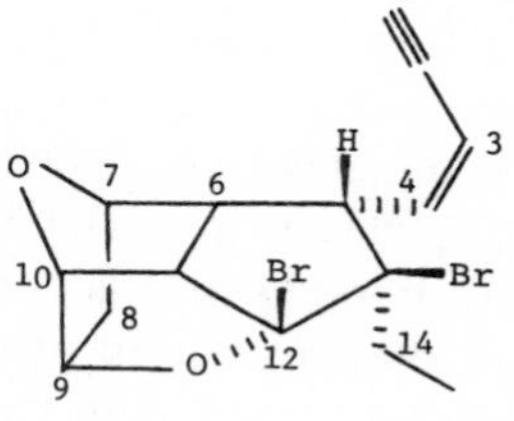

193

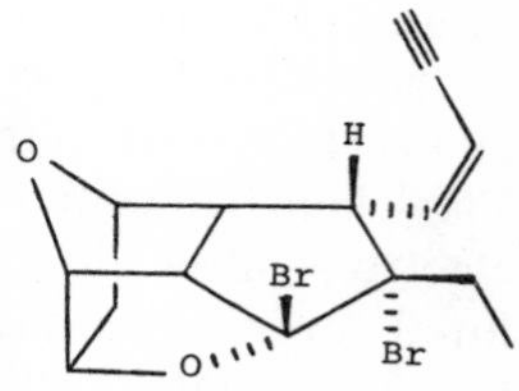

194

TABLE 23

13C nmr Chemical Shifts of the Maneonenes and Isomaneonenes[a]

	Compound					
Carbon	**189**[d]	**190**[e]	**191**[e]	**192**[e]	**193**[c,f]	**194**[c,f]
1	84.8	85.5	79.1	85.3	84.7	83.5
2	n.o.	n.o.[b]	n.o.	n.o.	79.5	79.9
3	111.9	111.7	113.3	109.9	109.2	108.7
4	142.9	141.4	142.2	141.9	144.5	146.9
5	58.3	58.3	63.8	54.4	49.7	43.2
6	58.3	58.3	58.2	54.0	50.5	50.0
7	78.3	78.1	78.0	77.9	77.6	77.6
8	38.6	38.6	39.0	36.4	41.4	41.1
9	80.4	79.4	80.7	81.2	83.3	81.4
10	82.2	82.8	82.6	83.5	82.9	84.6
11	46.1	44.0	45.1	45.0	63.1	58.7
12	n.o.	n.o.	n.o.	n.o.	110.6	109.5
13	n.o.	n.o.	n.o.	n.o.	89.7	85.2
14	26.7	29.2	29.2	27.9	26.8	38.0
15	13.0	14.0	14.0	13.6	10.0	10.6

[a] The δ values are in parts per million downfield from TMS in C_6D_6 solution.
[b] n.o., not observed.
[c] In $CDCl_3$ solution.
[d] Waraszkiewicz *et al.* (1976).
[e] Erickson (1977).
[f] Sun *et al.* (1976b).

The chemical shift of C-5 in the maneonenes is worthy of further mention. The *cis*- and *trans*-olefins **195** and **196** (Couperus *et al.*, 1976) are reasonable model systems for the environment about C-5. Using the model olefins and the branched alkyl substituent parameters from Table 1, one can obtain the calculated shift values for the maneonenes C-5 atom. The calculated chemical shift values for a chlorine and bromine atom adjacent to a *cis*-olefin are given on **195a** and **195b** and similar calculated values for a trans system are listed on **196a** and **196b**. Such calculations indicate that C-5 should be a bromomethine carbon. However, the chemical evidence supports the proposed structure, in which C-5 is a chloromethine carbon.

Specific proton decoupling experiments with isomaneonenes A (**193**) and B (**194**) verified the chemical shift assignments of C-7, C-8, and C-10. The chemical shift of C-12 (110.6 and 109.5 ppm) indicated that two electronegative atoms were bonded to this center. An identically substituted carbon in johnstonol (**90**) resonated at 113.6 ppm. The remaining low-field signals could then be assigned to the double bond (C-3, ~109 ppm; C-4, ~145 ppm) and to the terminal acetylene (C-1, ~94 ppm; C-2, ~79 ppm). Full carbon assignments are indicated in Table 23.

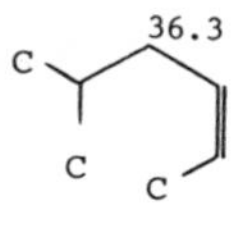

195 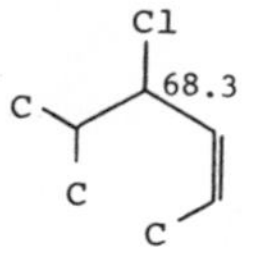 **195a** 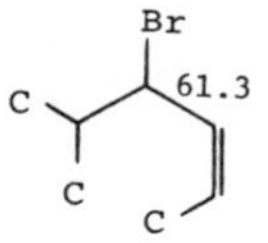 **195b**

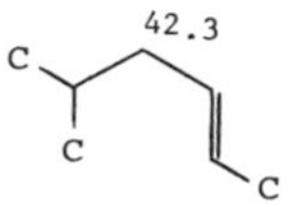

196 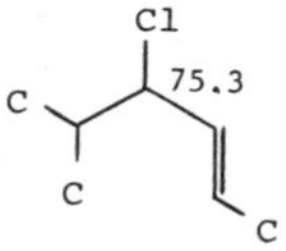 **196a** 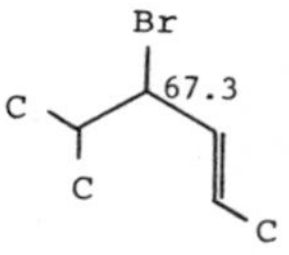 **196b**

Carbon-13 nmr data for two additional C_{15} red algal metabolites have been reported. The structure of chondriol (**197**), a metabolite of *Chondria oppositiclada* Dawson, was determined by X-ray diffraction analysis (Fenical *et al.*, 1974). Fenical (1974) proposed the structure of rhodophytin (**198**), a stable cyclic vinyl peroxide secondary metabolite of *Laurencia*, on the basis of chemical conversions and spectroscopic evidence. The ^{13}C nmr spectrum of rhodophytin (**198**) clearly showed it to be closely related to chondriol (**197**).

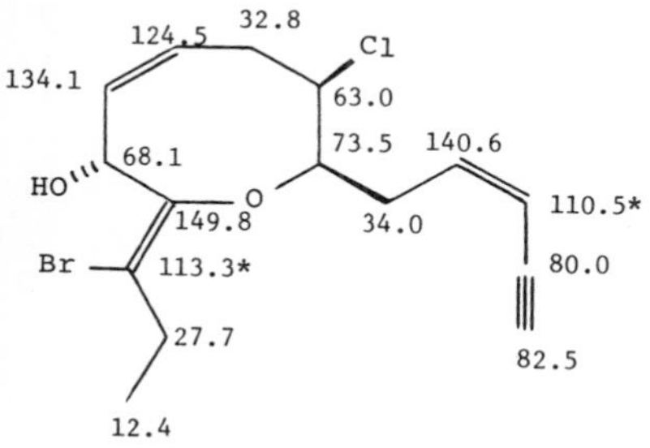

197 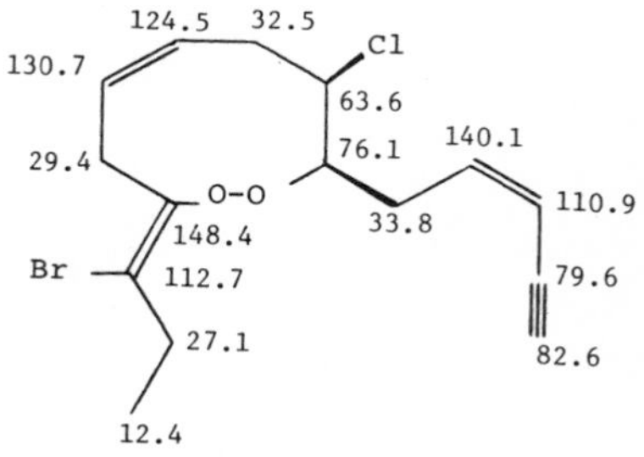 **198**

The ^{13}C nmr chemical shift assignments shown on the formulas of chondriol (**197**) and rhodophytin (**198**) deviate from those reported by Fenical (1974). The side chain carbons C-1 through C-4 are assigned by reference to the conjugated acetylene carbon shifts of the isomaneones (Table 23). The ring olefin shifts are not unlike those of model systems (Couperus *et al.*, 1976).

The methine carbons are assigned as follows. The ^{13}C nmr signal at 68.1 ppm of chondriol (**197**) is absent in rhodophytin (**198**), where a new methylene absorption at 29.4 ppm appears. This fixes the assignment of

C-11 in both compounds. The resonance at 73.5 ppm of chondriol (**197**) shifts downfield (+2.6 ppm) to 76.1 ppm in rhodophytin (**198**). The 2.6 ppm shift is consistent with an ether to peroxide functionality change and assigns the 73.5 ppm (and 76.1 ppm) resonance to C-6. The remaining signal near 63 ppm is therefore the chloromethine carbon, C-7.

B. Miscellaneous Compounds

Metabolites of the sponge *Dysidea herbacea* exhibited regional metabolite variation. A small sample of *D. herbacea* collected northeast of Cooktown on the Great Barrier Reef was freeze-dried and extracted with dichloromethane. Dysidenin (**199**) (Wells, 1977) was crystallized from hexane and its structure determined by spectroscopic methods. Dysidenin (**199**) was absent from a collection of *D. herbacea* from the Caroline Islands in the Pacific Ocean.

The ^{13}C nmr spectrum ($CDCl_3$) of dysidenin (**199**) accounted for all 17 carbon atoms as follows: singlets: 171.9, 171.2, 168.2, 105.5, and 105.1 ppm; doublets: 142.3 (dd, J = 185.5, 5.9 Hz), 118.9 (dd, J = 189.4, 15.6 Hz), 54.0, 51.9, 51.4, and 47.3 ppm; triplets: 37.4 and 31.0 ppm; quartets: 30.8, 21.8, 17.3, and 16.2 ppm. The singlets at 105.5 and 105.1 ppm were the two trichloromethyl groups.

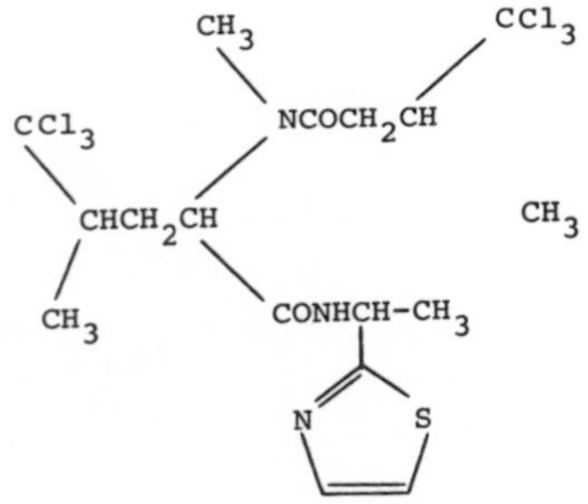

CH_3 CCl_3
CCl_3 $NCOCH_2CH$
$CHCH_2CH$ CH_3
CH_3 $CONHCH-CH_3$
N S

199

The ^{13}C nmr spectrum of 2-acetoxymethylthiazole displayed signals at 142.8 (dd, J = 187.5, 6.8 Hz) and 120.2 ppm (dd, J = 188.9, 14.6 Hz), confirming the thiazole ring in dysidenin (**199**) and assigning the ^{13}C nmr signals at 142.3 and 118.9 ppm to C-16 or C-17.

Pheromones are widely distributed in nature, but very few of these substances have so far been reported from the marine environment. Sleeper and Fenical (1977) reported the isolation and structural determination of navenone A (**200**), navenone B (**201**), and navenone C (**202**), which caused the sea slug *Navanax inermis* to terminate its trail-following

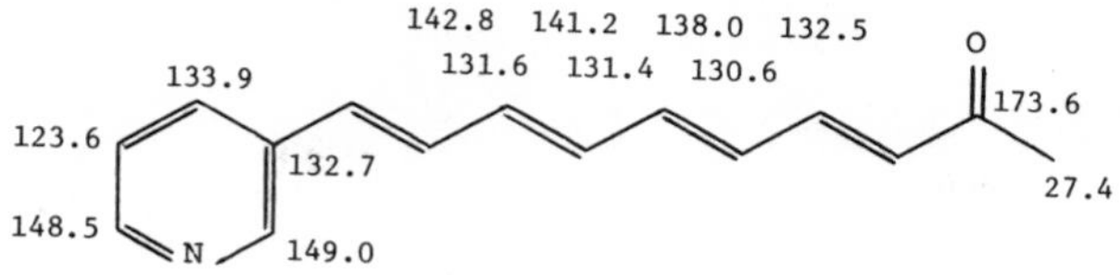

142.8 141.2 138.0 132.5
131.6 131.4 130.6
133.9
O
173.6
123.6
132.7
27.4
148.5
N
149.0

200

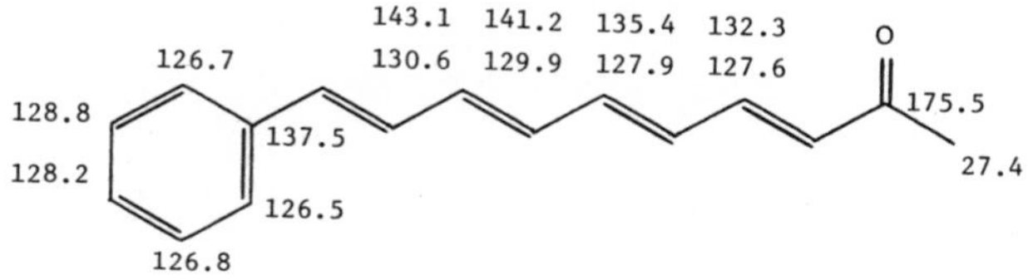

143.1 141.2 135.4 132.3
130.6 129.9 127.9 127.6
126.7
O
175.5
128.8
137.5
128.2
27.4
126.5
126.8

201

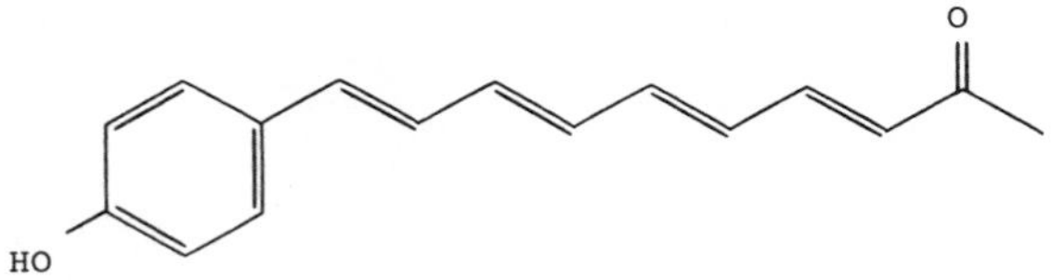

O
HO

202

behavior. The chemical shift assignments of the aromatic carbons were determined by comparison to 3-ethylpyridine [navenone A (**200**)] and styrene [navenone B (**201**)]. In the absence of additional data, these assignments are tenuous.

Although several toxins from marine organisms have been reported, few have received the attention of palytoxin (for a review of marine toxins, see Scheuer, 1977). Palytoxin, despite its lethality and water solubility, lacks repetitive amino acid units or sugar residues. Palytoxin(s) has been isolated from coelenterates of the genus *Palythoa*. To date, only a small partial structure of palytoxin has appeared in the literature (Moore *et al.*, 1975) and reports of metabolites of *Palythoa* are sure to draw attention.

Ito and Hirata (1977), in the course of their studies of *Palythoa tuberculosa*, isolated a water-soluble compound, mycosporine-Gly (**203**), which has an ultraviolet absorption maximum at 310 nm. The polyps of *Palythoa* are symbiotic with a blue-green alga, and it is possible that mycosporine-Gly (**203**) stimulates reproduction of the alga. The ^{13}C nmr chemical shifts of mycosporine-Gly (**203**), mycosporine-1 (**204**), and β-diketone (**205**) are indicated on the formulas.

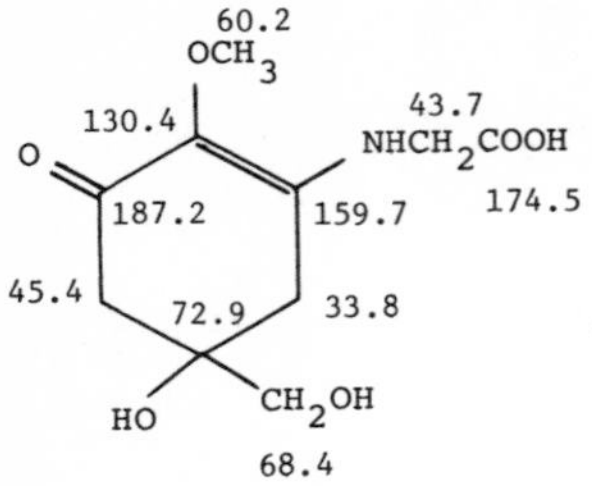

203 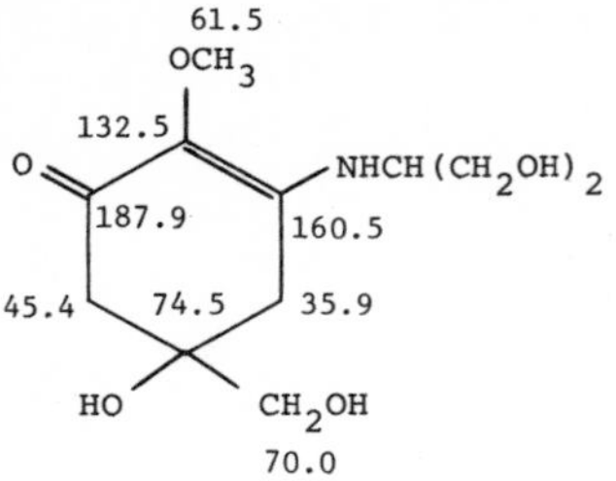 **204** 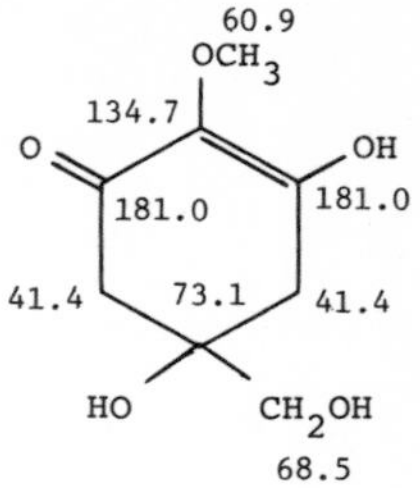 **205**

The causative agent associated with the lethality of "red tides" has been linked to a toxin(s) produced by dinoflagellates, particularly those of the genus *Gonyaulax*. The toxin(s) produced by these organisms are water soluble and are concentrated by filter-feeding animals. Saxitoxin (**206**), produced by *Gonyaulax catanella*, was initially isolated from the syphon tubes of the Alaskan butter clam *Saxidomus giganteus*. Although considerable effort was devoted to the structure elucidation of this toxin by chemical means (Wong *et al.*, 1971), the correct structure was determined by X-ray analysis of a *p*-bromobenzene sulfonate derivative (Schantz *et al.*, 1975). This structure was confirmed by X-ray analysis of saxitoxin ethyl hemiketal dihydrochloride and by ^{13}C nmr analysis of saxitoxin (**206a**) and a reduction product, dihydrosaxitoxin-13-d (**207**) (Bordner *et al.*, 1975). Shimizu *et al.* (1976) obtained gonyautoxin (**208**) from *Gonyaulax tamarensis*, the toxic dinoflagellate associated with shellfish poisoning along the North Atlantic coast. The ^{13}C nmr spectrum of gonyautoxin (**208**) clearly showed it to be related to saxitoxin (**206**) and was fully consistent with placement of a hydroxyl at C-12 of saxitoxin (**206**). The off-resonance ^{13}C nmr spectrum of gonyautoxin (**208**) required 111,499 scans, a considerable but worthwhile investment of instrument time.

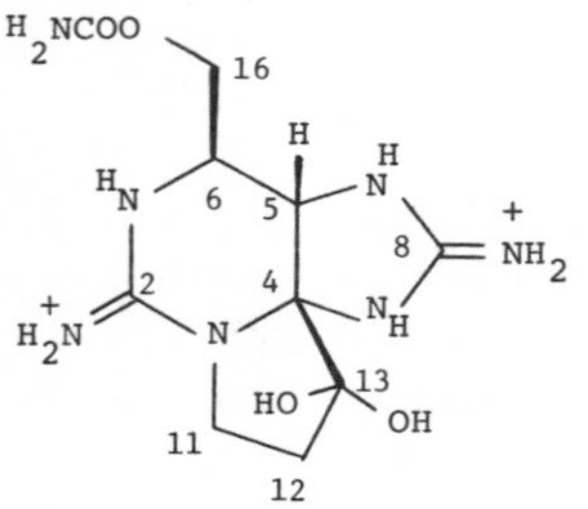

206 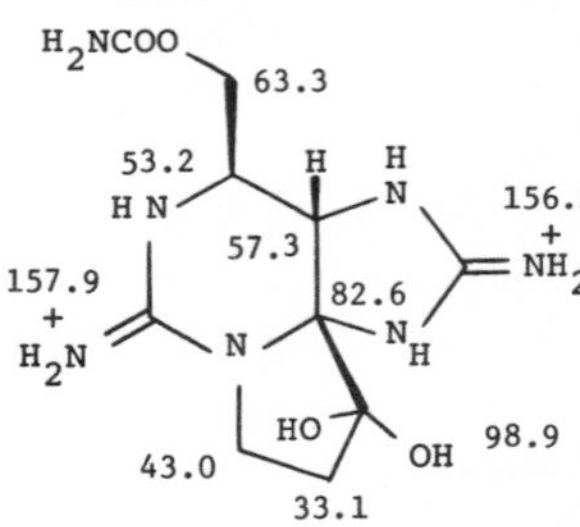 **206a**

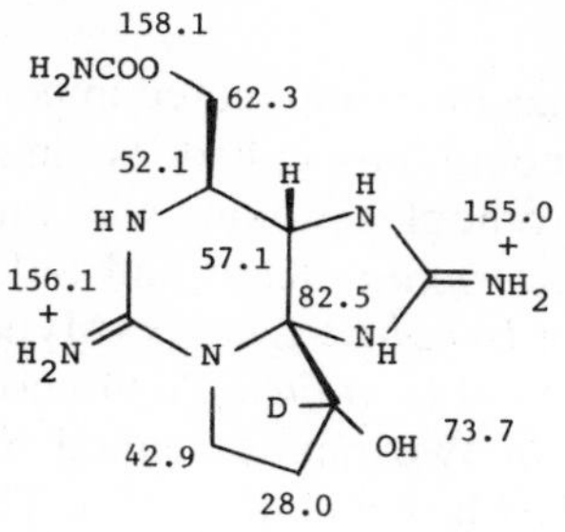

207

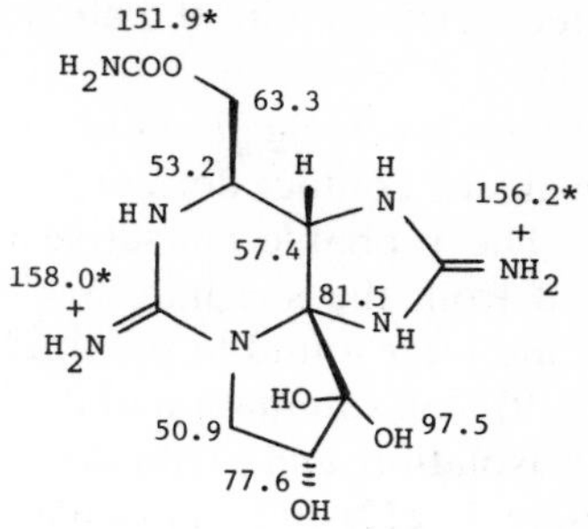

208

Chondria californica is a common filamentous red alga that is conspicuous in tide pools because of its striking blue iridescence. When removed from the water for only a short time, this noisome alga has a strong sulfurlike odor. Wratten and Faulkner (1976) reported the isolation and structure elucidation of seven sulfur-containing heterocyclic compounds from *C. californica*. Carbon-13 nmr data (DMSO-d_6) for 4-dioxo-1,2,4,6-tetrathiepane (**209**) were reported as shown on the structure.

Wells (1976) reported the characterization of chondrillin (**210**), an optically active, stable peroxy ketal metabolite of a sponge in the genus *Chondrilla*.

The thalli of *Delisea fimbriata* (Lamour.) Mont. collected near Palmer Station, Antarctic Peninsula, were unique because they were remarkably free of the normal organisms, both micro- and macroscopic, which colonize other algae in the immediate area. Pettus *et al.* (1977) isolated a series of acetoxyfimbrolides from the dichloromethane-soluble extract of the air-dried alga. Carbon-13 nmr data for the major lactone (**211**) were

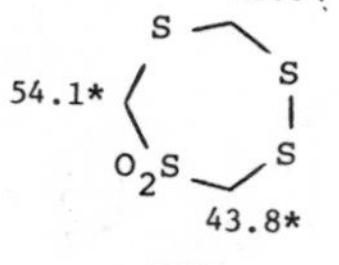

209

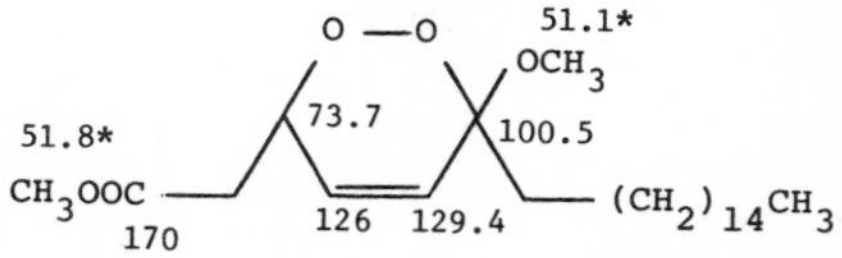

210

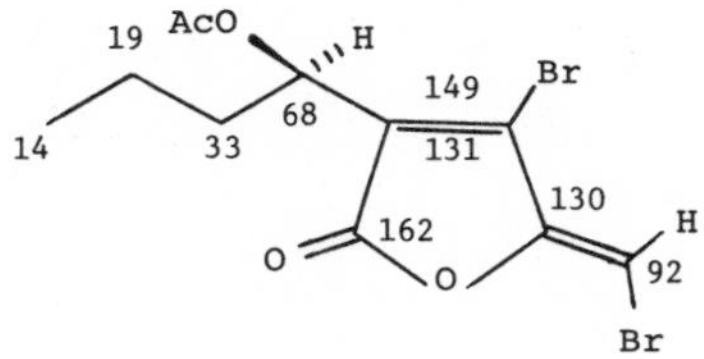

211

reported ($CDCl_3$) with the chemical shift assignments as indicated on the formula.

The blue-green alga *Lygnbya majuscula* has been implicated in periodic outbreaks of contact dermatitis in Hawaii commonly called "swimmers' itch." The probable causative agent, debromoplysiatoxin, was initially isolated from the sea hare *Stylocheilus longicauda* (Kato and Scheuer, 1975) and later found in the lipid extracts of *Lyngbya gracilis* (Mynderse *et al.*, 1977). Examination of the lipid extracts of *L. majuscula* Gomont led to the isolation and structure elucidation of two amides called majusculamide A (**212**) and majusculamide B (**213**) (Marner *et al.*, 1977). The ^{13}C nmr spectrum (DMSO-d_6) of majusculamide B (**213**) was complex when recorded at room temperature but was greatly simplified and showed all 28 carbon signals when the spectrum was recorded at 140°. The ^{13}C nmr chemical shifts for majusculamide B (**213**) are shown on the formula. The reported ^{13}C nmr data (Marner *et al.*, 1977) indicate that the resonances at

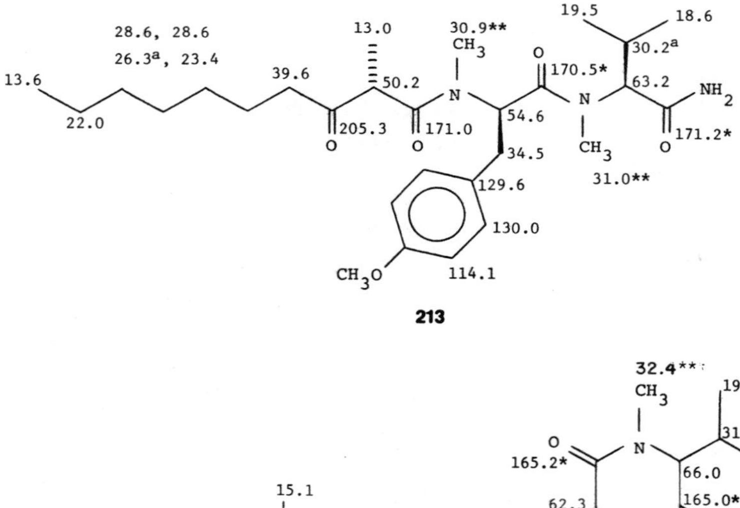

213

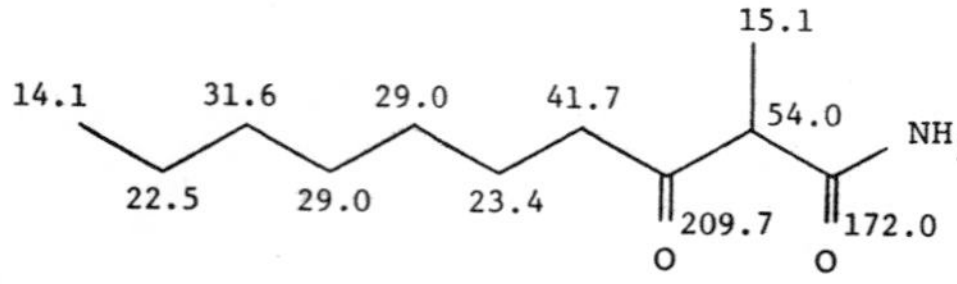

214

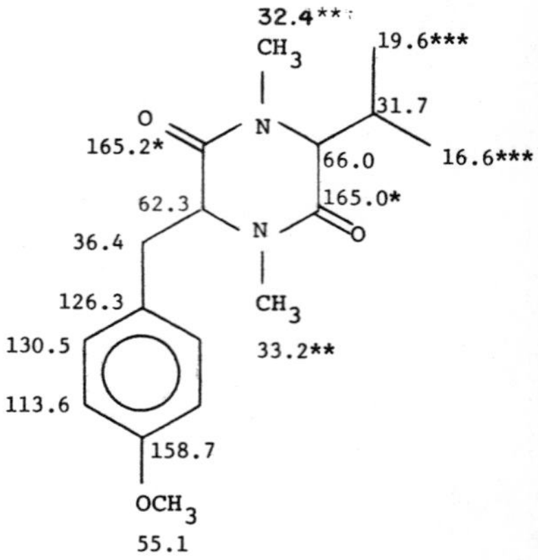

215

30.2 and 26.3 ppm are both off-resonance doublets. This must have been a typographical error, and one of these doublets (26.3? ppm) should have been a triplet.

Pyrolysis of majusculamide B (**213**) yielded racemic 2-methyl-3-oxodecanoic amide (**214**) and compound (**215**). The ^{13}C nmr chemical shift assignments are indicated on the formulas.

ACKNOWLEDGMENTS

This research was supported in part by NSF Grant CHE-74-13938. The Bruker WH90-D nmr Spectrometer was supported in part by NIH Biomedical Science Grant 5-SO5-RR07010-19 and NSF Grant MPS75-06138. We would also like to thank Sharon Herbert and Connie Watkins for assistance in preparation of this manuscript.

REFERENCES

Archer, R. A., Cooper, R. D. G., DeMarco, P. V., and Johnson, L. F. (1970). *Chem. Commun.* p. 1291.

Beechan, C. M., Djerassi, C., Finer, J. S., and Clardy, J. (1977). *Tetrahedron Lett.* p. 2395.

Bernstein, J., Shmeuli, U., Zadock, E., Kashman, Y., and Néeman, I. (1974). *Tetrahedron* **30**, 2817.

Blackman, A. J., and Wells, R. J. (1976). *Tetrahedron Lett.* p. 2729.

Bohlmann, F., Zeisberg, R., and Klein, E. (1975). *Org. Magn. Reson.* **7**, 426.

Bordner, J., Thiessen, W. E., Bates, M. A., and Rapoport, H. (1975). *J. Am. Chem. Soc.* **97**, 6008.

Breitmaier, E., and Voelter, W. (1974). *In* "Monographs in Modern Chemistry" (H. F. Ebel, ed.), Vol. 5, Verlag Chemie, Weinheim.

Burreson, B. J., Woolard, F. X., and Moore, R. E. (1975a). *Tetrahedron Lett.* p. 2155.

Burreson, B. J., Woolard, F. X., and Moore, R. E. (1975b). *Chem. Lett.* p. 1111.

Cafieri, F., Fattorusso, E., Magno, S., Santacroce, C., and Sica, D. (1973). *Tetrahedron* **29**, 4259.

Cimino, G. (1977). *In* "Marine Natural Products Chemistry" (D. J. Faulkner and W. H. Fenical, eds.), pp. 61–86. Plenum, New York.

Cimino, G., DeStefano, S., Minale, L., and Trivellone, E. (1975a). *Tetrahedron Lett.* p. 3727.

Cimino, G., DeStefano, S., and Minale, L. (1975b). *Experientia* **31**, 1117.

Coll, J. C., Mitchell, S. J., and Stokie, G. J. (1977). *Tetrahedron Lett.* p. 1539.

Corey, E. J., and Ursprung, J. J. (1956). *J. Am. Chem. Soc.* **78**, 5041.

Couperus, P. A., Clague, A. D. H., and van Dongen, J. P. C. M. (1976). *Org. Magn. Reson.* **8**, 426.

Crews, P. (1977). *J. Org. Chem.* **42**, 2634.

Crews, P., and Kho, E. (1974). *J. Org. Chem.* **39**, 3303.

Crews, P., and Kho, E. (1975). *J. Org. Chem.* **40**, 2568.

Crews, P., and Kho-Wiseman, E. (1977). *J. Org. Chem.* **42**, 2812.

Dalling, D. K., and Grant, D. M. (1972). *J. Am. Chem. Soc.* **94**, 5318.

Dalling, D. K., Grant, D. M., and Paul, E. G. (1973). *J. Am. Chem. Soc.* **95**, 3718.

Danise, B., Minale, L., Riccio, R., Amico, V., Oriente, G., Piattelli, M., Tringali, C., Fattorusso, E., Magno, S., and Mayol, L. (1977). *Experientia* **33**, 413.
de Rosa, S., Minale, L., Riccio, R., and Sodano, G. (1976). *J. Chem. Soc., Perkin Trans. 1* p. 1408.
Enzell, C., and Erdtman, H. (1958). *Tetrahedron Lett.* p. 361.
Erickson, K. L. (1977). Personal communication.
Fattorusso, E., Magno, S., Mayol, L., Santacroce, C., Sica, D., Amico, V., Oriente, G., Piattelli, M., and Tringali, C. (1976). *Chem. Commun.* p. 575.
Faulkner, D. J. (1976). *Phytochemistry* **15**, 1992.
Faulkner, D. J., and Stallard, M. D. (1973). *Tetrahedron Lett.* p. 1171.
Fenical, W. (1974). *J. Am. Chem. Soc.* **96**, 5580.
Fenical, W., and Sims, J. J. (1973). *J. Org. Chem.* **38**, 2383.
Fenical, W., Gifkins, K. B., and Clardy, J. (1974). *Tetrahedron Lett.* p. 1507.
Fenical, W., Howard, B. M., Gifkins, K. B., and Clardy, J. (1975). *Tetrahedron Lett.* p. 3983.
Fenical, W., Finer, J., and Clardy, J. (1976). *Tetrahedron Lett.* p. 731.
González, A. G., Darias, J., Martín, J. D., and Pérez, C. (1974). *Tetrahedron Lett.* p. 1249.
González, A. G., Darias, J., Martín, J. D., Pérez, C., Sims, J. J., Lin, G. H. Y., and Wing, R. M. (1975). *Tetrahedron* **31**, 2449.
González, A. G., Darias, J., Díaz, A., Fourneron, J. D., Martín, J. D., and Pérez, C. (1976). *Tetrahedron Lett.* p. 3051.
González, A. G., Martín, V. S., and Martín, J. D. (1977). Unpublished observations.
González, A. G., Arteaga, J. M., Martín, J. D., and Rodriguez, M. L. (1978). *Phytochemistry* **17**, 947.
Gray, G. A. (1975). *Anal. Chem.* **47**, 547a.
Grover, S. H., and Stothers, J. B. (1975). *Can. J. Chem.* **53**, 589.
Günther, H., and Jikeli, G. (1977). *Chem. Rev.* **77**, 599.
Hagaman, E. W. (1976). *Org. Magn. Reson.* **8**, 389.
Hall, S. S., Faulkner, D. J., Fayos, J., and Clardy, J. (1973). *J. Am. Chem. Soc.* **95**, 7187.
Higgs, M. D., Vanderah, D. J., and Faulkner, D. J. (1977). *Tetrahedron* **33**, 2775.
Hirschfeld, D. R., Fenical, W., Lin, G. H. Y., Wing, R. M., Radlick, P., and Sims, J. J. (1973). *J. Am. Chem. Soc.* **95**, 4049.
Howard, B. M., and Fenical, W. (1976a). *Tetrahedron Lett.* p. 41.
Howard, B. M., and Fenical, W. (1976b). *Tetrahedron Lett.* p. 2519.
Howard, B. M., and Fenical, W. (1976c). *Tetrahedron Lett.* p. 2519.
Howard, B. M., and Fenical, W. (1978). *J. Org. Chem.* (submitted for publication).
Iengo, A., Mayol, L., and Santacroce, C. (1977). *Experientia* **33**, 11.
Ireland, C., and Faulkner, D. J. (1976). *J. Am. Chem. Soc.* **98**, 4664.
Ireland, C., and Faulkner, D. J. (1977). *J. Org. Chem.* **42**, 3157.
Irie, T., Yasunari, Y., Suzuki, T., Imai, N., Kurosawa, E., and Masamune, T. (1965). *Tetrahedron Lett.* p. 3619.
Irie, T., Suzuki, M., Kurosawa, E., and Masamune, T. (1966). *Tetrahedron Lett.* p. 1837.
Irie, T., Fukuzawa, A., Izawa, M., and Kurosawa, E. (1969). *Tetrahedron Lett.* p. 1343.
Irie, T., Suzuki, M., Kurosawa, E., and Masamune, T. (1970). *Tetrahedron* **26**, 3271.
Ito, S., and Hirata, Y. (1977). *Tetrahedron Lett.* p. 2429.
Izac, R. R., and Sims, J. J. (1977). Unpublished observations.
Kashman, Y. (1977). *In* "Marine Natural Products Chemistry" (D. J. Faulkner and W. H. Fenical, eds.), pp. 17–21. Plenum, New York.
Kashman, Y., and Groweiss, A. (1977). *Tetrahedron Lett.* p. 1159.
Kato, T., Kanno, S., and Kitahara, Y. (1970). *Tetrahedron* **26**, 4287.
Kato, Y., and Scheuer, P. J. (1975). *Pure Appl. Chem.* **41**, 1.

Kazlauskas, R., Murphy, P. T., Quinn, R. J., and Wells, R. J. (1976a). *Tetrahedron Lett.* p. 4451.

Kazlauskas, R., Murphy, P. T., Quinn, R. J., and Wells, R. J. (1976b). *Aust. J. Chem.* **29**, 2533.

Kazlauskas, R., Murphy, P. T., Wells, R. J., Daly, J. J., and Oberhänsli, W. E. (1977). *Aust. J. Chem.* **30**, 2679.

Kazlauskas, R., Murphy, P. T., Wells, R. J., Oberhänsli, W. E., and Noack, A. (1978a). *Aust. J. Chem.* submitted.

Kazlauskas, R., Murphy, P. T., and Wells, R. J. (1978b). *Aust. J. Chem.* submitted.

Kurosawa, E. (1977). Personal communication.

Levy, G. C., and Nelson, G. L. (1972). "Carbon-13 Nuclear Magnetic Resonance for Organic Chemists," pp. 38–58. Wiley (Interscience), New York.

McEnroe, F. J., Robertson, K. J., and Fenical, W. (1977). *In* "NATO Conference on Marine Natural Products" (D. J. Faulkner and W. Fenical, eds.), pp. 179–189. Plenum, New York.

Marner, F. J., Moore, R. E., Hirotsu, K., and Clardy, J. (1977). *J. Org. Chem.* **42**, 2815.

Minale, L., and Riccio, R. (1976). *Tetrahedron Lett.* p. 2711.

Missakian, M. G., Burreson, B. J., and Scheuer, P. J. (1975). *Tetrahedron* **31**, 2513.

Moore, R. E. (1977). Personal communication.

Moore, R. E., Dietrich, R. F., Hatton, B., Higa, T., and Scheuer, P. J. (1975). *J. Org. Chem.* **40**, 540.

Mynderse, J. S., and Faulkner, D. J. (1974). *J. Am. Chem. Soc.* **96**, 6771.

Mynderse, J. S., Moore, R. E., Kashiwagi, M., and Norton, T. R. (1977). *Science* **196**, 538.

Norton, R. S., Warren, R. G., and Wells, R. J. (1977). *Tetrahedron Lett.* p. 3905.

Pettus, J. A., Jr., Wing, R. M., and Sims, J. J. (1977). *Tetrahedron Lett.* p. 41.

Robertson, K. J., and Fenical, W. (1977). *Phytochemistry* **16**, 1071.

Rose, A. F., and Sims, J. J. (1977a). *Tetrahedron Lett.* p. 2935.

Rose, A. F., and Sims, J. J. (1977b). Unpublished observations.

Rose, A. F., Sims, J. J., Wing, R. M., and Wiger, G. (1977). *Tetrahedron Lett.* p. 2533.

Schantz, E. J., Ghazarossian, V. E., Schnoes, H. K., Strong, F. M., Springer, J. P., Pezzanite, J. O., and Clardy, J. (1975). *J. Am. Chem. Soc.* **97**, 1238.

Scheuer, P. J. (1977). *Acc. Chem. Res.* **10**, 33.

Schmitz, F. J., and McDonald, F. J. (1974). *Tetrahedron Lett.* p. 2541.

Shimizu, Y., Alam, M., Oshima, Y., Buckley, L. J., Fallon, W. E., Kasai, H., Miura, I., Gullo, V., and Nakanishi, K. (1976). *J. Am. Chem. Soc.* **98**, 5414.

Sleeper, H. L., and Fenical, W. (1977). *J. Am. Chem. Soc.* **99**, 2367.

Stallard, M. O., and Faulkner, D. J. (1974). *Comp. Biochem. Physiol. B* **49**, 25.

Stierle, D. B., and Sims, J. J. (1978). *Tetrahedron* (submitted for publication).

Stierle, D. B., Wing, R. M., and Sims, J. J. (1976). *Tetrahedron Lett.* p. 4455.

Stothers, J. B. (1972). "Carbon-13 NMR Spectroscopy," Organic Chemistry Monograph Series, Vol. 24. Academic Press, New York.

Sun, H. H., Waraszkiewicz, S. M., and Erickson, K. L. (1976a). *Tetrahedron Lett.* p. 4227.

Sun, H. H., Waraszkiewicz, S. M., and Erickson, K. L. (1976b). *Tetrahedron Lett.* p. 585.

Sun, H. H., Waraszkiewicz, S. M., Erickson, K. L., Finer, J., and Clardy, J. (1977). *J. Am. Chem. Soc.* **99**, 3516.

Suzuki, M., Kurosawa, E., and Irie, T. (1970). *Tetrahedron Lett.* p. 4995.

Suzuki, M., Kurosawa, E., and Irie, T. (1974a). *Tetrahedron Lett.* p. 821.

Suzuki, M., Kurosawa, E., and Irie, T. (1974b). *Tetrahedron Lett.* p. 1807.

Suzuki, T., Suzuki, M., and Kurosawa, E. (1975). *Tetrahedron Lett.* p. 3057.

Suzuki, T., Furusaki, A., Hashiba, N., and Kurosawa, E. (1978). *Tetrahedron Lett.* (submitted for publication).

van Engen, D., Clardy, J., Kho-Wiseman, E., and Crews, P. (1978). *Tetrahedron Lett.* p. 29.
Waraszkiewicz, S. M., Sun, H. H., and Erickson, K. L. (1976). *Tetrahedron Lett.* p. 3021.
Waraszkiewicz, S. M., Erickson, K. L., Finer, J., and Clardy, J. (1977). *Tetrahedron Lett.* p. 2311.
Wehrli, F. W., and Wirthlin, T. (1976). ''Interpretation of Carbon-13 NMR Spectra,'' Heyden, London.
Wells, R. J. (1976). *Tetrahedron Lett.* p. 2637.
Wells, R. J. (1977). *Tetrahedron Lett.* p. 3183.
Wenkert, E., Buckwalter, B. L., Burfitt, I. R., Gasič, M. J., Gottlieb, H. E., Hagaman, E. W., Schell, F. M., and Woukulich, P. M. (1976). *In* ''Topics in Carbon-13 NMR Spectroscopy'' (G. C. Levy, ed.), p. 94. Wiley (Interscience), New York.
Wong, J. L., Oesterlin, R., and Rapoport, H. (1971). *J. Am. Chem. Soc.* **93**, 7344.
Woolard, F. X., and Moore, R. E., van Engen, D., Clardy, J. (1978). *Tetrahedron Lett.* p. 2367.
Wratten, S. J., and Faulkner, D. J. (1976). *J. Org. Chem.* **41**, 2465.
Wratten, S. J., and Faulkner, D. J. (1977a). *J. Am. Chem. Soc.* **99**, 7367.
Wratten, S. J., and Faulkner, D. J. (1977b). *J. Org. Chem.* **42**, 3343.
Yamamura, S., and Hirata, Y. (1963). *Tetrahedron* **19**, 1485.

Index

D

E

F

G

M

N

O

P

T